Handbook of Vascular Biology Techniques

Mark Slevin • Garry McDowell
Editors

Handbook of Vascular Biology Techniques

Springer

Editors
Mark Slevin
School of Healthcare Science
Manchester Metropolitan University
Manchester, UK

Garry McDowell
School of Healthcare Science
Manchester Metropolitan University
Manchester, UK

ISBN 978-94-017-9715-3 ISBN 978-94-017-9716-0 (eBook)
DOI 10.1007/978-94-017-9716-0

Library of Congress Control Number: 2015936145

Springer Dordrecht Heidelberg New York London

Printed on acid-free paper

Springer Science+Business Media B.V. Dordrecht is part of Springer Science+Business Media (www.springer.com)

The authors would like to dedicate this book to:
(Mark-Asha Shakira Slevin)
(GMcD: To Christine and Amy)

Preface

Within this book, a variety of techniques are described in detail pertaining to methods used in both basic and advanced vascular biology. Methodologies range from in vitro cell culture to in vivo manipulations, through cell signalling proteomics and genomics to patient imaging in disease. A number of novel, state-of-the-art methodologies are also included. Each chapter is fully inclusive, containing sections on trouble shooting and additional notes/links thus ensuring the reader has sufficient information to carry out the protocol without additional requirements. This book should appeal to students, researchers and medical professionals working in all vascular-linked fields such as cardio- and cerebro-vascular, cancer and dementia.

Manchester, UK

Mark Slevin
Garry McDowell

Preface

Contents

Part I
In Vitro Techniques

Chapter 1
In Vitro Angiogenesis Assay: Endothelial Migration, Proliferation, and Tube Formation

Kazuhide Hayakawa, Anna Chun-Ling Liang, Changhong Xing, Eng H. Lo, and Ken Arai

1.1 Introduction

Over the past four decades, angiogenesis has been well studied in both normal developmental processes and numerous pathologies including diabetes, cancer, myocardial infarction, brain injury and ischemic stroke [1–6]. Angiogenesis is a fundamental process to develop new blood vessels from pre-existing vessels and two different mechanisms have been described. One is an intussusceptive angiogenesis that is caused by the insertion of interstitial cellular columns into the lumen of pre-existing vessels [7]. The subsequent growth of these columns and their stabilization results in partitioning of the vessel and remodeling of the local vascular network [8]. The other is a sprouting angiogenesis induced by migration of endothelial cells toward the site of angiogenesis, proliferation of endothelial cells behind the front of migration, and lumen formation within the endothelial sprout and formation of loops by anastomoses of sprouts.

Most endothelial cells within the body are naturally quiescent and physiology of angiogenesis is well controlled to meet the tissue requirements. Under some pathological conditions, however, angiogenesis occurs in a disorderly sequence of cellular events. For example, moyamoya syndrome is an angiogenic disease caused by blocking blood flow through the vessel constriction, and blood clots. Newly formed brain vessels in moyamoya disease are small, weak and prone to hemorrhage, aneurysm and thrombosis [9]. Therefore, a greater understanding of the cellular/molecular mechanisms of angiogenesis in relevant diseases may be necessary to advanced therapeutic strategies. In this chapter, we will describe the methods to assess angiogenesis in vitro.

K. Hayakawa • A.C.-L. Liang • C. Xing • E.H. Lo • K. Arai (✉)
Neuroprotection Research Laboratory, Department of Radiology and Neurology, Massachusetts General Hospital and Harvard Medical School, MGH East 149-2401, Charlestown, MA 02129, USA
e-mail: karai@partners.org

M. Slevin, G. McDowell (eds.), *Handbook of Vascular Biology Techniques*,
DOI 10.1007/978-94-017-9716-0_1

1.2 Methodology

In vitro angiogenesis assays are well-established and useful tools to test efficacy of both pro- and anti-angiogenic agents. Major assessments of angiogenesis in vitro focus on endothelial migration, proliferation, and tubular- or capillary-like network formation on the biological materials including extracellular matrix. Here we will describe the endothelial scratch migration assay (wound healing assay), proliferation assay, and matrigel tube formation assay as principal methods commonly used in in vitro angiogenesis assays. Our protocol is outlined below.

1.3 Materials

1. Collagen I-coated culture plates (6 wells or 24 wells)
2. 5 % CO_2 incubator at 37 °C
3. EBM-2 basal medium (Lonza)
4. EGM-2 supplemental (Lonza)
5. Penicillin/streptomycin (GIBCO)
6. 0.05 % Trypsin-EDTA solution (GIBCO)
7. Phosphate buffer saline (GIBCO)
8. p200 or p1000 pipette tip
9. Cell counting kit-8 (WST) (Dojindo)
10. Growth factor reduced (GFR) Matrigel matrix (BD Bioscience)
11. Endothelial cells
    - Human brain microvascular endothelial cells (HBMEC, Cell Systems Corporation)
    - Rat brain endothelial cells (RBE.4)
    - Outgrowth endothelial progenitor cells (see the Sect. 1.4.4 in detail)
12. Two spleens isolated from 11 to 12 weeks old Sprague–Dawley (SD) rats.
13. Ficoll-Paque Plus (Amersham Biosciences Corp)
14. 40-μm nylon membrane
15. EPC lysis buffer (4 % BSA, 1 mM EDTA in PBS).
16. 4 % paraformaldehyde (PFA) (Wako)

1.4 Basic Protocol

1.4.1 Scratch Migration Assay

1. Use endothelial cells fully confluent in the flask, wash twice with PBS, then add 3 mL cell detachment solution and incubate at 37 °C for 5 min.
2. Add fresh 3 mL EGM, gently draw up the cell suspension, and spin down to obtain cell pellet.

3. Resuspend cells with EGM, and plate 2×10^5 cells per well in 6-well plate pre-coated with collagen-I, as 5×10^4 cells per well for 24-well plate.
4. Return plates with cells to incubator at 37 °C and 5 % CO_2. Change to fresh EGM every 2–3 day until the cells are almost confluent and make a monolayer.
5. Change to working medium, and incubate overnight for serum starvation.
6. Scrape the cell monolayer in a straight line with p200 or p1000 pipette tip. Remove the debris by washing twice with warm working medium, and then replace with fresh working medium.
7. Mark the reference points to obtain the same field during the image acquisition with a razor blade on the outer bottom of the culture plate or with an ultrafine tip marker.
8. Return the cells back to culture incubator for 8–24 h.
9. After the incubation, wash the cells in each well twice with PBS gently to remove detached or dead cells, and fix cells by 4 % PFA for 10 min at 4 °C.
10. Place the culture plate under a phase-contrast microscope, match the reference point, and acquire images at least four random areas.
11. Count the cells that cross into the scratched area from their reference points, these cells are determined as migrated cells.

1.4.2 Proliferation Assay

1. Use endothelial cells fully confluent in the flask, wash twice with PBS, then add 3 mL cell detachment solution and incubate at 37 °C for 5 min.
2. Add fresh 3 mL EGM, gently pipette to get the cell suspension, and spin down to get cell pellet.
3. Resuspend cells with EGM, and plate 2×10^4 cells per well in 24-well plate pre-coated with collagen-I.
4. Return plates with cells to incubator at 37 °C and 5 % CO_2.
5. Wait until 50 % confluent and change to working medium, and incubate overnight for serum starvation.
6. Stimulate endothelial cells with reagents or test drugs for 24 h.
7. The cells are incubated with 10 % WST solution for 1 h at 37 °C. Then the absorbance of the culture medium was measured with a microplate reader at a test wavelength of 450 nm and a reference wavelength of 630 nm.

1.4.3 Matrigel Tube Formation

1. Thaw the matrigel on ice or in cold room completely and slowly. Keep it on ice until use.
2. Keep the multi-well plate on ice. Use pre-cooled tips to add 200 µl soluble matrix into wells of 24-well plate, make sure it cover the bottom evenly, and no bubbles in the gel mixture.

3. Incubate the plates with matrix at 37 °C for 30–60 min, the matrix will be polymerized to insoluble gels and ready to use.
4. Endothelial cells at 80–90 % confluency were trypsinized, collected and suspended in culture medium with reduced serum and growth factors.
5. Count the cells with hemocytometer, diluting the cells with enough medium to 2.5×10^5 per mL.
6. Seed the endothelial cells on the surface of Matrigel-coated wells at 5×10^4 in each well in 200 μl.
7. Incubate in the regular cell incubator for 6–18 h, at 37 °C and 5 % CO_2.
8. Cells are observed under phase-contrast microscope, and 2–3 images for each well are acquainted under light microscope.
9. Images are imported and counted for the complete rings in each well. More rings mean higher capability of tube formation.

1.4.4 Outgrowth Endothelial Progenitor Cells Isolation

1. Under the hood, spleens were mechanically minced, placed at 37 °C for 15 min in 20 ml EPC lysis buffer.
2. Then run through a 40-um nylon membrane to obtain cell suspension.
3. Add 4.5 ml cell suspension gently onto 5 ml Ficoll-Paque Plus and centrifuge 1,500 rpm for 25 min to obtain mononuclear cells (MNCs).
4. Isolated MNCs are gently washed twice with complete growth media EBM-2 plus supplement.
5. 3×10^7 MNCs per well are seeded on collagen I-coated six-well plates and incubated in a 5 % CO_2 incubator at 37 °C.
6. Outgrowth EPCs are obtained after culturing for 1–1.5 months.

1.5 Sample Results

1.5.1 Scratch Migration Assay [10]

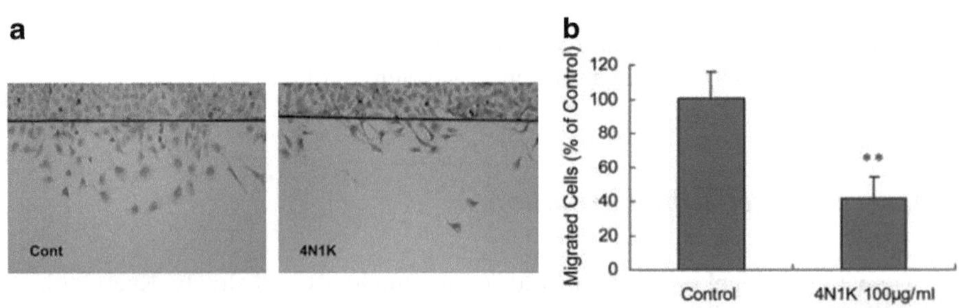

(**a**) Representative images showed decreased migration of human brain endothelial cells across a wound scratch line 24 h after treatment with a CD47-specific activator 4N1K (100 μg/ml) as compared to the control (untreated) cells. (**b**) Quantified cell counts showed a significant difference in cell migration between controls and 4N1K-treated cells (**$P<0.01$, $n=3$. Data are expressed as mean ± SD)

1.5.2 *Proliferation Assay*

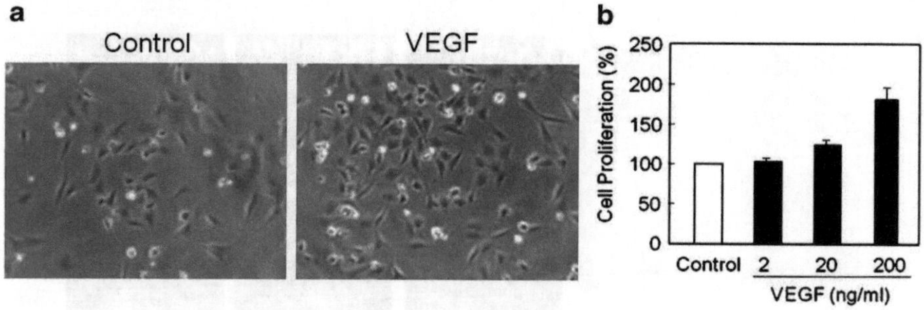

Rat brain endothelial cells were incubated with VEGF-A for 24 h. (**a**) Representative images of proliferation using rat brain endothelial cells. VEGF-A: 200 ng/ml (**b**) VEGF-A induced endothelial proliferation in a dose-dependent manner ($n=4$. Data are expressed as mean ± SEM)

1.5.3 *Tube Formation Assay [11]*

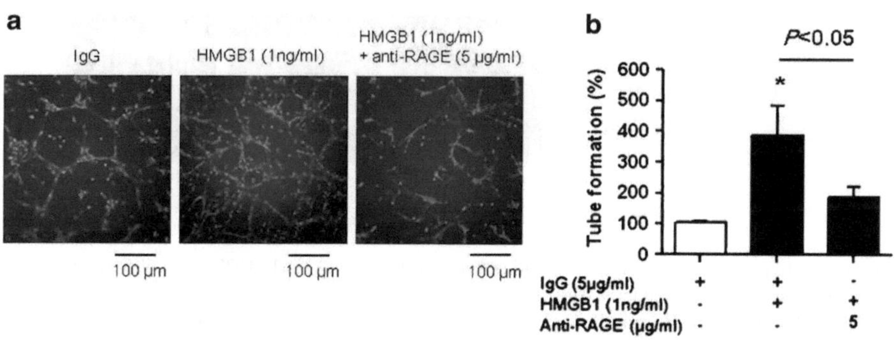

(**a**) Representative images of tube formation using outgrowth EPCs. HMGB1: 1 ng/mL, anti-RAGE: 5 ug/mL. (**b**) HMGB1 (1 ng/mL) significantly increased the number of tubes, and the HMGB1-induced tube formation was reduced by co-treatment with anti-RAGE neutralizing antibody (*$P<0.05$, $n=5$. Data are expressed as mean ± SEM)

1.5.4 *Outgrowth Endothelial Progenitor Cells*

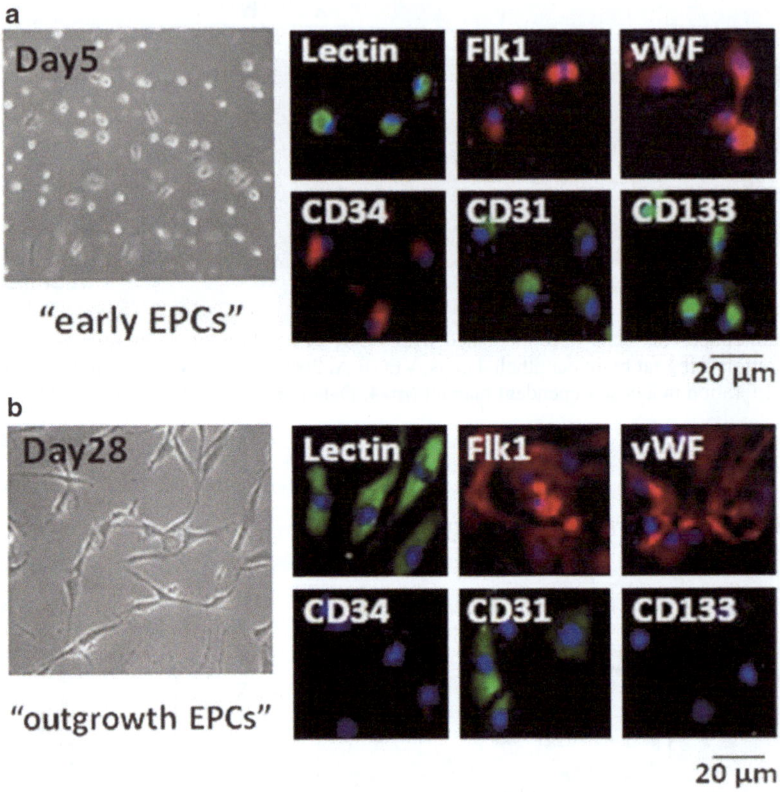

Representative images of early EPCs and outgrowth EPCs. (**a**) Early EPCs at day 5 after seeding bind to lectin (Ulex Europaeus Aglutinin) and showed positive staining for Flk1, vWF, CD34, CD31, and CD133. (**b**) When allowed to mature into late-phase EPCs (28 days), these cells show endothelial morphology, and positive for lectin, Flk1, vWF and CD31, but dim stain for the immature markers CD34 or CD133 (All figures are counter-stained blue for DAPI)

1.6 Troubleshooting

1.6.1 *Scratch Migration Assay*

1. The endothelial cells should not be overgrown to keep monolayer. Overgrown cells could be multi-layers which causes higher amount of cell debris or free floating cells after scratching.
2. Serum starvation is an important step to minimize background signal before cell stimulation.

3. To create scratches of approximately similar size is critical to minimize any possible variation caused by the difference in the width of the scratches. For 24 well, the p200 pipette tip can be used easier than p1000 pipette tip.
4. Choose shorter incubation time under faster migrating conditions.
5. Before imaging, washing step is important. Wash gently and try to remove all debris and free-floating cells to distinguish real migrating cells. Fixing cells helps to keep cell morphology and allows longer imaging time.

1.6.2 Proliferation Assay

1. Serum starvation is an important step to minimize background signal before cell stimulation. Sensitive endothelial cells such as endothelial progenitor cells may need to be supplemented by adding 0.5 % FBS.
2. Percentage of cell confluence is important. Please use appropriate cell confluence around 50–60 %.

1.6.3 Tube Formation Assay

1. In the case of no tube formation is observed in the positive control group, there are two possibilities. One is that cells may not be healthy or cells are too old. Please use healthy cells or a lower passage. Secondly, the cell density may be too high or low. Please use optimal cell density (e.g., 2×10^4 cells/well for 24-well plate).
2. After getting solid matrigel, do not disturb the gel layer. If you need to change the culture medium during the experiment, use a pipette to gently remove the medium rather than using aspiration.
3. To avoid generating bubbles, do not mix the liquid matrigel solution or pipette the solution up and down before coating the plate. If bubbles are generated, a few seconds of centrifugation using ~10,000 g would be helpful to remove bubbles.
4. For the thin coating of Matrigel matrix described here, the surface is not flat but crescent, which may make it difficult to get the pictures with good focus. The μ-Slide Angiogenesis from ibidi GmbH (Planegg/Martinsried, Germany) is used with only 10 μL of matrix, and makes a flat surface and bring all cells in one focal plane. Please see the Method variations/alternative.

1.7 Method Variations/Alternative

1.7.1 Transwell Migration Assay

Using a modified Boyden chamber [12], endothelial migration from upper chamber to lower part of membrane could be counted for migrated cells.

1. Use endothelial cells fully confluent in the flask and culture overnight with serum starvation, wash twice with PBS, then add 3 mL cell detachment solution and incubate at 37 °C for 5 min.
2. Add fresh 3 mL EGM, gently pipette to get the cell suspension, and spin down to obtain cell pellet.
3. Resuspend cells with EBM-2 basal media, and plate 2×10^4 cells/100 μl per upper chamber for 24-well plate.
4. Carefully place the upper chamber onto the lower wells with 600 μl EBM-2 basal media and test compounds.
5. Return plates with cells to incubator at 37 °C and 5 % CO_2.
6. Wait for 4 h, and gently wash upper chamber three times with PBS.
7. Incubate the upper chamber in 4 % PFA for 15 min to fix migrated cells.
8. Rinse in PBS again, cut the membrane, and place the membrane onto glass slides, mount onto glass slides with mounting media, cover with cover slips, and allow to dry overnight.
9. Image using a microscope to quantify the number of cells that have physically migrated from one side of the membrane to the other side.

1.7.2 BrdU Cell Proliferation Assay

5-bromo-2'-deoxyuridine (BrdU), a thymidine analog, enables detection of DNA replication in actively proliferating cells using a monoclonal antibody directed against BrdU and a fluorophore-conjugated secondary antibody. BrdU staining facilitates the identification of cells that have progressed through the S phase of the cell cycle during the BrdU-labeling period.

1. Use endothelial cells fully confluent in the flask, wash twice with PBS, then add 3 mL cell detachment solution and incubate at 37 °C for 5 min.
2. Add fresh 3 mL EGM, gently pipette to get the cell suspension, and spin down to obtain cell pellet.
3. Resuspend cells with EGM, and plate 2×10^4 cells per well in 24-well plate pre-coated with collagen-I.
4. Return plates with cells to incubator at 37 °C and 5 % CO_2.
5. Wait until 50 % confluent and change to working medium, and incubate overnight for serum starvation.
6. Stimulate endothelial cells with reagents or test drugs for 24 h.
7. Dilute BrdU with EBM-2 basal media to a final concentration 160 μM.
8. Incubate endothelial cells with BrdU solution for 60 min at 37 °C and 5 % CO_2.
9. Aspirate culture medium and add 100 μl of 70–80 % methanol for 15 min.
10. Aspirate fixation solution and wash plate twice with PBS.
11. Proceed with immunocytochemical staining to detect BrdU.

1.8 Additional Note

1.8.1 Scratch Migration Assay

1. RBE.4 cells could migrate into the scratched area under basal condition (with working medium). Four random areas from both edges of the scratched area could be photographed, providing large sample sizes for statistical analysis.
2. Acquired images could be further analyzed quantitatively with other indexes, such as the migrating distance from reference point from at least 100 cells for each condition, or the area occupied by migrated cells. Some free software, such as Image J (http://rsb.info.nih.gov/ij/) could be used for the measuring.

1.8.2 Tube Formation Assay

1. Recommended protein concentration of BD Matrigel matrix is 10 mg/ml or higher. The effect of Matrigel from a different lot on tube formation might be different, so it is better to use the Matrigel from same lot number for your whole experiment.
2. Thaw the soluble matrix slowly on ice, usually overnight. Once thawed, swirl the bottle slowly on ice to make the solution leveled. Try to avoid repeatedly freezing and thawing. The new bottle of Matrigel matrix could be dispensed into appropriate aliquots, and refreeze immediately.
3. The cells could elongate and align to one another as early as 3 h after seeding. The images can be acquired at early time (~6 h incubation time) or late time (~18 h incubation time) for analysis. After longer incubation time, the proteinases secreted from cells might digest the gels and destroy the supports for the tubes.
4. The parameters for quantitative image analysis could be numbers, areas or perimeters of the rings (tubes), or the numbers of branching points, and cell-covered areas.

1.9 Applications and Discussion

Angiogenesis plays important roles in tissue healing and in recovery from ischemic incidents, including heart attack or stroke. On the other hand, angiogenesis is also triggered in cancer, and solid tumors promote new vessel formation that allows supplying nutrients and oxygen to the growing mass. Thus, understanding cellular/molecular mechanisms of angiogenesis in relevant diseases may be critical to establish advanced therapeutic strategies.

In vitro angiogenesis assays are useful for screening potential targets and valuable tools to resemble the in vivo conditions, especially with supporting materials such as extracellular matrix. However, in vitro assays are only best viewed as inceptive data. In order to fully understand the effects of the test compounds in angiogenesis, multiple tests should be used to obtain maximum interpretations from in vitro tests. Also, angiogenesis pathways with either the relevant endothelial cells or multiple endothelial cell types are necessary to organ-specificity. Finally, combination assessments with in vivo models are required to fully investigate disease-related angiogenic mechanisms under pathological conditions.

Acknowledgments and Funding Supported in part by the National Institutes of Health. Materials including figures in this chapter have been extensively drawn from our previously published papers including: Xing et al., J Neurosci Res, 2009; Hayakawa et al., J Neurochem, 2012; Hayakawa et al., PNAS, 2012.

References

1. Hanahan D, Folkman J (1996) Patterns and emerging mechanisms of the angiogenic switch during tumorigenesis. Cell 86(3):353–364
2. Papetti M, Herman IM (2002) Mechanisms of normal and tumor-derived angiogenesis. Am J Physiol Cell Physiol 282(5):C947–C970
3. Watanabe D, Suzuma K, Matsui S, Kurimoto M, Kiryu J, Kita M et al (2005) Erythropoietin as a retinal angiogenic factor in proliferative diabetic retinopathy. N Engl J Med 353(8):782–792
4. Li J, Brown LF, Hibberd MG, Grossman JD, Morgan JP, Simons M (1996) VEGF, flk-1, and flt-1 expression in a rat myocardial infarction model of angiogenesis. Am J Physiol 270(5 Pt 2):H1803–H1811
5. Krupinski J, Kaluza J, Kumar P, Kumar S, Wang JM (1994) Role of angiogenesis in patients with cerebral ischemic stroke. Stroke 25(9):1794–1798
6. Carmeliet P, Jain RK (2000) Angiogenesis in cancer and other diseases. Nature 407(6801):249–257
7. Burri PH, Hlushchuk R, Djonov V (2004) Intussusceptive angiogenesis: its emergence, its characteristics, and its significance. Dev Dyn 231(3):474–488
8. Risau W (1997) Mechanisms of angiogenesis. Nature 386(6626):671–674
9. Scott RM, Smith ER (2009) Moyamoya disease and moyamoya syndrome. N Engl J Med 360(12):1226–1237
10. Xing C, Lee S, Kim WJ, Wang H, Yang YG, Ning M et al (2009) Neurovascular effects of CD47 signaling: promotion of cell death, inflammation, and suppression of angiogenesis in brain endothelial cells in vitro. J Neurosci Res 87(11):2571–2577. PMCID: 3712846
11. Hayakawa K, Miyamoto N, Seo JH, Pham LD, Kim KW, Lo EH, et al (2012) High-mobility group box 1 from reactive astrocytes enhances the accumulation of endothelial progenitor cells in damaged white matter. J Neurochem. PMCID: 3604050
12. Boyden S (1962) The chemotactic effect of mixtures of antibody and antigen on polymorphonuclear leucocytes. J Exp Med 115:453–466. PMCID: 2137509

Chapter 2
Endothelial Cell Tube Formation on Basement Membrane to Study Cancer Neoangiogenesis

Amelia Casamassimi, Filomena de Nigris, Concetta Schiano, and Claudio Napoli

2.1 Introduction

During angiogenesis, endothelial cells (ECs) undergo activation after binding of angiogenic factors to their receptors, release of proteases to dissolve the basement membrane, migration towards an angiogenic signal, proliferation, and an increase in cell number for new blood vessel formation. Finally, reorganization of ECs forms the three-dimensional vasculature. Tube-formation assay is one of the simple, but well-established in vitro angiogenesis assays based on the ability of ECs to form three-dimensional capillary-like tubular structures, when cultured on a gel of growth factor-reduced basement membrane extracts. During the assay, ECs differentiate, directionally migrate to align, branch, and form the tubular polygonal networks of blood vessels.

A. Casamassimi (✉) • F. de Nigris
Department of Biochemistry, Biophysics and General Pathology,
Second University of Naples, 80138 Naples, Italy

C. Schiano
Institute of Diagnostic and Nuclear Development (SDN), IRCCS, 80143 Naples, Italy

C. Napoli
Department of Biochemistry, Biophysics and General Pathology,
Second University of Naples, 80138 Naples, Italy

Institute of Diagnostic and Nuclear Development (SDN), IRCCS, 80143 Naples, Italy

U.O.C. Division of Immunohematology, Transfusion Medicine and Transplantation
Immunology (SIMT) and Regional Reference Laboratory of Immunogenetics
and Transplantation Immunology (LIT), Second University of Naples, Naples, Italy

M. Slevin, G. McDowell (eds.), *Handbook of Vascular Biology Techniques*,
DOI 10.1007/978-94-017-9716-0_2

The in vitro formation of capillary-like tubes by endothelial cells on a basement membrane matrix is a powerful in vitro method to screen for pro-angiogenic and anti-angiogenic factors. This assay can be used as a first screen before in vivo models and can be done also as a high throughput procedure. Since the first description in 1988, this assay have been used for many purposes both in cancer and cardiovascular fields, including analysis of pro-angiogenic or anti-angiogenic factors, definition of the signaling pathways involved in angiogenesis, identification of the genes regulating angiogenesis and endothelial progenitor cell characterization.

In our experimental settings we used in a model of osteosarcoma this assay to analyze the role of YY1 transcription factor and some cyclin-dependent kinases (CDK) on angiogenesis; to this aim, we performed tube formation assay to test both the effect of YY1 silencing and some inhibitors, like T22, a peptide affecting the chemokine receptor CXCR4 and roscovitine, a CDK inhibitor [1, 2].

2.2 Methodology

This angiogenesis assay consists of a quick measurement of the ability of ECs to form three-dimensional structures (tube formation) in vitro when they are plated on an appropriate extracellular matrix support. Endothelial tube formation on ECM gel mimics the in vivo environment and may be employed to test angiogenesis stimulators or inhibitors before in vivo analysis. Factors to be tested can be added exogenously to the medium, or they can be transfected or knocked down in the endothelial cells to determine their effects on angiogenesis.

As extracellular matrix support, the basement membrane extract obtained from a murine tumor rich in extracellular matrix proteins is generally used for this assay. The extract is in a liquid state at 4 °C or lower temperatures whereas it forms a gel at 16–37 °C. The routinely used basement membrane extracts are sold commercially; examples are BD Matrigel, Cultrex BME and other Engelbreth–Holm–Swarm (EHS) extracts. It is generally recommended the use of growth factor-reduced material especially to study stimulators of angiogenesis. Importantly, variability in the tube-formation activity of different preparations purchased from different vendors has also been observed. Thus, when possible, the same preparation should be used for the entire study.

Endothelial cells (ECs) typically used for this assay are obtained from human umbilical vein (HUVECs) or human aorta (HAECs), but other cells, such as lines established from mice (SVEC4-10) and humans (HMEC-1) also work well. It should be considered that ECs have considerable organ- and tissue-specific heterogeneity that may affect their response to specific factors and the time required for tube formation in vitro. The cells should be of low passages if they are primary; for instance, HUVECs should not be passaged more than 10 times, HAECs work well until the 7th–8th passage. The cells should be 80% confluent and passaged the day before the assay for optimum and consistent tube formation. The seeding density is another important point to consider with about 50,000 cells per cm^2 as recommended density for HUVECs and HAECs, but this number may vary depending on the cell source. Cells initially attach to the matrix, then migrate toward each other, align and finally form tubes. The time of the assay is very short for transformed cells compared to primary cells (3 h versus 16–20 h for primary cells), but should be determined for each EC type. Moreover, time is also dependent on the utilized matrix.

2.3 Materials (and Company Name)

Endothelial cell lines can be those with and without drug treatment or expressing the gene of interest.

2.3.1 Reagents

Growth factor-reduced BD Matrigel (BD Biosciences 354230)
Primary Human Aortic Endothelial Cell (HAEC) (Lonza CC2535)
Trypsin-EDTA
Endothelial Basal Medium-2 (EBM-2; Lonza, CC-3156)
EGM-2 SingleQuot BulleKit (Endothelial cell growth medium-2; Lonza, CC-3162)
Dulbecco's Phosphate-Buffered Saline, 1x
OPTI-MEM (Life technologies, Gibco 31985-047)
Cultrex Cell Staining Solution (Trevigen, 3437-100-01)
Methanol (Sigma, M3641)
96-well cell culture plates
15 ml conical centrifuge tubes-sterile
Tissue culture flasks, 25 cm^2, filter cap, 50 ml
Disposable sterile plastic pipettes
ImageJ software downloaded from the NIH website.

2.3.2 Equipment

Tissue culture setup
Inverted microscope with digital camera
Cell culture incubator (humidified, 5 % CO_2)
Biological hood with laminar flow and UV light
Pipette aid

Sterile micropipette
37 °C water bath
Centrifuge with a swing-bucket rotor, refrigerated
Hemocytometer (Burker chamber)

2.4 Basic Protocol

All procedures should be performed under sterile conditions in a biological safety cabinet using aseptic technique to prevent contamination. The procedure is standardized for HAEC (Lonza) as ECs and Matrigel (BD Biosciences) as matrix; however, other ECs or matrix extracts can also be used in substitution; in this case technique requires optimization. See also additional notes for reagent preparation.

2.4.1 Thawing and Passaging of Human Aortic Endothelial Cells (HAEC)

1. Prepare a bottle of complete EGM-2 as indicated. Note: the supplemented medium should be stored in the dark at 4 °C and should not be frozen. When stored in these conditions it is stable for 1 month.
2. Seed cryopreserved endothelial cells at 2.5×10^5 viable cells per a 25-cm^2 tissue-culture flask using 5mL EGM-2.
3. Change culture medium 24–36 h after seeding.
4. Change the medium every other day thereafter, until the culture is approximately 80 % confluent (5–6 days).
5. Using standard procedures to passage endothelial cells, split them when they are 80 % confluent. Usually, plating 5×10^5 to 1×10^6 cells in a 25-cm^2 flask works well.

The cells should be passaged at least twice after thawing before being used in the tube formation assay. However, the HAECs should not exceed passage 8–10.

2.4.2 Coating Plates with Matrigel

6. Remove Growth factor-reduced BD Matrigel from the freezer (−20 °C) and thaw in an ice bath in a refrigerator (4 °C) overnight. As other extracellular matrix preparations, Matrigel gels very easily; thus, it is important not to warm it during the thawing process and always to keep it on ice and to pre-chilled pipet-tips and plates.

7. Add 150–200 µl of Matrigel per cm^2 growth area and incubate plate for 30–60 min at 37 °C on a level surface to allow the gel to solidify. Note that 100 µl is necessary for each well in 96-well plates. It is very important to avoid bubble formation. If air bubbles get trapped in the wells, centrifuge the plate at 300xg for 10 min at 4 °C.

2.4.3 *HAEC Preparation for the Assay*

8. Starve cells with non-supplemented EBM-2 for 3–6 h prior to performing the assay. Then, harvest cells according to the producer instructions (Lonza).
9. Determine cell number and viability by counting cells in a Burker chamber.
10. Collect cells by centrifugation at 220xg for 5 min.
11. Aspirate supernatant and resuspend cell pellet in non-supplemented EBM-2 at a concentration of 2×10^5 cells/ml by gently pipetting up and down a few times to obtain a homogeneous single cell suspension. Be sure that cells are well mixed since cell density has an effect on tube formation (see also the point below and sect. 2.8).
12. To set the experimental points, dilute cells in non-supplemented EBM-2 in the presence or absence of angiogenesis inducers and inhibitors to be tested. Use appropriate negative and positive control samples (non-supplemented EBM-2 and complete EGM-2, respectively).

2.4.4 *Starting the Assay and Analyzing Data*

13. Prepare the appropriate number of HAEC cells in 1.5 ml tubes, according to the number of target cell lines to be used. Note: Per each test, one tube of HAEC will be suspended with a sufficient amount of cells to be dispensed in three wells. Thus, a triplicate will be done per each experimental point. Note that each well requires 100 µl of the HAEC cell suspension (corresponding to 2×10^4 cells/well).
14. Dispense 100 µl of the HAEC cell suspension obtained thorough mixing into the labeled wells of a 96-well plate. Be careful not to touch the surface of the gel when adding the cells to avoid injuring the gel. Incubate the plate at 37 °C, 5 % CO_2 for a period of 4–16 h, or until the desired result is achieved.
15. Visualize the cells every hour for tube formation under an inverted light microscope with 4× or 10× objective.
16. Photograph the capillary network in the wells using a digital camera attached to the inverted microscope (see also paragraph 8).
17. Analyze data using the ImageJ software (Fig. 2.1).

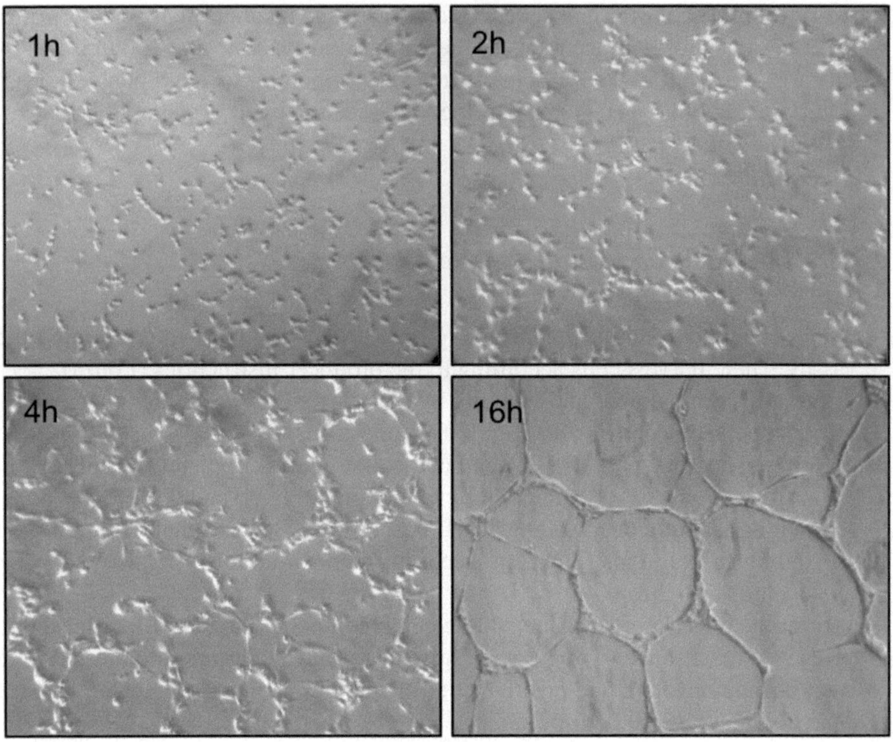

Fig. 2.1 Representative images of tube formation assay on the growth factor-reduced Matrigel by HAEC after incubation with supplemented EGM-2 for 1, 2, 4 and 16 h

2.5 Examples of Experimental Plans

This assay can be used to test in vitro a novel molecule and verify its potential as pro-angiogenic or anti-angiogenic factor. The angiogenic activity of one molecule at different concentrations or of more molecules can be easily tested simultaneously. To this purpose, cells (HAECs or alternatively HUVEC or other ECs) are diluted in non-supplemented EBM-2 medium in the presence or absence of the molecule to be assayed.

Alternatively, you can use conditioned medium from a cancer cell line to study its potential to activate angiogenesis through the release of angiogenic factors in the medium. In this setting you can test the cancer cell line under different experimental conditions. For instance, you can study the involvement of a particular gene in this process by utilizing the cancer cell line transfected to overexpress or silence that particular gene. Indeed, in our studies we have used an osteosarcoma cell line (SaOS) both un-transfected and silenced for the transcription factor YY1 [1, 2].

2.6 Sample Results

2.6.1 Data Interpretation

In this assay activated ECs form cellular networks from capillary tubes sprouting into the matrix. The formation of these networks is a dynamic process, which begins with cell migration and alignment, followed by the development and sprouting of capillary tubes, and ends with the formation of cellular networks. Although this is designed as a qualitative assay, it is also possible to quantitate the extent of the formed cellular networks. Several methods have been used for quantitation. Simple visual assessment by a blinded observer, including scoring of the tube quality or counting branch points, can provide an accurate evaluation even though often it is necessary to have an objective quantitative method, which is more rapid and easy to perform for the investigator. Many labs measure tube area, whereas others measure tube number, tube length or number of sprouts or combinations of these measurements. Some labs have adapted equipment or developed their own programs for tube formation analysis. Free software is also available, like ImageJ that can perform these measurements on the images obtained by microscopy. Similarly, website services, such as S.CORE, Angiosys and Wimasis, can also analyze data on acquired images (Fig. 2.2).

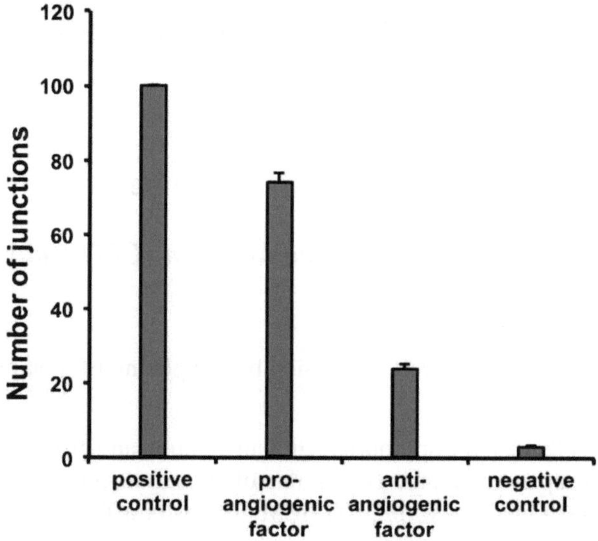

Fig. 2.2 Evaluation of the effect of pro-angiogenic and anti-angiogenic factors on HAEC tube formation. In this example of results analysis, number of junctions is represented as percent of positive control. Similar results can be plotted for tube length and for tube number

2.7 Troubleshooting

When Matrigel is too viscous, probably it is not thawed completely. Thus, keep it at 4 °C on ice until completely thawed. Another cause can be that the tube with matrigel has warmed up and started to gel. Keep the extract on ice or at 4 °C for a few hours or overnight before use.

If bubbles occur in the matrigel before or after coating, centrifuge the tube or the plate at 4 °C for 10 min at 300xg. To prevent the formation of an irregular matrigel surface, avoid touching the gel surface with the pipette tip.

Different cell density in the wells of the 96-well plate can result when cells have not been mixed properly. To avoid this problem, carefully invert the tube with cells or pipette cell suspension up and down for several times before loading the cells on the top of gelled matrigel.

When the quality of cell preparation is poor (i.e. cells from a late passage), cells may not adhere to the matrix or not to form tubes; in the last case also the number of cells used in the assay is a critical point. Indeed, if the number of cells per well is too low for the specific EC line used in the assay, no fully formed tubular structures can be observed also in the presence of pro-angiogenic factors and, hence, in the positive control. On the other hand, when the number of cells per well is too high for the specific EC line used in the assay, tube formation is observed also in the wells with a basal medium without any angiogenic factors (negative control). Thus, it would be useful to titrate the number of cells per well to establish optimal density for the cell line utilized before starting with the entire experiment. Moreover, to avoid tube formation in the negative control, be careful to properly starve cells before the assay.

2.8 Method Variations/Alternative Staining

2.8.1 Preparation of Conditioned Medium (CM) from Target Cell Line(s)

CM can be used to make the cell suspension before plating ECs on matrigel-coated plates.

1. Seed target cells with the appropriate growth medium and grow them to 30–40 % confluence (depending on the growth rate of the cell lines).
2. Replace growth medium with OPTIMEM (10 ml for a T75 tissue culture flask;) for 24–48 h.
3. Collect the CM, when cells reach 60–80 % confluence in a T75 tissue culture flask.
4. Spin at 200xg to eliminate cellular debris from the CM.
5. Make 0.5 ml aliquots of the CM and store at −80 °C if it is not immediately used after collection.

2.8.2 Fixation and Labeling with Cell Staining Solution

During all steps be careful to gently aspirate and load the following solutions in order to avoid damaging gel and the cellular network.

1. Remove the medium from the wells and rinse the wells three times with 100 μl of PBS per each well.
2. Add 100 μl of cold methanol per well and incubate the plate for 30 s–1 min. Note that longer times will result in the appearance of large precipitates of the matrix proteins that can interfere with the imaging process.
3. Rinse the wells three times with distilled water.
4. Add 100 μl of cell staining solution (a mixture of Azur A and Methylene Blue) per well and incubate the plate for 15–30 min at room temperature.
5. Rinse the wells three times with distilled water.
6. Photograph tubular networks and analyze images for quantitation.

2.9 Additional Notes

2.9.1 EGM-2 Preparation

Add all supplements and growth factors of the EGM-2 SingleQuots BulleKit to EBM-2 and store at 4 °C for up to 1 month.

2.10 Applications and Discussion

One of the most widely used in vitro assays to model the reorganization stage of angiogenesis is the tube formation assay [3]. The assay measures the ability of endothelial cells to rapidly form capillary-like structures in vitro when plated at sub-confluent densities on top of a reconstituted extracellular matrix support. Over the past several decades, researchers have typically employed this assay to determine the ability of various compounds to promote or inhibit tube formation. Anti-angiogenic factors could be useful in various diseases, including cancer, where tumors stimulate new blood vessel formation to receive oxygen and nutrients to grow.

The advantages of this assay are that it is quite easy and quick to perform, is quantifiable, and is suitable to high-throughput analysis. It also allows in vitro modeling of endothelial cell behavior, including survival, apoptosis, and the steps leading to capillary formation and invasion. A disadvantage of this assay is the great variation in the data obtained with different lots of ECs and support matrixes, thus rendering the choice of these resources crucial to get consistent and reliable data.

Recently growing evidence has shown that this assay is not limited to test vascular behavior for ECs, since certain non-endothelial cell types have also been reported to form capillary structures in vitro. Indeed, tube formation assay has also been used to test the ability of a number of tumor cells to develop a vascular phenotype, a process known as vasculogenic mimicry (VM). Tumor cell-mediated VM has been demonstrated to play a vital role in the tumor development, independent of EC-directed angiogenesis [4]. This vascular phenotype is also dependent on cell numbers plated on the Matrigel. Therefore, this assay may be useful to screen the vascular potential of a variety of cell types including vascular cells, tumor cells as well as other cells.

References

1. de Nigris F, Crudele V et al (2010) CXCR4/YY1 inhibition impairs VEGF network and angiogenesis during malignancy. Proc Natl Acad Sci U S A 107:14484–14489
2. de Nigris F, Mancini FP, Schiano C, Infante T, Zullo A, Minucci PB, Al-Omran M, Giordano A, Napoli C (2013) Osteosarcoma cells induce endothelial cell proliferation during neo-angiogenesis. J Cell Physiol 228:846–852
3. Arnaoutova I, George J, Kleinman HK, Benton G (2009) The endothelial cell tube formation assay on basement membrane turns 20: state of the science and the art. Angiogenesis 12:267–274
4. Seftor RE, Hess AR, Seftor EA, Kirschmann DA, Hardy KM, Margaryan NV, Hendrix MJ (2012) Tumor cell vasculogenic mimicry: from controversy to therapeutic promise. Am J Pathol 181:1115–1125

Key Web-Addresses for Further Information

Web site address to download ImageJ software. http://rsbweb.nih.gov/ij/

Chapter 3
Induction of Hypoxia in Vascular Endothelial Cell Culture

Hyun-Young Koo, Meredith Millay, and Tsutomu Kume

3.1 Introduction

O_2 is an essential molecule for the function and survival of cells. It is required for almost every cellular process, including cell metabolism, differentiation and division. In mammals, O_2 distribution to the body is regulated by the vascular system. Therefore, in vivo physiological O_2 concentrations can range from 1 % to 13 %. However, most in vitro studies are completed at atmospheric conditions (20–21 % O_2). Hypoxia is defined as the reduction or lack of O_2 in organs, tissues or cells. It generally occurs at 1 % O_2, even though each tissue and cell has a different ability to adapt to this condition [1]. Studies have shown differences in gene expression profiles in low O_2 conditions compared to atmospheric conditions [2]. Hypoxia is an important component of various ischemic diseases [3], including ischemic heart disease and ischemic limb. Hypoxic regions in these ischemic diseases are caused mainly by reduced blood supply into the tissue. Decreased blood supply limits O_2 and nutrient supply to the tissues and induces hypoxia. Hypoxia also occurs in solid tumours [4] because solid tumours grow very rapidly and require a large blood supply. When new blood vessel formation is not fast enough to meet the requirements of tumour growth, the central region of the solid tumours becomes hypoxic.

There are several ways to generate hypoxic conditions in cell cultures. Previous studies have shown that cobalt chloride ($CoCl_2$) mimics the hypoxic/ischemic condition, including the activation of hypoxia-inducible factor-1 (HIF-1) [5, 6]. Another method involves using a hypoxia chamber with a two-gas (carbon dioxide and nitrogen) supply. A cell culture incubator with a water tray needs to be placed

H.-Y. Koo • M. Millay • T. Kume (✉)
Feinberg Cardiovascular Research Institute, Feinberg School of Medicine,
Northwestern University, Chicago, IL 60611, USA
e-mail: t-kume@northwestern.edu

M. Slevin, G. McDowell (eds.), *Handbook of Vascular Biology Techniques*,
DOI 10.1007/978-94-017-9716-0_3

inside of the hypoxia chamber to maintain the temperature and humidity (Fig. 3.1). In this chapter, we will describe the latter method of inducing hypoxia in mammalian cell cultures with evaluation of the hypoxic status of these cells.

3.2 Methodology

3.2.1 Exposure of Cells to Ambient Atmospheric Hypoxia

Cells are grown in normal tissue culture conditions and placed in a hypoxia chamber (5 % CO_2, balanced N_2, 37 °C, humidified). Typically, O_2 levels in the chamber are maintained at 1 %. The manipulation of the cells and preparation of the cell extracts is completed inside of the chamber while maintaining a hypoxic environment. This manipulation is critical to exclude any possible complicating responses associated with reoxygenation. The oxygen partial pressure (PO_2) values of the medium equilibrate with the atmospheric values within approximately 20–30 min. Therefore, in studies of acute hypoxia, the medium is pre-equilibrated overnight in the hypoxia chamber. This hypoxic medium is applied to the cells in the hypoxia chamber at the beginning of the experiments, and in addition to the medium, PBS is pre-equilibrated in the hypoxia chamber. Our in-house protocol is outlined below.

3.2.2 Hypoxia Chamber

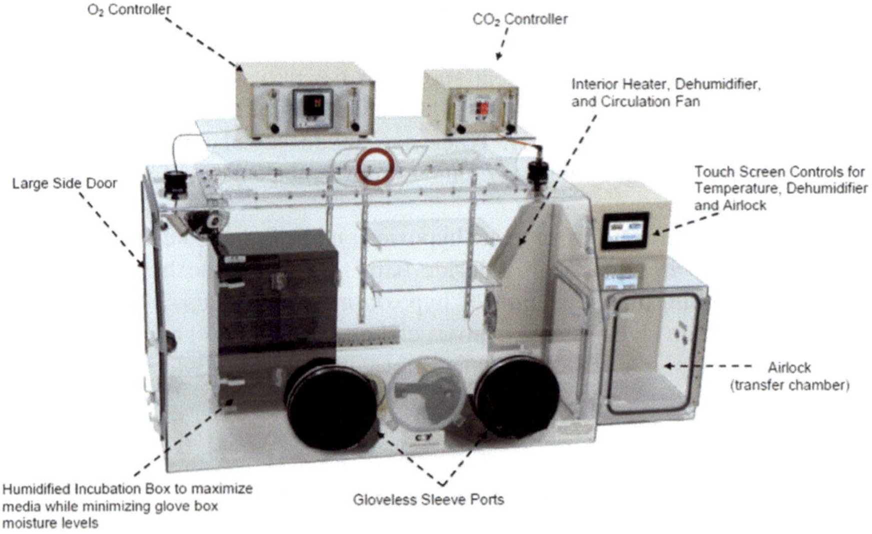

Fig. 3.1 Hypoxia chamber (Photo Courtesy Coy Laboratory Products Inc.)

3.3 Materials (and Company Name)

1. Incubator or oven at 37 °C (VWR, 2,350T)
2. Hypoxia chamber (Coy Laboratory Products, Inc., Custom built)
3. PBS (Corning, 21-040-CV)
4. EndoGro VEGF Complete Media kit (Millipore, SCME002)
5. Trypsin (Hyclone, SH30042.01)
6. Six well plates (BD, 3046)
7. 175-cm^2 polystyrene flasks (Corning, 431,080)
8. Conical tubes (Corning, 430,790)

3.4 Basic Protocol

3.4.1 *Cell Culture*

Human umbilical vein endothelial cells (HUVECs) are maintained in culture under atmosphere conditions (37 °C, 20 % O_2, and 5 % CO_2).

Procedure

1. Warm medium, trypsin and PBS to 37 °C in a bath.
2. Remove cells from incubator. Rinse two times with 3 ml PBS/trypsin, removing with suction.
3. Add 3 ml of trypsin to each plate and let sit for 5 min. It is best to check for cell detachment by shaking the flask sideways very gently.
4. Add 3 ml of medium, pipetting trypsin-medium mix in and out vigorously in the centre of the plate.
5. Transfer the mix to a conical tube.
6. Count cells.
7. Spin the tube at 1,000 rpm for about 5 min.
8. Remove the supernatant, resuspend the pellet in 1 mL of media and count the cells using a haemocytometer.
9. Plate 2×10^6 cells into each T175 flask.
10. Put the flasks in the incubator at 37 °C.

3.4.2 *Hypoxia Treatment*

Experiments are performed either 1 day or 2 days after plating cells.

1. (Day 1) Plate 150,000–200,000 cells per well in 2 mL of medium into a six-well plate.
2. (Day 2, optional) Change the medium to the EndoGRO medium with no growth factors and 0.5 % serum for 24 h.

3. Place medium and PBS in the hypoxia chamber to pre-equilibrate overnight. Caps on bottle/tube should be left loose to allow gas exchange.
4. (Day 3) Move plates from culture incubator to the hypoxia chamber.
5. Change medium to pre-equilibrated medium.
6. Incubate cells for desired time at 1 % O_2, 5 % CO_2, and 37 °C in the incubator inside the hypoxia chamber.
7. Harvest cells for further analysis (RNA or protein).

3.4.3 *Evaluation of Hypoxia*

Hypoxia-inducible factor-1 (HIF-1) is the major transcription factor specifically activated during hypoxia, and HIF-1α is targeted for proteasome degradation by ubiquitination in normoxia [7]. In hypoxia, HIF-1α is stabilized and translocates to the nucleus and binds its consensus DNA sequence (present in the promoter/enhancer elements of the target genes). Active HIF-1 induces the expression of different genes, such as vascular endothelial growth factor-A (VEGF-A) and different glycolytic enzymes, as well as plays an adaptive role in the new conditions induced by hypoxia [8]. The stabilization of this protein is a hallmark of hypoxia; therefore, detection of HIF-1α's protein levels is routinely used to evaluate for this condition. Alternatively expression of HIF-1α target genes, such as VEGF-A, can be measured.

3.5 Sample Results

To evaluate hypoxic effects in culture of HUVECs, VEGF-A expression was detected by real-time polymerase chain reaction (RT-PCR) (Fig. 3.2).

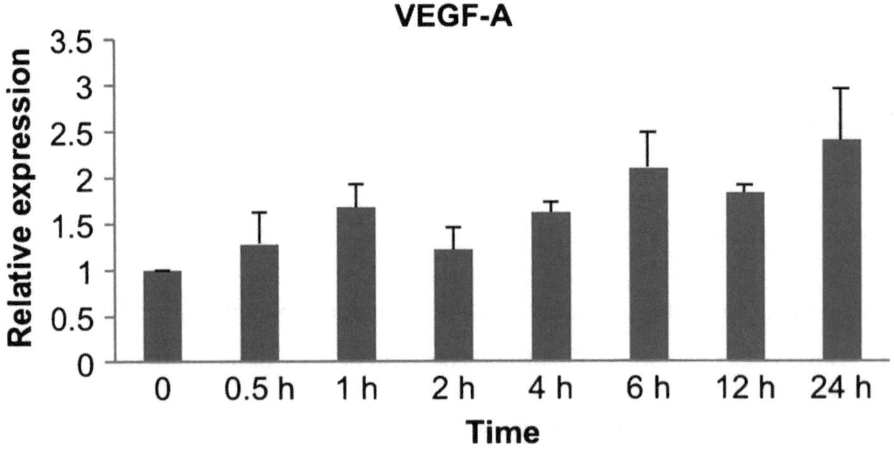

Fig. 3.2 VEGF gene expression under hypoxia in HUVECs (Unpublished data, Koo HY et al.)

HUVECs were treated with hypoxia (1 % O_2, 5 % CO_2) for 0.5, 1, 2, 4, 6, 12 and 24 h. Total RNA isolated with Trizol reagent (Invitrogen) was used for cDNA synthesis with an iScript cDNA synthesis kit (Biorad). cNDA transcripts were then amplified by RT-PCR with a human VEGF-A primer set. The result was normalized to human GAPDH mRNA.

References

1. Leach RM, Treacher DF (1998) Oxygen transport-2. Tissue hypoxia. BMJ 317(7169): 1370–1373
2. Wenger RH (2000) Mammalian oxygen sensing, signalling and gene regulation. J Exp Biol 203(Pt 8):1253–1263
3. Kim HA, Mahato RI et al (2009) Hypoxia-specific gene expression for ischemic disease gene therapy. Adv Drug Deliv Rev 61:614–622
4. Semenza GL (2007) Hypoxia and cancer. Cancer Metastasis Rev 26:223–224
5. Dai ZJ, Gao J et al (2012) Up-regulation of hypoxia inducible factor-1α by cobalt chloride correlates with proliferation and apoptosis in PC-2 cells. J Exp Clin Cancer Res 31:28
6. Torii S, Kurihara A et al (2011) Inhibitory effect of extracellular histidine on cobalt-Induced HIF-1alpha expression. J Biochem 149:171–176
7. Brahimi-Horn C, Mazure N et al (2004) Signalling via the hypoxia-inducible factor -1α requires multiple posttranslational modifications. Cell Signal 17:1–9
8. Semenza GL (1999) Regulation of mammalian O_2 homeostasis by hypoxia-inducible factor 1. Annu Rev 15:551–579

Chapter 4
Evaluating In Vitro Angiogenesis Using Live Cell Imaging

Elen Bray and Mark Slevin

4.1 Introduction

In today's research laboratories the use of microscopy most commonly involves the manual visualisation and capture of experimental examples at specific time points, normally at the beginning and end of a study. Although informative these approaches are often labour intensive, subjective and run the risk of vital information being missed or misinterpreted, as data from only a fraction of the experimental time period is collected. In addition, information on cellular movement, interactions and behaviour cannot be obtained using such techniques.

The ability to continually monitor and obtain kinetic experimental data has become a desired tool in many research fields, and as a result many microscopes have since added an environmental chamber to their platform, offering a form of live cell kinetic imaging. In practice however, sample throughput and analysis capabilities are often limited on these systems with environmental conditions not remaining optimal for long term experiments.

Recent advances in image acquisition technology and software algorithms have enabled specialised live cell imaging platforms to be developed. One such system is Cell-IQ™, which combines optimised incubation conditions with phase contrast/fluorescence imaging and analysis, allowing long term kinetic cell studies to be performed for extended periods of time.

The additional information that kinetic live cell imaging can provide makes it a suitable and appealing tool for a large number of fields including stem cell research [1–4], oncology [5], neuroscience research [6, 7], drug discovery [8] and angiogenesis [9].

E. Bray (✉)
CM Technologies Oy, Biokatu 12, 33520 Tampere, Finland

M. Slevin
School of Healthcare Science, Manchester Metropolitan University, Manchester, UK

M. Slevin, G. McDowell (eds.), *Handbook of Vascular Biology Techniques*,
DOI 10.1007/978-94-017-9716-0_4

Angiogenesis plays a critical role in the repair, replacement and cellular survival following tissue damage. The in-vitro angiogenic assays commonly used to assess the biological activity of potential pro- or anti-angiogenic therapies include cell migration and tubule formation assays. Traditionally cell migration is measured using a 2D wound healing assay, an in-vitro assay where a scratch is made through the centre of a confluent cellular layer and the rate of healing calculated via changes in wound width measurements and wound area. The formation of tubules is commonly aided by culturing the cells at specific densities on extracellular matrices such as matrigel.

Measurements for both assays traditionally involve the manual capture of images at specific time intervals, followed by manual measurements using basic software tools.

The following chapter aims to discuss how kinetic live cell imaging is an appropriate and valuable tool for studying the in–vitro angiogenic process in real time.

4.2 Methodology

4.2.1 Instrumentation- Cell-IQ®

Cell-IQ® is a fully integrated live cell imaging and analysis platform. The temperature controlled incubator, tailored gaseous environment and the use of light emitting diodes for both phase and fluorescence imaging creates an optimised cellular environment, allowing biological responses to be monitored in real time from a few hours to many weeks.

The incubator is integrated into the system (+/–0.5 °C) and houses two independent controlled gas lines. These gas lines feed directly into the culture vessel through specially designed cell secure™ lids. The gas mix is user defined and as the two lines are independent, different gas mixes can be fed to the two positions, allowing for example normoxic and hypoxic studies to run simultaneously.

Images are captured using a CCD camera, and a high precision motorised stage (XY = +/–1 μM Z = +/–0.4 μM) controls the movement of the culture vessel, allowing images to be acquired from any X, Y, Z position within the plate/dish/slide/flask.

Unlike many traditional microscopes the Cell-IQ® is able to image samples using variable sized Z-stacks (user defined), accommodating both 2D and 3D structures [10]. The final images are generated by creating a unique 'all-in-focus image'. This is achieved by collating together the best focused pixels from each Z level at every XY position (Diagram 4.1). The generated high quality single image is ideal for data processing [11].

Additional imaging features include 'auto focus', a focusing tool which will automatically focus the region of interest for the user and 'auto-adjust focus'.

Independent to the 'auto-focus' the 'auto-adjust focus' enables the defined Z-stack to re-adjust its position over time as the cells/structures change position. The benefit of such technology is not only can cells/structures which exist in different planes be focused simultaneously, but they will also remain in focus as their positions move through the different Z planes.

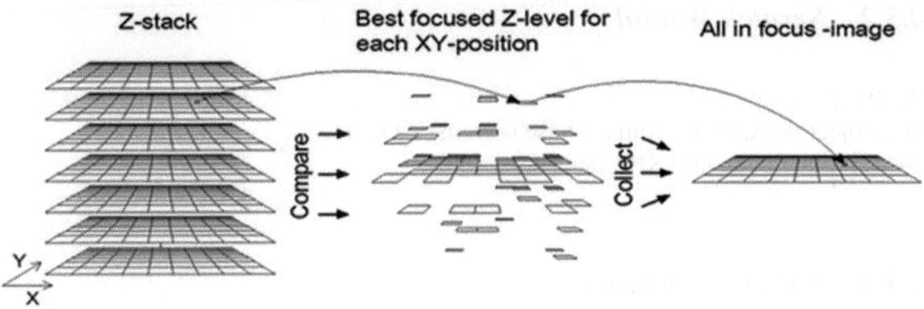

Diagram 4.1 A schematic diagram detailing the generation of an' all in focus image' created by Imagen software

Accompanying the hardware is the Cell-IQ® analyser™ software programme. This software utilises 'machine vision technology', pattern recognition software which is traditionally used in medical imaging and precision robotics [12].

This software enables the user to teach the system to recognise different patterns, allowing user defined protocols to be created that distinguish between morphologically different structures, for example different cell types or different phases such as live and dead cells. Once taught this cell counting and classification tool is able to automatically identify and quantify the specified cellular structures in each of the captured images.

Additional tools based on threshold measurements are also provided as separate tools in the software. These tools automatically measure changes in area or wound widths, fluorescence intensity and area, which in conjunction with the machine vision capabilities facilitates all aspects of the in vitro angiogenesis process to be quantified automatically.

4.3 Materials

4.3.1 General

Cell-IQ® Imaging and analysis platform (including appropriate objective)
Cell Secure™ lids
Sterile phosphate buffered saline (PBS)
Pressure Regulator

- Single stage cylinder regulator
- Outlet pressure 0.2–3 bar
- Valve connection DIN 10 and 6 mm (tubing)

Electrical Tape
0.2 μM Filters
Premixed gaseous mixture (e.g. 5 % CO_2 mixed with air)

4.3.2 Scratch Wound

P200 pipette tip
10x objective (CFI Achromat Phase contrast ADL 10x)
24 well plate (standard SBS plate footprint)

4.3.3 Tubule Formation

Growth factor reduced matrigel (BD Biosciences)
10x objective (CFI Achromat Phase contrast ADL 10x)
48 well plate (Standard SBS plate footprint)

4.4 Basic Protocol

4.4.1 Scratch Wound (Mouse Embryo Fibrosarcoma Cells)

Nine wells of a twenty-four well plate were seeded with mouse embryo fibrosarcoma cells (generated from single VEGF isoform expressing mice, specifically the VEGF164 isoform) [13], and incubated at 37 °C/5 % CO_2 until the cells achieved 100 % confluence. The cells were then washed with PBS and an artificial wound created using a commercially available automated wound maker. Prior to imaging, the cells were washed with PBS (Invitrogen) and the media replenished with, (i) fresh culture medium (high glucose DMEM (Invitrogen) medium containing L-glutamine, Fetal Calf Serum (FBS), and the antibiotics G-418 and puromycin), (ii) fresh culture medium plus drug X, (iii) fresh culture medium plus drug Y. Triplicate wells of the treatment were set up.

The plate was subsequently prepared for imaging. In order to prevent media evaporation, PBS was added to all the empty wells to help humidify the flowing gas. The plate was then secured with a cell Secure™ lid and sealed using electrical tape, creating an enclosed environment.

Using the Cell IQ™ Imagen software, two regions of interest (ROI) were selected from each well and imaged automatically every 30 min for a period of 65 h. Images were captured using a Z stack of 20 μM with the 'auto-adjust focus' activated.

Data is presented by the change in wound width over time and by the percentage by which the original scratch wound width has decreased.

The captured images were subsequently analysed using the Analyser™ software scratch wound analysis tool. Using threshold measurements, the tool is able to calculate the closure of the wound automatically over time.

4.4.2 Co-culture Scratch Wound (HaCaT and L929 Cells)

Tissue culture plates (24 well) were collagen-coated by adding 0.2 mg/ml of collagen type I solution (Sigma, Dorset UK) for 2 h at 37 °C before rinsing with PBS (Invitrogen).

Each well was seeded with HaCaT (keratinocyte cells) and L929 (Fibroblasts cells) to give a final density of 100,000 cells per well (equal numbers of each cell type) and maintained at 37 °C and 5 % CO_2 for 24 h to permit cell adhesion and the formation of a confluent monolayer. These confluent monolayers were then scored with a sterile pipette tip to leave a scratch of approximately 0.4–0.5 mm in width. Culture medium was then immediately removed (along with any dislodged cells) and replaced with (i) fresh serum supplemented culture medium (10 % FCS), or (ii) conditioned medium which had been generated from culturing Mesenchymal stem cells under serum free conditions (MSC-CM). Each treatment was performed in quadruplicate [14].

Prior to imaging with Cell-IQ®, the plates were prepared as described above. Pre-selected fields were imaged continually (every 6 min) until wound closure was complete.

Data is presented as the percentage by which the original scratch wound width has decreased for each given time point. In addition, for the duration of the closure, both L929 and HaCaT cell numbers were quantified using the automated cell counting and classification tool to try and establish if both or just one cell type was contributing to the closure of the wound.

4.4.3 Tubule Formation Assay

A pre-chilled 48 well plate was coated with 100 μl of undiluted growth factor reduced matrigel (BD Biosciences) on ice. In order to allow the matrigel to polymerise, the plate was incubated for an hour at 37 °C. Human umbilical vein endothelial cells (HUVECs) were seeded at a density of 4×10^4cells/well and the plate prepared for live cell imaging as previously described.

In order to monitor larger areas, a large field of view can be generated by adding a number of positions as a grid and stitching them into a single image. For this experiment a 3×3 grid was added and the cells were imaged every 30 min for a period of 22 h. In order to ensure the three dimensional tubule structures remained in focus over time a 50 μM Z-stack with the 'auto-adjust feature' activated was applied in the imaging settings.

Images were subsequently analysed using the tubule and branch-point Analyser™ tool, and the data is presented as total tubule length and the total number of different branch points over time.

4.5 Results

4.5.1 Scratch Wound (Mouse Embryo Fibrosarcoma Cells)

The data generated clearly illustrates that both treatments (drug X and Y) enhance wound healing, accelerating the rate of closure (Figs. 4.3 and 4.4). The automated results allow closure times to be easily and accurately identified, allowing the exact difference between the rates of closure for different treatment groups to be documented.

4.5.2 Wound Closure: Co-culture (HaCaT and L929 Cells)

Below are sample results showing scratch wound closure assay at increasing time points culminating in complete resolution.

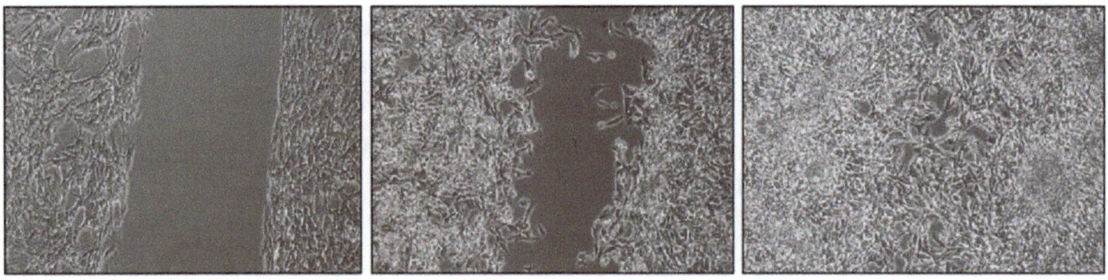

Fig. 4.1 Example of the captured scratch wound using the Cell-IQ Imagen acquisition software

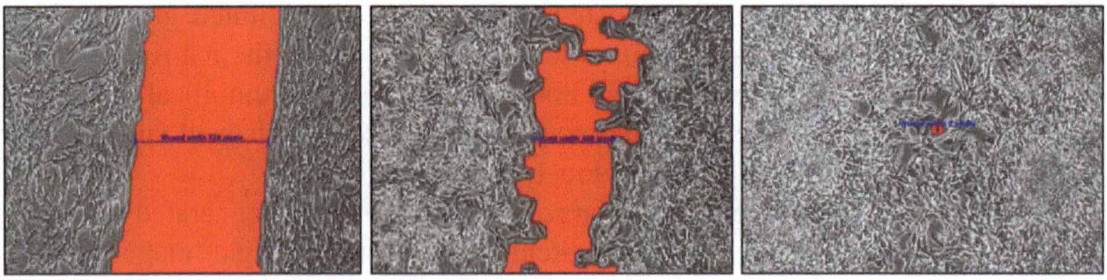

Fig. 4.2 Example of the analysed scratch wound using Analyser software

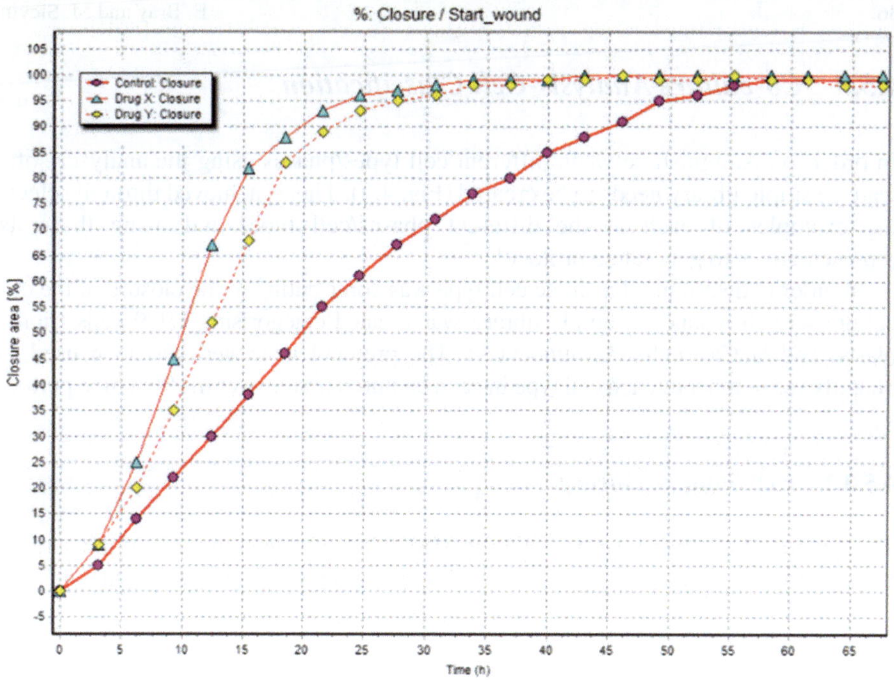

Fig. 4.3 Graphical results illustrating the percentage closure of the scratch wounds over time (Data obtained from two positions in three replicate wells ($n=6$))

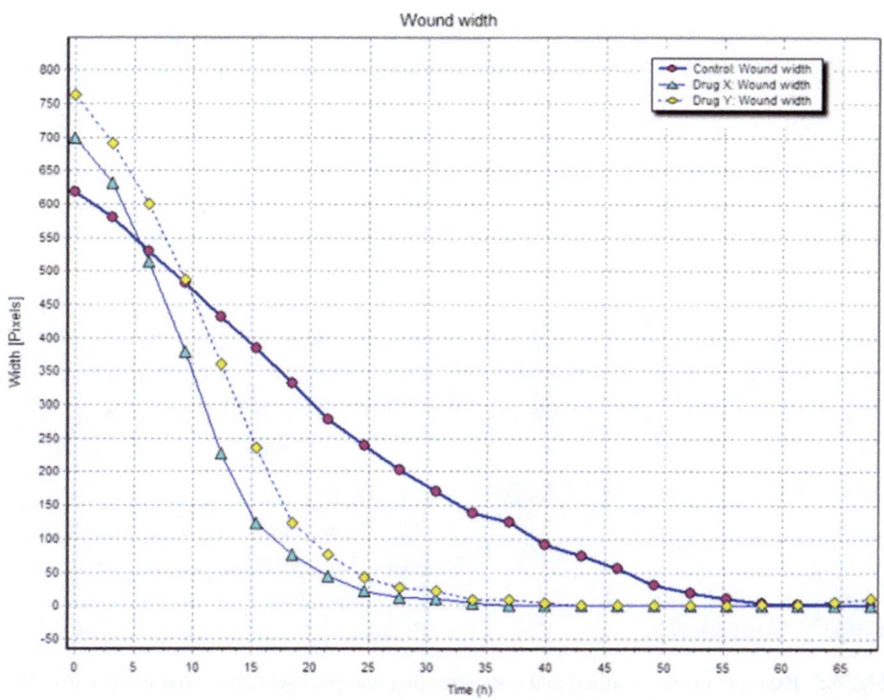

Fig. 4.4 Graphical results illustrating the change in wound width over time (Data obtained from two positions in three replicate wells ($n=6$))

4.5.3 Co-culture Analysis-Cell Classification

In order to distinguish between different cell types/phases using the analyser software a sample library needs to be created (Fig. 4.7). This is achieved through selecting examples of each of the different phases/cell types and using these as representatives to generate a protocol.

In order to determine if a single cell type was contributing to the closure of the co-culture scratch wound, a sample library was created categorising L929 cells (fibroblasts) and HaCaT cells (keratinocytes). The protocol generated was then used to quantify the number of each cell type in each captured image over the 96 h time period.

4.5.3.1 Cell Sample Library

The generated results illustrated that the MSC-CM enhanced the rate of the scratch wound closure, accelerating the closure time by approx. 30 h (Figs. 4.5 and 4.6).

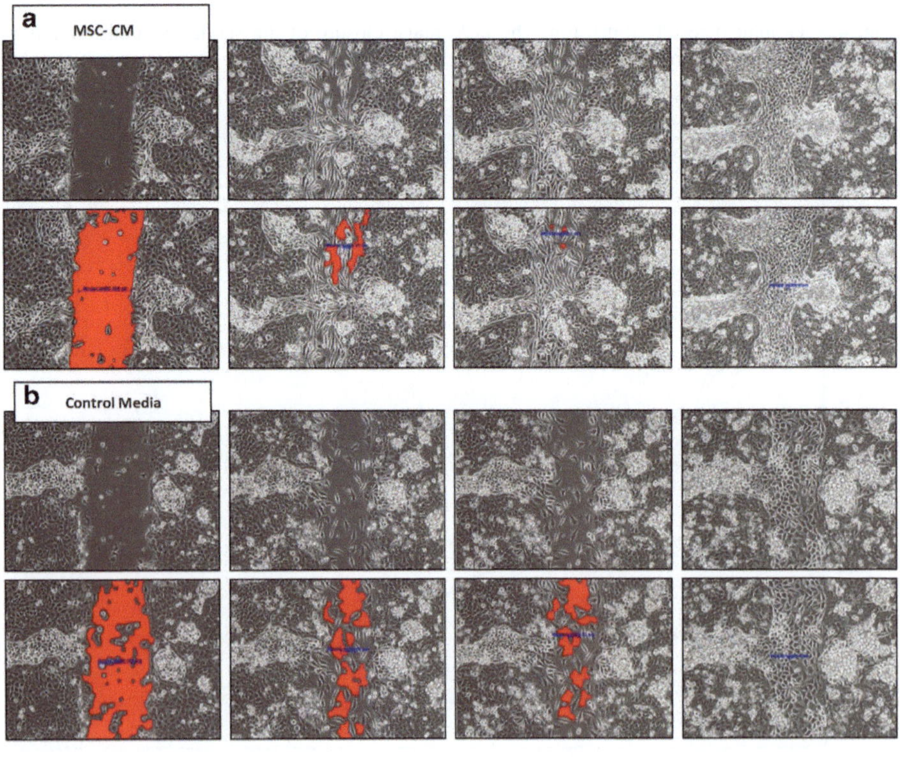

Fig. 4.5 Example of the captured and corresponding analysed scratch wound images over 96 h using (**a**) conditioned MSC media (**b**) control media

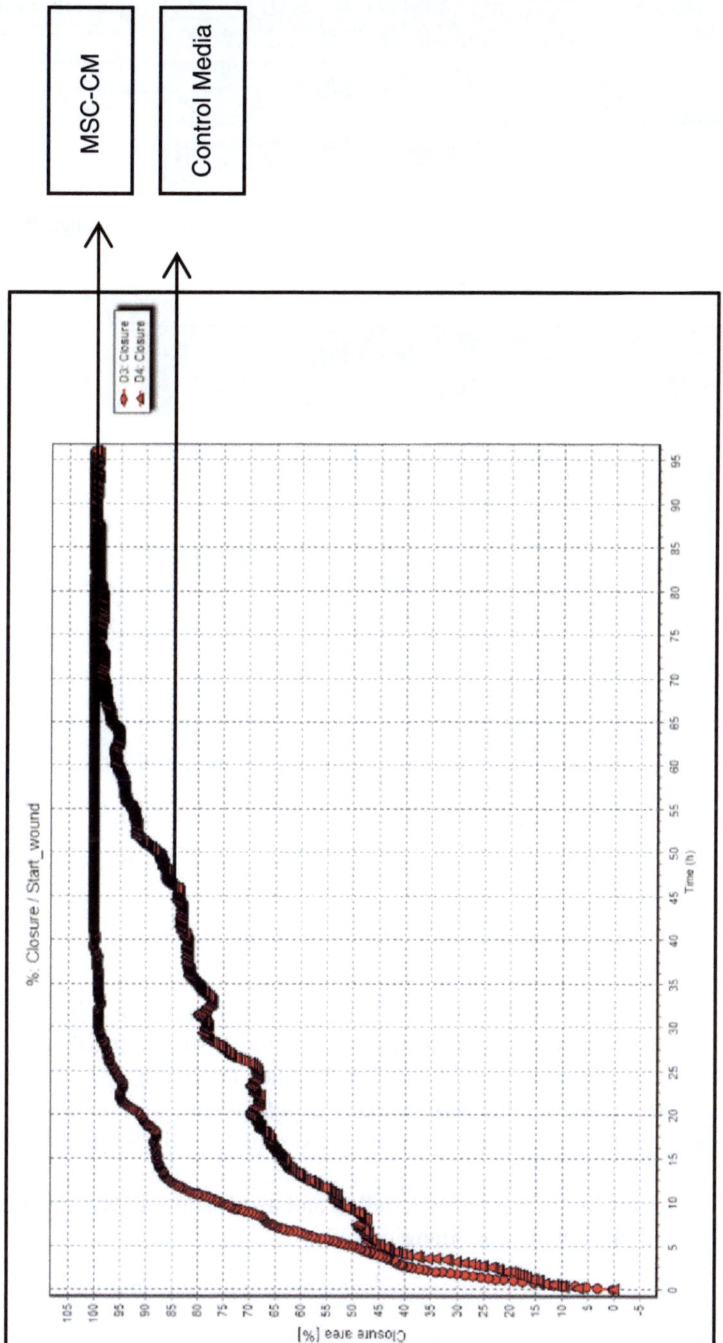

Fig. 4.6 Graphical results illustrating the percentage closure of the scratch wounds over a 96 h time period

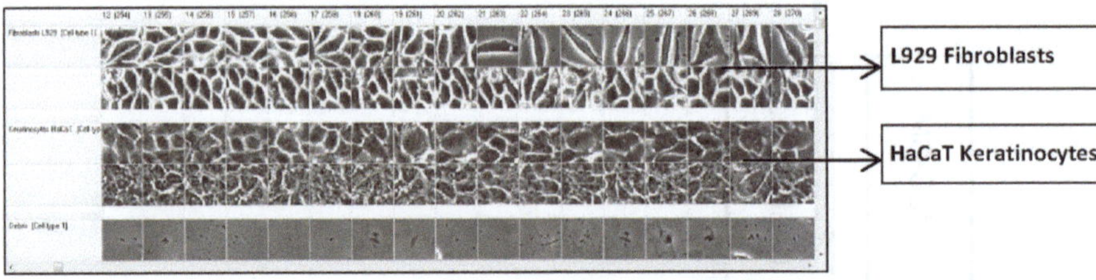

Fig. 4.7 A snapshot of the sample library used to generate the protocol to distinguish between L929 and HaCaT cells

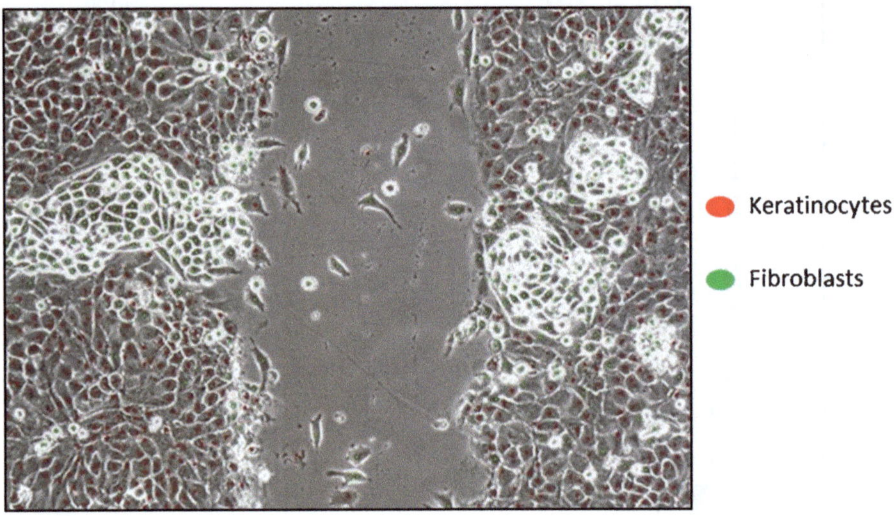

Fig. 4.8 Analysed image demonstrating the identification of the different cell types in the absence of any fluorescent labelling. The *red dots* indicate the identification of HaCaT keratinocyte cells, and the *green dots* represent the L929 fibroblast cells

The quantification of the different cell types revealed an increase in the number of L929 fibroblast cells during the closure of the wound, suggesting that it was the L929 fibroblast cells that migrated into the wound, facilitating its closure. This result is seen irrespective of the media used (Fig. 4.9).

The above detailed classification results were subsequently verified by fluorescent data.

Both cell types were fluorescently labelled and a snapshot of their location over the 96 h obtained through manual imaging [12].

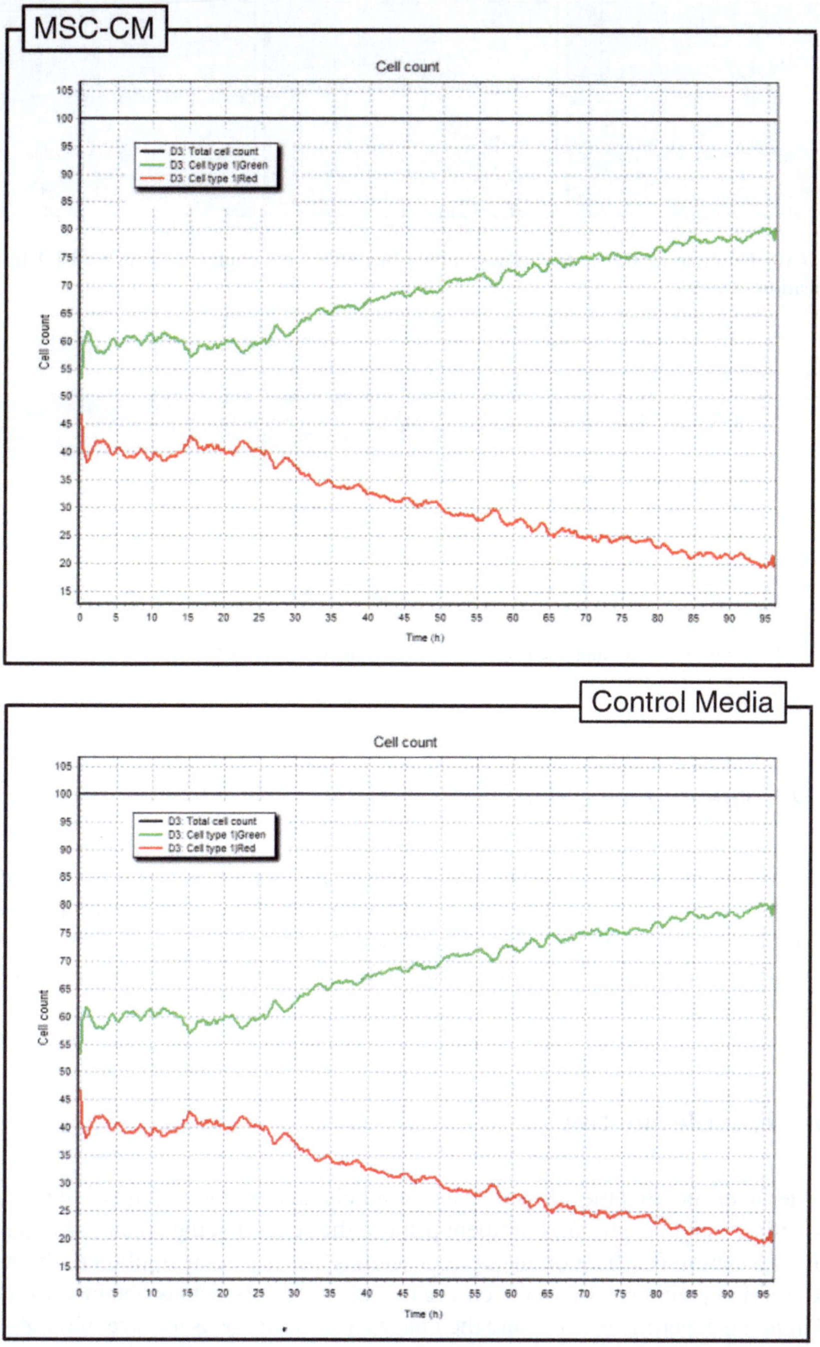

Fig. 4.9 Quantification data demonstrating the percentage of each cell type identified during the closure of the wound over 96 h

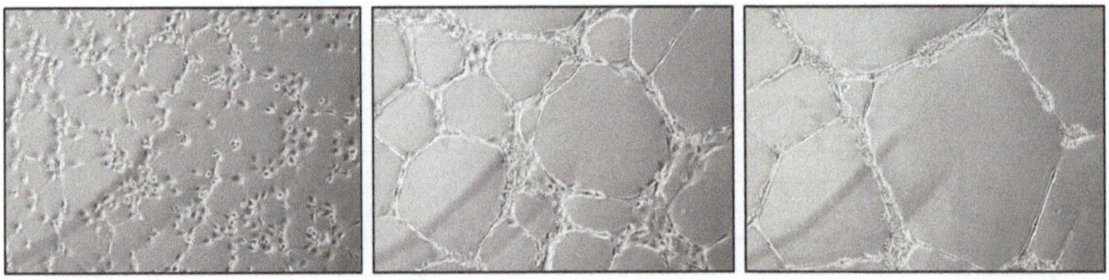

Fig. 4.10 Example of the cells forming tubule like structures captured using Cell-IQ Imagen acquisition software

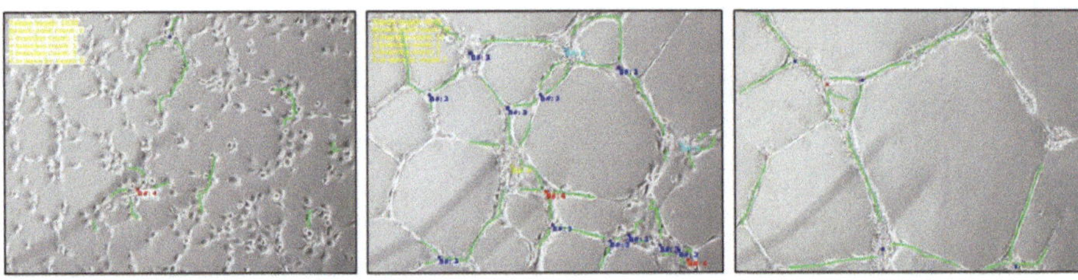

Fig. 4.11 Example of the analysed images using the analyser tubule and branch-point measurement tool

4.5.4 Tubule Formation

As the captured images illustrate, the increased Z-stack of 50 μM ensures that the tubule like structures remain in focus as they form and move in different Z planes over time (Fig. 4.10). As the tubule like structures form the tubule length increases, followed by a small decrease as the tubule network expands forming longer network extensions (Fig. 4.12).

4.6 Troubleshooting

In order to ensure that the software can accurately analyse the closure of the scratch, it is critical that the cells are confluent prior to the scratch being created. In addition tubule formation is reliant on the correct seeding density being applied to the wells.

Careful planning is required to ensure that the necessary data is obtained from live cell imaging experiment. To ensure the images remains in focus the size of the Z-stack must be appropriate for the samples being imaged. The number of regions imaged must also be carefully selected; the addition of too many could result in the time taken to return to the same region to be too long, resulting in key events being missed.

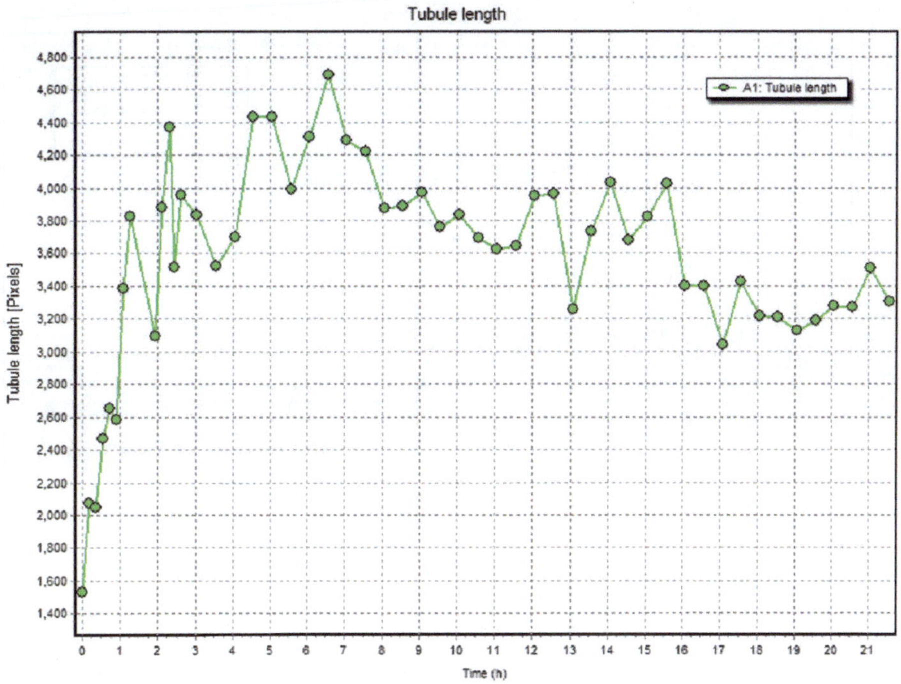

Fig. 4.12 Graphical results illustrating the total tubule length over time as the tube -like structures form in matrigel

If imaging with fluorescence, care must be taken not to damage the cells by imaging with too high an exposure or too frequently.

4.7 Applications and Discussions

Traditionally the assays used to assess in vitro angiogenesis are end point and provide only a snapshot of the biological response. By employing live cell imaging a complete over view of the response can be obtained, which is especially important when investigating the effects of specific factors (pro or anti) on a dynamic processes such as angiogenesis.

Not only does this powerful technique facilitate automated capture and results processing, reducing time and labour costs, it will also confirm the appropriateness of previously chosen time points in former studies. In addition, the ability to collect a series of data points from the same population of cells enables more detailed studies to be performed with fewer samples, and as the imaging process is non-destructive these samples can subsequently undergo further analysis e.g. RNA or protein extractions.

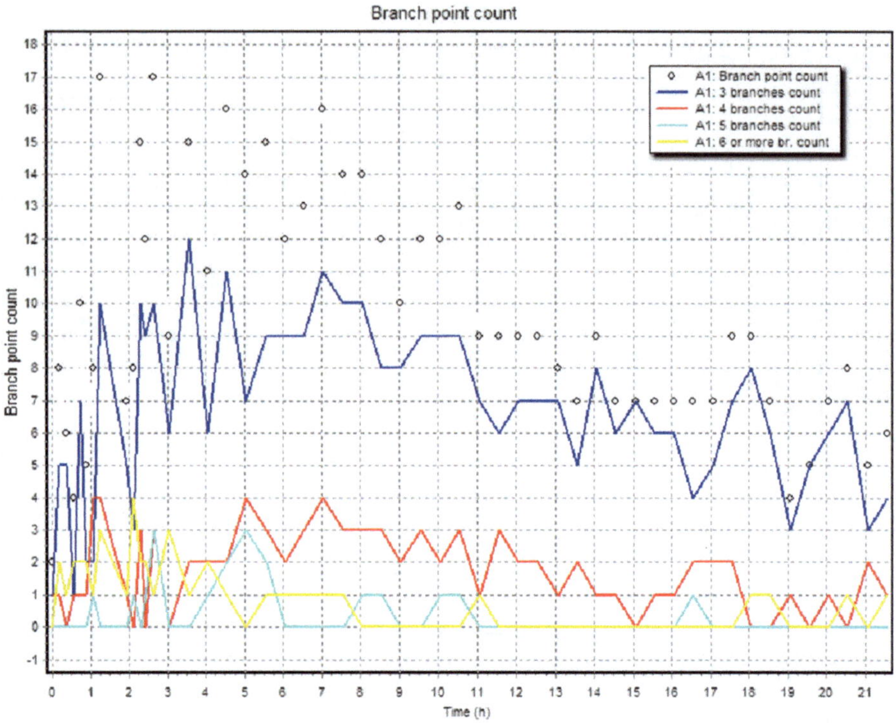

Fig. 4.13 Graphical results illustrating the different branch-point count over time as the tubule like structures form in the matrigel

The use of machine vision to classify and quantify different cell populations provides the user with an intelligent but achievable approach to analysing various experimental data sets, including complex models such as co-culture experiments. The ability to identify different cell types and phases from phase contrast images alone eliminates the need for fluorescent dyes in many experimental protocols. For those experiments where protein expression is of interested and can be determined by a fluorescent signal, the fluorescence image acquisition and analysis tool enables information on specific expression patterns to be obtained.

Initially developed in the seventeenth century, microscopes have since become an integral part of cell biology laboratories. Recent developments have resulted in the availability of live cell imaging platforms, allowing numerous biological processes to be studied in real time.

Although still developing, this informative technique is already enriching the data sets of many studies, demonstrating compatibility with both traditional techniques such as PCR, western blotting and more complex genomic methods such as microarray studies [5].

References

1. Huttunen TT, Sundberg M et al (2011) An automated continuous monitoring system: a useful tool for monitoring neuronal differentiation of human embryonic stem cells. Stem Cell Stud 1(e10):71–77
2. Mikkola M, Toivonen S et al (2013) Lectin from Erythrina cristagalli supports undifferentiated growth and differentiation of human pluripotent stem cells. Stem Cells Dev 22(5):707–716
3. Toivonen S, Ojala M et al (2013) Comparative analysis of targeted differentiation of human induced pluripotent stem cells (hiPSCs) and human embryonic stem cells reveals variability associated with incomplete transgene silencing in retrovirally derived hiPSC lines. Stem Cell Trans. doi: 10.5966/sctm.2012-0047
4. Vuoristo S, Toivonen S et al (2013) A novel feeder-free culture system for human pluripotent stem cell culture and induced pluripotent stem cell derivation. PLoS One 8(10):e76205
5. Kivi N, Greco D et al (2008) Genes involved in cell adhesion, cell motility and mitogenic signalling are altered due to HPV 16 ES protein expression. Oncogene 27:2532–2541
6. Talmana V, Amadiob M et al (2013) The C1 domain-targeted isophthalate derivative HMI-1b11 promotes neurite outgrowth and GAP-43 expression through PKC activation in SH-SY5Y cells. Pharmacol Res 73:44–54
7. Palazzolo G, Horvath P et al (2012) The flavonoid isoquercitrin promotes neurite outgrowth by reducing Rho A activity. PLoS One 7(11):e49979
8. Talman V, Tuominen RK et al (2011) C1 Domain-targeted isophthalate derivatives induce cell elongation and cell cycle arrest in HeLa cells. PLoS One 6(5):e20053. doi:10.1371/journal.pone.0020053
9. Bosutti A, Qui J et al (2013) Targeting p35/cdk5 signalling via CIP-peptides promotes angiogenesis in hypoxia. PLoS One 8(9):e75538
10. Ball SG, Worthington JJ et al (2013) Mesenchymal stromal cells: inhibiting PDGF receptors or depleting fibronectin induces mesodermal progenitors with endothelial potential. Stem Cells 10. doi: 10.1002/stem.1538
11. Sonka M, Hlavac V et al (1999) Image processing, analysis, and machine vision, 2nd edn. PWS Pub, Pacific Grove
12. Tarvainen J, Saarinen M et al (2002) Creating images with high data contents for microworld applications. Ind Sys Rev:17–23
13. Tozer GM, Akerman S et al (2008) Blood vessel maturation and response to vascular-disrupting therapy in single vascular endothelial growth factor-A isoform–producing tumors. Cancer Res 68(7):2301–2311
14. Walter MNM, Wright KT et al (2010) Mesenchymal stem cell-conditioned medium accelerates skin wound healing: an in vitro study of fibroblast and keratinocyte scratch assays. Exp Cell Res 316(7):1271–1281

Chapter 5
Isolation of Endothelial Progenitor Cells (EPCs)

Aaron Liew and Timothy O'Brien

5.1 Introduction

Endothelial progenitor cells (EPCs) are non-endothelial cells that are capable of differentiating into cells with endothelial phenotypes. They secrete angiogenic factors and enhance in vitro and in vivo vascular regeneration [1]. We defined EPC based on the following criteria: (a) cells derived from a non-endothelial source; (b) adherence to matrix molecule (our protocol utilises fibronectin); (c) demonstrate acetylated Low Density Lipoprotein (acLDL) uptake; and (d) positive staining with lectin. While the precise definition of EPCs is controversial, this terminology is often used to denote a heterogenous population of cells, consisting of two distinct cell types: (a) early EPCs or circulating angiogenic cells (CACs) and (b) late EPCs or outgrowth endothelial cells (OECs). These two different subpopulations have distinct cell growth patterns and the ability to secrete angiogenic factors [2, 3]. The former are spindle-shaped cells with a peak growth in culture at 2–3 weeks and die after approximately 4 weeks in vitro [3]. Late EPCs usually appear following 2–3 weeks of in vitro culture. They are cobblestone-shaped and have an exponential growth rate at 4–8 weeks which can be maintained for up to 12 weeks in culture [2, 3].

EPCs can be isolated from various sources including bone marrow, peripheral blood, and umbilical cord blood [2, 4, 5]. They can also be isolated by using either a single or a combination of haematopoetic (CD34) and endothelial (KDR) markers by using either FACS or a magnetic bead sorter [6]. The majority of early EPCs and late EPCs arise from a CD14 positive and CD14 negative subpopulation of peripheral blood mononuclear cells, respectively [2, 7]. The discussion regarding the surface markers of EPC is beyond the scope of this article. The late EPCs secrete a

A. Liew • T. O'Brien (✉)
Regenerative Medicine Institute (REMEDI), National Centre for Biomedical Engineering Science (NCBES), National University Ireland Galway (NUIG), Galway, Ireland
e-mail: Timothy.obrien@nuigalway.ie

M. Slevin, G. McDowell (eds.), *Handbook of Vascular Biology Techniques*,
DOI 10.1007/978-94-017-9716-0_5

smaller array of cytokines as compared with the early EPCs [3, 8]. Even though late EPCs are more capable of in vitro tube formation than early EPCs, the data regarding their in vivo functional effects are conflicting [2, 3, 7].

EPC number and function inversely correlate with the number of cardiovascular risk factors and have been utilised as a surrogate marker for endothelial dysfunction [9–12]. Preclinical data showed that EPC based therapy has a therapeutic potential for the treatment of a variety of vascular disease states including ischaemia, restenosis and pulmonary hypertension [13–15]. Early human trials showed that EPC-based therapy is safe and feasible [16, 17].

Herein, we describe the protocol for isolation and characterisation of EPCs (CACs and OECs) from peripheral mononuclear cells. This protocol can be used for (a) direct quantification of EPC number as a surrogate marker of cardiovascular risk factor; (b) in vitro assessment of EPC biology; or (c) in vivo assessment of EPC biology by administration into animal model.

EPCs can also be isolated by two other methods which involved: (a) plating the mononuclear cells on a fibronectin-coated plate for 48 h and replating the non-adherent cells onto a fresh fibronectin-coated plate (These cells are termed colony forming unit-endothelial cells or CFU-ECs) [10]; and (b) plating the mononuclear cells onto a collagen-coated plate instead of fibronectin-coated plate (These cells are termed endothelial colony forming cells or ECFCs) [18].

5.2 Methodology

Our in-house protocol outlined below describe the protocol for isolation and characterisation of EPCs (CACs and OECs) derived from peripheral mononuclear cells. All research involving human subjects should be performed following the approval of relevant institutional ethical committee and blood samples should be obtained following informed consent. All experiments involving live animals must only be performed in accordance with the Guide for the Care and Use of Laboratory Animals or relevant regulations. Standard precautions for safe handling of blood products should be followed. All procedures described below are performed under strict aseptic technique in a Class II biohazard flow hood.

5.3 Materials (and Company Name)

5.3.1 Collection of Blood (60 mls)

5.3.1.1. Eight to ten 8 ml Venous blood collection tubes (vacutainer) coated with K3 EDTA 15 % solution (BD Biosciences)

5.3.1.2. Standard phlebotomy set

5.3.2 Isolation of Mononuclear Cells

5.3.2.1. Fourteen 50 ml conical tubes
5.3.2.2. One 10 ml Pasteur pipette
5.3.2.3. Hanks' Balanced Salt solution (SIGMA)
5.3.2.4. Bovine Serum Albumin (SIGMA)
5.3.2.5. EDTA pH8 (SIGMA)
5.3.2.6. Endothelial Growth Medium 2 (EGM-2) with Bullet Kit (Lonza)
5.3.2.7. Ficoll-Paque PLUS (GE Healthcare Life Sciences)
5.3.2.8. Red Cell Lysis Buffer (SIGMA)
5.3.2.9. Phosphate Buffered Saline (PBS) (Gibco)

5.3.3 Counting and Plating of Mononuclear Cells

5.3.3.1. Haemocytometer
5.3.3.2. Trypan blue (SIGMA)
5.3.3.3. The BioCoat™ fibronectin-coated six well plates (BD Bioscience)

5.3.4 Cell Culture

5.3.4.1. Endothelial Growth Media 2 (EGM-2) with Bullet kits (Lonza)
5.3.4.2. Incubator with 5 % CO_2 at 37 °C

5.3.5 acLDL Uptake and Lectin Staining

5.3.5.1. 1,1′-dioctadecyl-3,3,3′,3′-tetramethylindocarbocyanine-labeled Ac-LDL (DiI-Ac-LDL) (Molecular Probes)
5.3.5.2. Phosphate Buffered Saline (PBS) (Gibco)
5.3.5.3. 3 % Paraformaldehyde (PFA)
5.3.5.4. FITC-labeled Ulex europaeus agglutinin-I lectin (SIGMA)

5.4 Basic Protocol

5.4.1 Isolation of EPCs

5.4.1.1. Prepare 3 separate 50 mls tubes containing 20 mls of Hank's Complete Solution each (Please see Sect. 5.7 for the preparation of Hank's Complete Solution)

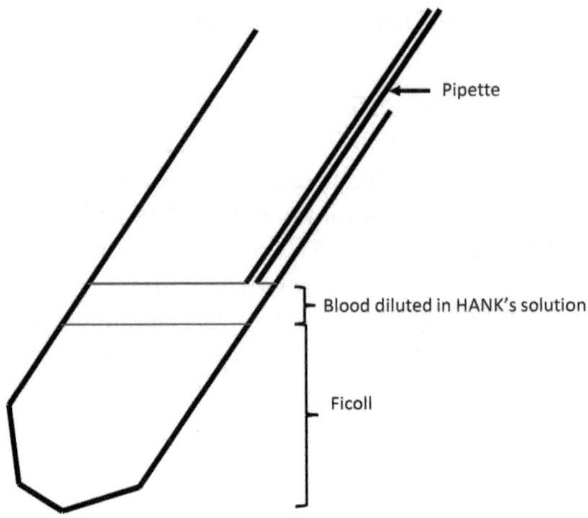

Fig. 5.1 The diluted blood can be gently layered on the Ficoll layer by the 'tilting' method

5.4.1.2. Prepare 8 separate 50 mls tubes containing 21 mls of Ficoll each
5.4.1.3. Prepare 2 separate 50 mls tubes containing 25 mls of PBS each
5.4.1.4. Add 20 mls of blood into each 50 mls tube containing 20 mls of Hank's solution prepared in step 5.4.1.1.
5.4.1.5. Gently layer 15 mls of the diluted blood from step 5.4.1.4. into each 50 mls tube containing 21 mls of Ficoll prepared in step 5.4.1.2. using the 'tilting' method as shown in Fig. 5.1 (Care should be taken to avoid mixing the diluted blood with Ficoll)
5.4.1.6. Centrifuge for 30 min at 1,800 rpm WITH ZERO DECELERATION (without the break following a deceleration phase)
5.4.1.7. Remove cloudy buffy coat containing mononuclear cells with Pasteur pipette and transfer it equally into two separate 50 mls tubes (approximately 5–7 mls of cloudy buffy coat containing mononuclear cells from each 4 tubes into a new tube) containing 25 mls of PBS (approximately 1:1 dilution) prepared in step 5.4.1.3. (Care should be taken to avoid collecting serum or Ficoll) as shown in Fig. 5.2
5.4.1.8. Centrifuge for 10 min at 1,600 rpm without zero deceleration (with the break following a deceleration phase)
5.4.1.9. Aspirate and discard the supernatant
5.4.1.10. Add 5 mls of Red Blood Cell Lysis Buffer to the pellet
5.4.1.11. Gently resuspend the pellet with the Red Blood Cell Lysis Buffer
5.4.1.12. Centrifuge for 10 min at 1,600 rpm without zero deceleration (with the break following a deceleration phase)
5.4.1.13. Repeat step 5.4.1.9 if residual RBC remains (If the pellet still red in colour)
5.4.1.14. Aspirate and discard the supernatant
5.4.1.15. Add 25 mls of PBS in to each 50 mls tube
5.4.1.16. Gently resuspend pellet with PBS
5.4.1.17. Transfer the resuspended pellets with PBS from two 50 mls tubes into a new 50 mls tube

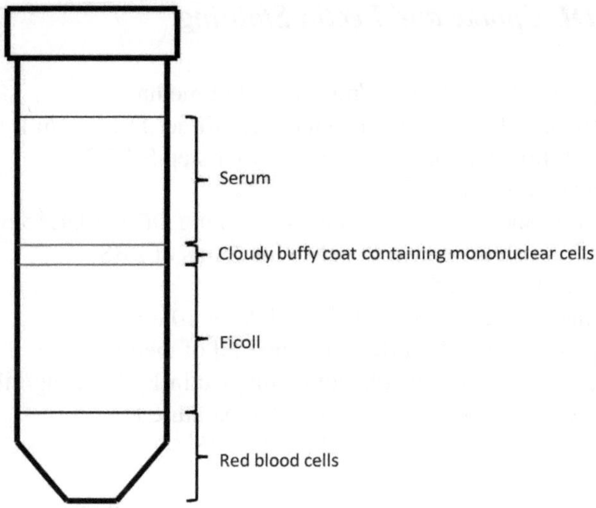

Fig. 5.2 Following Ficoll density centrifugation step, the cloudy buffy coat containing mononuclear cells can be seen between the serum and Ficoll layers

5.4.1.18. Centrifuge for 10 min at 1,600 rpm with no zero deceleration (with the break following a deceleration phase)
5.4.1.19. Aspirate and discard the supernatant
5.4.1.20. Gently resuspend pellet with 50 mls of PBS
5.4.1.21. Centrifuge for 10 min at 1,600 rpm (with no zero deceleration)
5.4.1.22. Repeat steps 5.4.1.19–5.4.1.21
5.4.1.23. Aspirate and discard the supernatant
5.4.1.24. Gently resuspend the pellet in EBM-2 media supplemented with growth factors
5.4.1.25. Count the mononuclear cells with haemocytometer with trypan blue exclusion
5.4.1.26. Gently resuspend the mononuclear cells at a concentration of 10 million mononuclear cells per 2 mls of EBM-2 media supplemented with growth factors
5.4.1.27. Gently transfer the resuspended mononuclear cells into each well of the 6-well plate
5.4.1.28. Incubate for 4 days at 37°C in 5 % CO_2 incubator
5.4.1.29. Gently add 2 mls of EBM-2 media supplemented with growth factors 2 days after plating the mononuclear cells
5.4.1.30. Change media every second day from Day 4 onwards
5.4.1.31. The number of adherent cells can be assessed from Day 4 onward (step 5.4.2.1.)
5.4.1.32. To isolate late EPCs or OECs, continue the culture for up to 3 weeks from step 5.4.1.30. for the appearance of colonies of late EPCs or OECs, which may start to appear from week 1 onwards (However, this is only seen in 5–10% of cultures)

5.4.2 acLDL Uptake and Lectin Staining

5.4.2.1. Prepare DiI-acLDL (5ug/ml) in 1 ml of media
5.4.2.2. Gently add 1 ml of media containing DiI-acLDL (5ug/ml) to the adherent cells, following the media change after step 5.4.1.31.
5.4.2.3. Incubate for 3 h at 37°C
5.4.2.4. Aspirate and discard the media containing DiI-acLDL (5ug/ml)
5.4.2.5. Gently wash the adherent cells with 5 mls of PBS
5.4.2.6. Repeat step 5.4.2.5.
5.4.2.7. Fix the adherent cells with 3 % PFA for 10 min
5.4.2.8. Prepare lectin-FITC (10 ug/ml) in 1 ml of media
5.4.2.9. Gently add 1 ml of media containing lectin-FITC (10ug/ml)
5.4.2.10. Incubate in a dark room at room temperature for 1 h
5.4.2.11. Gently wash the adherent cells with PBS
5.4.2.12. Repeat step 5.4.2.11. twice
5.4.2.13. Visualise adherent cells under fluorescent microscope (Adherent cells which are dual positive for acLDL and lectin are considered EPCs)

5.5 Results

Both the early EPCs and late EPCs are adherent cells of non-endothelial origin, which demonstrated dual positivity for acLDL and lectin. The phenotypes for early EPCs and late EPCs are spindle-shaped and cobblestone appearance, respectively, as demonstrated in Fig. 5.3.

5.6 Troubleshooting

5.6.1. Perform phlebotomy at the same time in the morning
5.6.2. Mix blood with EDTA within the vacutainer tube by gently shaking the tube immediately following phlebotomy
5.6.3. Process the blood within an hour following phlebotomy
5.6.4. Gently layer the diluted blood on the Ficoll as shown in Fig. 5.1
5.6.5. Ensure that the deceleration break is off so that a distinct buffy coat of mononuclear cells is obtained
5.6.6. Care should be taken to avoid leaving the mononuclear cells in red cell lysis buffer for too long
5.6.7. Care should be taken to avoid agitating the cells too much when washing them
5.6.8. Gently layer the mononuclear cells on fibronectin-coated plate to prevent bubble formation

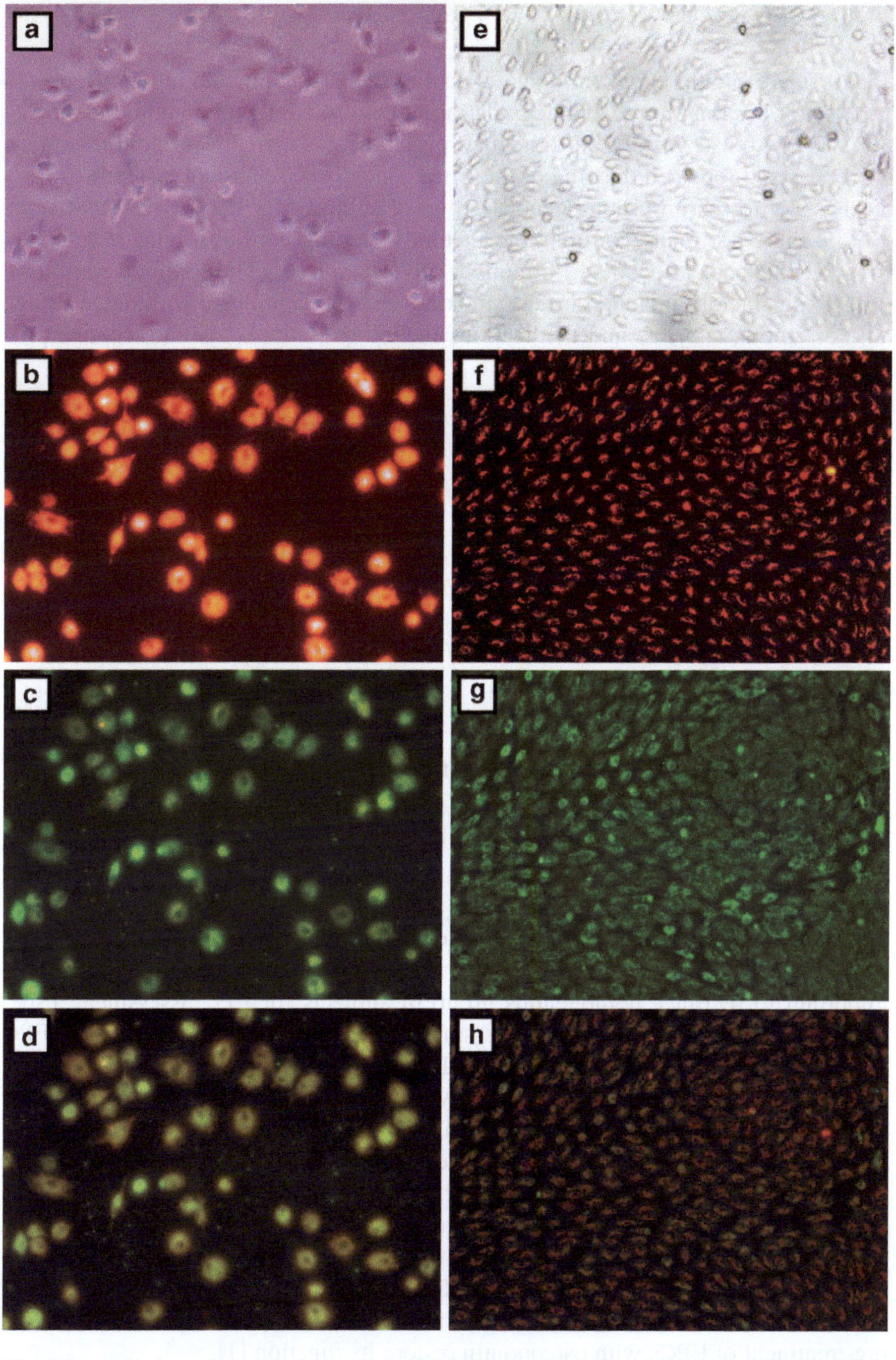

Fig. 5.3 (**a**) Early endothelial progenitor cells (EPCs) under light microscopy; (**b**) early EPCs stained with 1,1′-dioctadecyl-3,3,3′,3′-tetramethylindocarbocyanine-labeled Ac-LDL acetylated low-density lipoprotein (diI-acLDL) (*red*); (**c**) early EPCs stained with fluoroscein isothiocyanate (FITC)-lectin (*green*); (**d**) early EPCs stained with both diI-acLDL and FITC-lectin; (**e**) late EPCs under light microscopy; (**f**) late EPCs stained with diI-acLDL (*red*); (**g**) late EPCs stained with FITC-lectin (*green*); (**h**) late EPCs stained with both diI-acLDL and FITC-lectin (Adapted from Liew et al. [19] with permission from Wiley Materials)

5.6.9. Care should be taken to avoid leaving the EBM-2 media in the water bath for too long to prevent degradation of the growth factors within the media
5.6.10. Try to use fresh EBM-2 media if possible (less than 3 weeks old)
5.6.11. Avoid repeated thawing of the EBM-2 media (Use aliquot instead)
5.6.12. Try to recruit a homogenous group of subjects for EPC assessment since there are many confounding factors including cardiovascular risk factors which can affect the EPC number and function [6] (The common confounding factors include current smoking history, significant difference in age among subjects and medications including statin, oral contraceptive pills and ACE inhibitors)

5.7 Additional Notes

Preparation of Hank's Complete Solution can be prepared by adding 2.5 g of Bovine Serum Albumin (BSA) and 1 ml of 500 mM EDTA pH8 to 500 mls of Hank's Balanced Salt Solution. Filter and aliquot the Hank's Complete Solution and store in fridge at 3°C. Thaw in water bath at 37°C prior use in step 5.4.1.1.

5.8 Applications and Discussion

Direct assessment of EPC number can be used as a surrogate marker for endothelial dysfunction [1, 9–12]. Many confounding factors including cardiovascular risk factors can affect the EPC number and function [6]. The common confounding factors are current smoking history, significant difference in age among subjects and medications including statin, oral contraceptive pills and ACE inhibitors. It is important to recruit a homogenous group of subjects for EPC assessment, especially, if the proposed sample size is small. The timing for phlebotomy for EPC isolation needs to be standardised since EPC level exhibits diurnal variation [20]. We recommend performing phlebotomy between nine and ten o'clock in the morning and proceed to immediate EPC isolation within 1 h of the phlebotomy.

Preclinical data showed that EPC based therapy has a therapeutic potential for the treatment of a variety of vascular disease states including ischaemia, restenosis and pulmonary hypertension [13–15]. Early human trials showed that EPC based therapy is safe and feasible [16, 17]. These cells can be genetically modified to enhance their therapeutic potential [21]. Furthermore, we have shown that patients with diabetes mellitus have less EPCs and that their EPCs are dysfunctional, and pre-treatment of EPCs with osteopontin restore its function [1].

This protocol can be used for (a) direct quantification of EPC number as a surrogate marker of cardiovascular risk factor; (b) in vitro assessment of EPC biology; or (c) in vivo assessment of EPC biology by administration into animal model [1].

References

1. Vaughan EE, Liew A, Mashayekhi K, Dockery P, McDermott J, Kealy B, Flynn A, Duffy A, Coleman C, O'Regan A et al (2012) Pretreatment of endothelial progenitor cells with osteopontin enhances cell therapy for peripheral vascular disease. Cell Transplant 21(6):1095–1107
2. Gulati R, Jevremovic D, Peterson TE, Chatterjee S, Shah V, Vile RG, Simari RD (2003) Diverse origin and function of cells with endothelial phenotype obtained from adult human blood. Circ Res 93(11):1023–1025
3. Hur J, Yoon CH, Kim HS, Choi JH, Kang HJ, Hwang KK, Oh BH, Lee MM, Park YB (2004) Characterization of two types of endothelial progenitor cells and their different contributions to neovasculogenesis. Arterioscler Thromb Vasc Biol 24(2):288–293
4. Lin Y, Weisdorf DJ, Solovey A, Hebbel RP (2000) Origins of circulating endothelial cells and endothelial outgrowth from blood. J Clin Invest 105(1):71–77
5. Nagaya N, Kangawa K, Kanda M, Uematsu M, Horio T, Fukuyama N, Hino J, Harada-Shiba M, Okumura H, Tabata Y et al (2003) Hybrid cell-gene therapy for pulmonary hypertension based on phagocytosing action of endothelial progenitor cells. Circulation 108(7):889–895
6. Liew A, Barry F, O'Brien T (2006) Endothelial progenitor cells: diagnostic and therapeutic considerations. Bioessays 28(3):261–270
7. Urbich C, Heeschen C, Aicher A, Dernbach E, Zeiher AM, Dimmeler S (2003) Relevance of monocytic features for neovascularization capacity of circulating endothelial progenitor cells. Circulation 108(20):2511–2516
8. He T, Smith LA, Harrington S, Nath KA, Caplice NM, Katusic ZS (2004) Transplantation of circulating endothelial progenitor cells restores endothelial function of denuded rabbit carotid arteries. Stroke 35(10):2378–2384
9. Vasa M, Fichtlscherer S, Aicher A, Adler K, Urbich C, Martin H, Zeiher AM, Dimmeler S (2001) Number and migratory activity of circulating endothelial progenitor cells inversely correlate with risk factors for coronary artery disease. Circ Res 89(1):E1–E7
10. Hill JM, Zalos G, Halcox JP, Schenke WH, Waclawiw MA, Quyyumi AA, Finkel T (2003) Circulating endothelial progenitor cells, vascular function, and cardiovascular risk. N Engl J Med 348(7):593–600
11. Ghani U, Shuaib A, Salam A, Nasir A, Shuaib U, Jeerakathil T, Sher F, O'Rourke F, Nasser AM, Schwindt B et al (2005) Endothelial progenitor cells during cerebrovascular disease. Stroke 36(1):151–153
12. Schmidt-Lucke C, Rossig L, Fichtlscherer S, Vasa M, Britten M, Kamper U, Dimmeler S, Zeiher AM (2005) Reduced number of circulating endothelial progenitor cells predicts future cardiovascular events: proof of concept for the clinical importance of endogenous vascular repair. Circulation 111(22):2981–2987
13. Chang ZT, Hong L, Wang H, Lai HL, Li LF, Yin QL (2013) Application of peripheral-blood-derived endothelial progenitor cell for treating ischemia-reperfusion injury and infarction: a preclinical study in rat models. J Cardiothorac Surg 8:33
14. Kong D, Melo LG, Mangi AA, Zhang L, Lopez-Ilasaca M, Perrella MA, Liew CC, Pratt RE, Dzau VJ (2004) Enhanced inhibition of neointimal hyperplasia by genetically engineered endothelial progenitor cells. Circulation 109(14):1769–1775
15. Yip HK, Chang LT, Sun CK, Sheu JJ, Chiang CH, Youssef AA, Lee FY, Wu CJ, Fu M (2008) Autologous transplantation of bone marrow-derived endothelial progenitor cells attenuates monocrotaline-induced pulmonary arterial hypertension in rats. Crit Care Med 36(3):873–880
16. Zhu JH, Wang XX, Zhang FR, Shang YP, Tao QM, Zhu JH, Chen JZ (2008) Safety and efficacy of autologous endothelial progenitor cells transplantation in children with idiopathic pulmonary arterial hypertension: open-label pilot study. Pediatr Transplant 12(6):650–655
17. Losordo DW, Henry TD, Davidson C, Sup Lee J, Costa MA, Bass T, Mendelsohn F, Fortuin FD, Pepine CJ, Traverse JH et al (2011) Intramyocardial, autologous CD34+ cell therapy for refractory angina. Circ Res 109(4):428–436

18. Mead LE, Prater D, Yoder MC, Ingram DA (2008) Isolation and characterization of endothelial progenitor cells from human blood. Curr Protoc Stem Cell Biol. Chapter 2:Unit 2C 1
19. Liew A, McDermott JH, Barry F, O'Brien T (2008) Endothelial progenitor cells for the treatment of diabetic vasculopathy: panacea or Pandora's box? Diabetes Obes Metab 10(5):353–366
20. Thomas HE, Redgrave R, Cunnington MS, Avery P, Keavney BD, Arthur HM (2008) Circulating endothelial progenitor cells exhibit diurnal variation. Arterioscler Thromb Vasc Biol 28(3):e21–e22
21. Kealy B, Liew A, McMahon JM, Ritter T, O'Doherty A, Hoare M, Greiser U, Vaughan EE, Maenz M, O'Shea C et al (2009) Comparison of viral and nonviral vectors for gene transfer to human endothelial progenitor cells. Tissue Eng Part C Methods 15(2):223–231

Chapter 6
Culture and Maintenance of Human Embryonic Stem Cells: A Potential Source for Vasculargenesis

Michael Carroll and Clare Nevin

6.1 Introduction

Human Embryonic Stem cells (hESCs) are pluripotent stem cells derived from the inner cell mass of the blastocyst. Their propensity to differentiate in to the cells representative of all three germ layers has evoked great interest. Since their derivation they have provided a source for investigations in to human developmental biology, regenerative medicine and cell-based therapies [1, 2].

Ischemic diseases are considered one of the major causes of morbidity and mortality throughout the world. This has provoked much interest in the development of cell-replacement therapeutics and vascular regenerative medicine [3, 4]. hESCs have proven to be a potential and unlimited source for generating cardiomyocytes [5] and ES cell-derived vascular progenitor cells and could provide a potential source of cells for vascular tissue engineering [6].

The maintenance and differentiation of hESCs form the basis for significant research in human cell and developmental biology, and in the potential clinical application of cell replacement therapies. However, due to their inherent capacity for differentiation, the maintenance of undifferentiated cultures of hESCs is not as simple as growing other types of mammalian cells.

This chapter will outline the procedures required to maintain hESCs on both mouse embryonic fibroblasts (MEFs) and in feeder-free conditions.

M. Carroll (✉) • C. Nevin
School of Healthcare Science, Manchester Metropolitan University, Manchester, UK
e-mail: michael.carroll@mmu.ac.uk

M. Slevin, G. McDowell (eds.), *Handbook of Vascular Biology Techniques*,
DOI 10.1007/978-94-017-9716-0_6

6.2 Methodology

Maintaining hESC culture can be demanding and to ensure the cells retain their pluripotency status, proper cell culture procedures must be carried out. It is imperative that all media manufacturing and cell culture is performed under aseptic conditions. Cells must be observed at least once a day to ensure spontaneous differentiation is avoided, and the media must be changed daily.

All cells are cultured in 6-well cell culture plates at 37 °C, 5 % CO_2, relative humidity (85–100 %).

6.3 Materials

KO-DEMEM	Invitrogen
DMEM	Invitrogen
L-Glutamax	Gluta-Max™, Invitrogen
Non-essential amino acids	Invitrogen
Serum Replacement	Invitrogen
β–Mercaptoethanol	Gibco
FBS (ES cell defined)	Gibco
Plasmanate	Bayer
bFGF[2]	Invitrogen
Pen/Strep	Invitrogen
6 Well Cell Culture Microplates	Coring Life Science
DMSO	Sigma
Trypsin/EDTA – 0.05 %	Invitrogen
Triton X100	Sigma
Mouse Embryonic Fibroblasts – PMEFs	Speciality Media
Matrigel™ Basement Membrane Matrix	BD Bioscience
FBS	Valley Biomedical Inc.
Gelatin	Sigma
Tri-Reagent	Sigma
Stratagen SuperScript III.	Life Technologies
Alkaline phosphatase detection kit	Millipore
Antibodies:	
Oct4	Santa cruze biotech
SOX2	Chemicon
Nanog	R&D systems
TRA1-60 & TRA1-81	Gifted by P. Andrews (Sheffield University)

6.4 Cell Culture

hESCs can be maintained on a layer of mouse embryonic fibroblasts (MEF), in hESC media or in feeder-free conditions using MEF-conditioned media. It is advisable, when first culturing hESCs to grow them on MEFs initially. Then after establishing good, undifferentiated colonies the hESCs can be cultured on Matrigel™ under feeder-free conditions.

The following protocols have been adapted for culture and maintenance of the *HUES* ES Cell lines [7].

*6.4.1 Preparation of MEF Feeders (*PMEF-P3; Speciality Media*)*

1. Prepare a 6 well tissue dish (more if required) by adding 1.5 ml of 0.1 % gelatine and incubate at 37 °C for at least 30 min
2. Pre-warm MEF media to 37 °C in water bath
3. Add 10 ml of pre-warmed MEF media[1] to a 15 ml falcon tube
4. Remove MEFs from liquid nitrogen storage and thaw by half immersing in 37 °C water bath
5. Transfer cells to a sterile 15 ml falcon tube containing 10 ml of media
6. Spin cell-suspension at 500 g for 5 min.
7. While cells are centrifuging; prepare the 6-well plate by aspirating the gelatine off and adding 1 ml of preheated MEF media to each well. Leave to one side.
8. Remove cells from centrifuge and carefully aspirate supernatant.
9. Resuspend in 2–3 ml of preheated media (taken from the 15 ml tube from step 3).
10. Transfer cell suspension to the remaining media in 15 ml falcon tube (step 3), pipetting up and down gently to ensure an even cell suspension
11. Add 1 ml of cell suspension drop wise in to each well of the 6 well plate, distribute the cells evenly about the wells. Total volume per well should be 2 ml.
12. Place plate in incubator at 37 °C/5 % CO_2
13. The MEFs will be ready as feeders for hESC the following day. **Do not** use MEF plates older than 4 days.

[1] MEF Media: to make MEF Media, add 5.5 ml of Pen/Strep and 55 ml of FBS to 500 ml of DMEM, store at 4 °C. Filter sterilize at 0.2 µm before use.

6.4.2 hESC Culture

6.4.2.1 hESC Media

To prepare hESC Media add the following components, filter sterilized at 0.2 μm and store at −20 °C in 45 ml aliquots.

500 ml KO-DEMEM
6.5 ml L-Glutamax
6.5 ml Non-essential amino acids
65 ml Serum Replacement
0.65 ml β–Mercaptoethanol

When required, thaw an aliquot of hESC media and warm to 37 °C. Add 5 ml of Plasmanate and bFGF[2] (10 ng/ml, final). The media is ready for hESC culture.

6.4.2.2 Thaw from a Frozen Stock

HUES ES cells obtained from the Harvard Stem Cell Facility are delivered on dry ice. They can be stored in liquid nitrogen until required or thawed and cultured upon arrival. The following protocol can be adapted for most hES cell lines. However, ensure you are familiar with the specific culture requirements of whichever cell line you obtain.

1. Ensure that the MEF plate is prepared, properly plated and in good condition. DO NOT use a less than desirable plate, or one that is older than 4 days.
2. Pre-warm hES media to 37 °C.
3. Remove hESC vial/s from cryostorage and immediately submerge the bottom half of the tube in a 37 °C water bath. It should take about 45–60 s before the cells are 80 % thawed (small frozen portion left).
4. Quickly bring the tube to the laminar flow hood, spray down with 70 % alcohol, and gently transfer cells to the of pre-warmed media.
5. Spin the tube, remove media, and resuspend in an appropriate volume of pre-warmed hESC media.
6. Aspirate off the MEF media from the wells used for hESC culture.
7. Quickly, aliquot pre-warmed hESC media back into each well of the plate, being careful not to disturb the attached MEFs.
8. Drop the hESCs carefully on to the MEF wells. Carefully return the plate to a 37 °C incubator overnight to allow the hESCs to seed the MEFs. Change media ~48 h after thaw, replacing with fresh hESC media.

6.4.3 *Passaging hESC*

1. After 4–7 days post-thaw, hESCs should be near confluence. When deciding when to passage hESCs, plate MEFs (MEFs will be good to seed hESCs for up to 3 days post plating; **DO NOT** use MEFs older than 4 days).
2. For a 1:3 split ratio use 3 wells of a 6 well plate.
3. Pre-warm hESC media and 0.05 % Trypsin/EDTA at 37 °C in water bath.
4. Aspirate MEF media from 3 MEF wells and replace with 1 ml of pre-warmed hESC media, leave to one side.
5. Aspirate hESC media from cultures to be split and gently wash the well with 1× sterile PBS.
6. Aspirate PBS and add 0.3 ml of pre-warmed Trypsin/EDTA.
7. Observe cells under 4× objective, noting when the hESCs and MEFs begin to retract (should take about 3 min)
8. When MEFs have sufficiently rounded up and the borders of the hESC colonies are rough, return plate to hood.
9. Add 2 ml of pre-warmed hESC media to trypsinised cells.
10. Gently pipette up and down, washing the bottom of the well, until the MEF monolayer has completely detached (Careful not to produce too many bubbles!)
11. Transfer the cell suspension to a sterile 15 ml conical tube. Wash plate with another 1 ml of hESC media to remove any remaining cells.
12. Total volume of hESC media in 15 ml conical flask should be 3 ml. Resuspend the cells by gentle pipetting up and down.
13. Add 1 ml of cell suspension drop wise to each well, ensuring even distribution about the well.
14. Place at plate in a 37 °C/5 % CO_2 incubator.

6.4.4 *Cryopreservation of hESC*

Once sufficient colonies are grown it may be necessary to bank cells for future culture and experimentation.

1. Harvest cells as for passaging
2. When cells are in suspension after Tryspsin/EDTA treatment (step 9 of **Passaging hESC** steps) centrifuge at 500 g for 5 min.
3. Carefully aspirate supernatant and resuspend the pellet in Freeze-Media[2] by pipetting up and down gently.
4. Aliquot the cell suspension in well labelled cryovials.

[2] Freeze Media: 90 % FBS (ES cell defined), 10 % DMSO.

5. Place cryovials in Freeze container, ensuring the appropriate level of Isoproanol is accurate.
6. Place freeze container in −80 for 24 h, after which place in Liquid Nitrogen storage.

6.4.5 *Characterisation of hESCs*

6.4.5.1 Morphology

It is important to monitor for undifferentiated hESCs. Undifferentiated hESCs grow as compact, multicellular colonies with a distinct border. They also have a high nuclear-to-cytoplasm ratio and prominent nucleoli. Good hESC colonies will be multi- layered in the centre, resulting in clusters of phase-bright cells when viewed under phase contrast. Differentiation is characterised by loss of border integrity, gross non-uniformity of cell morphology within a colony, and the emergence of obvious differentiated cell types (Fig. 6.1a, j).

6.4.5.2 Immunocytochemistry for hESC-Specific Markers

Often is will be required to characterise the hESC cultures by immunocytochemical assessment of surface and nuclear makers (Fig. 6.1).

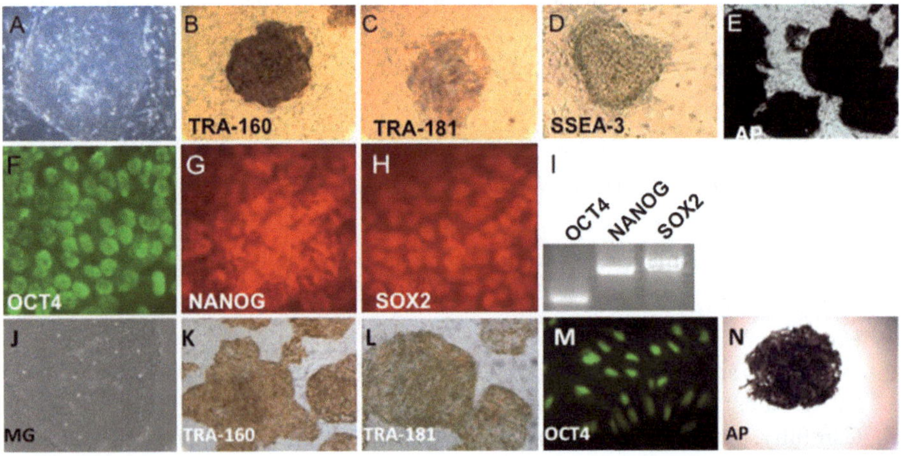

Fig. 6.1 Characterisation of hESCs: Undifferentiated hESCs maintained under feeder cell conditions display a characteristic morphology (**a**) and express several stem cell-specific markers: **TRA-160** (**b**), **TRA-181** (**c**), **SSEA-3** (**d**), **Alkaline Phosphatase, AP** (**e**), **OCT4** (**f**), **SOX2** (**g**) and **NANOG** (**h**) OCT4, SOX2 and NANOG (also detected by RT-PCR) (**i**). Undifferentiated hESCs grown in feeder-less culturing conditions (on Matrigel-coated plates) retain their morphological characteristics (**j**) and express the stem cell markers **TRA-160** (**k**) and **TRA-181** (**l**), **OCT4** (**m**) and **AP** (**n**)

1. hESCs were maintained in hESC culture media on 6-well plates, previously plated with mouse embryonic fibroblasts (MEFs) or coated with Matrigel (feeder-free).
2. Fix cells with fixative (acetone:methanol:water; 2:2:1), ethanol-dehydrated and stored at 4 °C until required.
3. After rehydration, wash cells in PBS/Triton X100 and incubated with primary antibody (Oct4, 1:500 (Santa cruze biotech.)); SOX2, 1:150 (Chemicon); Nanog, 1:20 (R&D systems); TRA1-60 and TRA1-81, 1:50 (gifted by P. Andrews) at 4 °C ON.
4. Incubate with the appropriate fluorophore – tagged secondary antibodies and image using an inverted *Zeiss* fluorescent microscope.

Detection of Alkaline phosphatase can be carried out as instructed in the Alkaline phosphatase detection kit (Millipore).

6.4.5.3 PCR for hESC-Specific Transcripts

5. Total RNA was isolated from cells using Tri-Reagent (Sigma) following the manufactures protocol and cDNA was synthesised using *Stratagen SuperScript III*.

   Primer sequence for PCR

Oct4 : Forward, 5′-TGGAGA AGGAGA AGCTGGAGCA A A A-3′
Reverse, 5′-GGCAGAGGTCGTTTGGCTGA ATAGACC-3′

Nanog : Forward, 5′-TCCTCCTCTTCCTCTATACTA AC-3′
Reverse, 5′-CCC ACAATCACAGGCATAG-3′

Sox-2 : Forward, 5′-GGGAAATGGAGG GGTGCA A A AGAGG-3′
Reverse, 5′-TTGCGTGAGTGT GGATGG GATTGGTG-3′

6.4.6 Feeder-Free Culture of hESC

Once pluripotent hESC colonies have been established on feeder layers, it is often necessary to transfer them to feeder-free culture to avoid feeder cell contamination of downstream applications. Matrigel is a gelatinous protein mixture that acts as a basement membrane matrix and maintains stem cells in an undifferentiated state.

6.4.6.1 Preparation of Matrigel-Coated Plates

1. Slowly thaw Matrigel stock solution on ice at 4 °C for at least 2 h to avoid gel formation. Shake well before use.
2. Dilute Matrigel stock solution according to the manufacturer's instructions in cold knockout DMEM. Coat each well of a 6 well plate with 1 ml Matrigel work-

ing solution. Keep culture plates on ice throughout coating process. Tilt the plate to ensure even covering of the wells.

3. Incubate Matrigel-coated plates for 1–2 h at room temperature (15–25 °C) or at 4 °C overnight.

   Plates can be stored at 4 °C for up to 1 week providing that plates are sealed to prevent dehydration (e.g. Parafilm)

4. Aspirate remaining liquid from the culture plate immediately before use.

6.4.6.2 Culturing hESCs on Matrigel from Feeder Cells

1. Thaw hESC vials at 37 °C in a water bath as stated in Sect. 6.4.2.2 (directions 1–5). Alternatively, if transferring directly from feeder culture, follow Sect. 6.4.3 (directions 5–12) to obtain a cell suspension.
2. Remove Matrigel coating solution from wells by tilting the plate and aspirating from one side. Do not scratch the surface of the wells. If plate has been stored at 4 °C allow it to come to room temperature (30 min) before removing the solution.
3. Wash Matrigel-coated plates twice with 1× PBS.
4. Place 2 ml of thawed or trypsinized cell suspension into each well of a 6 well plate. Gently shake the plate to ensure an even distribution of cells.
5. Culture cells in a humidified chamber at 37 °C with 5 % CO_2.
6. Change hESC media daily on days 3–7. Aspirate spent medium. Add 2 ml pre-warmed hESC medium to each well and return plate to incubator.

6.4.6.3 Passaging hESCs from Matrigel

hESC colonies are typically ready for passaging after 5–7 days of feeder-free culture. Undifferentiated colonies that are ready to be passaged have a characteristic dense centre. Colonies should be passaged at 1:3 or 1:6 ratio.

1. Aspirate hESC media from wells to be passaged. Rinse twice with DMEM.
2. Add 0.3 ml of pre-warmed Trypsin/EDTA to each well.
3. Incubate until the edges of colonies start to curl up when observed under a microscope. Add 1 ml hESC media per split (for 1:3 split add 3 ml media).
4. Gently disrupt colonies by scraping with a sterile serological pipette or cell scraper. Use a phase contrast microscope to identify undifferentiated cells. Pipette up and down but do not create a single cell suspension.
5. Collect suspension of cell aggregates in a 15 ml conical tube.
6. Add 1 ml of cell suspension to fresh culture plates coated with Matrigel and rinsed previously. Tilt the plate in all directions to ensure even distribution of cell aggregates.
7. Culture cells in a humidified chamber at 37 °C with 5 % CO_2.
8. Feed cells with 2 ml hESC supplemented with bFGF media every day.

6.5 Troubleshooting

Problem	Reason	Solution
No viable cells after thawing stock	Incorrect storage	Store cells in liquid nitrogen and follow freezing requirements as recommended by the supplier at the correct cell density
	Incorrect thawing	Ensure frozen cells are thawed quickly but dilute them slowly in pre-warmed media
	Cells are too dilute	Plate thawed cells at higher density to optimise recovery
	Cells not handled gently	Do not freeze-thaw, vortex or centrifuge at very high speeds
Poor colony growth in feeder-free culture	Improper matrigel coating	Ensure sufficient volume per well. If storing plates in fridge, ensure this is on an even surface. Cover plates with Parafilm to prevent drying out
	Improper matrigel handling	Keep vials of Matrigel and culture plates on ice at all times when coating to prevent gel formation
	Air bubbles present in Matrigel	Centrifuge plate at 300×g for 10 min at 4 °C after Matrigel coating. Ensure centrifuge is pre-cooled
Cells are differentiating	Faulty growth factor	Check quality of bFGF. Use fresh
	Uneven distribution of cells/aggregation of colonies	When culturing/passaging colonies onto a new plate, tilt the plate in all directions to get an even distribution. Do not swirl the plate as this will cause cells to aggregate in the centre
	Transfer of differentiating regions during passaging	Use a picking tool to cut and remove differentiating cells from wells before passaging. They tend to appear at the edges of the wells first
	Cells seeded at too high concentration	Use a lower split ratio when passaging (1:1 to 1:3)
Slow cell growth	Dissociation-induced apoptosis	Supplement hESC media or matrigel with ROCK inhibitor (10 μM)
	Incorrect growth medium	Always use pre-warmed medium. Check colour of medium for change in pH
	Contamination	Aseptic technique is imperative: wash hands with soap and water; wear protective gloves; spray gloves with 70 % ethanol
		Working within a microbiological safety cabinet: stock the cabinet with materials required for an experiment but keep it clutter-free so as not to block the air flow. All items that enter the cabinet must be sprayed with 70 % ethanol
		Clean spills that arise immediately with 70 % sterile ethanol

References

1. Thomson JA, Itskovitz-Eldor J, Shapiro SS, Waknitz MA, Swiergiel JJ, Marshall VS, Jones JM (1998) Embryonic stem cell lines derived from human blastocysts. Science 282(5391):1145–1147
2. Daley GQ (2010) Stem cells: roadmap to the clinic. J Clin Invest 120:8–10
3. Cheung C, Sinha S (2011) Human embryonic stem cell-derived vascular smooth muscle cells in therapeutic neovascularisation. J Mol Cell Cardiol 51(5):651–664
4. Takehara N (2013) Cell therapy for cardiovascular regeneration. Ann Vasc Dis 6(2):137–144
5. Ardehali R, Ali SR, Inlay MA, Abilez OJ, Chen MQ, Blauwkamp TA, Yazawa M, Gong Y, Nusse R, Drukker M, Weissman IL (2013) Prospective isolation of human embryonic stem cell-derived cardiovascular progenitors that integrate into human fetal heart tissue. Proc Natl Acad Sci U S A 110(9):3405–3410
6. Ferreira LS, Gerecht S, Shieh HF, Watson N, Rupnick MA, Dallabrida SM, Vunjak-Novakovic G, Langer R (2007) Vascular progenitor cells isolated from human embryonic stem cells give rise to endothelial and smooth muscle like cells and form vascular networks in vivo. Circ Res 101(3):286–294
7. Cowan CA, Klimanskaya I, McMahon J, Atienza J, Witmyer J, Zucker JP, Wang S, Morton CC, McMahon AP, Powers D, Melton DA (2004) Derivation of embryonic stem-cell lines from human blastocysts. N Engl J Med 350(13):1353–1356

Chapter 7
Assessment of Vascular Function and Contractility, *Ex Vivo*

May Azzawi

7.1 Introduction

In this chapter, I will describe two useful techniques utilised in the laboratory to assess vascular function and contractility. An important function of the vasculature is the regulation of blood pressure to ensure that blood perfusion is matched to tissue demand. One essential factor controlling arterial blood pressure is peripheral resistance. As defined by Poiseuille's law of fluid dynamics, resistance to flow is inversely proportional to radius (hence diameter). Therefore, small changes in vessel diameter will have a very large effect on vascular resistance. Blood vessels are made up of three layers: an outer adventitial layer, a medial layer composed mainly of vascular smooth muscle cells (VSMCs) and an inner endothelial layer. Within small arteries, the VSMCs constitute a major component of the vessel wall and through contraction and relaxation, are able to alter vessel diameter [1]. The endothelial cells, on the other hand, are important modulators of these responses, through the release of vasoconstrictor and vasodilator substances, in response to both mechanical and pharmacological stimuli [2].

It is possible to examine aspects of vessel function and to explore the role of endothelial derived mediators in modulating this function, by using an *ex vivo* system that enables the maintenance of isolated vessels in a viable state. A number of these exist [3], and they include, the organ bath system, wire and pressure myography. Here, I shall introduce you to the 'Organ bath' system, utilised for conduit and large size vessel studies, followed by 'Pressure myography', a state-of-the-art system that more closely mimics the physiological state, by enabling the maintenance of viable small size vessels (40–450 um diameter) under pressure, flow, and temperature. Using these *ex vivo* systems, it is possible to examine the following:

M. Azzawi (✉)
School of healthcare Science, Manchester Metropolitan University,
Manchester M1 5GD, UK
e-mail: m.azzawi@mmu.ac.uk

M. Slevin, G. McDowell (eds.), *Handbook of Vascular Biology Techniques*,
DOI 10.1007/978-94-017-9716-0_7

(i) The vasoconstrictor responses elicited by agonists, such as phenylephrine (Phe, an alpha adrenergic receptor agonist) and high K+ solution. (ii) The role of the endothelium in modulating vascular responses. Acetylcholine (ACh) acts directly on endothelial cells to induce the release of endothelial derived relaxants, including nitric oxide and has hence been used as a marker of endothelial-dependent dilation. In contrast, nitric oxide donors, such as sodium nitroprusside (SNP) can act directly on VSMCs to cause relaxation, in the absence of a functional endothelial layer (endothelial-independent dilation). (iii) The effect of both mechanical (pressure/flow) and pharmacological stimuli (agonists/cytokines/mediators) on constrictor and dilator responses.

7.2 Methodology

7.2.1 Solutions and Agonists

Physiological Salt Solution [(PSS) (mM): NaCl 119, KCl 4.7, MgSO4_7H20 1.2, NaHCO3 25, KHPO4 1.17, K2EDTA 0.03, glucose 5.5, CaCl2_2H2O 1.6, at 37 °C, gassed with 95 % air/5 % CO2, pH 7.4] (see Table 7.1 for preparation).

Calcium-free PSS (See Table 7.1 for preparation).
High potassium Physiological Salt Solution (KPSS- 60 mM, see Table 7.1 for preparation).
Acetylcholine Chloride (ACh).
Sodium Nitroprusside (SNP).
L-Phenylephrine Hydrochloride (Phe).

7.2.2 Materials and Equipment

Dissection tools, including fine forceps and fine scissors for micro-dissection (World precision Instruments, UK)
Stereomicroscope for micro-dissection (Olympus, UK).
Data acquisition system: PowerLab System and Labchart Software (AD Instruments, UK)
Water bath and heater (World precision Instruments, UK).
Organ bath system: Includes glass tissue bath (Aimer Products Ltd.); micrometer (World precision Instruments, UK); signal transducer and amplifier (Harvard Apparatus Ltd., UK).
Pressure myography system: Includes vessel chamber with self-heating system, superfusion pump, pressure servo control unit, video dimension analyser, CCD camera, microfusion pump; TV monitor; pressure transducer (all from Living systems Instrumentation, USA); Inverted Microscope (Nikon, UK).

7.3 The Organ Bath System

A number of organ bath systems (with single, dual or multiple tissue baths) are available commercially. Below is a step by step procedure for setting up and calibrating one such set up, as illustrated in Fig. 7.1.

7.3.1 Preparing and Calibrating the Organ Bath Setup

(i) Switch on the water bath heater to 35–37 °C.
(ii) Add PSS solution to the organ bath, using a measuring cylinder to measure an exact amount (e.g. 20 ml)
(iii) Gas the solution using the gas source (95 % oxygen + 5 % CO_2)
(iv) Allow to equilibrate.
(v) Using PowerLab, open 'Chart'. Go to 'Input amplifier' using the drop-down menu on the right hand side.
(vi) Adjust the level to 0 g, using the 'DC level' knob on the Transducer Amplifier.
(vii) Using the known weights, add 1 g and then 2 g to the transducer hook. Check that the level on the input amplifier box goes to 1 and 2 g readings, respectively.
(viii) Remove the weights. Level should go back to 0 g. Click O.K. on the Input Amplifier box.

Fig. 7.1 The organ bath setup. Vessels are mounted between two steel wires and placed in a glass tissue bath, immersed in gassed PSS (*left*). The upper wire is connected to an isometric force transducer, while the lower U shaped wire is screwed to a support leg (vessel holder) that is connected to a micrometer (illustration; *right*)

7.3.2 *Aortic Vessel Dissection and Mounting*

(i) Isolate the heart and lung tissue en-block (including the connecting aorta) and place promptly in ice cold oxygenated PSS.
(ii) Pin down into a petri dish (covered in clear silicone [Sylgard]) and cover in cold PSS. Gas continuously. Place the petri dish over a larger dish filled with ice (See Fig. 7.2).
(iii) Using a stereomicroscope and fine forceps, identify the aortic vessel and dissect away the surrounding perivascular adipose tissue very carefully.
(iv) Cut 3–4 mm size vessel ring and pass through the U shaped stainless steel wire very carefully.
(v) Fix this onto the vessel holder, using a screw driver. Now pass through the steel wire at one end of the long vessel hook, very gently, through the vessel ring, avoiding any damage to the tissue. Make sure that the vessel ring is not stretched when you do this.
(vi) Immerse the mounted vessel holder into the glass tissue bath. Do this by MOVING the organ bath UPWARDS onto the vessel holder.
(vii) Make sure that the steel wires are very close to each other (i.e. no tension), by moving the micrometer gently.
(viii) Allow the vessel to equilibrate for 30 min.

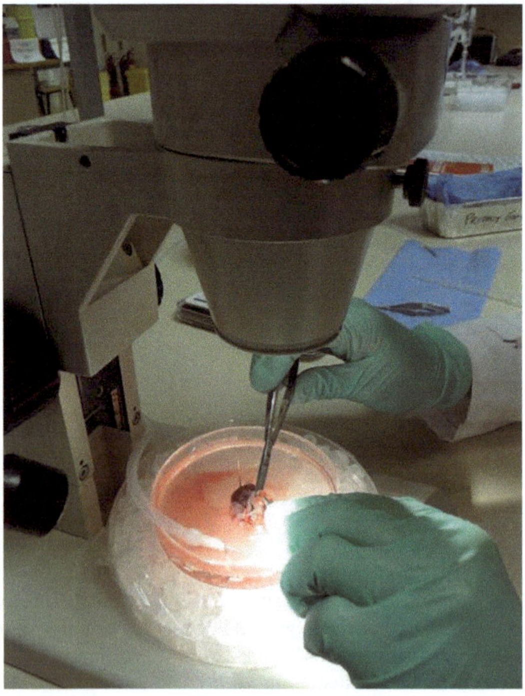

Fig. 7.2 Aortic vessel dissection procedure using a stereomicroscope. The tissue is pinned to a petri dish filled in gassed PSS and placed over a bed of ice

(ix) Go to the PowerLab recording. Click 'start'. Now adjust the tension to read 0, using the 'DC level' knob on the transducer amplifier.
(x) Start moving the lower steel wire away from the upper wire, using the micrometer until the tension reads 2 g. For this experimental set up, this tension value has been predetermined to be the optimal passive distending force that is necessary to allow for a maximal active force [see trouble shooting section].
(xi) Allow the vessel to equilibrate by stretching it to 2 g tension over a further period of 30 min. Adjust the tension using the micrometer as required, until the reading is stable on 2 g tension.

7.3.3 Standard Experimental Protocol

(i) Set up the vessel as above and check that the vessel tension has been stable over the last 15 min.
(ii) Now return the tension to 0, using the 'DC level' knob. This is for measurement purposes only, as it will allow you to register changes in tension starting at 0 value.
(iii) *Test vessel viability*: Constrict the vessel by adding KPSS (60 mM) to the tissue bath. Observe the increase in active tension. Allow the amplitude of this contraction to reach a steady state (i.e. stabilize). This will take approx. 5–10 min. Record this tension value. A contraction of 1 g or greater is considered viable.
(iv) Wash off the vessel by adding PSS and allow the vessel to relax (tension should return to 0 g).
(v) the procedure can be repeated a second time.
(vi) *Examine endothelial integrity*: Constrict the vessel again, by adding KPSS or an agonist (e.g. Phe). Allow the amplitude of contraction to stabilize. Record this tension value.
(vii) Add a specified concentration of ACh from stock.
(viii) Observe any change in tension and allow to stabilize over 2–3 min.
(ix) Wash off in PSS, and allow the vessel to relax completely (tension should return to 0 g).
(x) Calculate % dilation to ACh (see example overleaf). An intact endothelial layer is indicated by a % dilation greater than 60 %.
(xi) You may continue with the ACh dose response curve, by adding ACh cumulatively (e.g. 10^{-8}–10^{-4}). Allow to stabilise over 2–3 min at each concentration. This will provide you with information on endothelial-dependent dilator responses.
(xii) *Examine VSMC responsiveness*: Constrict the vessel again, by adding KPSS or an agonist (e.g. Phe). Allow the amplitude of contraction to stabilize. Record this tension value.

(xiii) Add a specified concentration of SNP. Observe any change in tension and allow to stabilize over 5–10 min.
(xiv) Wash off in PSS, and allow the vessel to relax completely (tension should return to 0 g).
(xv) Calculate % dilation to SNP (see example below). At least a 70–80 % dilator response should be AChievable for a responsive VSMC layer.
(xvi) You may continue with the SNP dose response curve, by adding SNP cumulatively (e.g. 10^{-9}–10^{-4}). Allow to stabilize over 2–3 min at each concentration. This will provide you with information on endothelial-independent dilator responses.

In order to examine the effects of a given substance/mediator/cytokine, on vascular function, incubate the vessel over a given period of time, as relevant, then repeat the steps above.

7.3.4 Calculating % Dilation

$\%\,\text{dilation} = \text{Maximum relaxation} / \text{maximum constriction} \times 100$. *Therefore:*

$$\%\,dilation = \frac{(Maximum\ tension\ after\ constriction - stable\ tension\ after\ dilation)}{(Maximum\ tension\ after\ constriction - tension\ before\ constriction)} \times 100$$

Example: $\%\,\text{Dilation} = (\text{b} - \text{c}) / (\text{b} - \text{a}) \times 100$

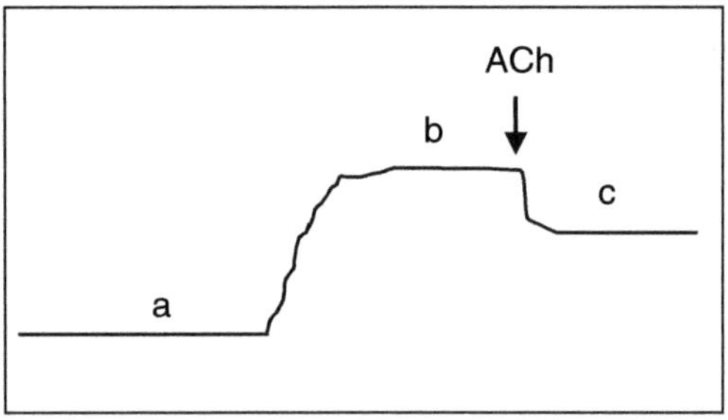

7.3.5 Sample Results

The labchart software provides trace outputs of each experiment conducted (see Fig. 7.3 below). One useful function is the insertion of 'comments' at each intervention. This is very helpful for calculations and data interpretation.

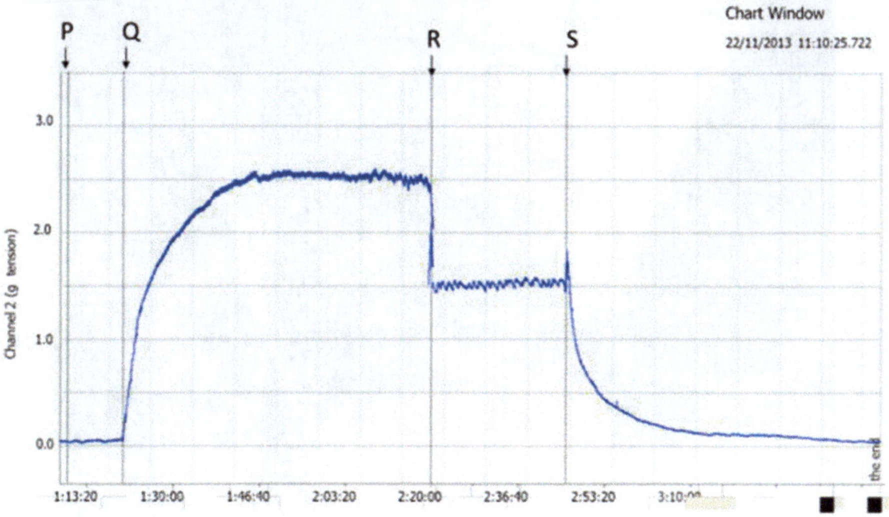

Fig. 7.3 Sample trace, illustrating the change in tension (g) over time (min) in a mounted aortic vessel, *ex vivo*, using lab Chart (P, PSS; Q, KPSS; R, ACh; S, PSS)

7.4 The Pressure Myograph System

For small size arteries, the pressure myograph system is ideal for investigating the effects of pharmacological and mechanical stimuli on vascular function and contractility. It has a number of advantages over the wire myograph system, as it preserves the geometry of the vessel and enables its maintenance under physiological conditions of temperature, pressure and flow. Vessel diameter measurements can be constantly monitored, in real time, under isobaric (constant pressure) conditions. Using this system, vessels within approximately 40–450 μm internal diameters can be studied. In the example below, I shall describe a step by step procedure for studying endothelial dependent and independent dilator responses in a murine mesenteric artery under static (zero) flow.

7.4.1 Vessel Dissection and Mounting

(i) Isolate the mesentery and place in ice cold PSS. Pin down in a petri dish with sylgard base and cover in gassed PSS. Place on ice and gas the PSS constantly (Fig. 7.4).
(ii) Using a stereomicroscope and fine forceps, identify the artery and dissect away any surrounding tissue very carefully. Arteries can easily be distinguished from veins by visualising their striped wall structure under the ×40 objective

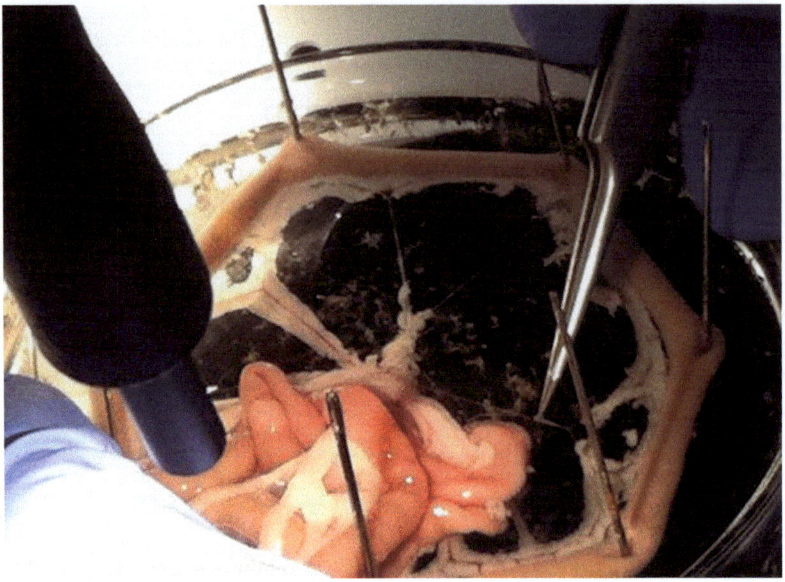

Fig. 7.4 Picture illustrating the preparation of the mesentery for dissection

lens. The walls are thicker than the walls of veins and they have a sharp v shape at bifurcations (as opposed to the more rounded bifurcation of veins).

(iii) Prepare the vessel chamber by infusing the cannulae and tubing with gassed PSS (Fig. 7.5). Ensure there are no air bubbles and close off the three way taps.

(iv) Cut a 2–3 mm length of vessel and mount between the two glass cannulae within the vessel chamber. Tie both ends of the vessel using polyester thread. Ensure that the vessel is covered in gassed PSS.

(v) Adjust the distance between the cannulae to ensure that the vessel is sufficiently stretched with no bends.

7.4.2 *Preparing the Setup*

Setting up of the system and its calibration can vary depending on the type of myograph available and its make. The information below, relates to use of the living systems model. Additional information is available on their website (http://www.livingsys.com/)

(i) Switch on the water bath heater to 37 °C.

(ii) Add PSS solution to the superfusion system and gas the solution using the gas source (95 % oxygen + 5 % CO_2). Allow to equilibrate.

Fig. 7.5 A close view of a vessel chamber. Two glass cannulae are aligned and infused with PSS prior to vessel mounting

(iii) Place the vessel chamber over an inverted microscope and connect the tubing to allow for constant superfusion of the vessel.

(iv) Connect one side of the vessel to the microfusion pump and start infusing with gassed PSS. Ensure that there are no air bubbles or leaks.

(v) Pressurise the vessel to an initial intravascular (Lumenal) pressure (IvP) of 60 mmHg, using the pressure servo control unit. Ensure that the vessel is not leaky by switching the servo control unit from 'auto' to 'manual' mode. The pressure reading should remain constant.

(vi) Switch on the 'video dimension analyser' and TV screen. Check that the vessel is fully visible and in focus, by focusing the microscope and adjusting the microscope stage.

(vii) Using the 'video dimension analyser' knobs, ensure that the 'scan line' is appropriately positioned and that the vessel walls are clearly identified using the 'level' knobs. This will allow constant measurement of the lumen diameter.

(viii) Switch on the powerlab system and open 'chart'. A live recording of the lumen diameter and right and left vessel wall thickness will be displayed as a function of time (Fig. 7.6). Additional parameters may be monitored (e.g. Pressure and temperature) depending on the data acquisition system available.

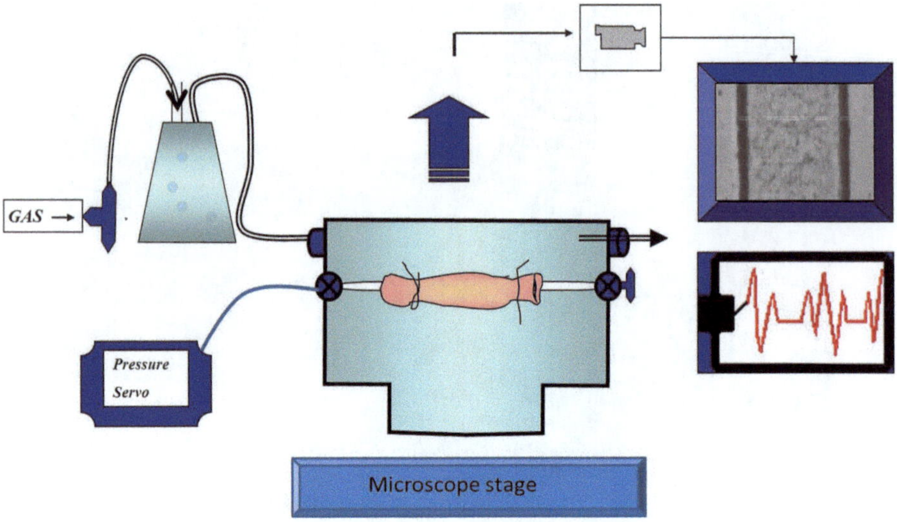

Fig. 7.6 Diagrammatic representation of the pressure myograph setup

7.4.3 Experimental Protocols

A number of experimental protocols can be followed, depending on the objectives of the experiment. These can be carried out either in isolation or in combination, to achieve the desired outcomes.

(a) **Test vessel viability**: Once the vessel has equilibrated (approx. 30 min), constrict it by superfusing it in KPSS (60 mM). Observe the reduction in diameter. Allow the amplitude of this contraction to reach a steady state (i.e. stabilize). This will take approx. 5–10 min. Wash off the vessel by superfusing in PSS and allow the vessel to relax and return to its original diameter. This procedure can be repeated a second time.

(b) **Examine endothelial integrity (endothelial dependent dilator responses)**:

  (i) Constrict the vessel again, by superfusing it in KPSS or by adding an agonist (e.g. Phe). Allow the amplitude of contraction to stabilize.
  (ii) Add a specified concentration of ACh from stock. Allow to stabilize over 2–3 min.
  (iii) Wash off by superfusing in PSS, and allow the vessel to relax completely.
  (iv) You may continue with the ACh dose response curve, by adding ACh cumulatively. Allow to stabilize over 2–3 min at each concentration. This will provide you with information on endothelial-dependent dilator responses.

(c) **Examine VSMC responsiveness (endothelial independent dilator responses)**.

  (i) Constrict the vessel again, by superfusing it in KPSS or by adding an agonist (e.g. Phe). Allow the amplitude of contraction to stabilize.

(ii) Add a specified concentration of SNP from stock. Allow to stabilize over 2–3 min.
(iii) Wash off by superfusing in PSS, and allow the vessel to relax completely.
(iv) You may continue with the SNP dose response curve, by adding it cumulatively. Allow to stabilize over 2–3 min at each concentration. This will provide you with information on endothelial-independent dilator responses.

(d) **Myogenic tone**: To investigate the effects of a physical parameter (e.g. pressure), or agonist on myogenic tone in small arteries: Firstly, set up the vessel at an initial IvP of 60 mmHg at 37 °C and allow the vessel to develop myogenic tone. This will take approximately 30 min to 1 h depending on the vascular bed and original resting arterial diameter. For investigating the effects of pressure or flow on myogenic tone, alter IvP using the pressure servo control, and note the change in diameter and vessel wall thickness. At the end of the experiment, superfuse the vessel in Calcium free PSS (containing 2 mM EGTA). The vessel will dilate and return to its passive resting diameter after 30 min. Myogenic tone is taken as the difference between intralumenal diameters in the presence and absence of calcium.

At the end of a protocol, the vessel is normally superfused in calcium-free PSS for 30 min to allow it to return to its passive diameter. The effects of a given substance/ mediator/ cytokine, on vascular function, can be examined by adding it either to the superfusate, or by directly infusing it through the vessel intravascularly, over a given period of time, then repeating the steps above.

7.4.4 Calculating % Dilation

$$\begin{aligned}\%\,\text{dilation}: &(\text{Maximum relaxation} - \text{Maximum constriction})/\\ &(\text{Original Diameter} - \text{Maximum constriction}) \times 100.\\ &= (a - c)/(b - c) \times 100\end{aligned}$$

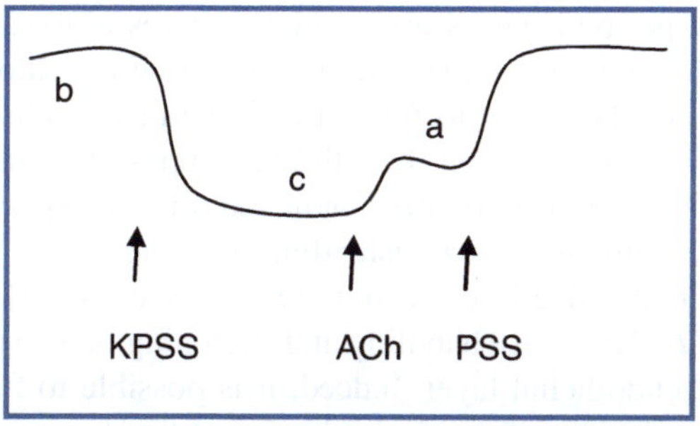

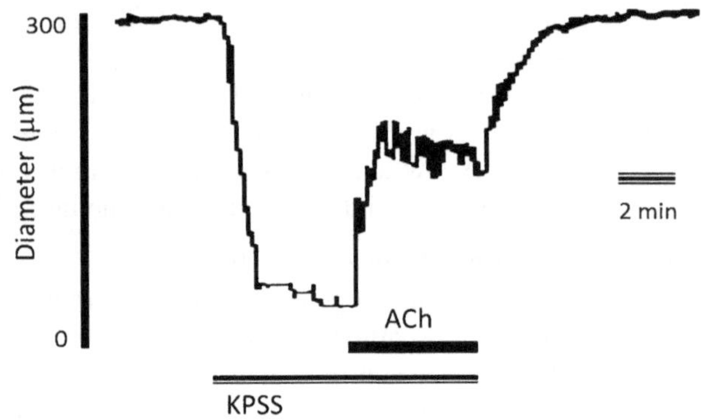

Fig. 7.7 Sample trace, illustrating the change in diameter over time (min) in a mesenteric artery (300 µm diameter). The artery was initially constricted in KPSS (60 mM), then dilated using 10^{-6} M ACh and finally washed in PSS

7.4.5 Sample Results

The labchart software provide trace outputs of each experiment conducted. One useful function is the insertion of 'comments' at each intervention. This is very helpful for data interpretation and calculations. The output trace can be imported into a drawing package software for presentation purposes (see Fig. 7.7).

7.5 Trouble Shooting

Preparation of agonists: Care must be taken in the preparation of agonists and the calculation of appropriate dosages. For example, ACh is hygroscopic and will absorb moisture, so speed in weighing out is necessary to limit exposure to the air. Also, SNP is very light sensitive, so solutions must be prepared freshly in the dark and placed in opaque tubing.

Vessel viability: Having a non-viable vessel is often due to lengthy time periods of dissection and/or excessive handling of the vessel. It can also be due to poor oxygenation and perfusion in PSS. An insufficient equilibration period can lead to reduced viability (KPSS responses). In the case of the organ bath experiments, the equilibration period is necessary to allow the vessel to reach its resting passive tension. Allowing for a longer equilibration time will allow the intermediate elastic filaments of the vessel to fully stretch, so that a subsequent contraction (active tension) can be observed. For different types of vessels, a passive force normalisation procedure is normally conducted prior to experimentation in order to determine the optimal passive distending force necessary to achieve a maximal, or submaximal active force, as required [4] [see also [3], for details].

Endothelial integrity: Excessive handling and stretching of the vessel can also damage the delicate endothelial layer. Indeed, it is possible to fully denude a large artery by rubbing the inner layer using the tips of fine forceps. Small vessels can be fully denuded by passing a number of air bubbles through the lumen. An ACh response <10 % confirms denudation.

Limiting variation in results: Use tissue from similar aged animals and from the same region (e.g. thoracic or abdominal aorta; 1st or 3rd order of mesenteric vessel). Use a two sided blade to cut a precise length of vessel.

Non-pressurised vessels (for pressure myography): This is often a result of a leaky vessel. Care should be taken in dissecting and mounting the vessel to avoid cutting through the vessel wall too. Stretching can cause a tear in the vessel wall. A very small side branch in the vessel may also cause loss of pressure. It is worth eliminating any pressure loss from the system itself. Check that the system is not leaking (loose connections etc.) by closing off the vessel from the pressure servo control. Turn the pressure control valve from 'auto' to 'manual' mode and check for any fall in pressure.

7.6 Applications and Discussion

A major application of the *ex vivo* techniques described herein, is in the study of endothelial dysfunction, a known predictor of coronary artery disease. Examining endothelial dependent dilator responses in vessels remains the gold standard for assessing endothelial dysfunction, both *ex vivo* and *in vivo*. Hence, there are wide ranging applications for both the organ bath and pressure myograph techniques in both health and disease, subsequent to medical intervention, and in assessing the influence of novel technologies (such as nanoparticles) for prospective therapies [5, 6]. The direct influence of the endothelial layer can be investigated by carrying out experiments on denuded vessels (as described above) [7]. Equally, it is possible to investigate the role of the adventitial layer by digesting in collagenase and then gently removing it using fine forceps [8]. The role of the surrounding perivascular fat has also attracted attention in relation to its anti-contractile function [9, 10]. Both techniques, in particular, the organ bath technique, is very useful for teaching and demonstrating purposes in the laboratory, at both undergraduate and postgraduate levels, to illustrate the basic principles of vascular biology [11, 12].

A number of adaptations of the pressure myography system exist, which allow the measurement of various parameters, including extravascular pressure (EvP), relevant to heart disease [13], and for the study of flow-mediated dilation [14]. Additional information from these vascular function experiments can also be gained by isolating RNA/DNA from the vessels for molecular biology studies. It is also possible to fix the vessels, at the end of an experiment, for immunohistochemical and morphological studies (light microscopy and electron microscopy), as appropriate.

Appendix

Table 7.1 Composition of the physiological salt solutions (Courtesy of Dave Maskew, School of HCS)

Salt *(Mol.Wt.)*	PSS (Litre)	PSS (5 l) 10×Conc	mM	Ca2+ Free PSS[a] (Litre)	Ca2+ Free PSS[a] (5 l) 10×Conc	mM	KPSS (60 mM) (Litre)	KPSS (60 mM) (5 l) 10×Conc	mM
	Weight (g)	Weight (g)		Weight (g)	Weight (g)		Weight (g)	Weight (g)	
NaCl *(58.44)*	6.95	347.7	119	6.95	347.7	119	4.57	228.5	78.2
KCl *(74.55)*	0.35	17.52	4.7	0.35	17.52	4.7	4.47	223.65	60.0
$MgSO_4.7H_2O$ *(246.48)*	0.30	14.79	1.2	0.30	14.79	1.2	0.30	14.79	1.2
KH_2PO_4 *(136.09)*	0.16	7.96	1.17	0.16	7.96	1.17	0.16	7.96	1.17
$CaCL_2.2H_2O$ *(147.02)*	0.24	11.76	1.6	–	–	–	0.24	11.76	1.6
$K_2EDTA.2H_2O$ *(404.45)* (add last)	0.012	0.61	0.03	0.012	0.61	0.03	0.012	0.61	0.03

Volume of glucose and $NaHCO_3$ to be added to 1 l of stock solution
Glucose (100 g/L) (Mol. Wt. = 180.16)
Add **9.91** ml/l of stock solution to give a working concentration of **5.5 mM**
NaHCO3 (50 g/L) (Mol. Wt. = 84.01)
Add **42** ml/l of stock solution to give a working concentration of **25 mM**
[a]Add 2 mM EGTA last

References

1. Webb RC (2003) Smooth muscle contraction and relaxation. Adv Physiol Educ 27:201–206
2. Furchgott RF, Zawadzki JV (1980) The obligatory role of endothelial cells in the relaxation of arterial smooth muscle by acetycholine. Nature 288:373–376
3. Angus JA, Wright CE (2000) Techniques to study pharmacodynamics of isolated large and small blood vessel. J Pharmacol Toxicol Methods 44:395–407
4. Mulvany MJ, Halpern W (1977) Contractile properties of small arterial resistance vessels in spontaneously hypertensive and normotensive rats. Circ Res 41:19–26
5. Akbar N, Mohamed T, Whitehead D, Azzawi M (2011) Biocompatibility of amorphous silica nanoparticles: size and charge effect on vascular function, in vitro. Biotechnol Appl Biochem 58:353–362
6. Shukur A, Rizvi S, Whitehead D, Seifalian A, Azzawi M (2013) Altered sensitivity to nitric oxide donors, induced by intravascular infusion of quantum dots, in murine mesenteric arteries. Nanomedicine 9:532–539
7. Gao YJ, Lu C, Sharma AM et al (2007) Modulation of vascular function by perivascular adipose tissue: the role of endothelium and hydrogen peroxide. Br J Pharmacol 151:323–331

8. Gonzalez MC, Arribas SM, Molero F et al (2001) Effect of removal of adventitium on vascular smooth muscle contraction and relaxation. Am J Physiol Heart Circ Physiol 280:H2876–H2881
9. Lohn M, Dubrovska G, Lauterbach B et al (2002) Periadventitial fat releases a vascular relaxing factor. FASEB J 16:1057–1063
10. Withers SB, Passi N, Williams AS et al (2013) Erythropoietin has a restorative effect on the contractility of arteries following experimental hypoxia. J Cardiovasc Dis Res 4:164–169
11. Gonzales RJ, Carter RW, Kanagy NL (2000) Laboratory demonstration of vascular smooth muscle function using rat aortic ring segments. Adv Physiol Educ 24:13–21
12. Vandier C, Le Guennec J-Y, Bedfer G (2002) What are the signalling pathways used by norepinephrine to contract the artery? A demonstration using guinea pig aortic ring segments. Adv Physiol Educ 26:195–203
13. Azzawi M, Austin C (2006) Extravascular pressure modulates responses of isolated rat coronary arteries to vasodilator, but not vasoconstrictor stimuli. Am J Physiol Heart Circ Physiol 290(3), H11516
14. Azzawi M, Austin C (2007) Transient vasodilatory responses to intralumenal flow in isolated pressurised rat coronary arteries: influences of different endothelial factors. J Vasc Res 44:223–233

JoVE Videos

Shahid M, Buys ES (2013) Assessing murine resistance artery function using pressure myography. J Vis Exp (76):e50328. doi:10.3791/50328. http://www.jove.com/video/50328/assessing-murine-resistance-artery-function-using-pressure-myography

Web Links

AD Instruments (PowerLab System). http://www.adinstruments.com/products/powerlab
Living Systems Instrumentation. http://www.livingsys.com/
World Precision Instruments. http://www.wpiinc.com/

Chapter 8
Measurement of Aβ Uptake by Cerebrovascular Smooth Muscle Cells

Wan Adriyani Wan Ruzali and Seth Love

8.1 Introduction

The accumulation of amyloid-β peptide (Aβ) is a pathological hallmark of Alzheimer's disease (AD), reflecting a dynamic imbalance between Aβ production and removal that is thought to play a central role in the pathogenesis of this disease, initiating a cascade of metabolic processes that lead to the formation of neurofibrillary tangles and to neuronal dysfunction and damage. Aβ accumulates within the brain parenchyma as plaques, and in the walls of cerebral blood vessels as cerebral amyloid angiopathy (CAA). The Aβ within blood vessels is thought to be largely if not solely of neuronal origin but its progressive accumulation within the tunica media leads to the death of cerebrovascular smooth muscle cells (CVSMCs). These cells are important for vascular contractility and tensile strength. Their loss affects cerebral vasomotor function and can lead to cerebral haemorrhage.

The pathogenesis of the death of CVSMCs in CAA is still unclear. Several studies have shown that CVSMCs are capable of Aβ uptake and that this is mediated at least partly through the binding of Aβ to lipoprotein receptors, particularly low density lipoprotein receptor-related protein 1 (LRP1) [1, 2]. It has been suggested that the death of CVSMCs in CAA is mediated by at least two mechanisms, one that relates to fibril formation on the CVSMC surface and another to the uptake of Aβ into these cells [2, 3]. There is evidence that apolipoprotein E (*APOE*) genotype also influences the death of CVSMCs in CAA and the development of cerebral haemorrhage [4, 5]. We therefore recently studied the influence of *APOE* genotype and contribution of LRP1-mediated cellular uptake of Aβ on the cytotoxic effects of

W.A.W. Ruzali
Institute of Biological Sciences, University of Malaya, Kuala Lumpur 50603, Malaysia

S. Love (✉)
Dementia Research Group, Institute of Clinical Neurosciences, Frenchay Hospital, Bristol BS16 1LE, UK
e-mail: seth.love@bristol.ac.uk

M. Slevin, G. McDowell (eds.), *Handbook of Vascular Biology Techniques*,
DOI 10.1007/978-94-017-9716-0_8

Aβ on CVSMCs [6, 7]. In this chapter we describe the image-based technique that we used to measure Aβ uptake by CVSMCs whilst excluding adherent fibrillar deposits of Aβ on the surface of the cells.

8.2 Methodology

The uptake of Aβ at different time points can be measured by immunofluorescent labelling of unlabelled Aβ or by using fluorescently tagged Aβ. In both cases the level of uptake is determined by measuring the intensity of intracellular fluorescence, with the help of imaging analysis software. The use of fluorescently tagged Aβ allows the detection of lower levels of uptake and is the method reported here. To illustrate the utility of this image-based approach, we describe its application to measurement of the uptake of fluorescently tagged $A\beta_{1-42}$, as measurement of the uptake of this longer isoform of Aβ is particularly susceptible to precipitation artefact. The method can, of course, be modified for measuring the uptake of a wide range of other molecules.

8.3 Materials

8.3.1 Cells and Cell Culture Reagents

- Human fetal primary brain vascular smooth muscle cells (ScienCell; from Caltag Medsystems, Buckingham, UK)
- Fetal bovine serum (ScienCell)
- Smooth muscle cell growth supplement (ScienCell)
- Penicillin/streptomycin solution (ScienCell)
- Albumin from bovine serum (Sigma-Aldrich, Gillingham, UK)
- HiLyteFluor 488-labelled $A\beta_{1-42}$; lyophilized powder (Cambridge Biosciences, Cambridge, UK)
- Paraformaldehyde/ PFA 16 % (Alfa Aesar, Heysham, UK)
- Vectashield mounting medium for fluorescence with 4′,6-diamidino-2-phenylindole (DAPI) (Vector Laboratories, Peterborough, UK)
- Poly-L-Lysine (PLL) (ScienCell)
- Acetonitrile (ACROS Organics, from Fisher Scientific, Loughborough, UK)
- Cellstar tissue culture flasks (Greiner Bio-One, Stroudwater, UK)
- 4-well plates (Nunc, from Fisher Scientific, Loughborough, UK)

8.3.2 Preparation of Solutions

- Smooth muscle cell medium: 100 % fetal bovine serum, 1 % smooth muscle cell growth supplement, 1 % penicillin/streptomycin solution

- Serum-free medium: 1 % smooth muscle cell growth supplement, 1 % penicillin/streptomycin solution, 0.1 % albumin solution from bovine serum
- 35 % acetonitrile: dilute acetonitrile into ddH_2O to a concentration of 35 %.
- 4 % PFA: dilute 16 % PFA in sterile PBS to a concentration of 4 %.

8.3.3 Equipment and Software

- Leica DMR fluorescent microscope (Leica Microsystems, Milton Keynes, UK)
- QICAM digital camera (Q Imaging, Surrey, BC, Canada)
- Andor IQ Live Cell Imaging software (Andor Technology, Belfast, UK)
- Simple PCI 6 software (Hanamatsu Photonics, Welwyn Garden City, UK)

The analysis is, of course, not dependent on these particular items of hardware and software and can be performed using other fluorescent microscopes, digital cameras and imaging software.

8.4 Protocol

The experiments are performed under low-light conditions to minimise photobleaching of the fluorescein.

8.4.1 Culture of CVSMCs

1. CVSMCs are grown in PLL-coated flasks at 37 °C under 5 % CO_2.
2. Grow the cells to confluence with a change of smooth muscle cell culture medium every 2 days.

8.4.2 Aβ Uptake Experiments

1. Once cells are confluent, plate CVSMCs at equal densities onto PLL-coated coverslips in 4-well plates: 1.4×10^4 to 1.8×10^4 CVSMCs for a coverslip of 1.0 cm diameter.
2. Allow the cells to grow (at 37 °C under 5 % CO_2) until they are 80 % confluent.
3. Remove the smooth muscle cell culture medium, replace it with an equal volume of serum-free medium and leave for 3 h in the incubator.
4. Dilute HiLyteFluor 488-labelled $A\beta_{1-42}$ in 35 % acetonitrile to a stock concentration of 80 μM and store at −20 °C.

5. Further dilute the stock concentrations of $A\beta_{1-42}$ in serum-free medium to working concentrations of 0.1, 1.0 and 10.0 μM and keep at −20 °C until use.
6. Remove the serum-free medium from the CVSMCs and add the working concentrations of $A\beta_{1-42}$ or acetonitrile in serum-free medium as a control.
7. Incubate CVSMCs at 37 °C under 5 % CO_2 for appropriate periods (e.g. 8 h, 1 day and 3 days).
8. After the period of exposure, wash the CVSMCs twice with PBS.
9. Fix the CVSMCs in 4 % PFA for 10 min.
10. Wash again twice with PBS.
11. Mount the coverslips with attached CVSMCs onto glass microscope slides in Vectashield mounting medium with DAPI.

8.4.3 Quantifying Aβ-Uptake Using Simple PCI 6

1. Under a ×20 objective, acquire images of approximately 30 non-overlapping cells. It is important to have the same illumination and exposure settings for all images and preferably to acquire all of the images in a single session.
2. In Simple PCI 6, load each image and the corresponding calibration file (i.e. for the ×20 objective).
3. Identify and select the region of interest (ROI), making sure not to select any precipitate (which is easily recognizable as clumps of intensely fluorescence material).
4. Manually exclude any superimposed precipitate in the Advanced Detection dialog box, by use of a mouse or a drawing tablet.
5. The Qualify Objects dialog box can be used to further exclude areas that are not of interest.
6. The final ROI is thus selected by a combination of automatic software-based selection of the ROI and manual exclusion of precipitate.
7. In the Select Measurements dialog box, select Area, Total Green and Mean Green (HiLyteFluor 488-labelled $A\beta_{1-42}$ fluoresces at 488 nm).
8. Repeat steps 4–7 above for each cell (i.e. measure twice for each cell).
9. Repeat steps above for measurement of Aβ-uptake into the other imaged cells.
10. Calculate the final values for the mean fluorescence, range and standard deviation (SD) for each coverslip.
11. Repeat the experiment and calculate overall mean fluorescence, range and SD for each time point (i.e. 8 h, 1 day and 3 days).

8.5 Sample Results

Images of CVSMCs incubated with HiLyteFluor 488-labelled $A\beta_{1-42}$ are shown in Fig. 8.1, which illustrates the way in which the Aβ can precipitate out of solution, not only onto the coverslip but also on to the surface of the cells. Aβ that has been

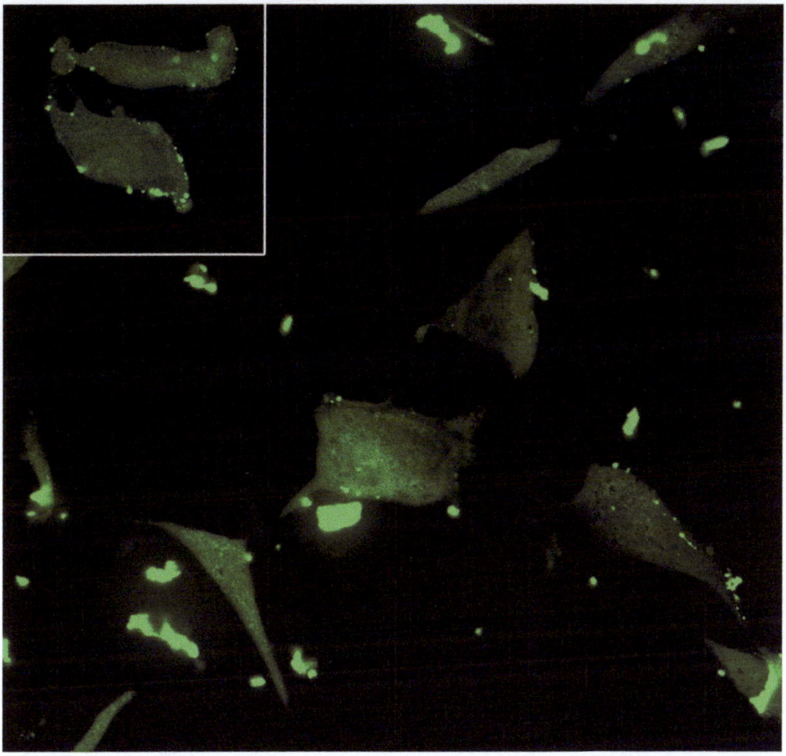

Fig. 8.1 Distinguishing between uptake and precipitation of $A\beta_{1-42}$. The main image illustrates CVSMCs after incubation with HiLyteFluor 488-labelled $A\beta_{1-42}$ for 8 h. The finely granular green fluorescence within the cells reflects uptake of Aβ (confirmed immunohistochemically). The large brightly fluorescent clumps are precipitates of Aβ on the coverslip. There are also small precipitates of Aβ on the surface of some of the cells. The inset shows cells incubated with HiLyteFluor 488-labelled $A\beta_{1-42}$ for 1 day, at which time there are more numerous clumps of Aβ on the surface of the cells

taken up by the cells produces a finely granular pattern of fluorescence. In contrast, there are relatively large, densely fluorescent precipitates of Aβ on the coverslip, and also smaller but still intense fluorescent deposits of Aβ on the surface of some of the cells. These are clearly seen in the inset. Direct visualisation of the cells allows the different patterns of fluorescence to be distinguished. The clumps of bright fluorescent material can be excluded from measurement by the image analysis software. Their inclusion overstates the uptake of Aβ.

As shown in Fig. 8.2, uptake of $A\beta_{1-42}$ by CVSMCs is relatively rapid and there is no significant increase beyond 8 h. By 72 h the uptake appears to have declined; however, many of the cells have degenerated after this period of exposure to $A\beta_{1-42}$, and the 'decline' is probably a reflection of bias towards cells that have survived because they have not taken up as much of the peptide. The cytotoxicity of $A\beta_{1-42}$ to CVSMCs is illustrated in Fig. 8.3, which indicates another advantage of image-based measurement of Aβ uptake – the ability to assess relationships between function

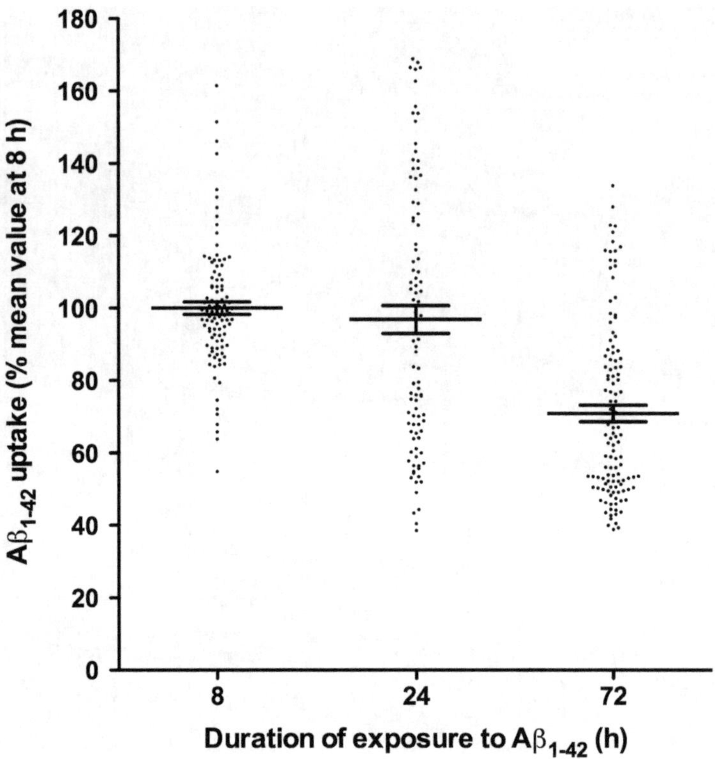

Fig. 8.2 Measurement of the uptake of HiLyteFluor 488-labelled $A\beta_{1-42}$ by CVSMCs at 8, 24 and 72 h. The values are expressed as the percentage of the mean level of intracellular fluorescence at 8 h. Each dot represents the mean of two measurements on a single cell. The horizontal bars show the mean and standard error of the mean. There is little change between 8 and 24 h. The decline at 3 days probably reflects sampling bias as many of the cells have degenerated after exposure to $A\beta_{1-42}$ for this length of time

and structure. In this case it can be seen that the uptake of $A\beta_{1-42}$ by CVSMCs is associated with marked cytoskeletal disruption, as evidenced by the subsarcolemmal accumulation and clumping of smooth muscle actin, with loss of its normal, elongated filamentous structure.

8.6 Reproducibility

This section describes experiments we performed to evaluate the reliability and reproducibility of the assay. Researchers are encouraged to assess the reliability of the assay for their own experiments, by an approach similar to that described below.

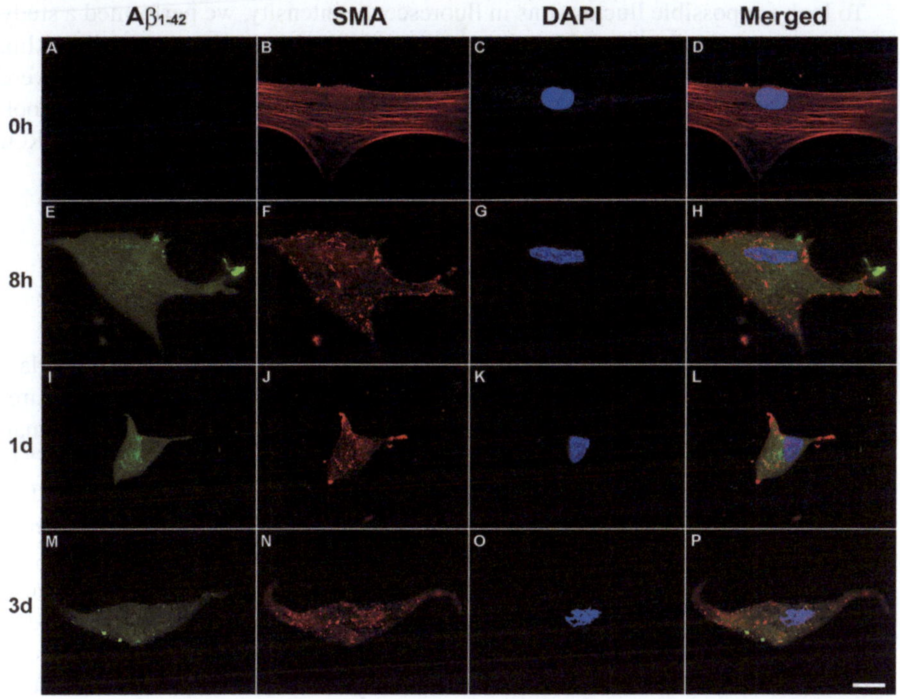

Fig. 8.3 Cytoskeletal disruption associated with the uptake of $A\beta_{1-42}$. After exposure of CVSMCs to HiLyteFluor 488-labelled $A\beta_{1-42}$ for up to 3 days, the cells were labelled with rabbit polyclonal antibody to smooth muscle actin (SMA; ab58619, Abcam, Cambridge, UK) followed by Alexa Fluor 568-tagged donkey anti-rabbit IgG (Invitrogen, Paisley, UK), and the nuclei counterstained with DAPI. (**a–d**) In the absence of Aβ, SMA (*red*) is labelled in a filamentous pattern, with most filaments orientated along the long axis of the cell. Nuclei are stained with DAPI (*blue*). (**e–h**) After incubation with $A\beta_{1-42}$ for 8 h, the cells have lost their strap-like shape and there is clumping of SMA, particularly in the subsarcolemmal region. (**i–l**) After incubation for 1 day, the cells appear shrunken. SMA is still clumped, particularly in the subsarcolemmal region, but the antibody also labels irregularly shaped granules deep within the cytoplasm. (**m–p**) By 3 days many of the cells incubated with $A\beta_{1-42}$ have degenerated and most of the SMA is in the form of cytoplasmic granules. Bar = 40 μm (Reprinted from Journal of Alzheimer's Disease [7] with permission from IOS Press)

The first step we took to assess the reproducibility of this method of measuring Aβ uptake, was simply to repeat the experiments on three separate occasions, which yielded similar findings, albeit with quite a wide scatter of measurements about the mean. We then used a statistical approach to distinguish variability resulting from the method of acquisition of imaging data (particularly from possible fluctuations in fluorescence intensity) from that caused by other experimental variables (e.g. variations in cell plating density or passage).

To look for possible fluctuations in fluorescence intensity, we performed a study in which three separate sets of images were acquired of the cells on each coverslip. We recorded the total time taken to acquire the images. The replicate images were acquired in three successive sequential sessions (S1, S2 and S3). Each session took 45 min to 1 h. The individual images and cells were matched up and the same ROI was used for the measurements of the same cell.

8.6.1 Results

Our initial inspection of the data included an assessment of whether all of the data points fell within the 95 % confidence interval (CI) of the means of the measurements of each set of data. Further analysis was performed by use of Bland-Altman plots, Passing and Bablok regression and mountain plots, with the help of MedCalc (v11.6; Mariakerke, Belgium). The Bland-Altman plot is also known as the difference plot and is used to compare the differences between two measurement techniques or to analyse the repeatability of a method. Passing and Bablok regression shows possible systematic or proportional bias between different measurements. The mountain plot looks at the distribution of differences in each measurement, comparing up to two methods against a reference method. These statistical tests have been used in several studies, such as those by Rosario et al. and Donadio et al. [8, 9], to look at the reliability of different methods and diagnostic markers. Krouwer and Monti [10] recommended that these tests be used together to provide a comprehensive perspective of the data.

Graphical representations of the analysis of the possible contribution of fluctuations in fluorescence intensity to the variation in measurements within our samples are shown in Fig. 8.4. All data points in S1 and S3 were well within the 95 % CI of both upper and lower limits of differences (blue line) in the Bland-Altman plot. Further analysis using Passing and Bablok regression revealed a slight systematic bias (data with regression intercept and slope values with 95 % confidence intervals are not shown) in measurements of fluorescence in S3 compared to S1. In the mountain plot, the median of differences between S3 and S1 was larger than the median of differences between S2 and S1 (which was close to zero).

The results of all of these analyses suggest that fluorescent intensity measurements change slightly over time despite our use of constant exposure settings. The longer the fluorescence microscope was used continuously, the greater the median of differences between measurements of the same cell. Since the whole image-taking session was carried over a period of about 2.5 h, these results show that the fluctuations in fluorescent intensity may affect the accuracy of measurements over this period. To maintain a fairly consistent level of fluorescent intensity, the acquisition of images for measurement of Aβ uptake by CVSMCs was limited to a maximum of 1 h per sitting.

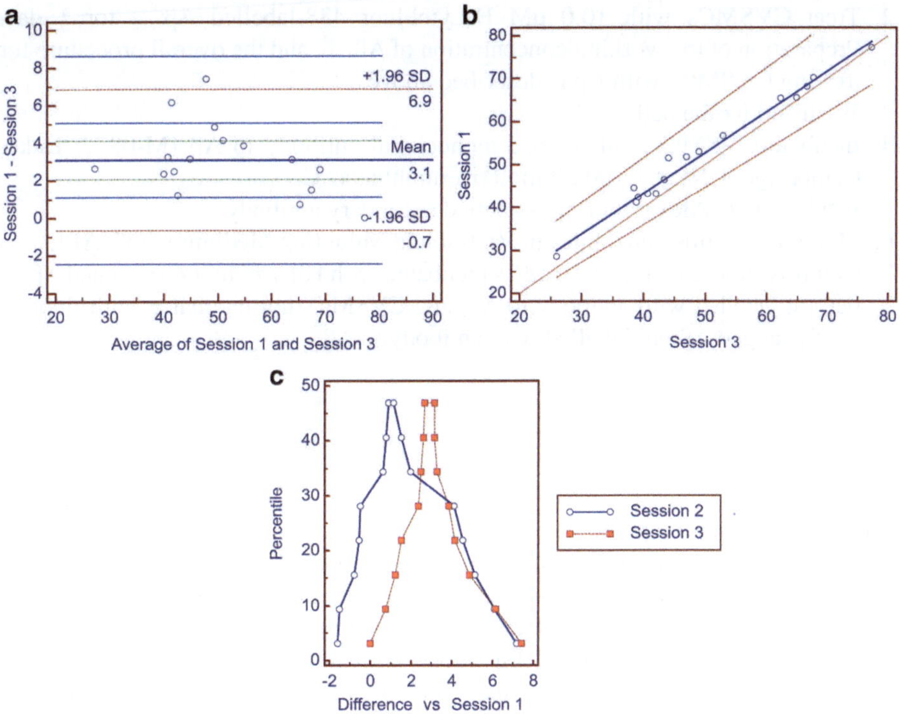

Fig. 8.4 Variation in measurement between sessions. In this experiment we measured fluorescence (i.e. Aβ uptake) in the same CVSMCs in three sequential imaging sessions (S1, S2 and S3), each lasting 45 min to 1 h. (**a**) Bland-Altman plot comparing measurements in S1 and S3. (**b**) Passing and Bablok regression graph comparing measurements in S1 and S3. (**c**) Mountain plot comparing measurements in S2 and S3 with those in S1

8.7 Control Experiments

The accuracy of this method of measurement depends on the retention of the fluorescein label by the protein/peptide of interest up to the time of measurement. We used immunofluorescence with Aβ-specific antibody to confirm the correspondence between the HiLyteFluor 488 fluorescence and $A\beta_{1-42}$ within the CVSMCs. We also used confocal imaging to confirm that the HiLyteFluor 488 fluorescence was intracellular and not solely on the surface of the CVSMC.

8.7.1 Correspondence Between HiLyteFluor 488 Fluorescence and $A\beta_{1-42}$

1. Grow CVSMCs as usual and plate them on coverslips in wells, as described above.

2. Treat CVSMCs with 10.0 μM HiLyteFluor 488-labelled $A\beta_{1-42}$ for 1 day. Preparation of the working concentration of $A\beta_{1-42}$, and the overall procedure for treating CVSMCs with Aβ is described above.
3. Wash and fix the cells.
4. Incubate CVSMCs with mouse monoclonal antibody to Aβ (M-0872, Dako, Cambridge, UK) at 1:100 dilution overnight at 4 °C.
5. Incubate with AlexaFluor 555-labelled secondary antibody.
6. Mount coverslips onto slides in Vectashield Mounting Medium with DAPI.
7. Controls should include CVSMCs incubated with HiLyteFluor 488-tagged $A\beta_{1-42}$ but not labelled with antibody to Aβ, and CVSMCs not incubated with fluorescently tagged Aβ but labelled with antibody to Aβ.

8.7.2 Results

Immunofluorescent labelling of CVSMCs that had been exposed to HiLyteFluor 488-tagged Aβ1-42 showed that the distribution of green fluorescence corresponded closely to that of the red immunofluorescent labelling of Aβ. Although the fluorescein signal tended to be more diffuse, less granular and sometimes more extensive than the immunolabelling of Aβ (Fig. 8.5a–l), the strength of immunolabelling of Aβ with antibody corresponded to the intensity of the fluorescein signal, i.e. cells that had less green fluorescence also had less red immunolabelling of Aβ. No immunofluorescent labelling for Aβ could be demonstrated within CVSMCs that had not been incubated with HiLyteFluor 488-tagged Aβ1-42 (Fig. 8.5m–p), confirming that measurement of the green fluorescein signal was a reliable way to assess the uptake of Aβ.

Examination of Z-stacks obtained by confocal laser scanning of CVSMCs treated with HiLyteFluor 488-tagged Aβ1-42 showed the diffuse green fluorescein signal to be intracellular (Fig. 8.6), clearly distinguishable from the brightly fluorescent precipitate that was present outside the cells and that was excluded from the ROIs delineated for measuring Aβ uptake.

8.8 Additional Comments on Methodology

For accurate measurements, it is important to be able to delineate the periphery of the cells to be subjected to image analysis. If the periphery of many of the cells cannot clearly be identified, the coverslip is unsuitable for measuring Aβ uptake. Excessive cell density, in particular, reduces the accuracy of the measurements by obscuring cell margins (as well as affecting the biological behaviour of the cells).

We found a slight decline in the uptake of Aβ with cell passage number. Many other studies have shown that differences in cell passage number and confluence may affect the morphology and biological behaviour [11–14] of cells, and their

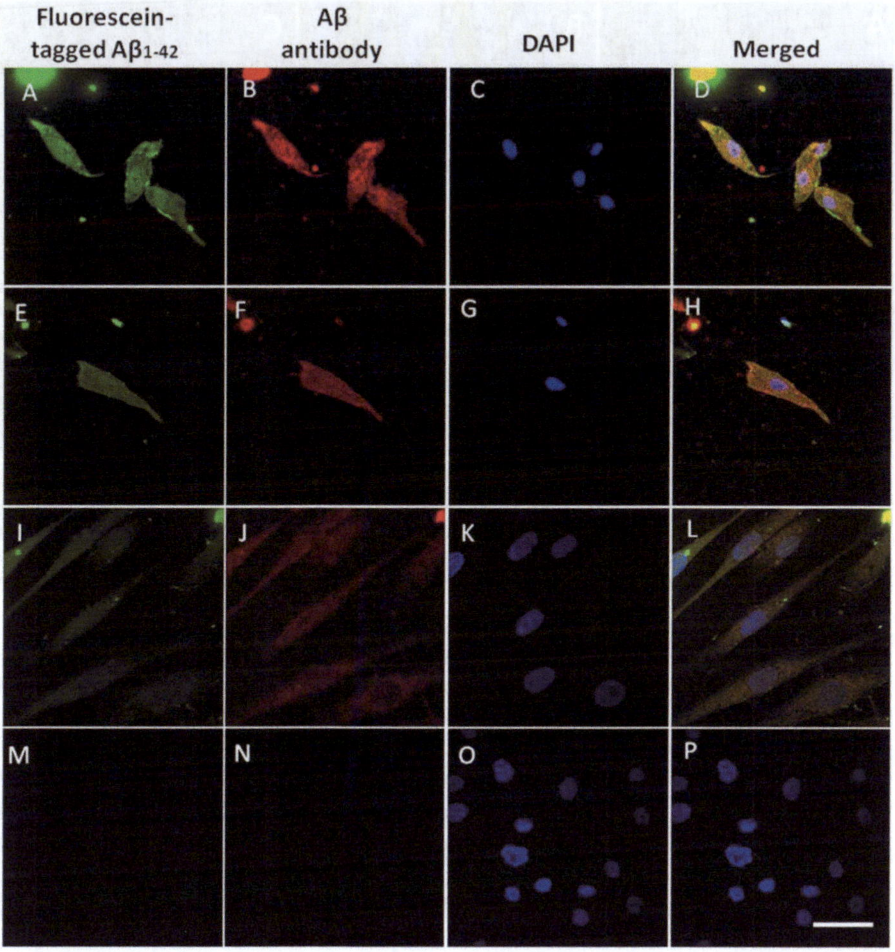

Fig. 8.5 (**a–d**) CVSMCs exposed to 10.0 μM HiLyteFluor 488-tagged $A\beta_{1-42}$ for 1 day, where the distribution of green fluorescence corresponded closely to that of immunolabelling of Aβ (*red*). Nuclei were stained with DAPI (*blue*). (**e–h**) Another example of CVSMCs exposed to 10 μM fluorescently tagged $A\beta_{1-42}$ for 1 day. (**i–j**) In CVSMCs exposed to fluorescently tagged $A\beta_{1-42}$ for 1 day, the intensity of $A\beta_{1-42}$ immunolabelled in red corresponded to the intensity of fluorescently tagged $A\beta_{1-42}$. (**m–p**) In CVSMCs that had not been exposed to fluorescently tagged $A\beta_{1-42}$, but immunostained with antibody to Aβ, there was neither *green* fluorescent signal nor *red* immunofluorescent labelling for $A\beta_{1-42}$. Bar = 20 μm (Part of this figure is reprinted from Journal of Alzheimer's Disease [7] with permission from IOS Press)

susceptibility to the cytotoxic effects of Aβ [15]. We used meta-analysis rather than simple pooling of data to examine the effects of various experimental manipulations in repeat experiments.

Use of fluorescently tagged Aβ (rather than immunofluorescent labelling of the peptide) increased the sensitivity of our assays and facilitated combined fluorescent-immunofluorescent labelling of other proteins of interest. However, in using a

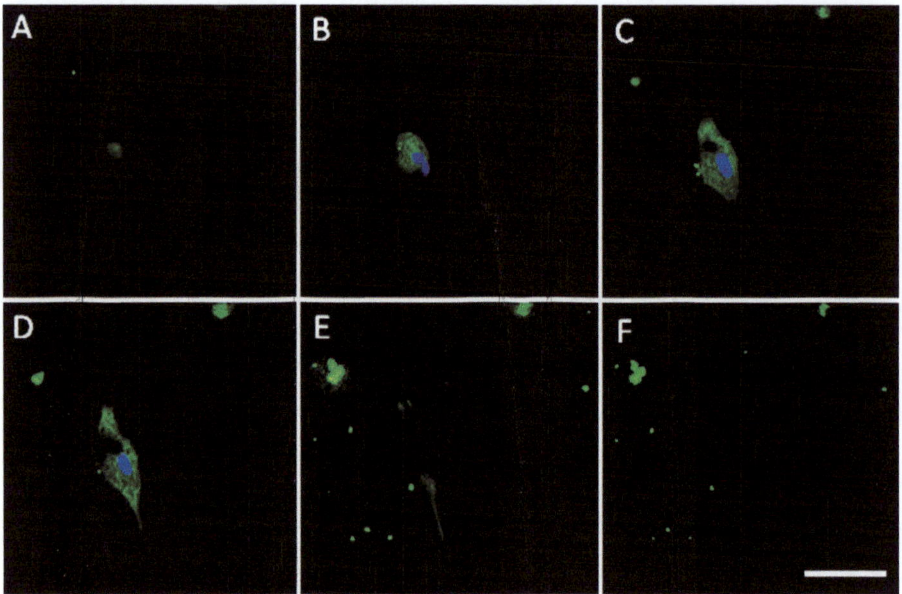

Fig. 8.6 Confocal imaging of fluorescently tagged $A\beta_{1-42}$ within a cell. Images (**a–f**) show sequential slices at multiple levels through the same CVSMC after exposure to 10.0 μM HiLyteFluor 488-tagged $A\beta_{1-42}$ for 1 day. There is diffuse fluorescence through the cytoplasm of the cell, as well as some brightly fluorescent clumps of extracellular material. Bar = 20 μm

fluorescently tagged protein, it is important to use a stable fluorescent tag that will not interfere with the experiment. In our case we used a HiLyteFluor 488 tag attached to the amino terminal of Aβ whereas it is mainly the carboxyl terminal that is involved in uptake of this peptide [16]. HiLyteFluor 488 added only 0.5 kDa to the molecular weight to Aβ and is said to be more resistant to change in pH and less prone to aggregation than fluorescein isothiocyanate.

8.9 Discussion and Applications

Several studies have assessed the uptake of Aβ by CVSMCs, although by use of other methods from the one described here. Urmoneit et al. [1] examined the uptake of soluble Aβ1-40 and Aβ1-42 into human and canine CVSMCs by immunohistochemical labelling of Aβ. Kanekiyo et al. [17] measured the uptake of Aβ into CVSMCs by FACS (fluorescence-activated cell sorting). We investigated several techniques for studying the uptake of Aβ, including these previously described methods and the use of a fluorescent plate reader to quantify total cellular fluorescence. We found that none of these methods allowed us to reliably distinguish genuine uptake of Aβ from the presence of aggregates of Aβ that were firmly adherent to the external surface of many cells and resisted attempts at removal by washing and by use of proteases.

The accumulation of insoluble aggregates of Aβ on the cell surface may simply reflect their precipitation out of solution. However, it has also been suggested that Aβ taken up by cells in soluble form may subsequently be released extracellularly as fibrils into cell culture medium [18]. In CAA, deposits of Aβ on the surface of CVSMCs may contribute to cell death [3, 19]. To study the relative contributions of extracellular and intracellular Aβ to the dysfunction and death of CVSMCs, it is clearly important to incorporate measurement techniques, such as those described here, that are able to distinguish between Aβ on the surface of the cells and that in the cytoplasm.

However, image-based measurement does have a number of limitations. It is more labour-intensive and time-consuming than most other methods. As noted above, there are also constraints relating to the consistency of fluorescent excitation and measurement over prolonged periods, and the manual editing of images is user-dependent, introducing another source of potential variability and highlighting the desirability of repeating measurements and experiments to confirm the findings, perhaps even using different investigators to acquire and analyse the images.

Regardless of the limitations, we suggest that image-based measurement techniques should form at least part of the analytical approach to measuring the uptake of Aβ, whether by CVSMCs or by other relevant types of cell, such as neurons, microglia and endothelial cells. And although we have focused on the uptake of Aβ, it is likely that most of the considerations that we have outlined above will be applicable to measurement of the uptake of proteins and other molecules in a wide range of biological contexts.

Acknowledgements WAWR was supported by a grant from the Ministry of Higher Education and University of Malaya, Malaysia. This research was also supported by grants from Alzheimer's Research UK and BRACE (Bristol Research into Alzheimer's and Care of the Elderly).

References

1. Urmoneit B, Prikulis I, Wihl G et al (1997) Cerebrovascular smooth muscle cells internalize Alzheimer amyloid β protein via a lipoprotein pathway: implications for cerebral amyloid angiopathy. Lab Invest 77:157–166
2. Wilhelmus MM, Otte-Holler I, van Triel JJ et al (2007) Lipoprotein receptor-related protein-1 mediates amyloid-β-mediated cell death of cerebrovascular cells. Am J Pathol 171:1989–1999
3. Van Nostrand WE, Melchor JP, Ruffini L (1998) Pathologic amyloid β-protein cell surface fibril assembly on cultured human cerebrovascular smooth muscle cells. J Neurochem 70:216–223
4. McCarron MO, Nicoll JA, Ironside JW et al (1999) Cerebral amyloid angiopathy-related hemorrhage. Interaction of *APOE* ϵ2 with putative clinical risk factors. Stroke 30:1643–1646
5. McCarron MO, Nicoll JA, Stewart J et al (1999) The apolipoprotein E ϵ2 allele and the pathological features in cerebral amyloid angiopathy-related hemorrhage. J Neuropathol Exp Neurol 58:711–718
6. Ruzali WA, Kehoe PG, Love S (2012) LRP1 expression in cerebral cortex, choroid plexus and meningeal blood vessels: relationship to cerebral amyloid angiopathy and *APOE* status. Neurosci Lett 525:123–128

7. Ruzali WA, Kehoe PG, Love S (2013) Influence of LRP-1 and apolipoprotein E on amyloid-β uptake and toxicity to cerebrovascular smooth muscle cells. J Alzheimers Dis 33:95–110
8. Rosero EB, Peshock RM, Khera A et al (2009) Agreement between methods of measurement of mean aortic wall thickness by MRI. J Magn Reson Imaging 29:576–582
9. Donadio C (2010) Serum and urinary markers of early impairment of GFR in chronic kidney disease patients: diagnostic accuracy of urinary β-trace protein. Am J Physiol Renal Physiol 299:F1407–F1423
10. Krouwer JS, Monti KL (1995) A simple, graphical-method to evaluate laboratory assays. Eur J Clin Chem Clin Biochem 33:525–527
11. Chang-Liu CM, Woloschak GE (1997) Effect of passage number on cellular response to DNA-damaging agents: cell survival and gene expression. Cancer Lett 113:77–86
12. Briske-Anderson MJ, Finley JW, Newman SM (1997) The influence of culture time and passage number on the morphological and physiological development of Caco-2 cells. Proc Soc Exp Biol Med 214:248–257
13. Esquenet M, Swinnen JV, Heyns W et al (1997) LNCaP prostatic adenocarcinoma cells derived from low and high passage numbers display divergent responses not only to androgens but also to retinoids. J Steroid Biochem Mol Biol 62:391–399
14. Yu H, Cook TJ, Sinko PJ (1997) Evidence for diminished functional expression of intestinal transporters in Caco-2 cell monolayers at high passages. Pharm Res 14:757–762
15. Balcells M, Wallins JS, Edelman ER (2008) Amyloid β toxicity dependent upon endothelial cell state. Neurosci Lett 441:319–322
16. Deane R, Bell RD, Sagare A et al (2009) Clearance of amyloid-β peptide across the blood-brain barrier: implication for therapies in Alzheimer's disease. CNS Neurol Disord Drug Targets 8:16–30
17. Kanekiyo T, Bu G (2009) Receptor-associated protein interacts with amyloid-β peptide and promotes its cellular uptake. J Biol Chem 284:33352–33359
18. Chung H, Brazil MI, Soe TT et al (1999) Uptake, degradation, and release of fibrillar and soluble forms of Alzheimer's amyloid β-peptide by microglial cells. J Biol Chem 274:32301–32308
19. Melchor JP, Van Nostrand WE (2000) Fibrillar amyloid beta-protein mediates the pathologic accumulation of its secreted precursor in human cerebrovascular smooth muscle cells. J Biol Chem 275:9782–9791

Web-Addresses for Further Information

Simple PCI 6 analysis software. www.cimaging.net/simplepci.htm
MedCalc statistical software. www.medcalc.org

Chapter 9
Measurement of Intracellular Ca^{2+} in Human Endothelial Cells

Sarah Jones

9.1 Introduction

Endothelial cells perform a plethora of physiological roles including regulation of immune responses, angiogenesis, wound healing, vascular permeability, coagulation and vascular tone [1, 2]. Integral to many of these functions is an elevation in intracellular calcium concentration $[Ca^{2+}]_{i,}$ which can occur through mobilisation of intracellular stores or influx from the extracellular milieu (Fig. 9.1).

Stored Ca^{2+} in endothelial cells largely resides in the endoplasmic reticulum (ER) and is released into the cytoplasm following activation of the inositol-1,4,5-triphosphate receptor (IP_3R) [3]. There are two mechanisms by which inositol-1, 4, 5-trisphosphate (IP_3) is generated in endothelial cells. These are through the activation of guanosine nucleotide binding protein-coupled receptors (GPCRs) or through stimulation of receptor tyrosine kinases (RTKs). Endothelial cells possess a variety of GPCRs, which respond to locally released vasoactive mediators, hormones and neurotransmitters, including endothelin-1, angiotensin II, ATP and bradykinin [4]. Stimulation of these membrane receptors leads to the activation of phospholipase Cβ (PLCβ) and the subsequent hydrolysis of phosphatidylinositol-4,5-bisphosphate (PIP_2) to IP_3 and diacylglycerol (DAG). Receptor tyrosine kinases (RTKs) including the receptors for platelet-derived growth factor (PDGF) [5], fibroblast growth factor (FGF) and vascular endothelial growth factor (VEGF) [6, 7] stimulate IP_3 and DAG production via the phosphorylation of phospholipase γ (PLCγ). Generation of IP_3 and DAG leads to a transient increase in $[Ca^{2+}]_i$ and the activation of protein kinase C (PKC) respectively. A small portion of stored Ca^{2+} in endothelial cells resides in the mitochondria. The mechanisms, which release mitochondrial Ca^{2+} from

S. Jones (✉)
School of Healthcare Science, Manchester Metropolitan University,
Chester Street, Manchester, M1 5GD, UK

M. Slevin, G. McDowell (eds.), *Handbook of Vascular Biology Techniques*,
DOI 10.1007/978-94-017-9716-0_9

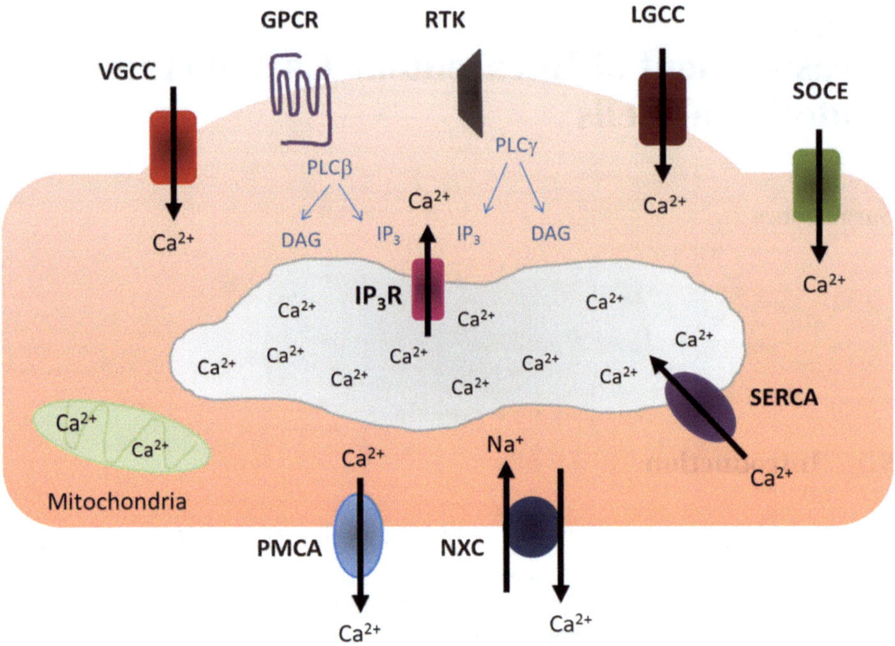

Fig. 9.1 Regulation of intracellular Ca2+ in endothelial cells. Many of the functions performed by endothelial cells are dependent on an increase in intracellular concentration ($[Ca^{2+}]_i$). Ca^{2+} levels can be elevated by the inositol 1, 4, 5, trisphosphate (IP_3)–dependent release from endoplasmic reticulum stores or through influx across the plasma membrane. Activation of G-protein coupled receptors (GPCRS) or receptor tyrosine kinases (RTKs) stimulates the phospholipase C (PLC)-dependent production of IP_3, which opens IP_3 receptors (IP_3Rs) on the ER membrane releasing Ca^{2+} into the cytosol. Store depletion subsequently leads to store-operated calcium entry (SOCE). Ca^{2+} influx can also occur through voltage-gate calcium channels (VGCCs) upon membrane depolarization or ligand-gated ion channels upon ligand binding. Low resting intracellular Ca^{2+} levels are maintained by the sarcoendoplasmic reticulum ATPase (SERCA), which pumps Ca^{2+} back into stores, the plasma membrane Ca^{2+} ATPase (PMCA) which extrudes Ca^{2+} across the plasma membrane in to the extracellular milieu and the Na^+, Ca^{2+} exchanger (NXC), which removes one Ca^{2+} ion from the cell in exchange for three Na^+ ions

endothelial cells remain unclear, but it is thought to require Ca^{2+} and occur following Ca^{2+} release from the ER (Mitochondrial Ca^{2+} induced Ca^{2+} release) [8, 9].

Agonist induced Ca^{2+} responses in endothelial cells are usually biphasic, with an initial transient response, evoked by release from intracellular stores, followed by a large sustained Ca^{2+} elevation, which represents store operated calcium entry (SOCE) or capacitive calcium entry (CCE) [10]. The molecular identity of the store-operated channels and mode of activation however, remained elusive for over 15 years, despite numerous candidates being suggested [11]. TRP channels received the most attention, due to their well-documented association with PLC signalling. The theory however did not gain general acceptance with many groups unable to demonstrate TRCP channel activation following store depletion [12–15].

More recently, STIM-1 and Orai 1 have been identified as pivotal components facilitating SOCE. STIM-1 is located in both the plasma membrane and the ER and is thought to act as a Ca^{2+} sensor, which interacts with Orai, following Ca^{2+} depletion in the ER [16, 17] while Orai, is a pore-forming channel in the plasma membrane [18–20].

Other mechanisms of Ca^{2+} entry include voltage-gated Ca^{2+} channels (VGCCs), mechanosensitive Ca^{2+} channels (stretch or shear stress activated receptors) and ligand-gated ion channels. Despite being considered non-excitable cells, endothelial cells in the microvasculature have been shown to express both L-type and T-type VGCCs [21–23]. At least two distinct types of ionotropic receptor have also been demonstrated; the nicotinic acethylcholine receptor nAChR and the purinergic P2X receptor, which are gated by acetylcholine (Ach) and ATP respectively. nAchRs are pentameric channels comprising of five α-subunits or a combination of different α- and β- subunits. Several different α- (α2–α6, α7) and β- (β2 and β4) subunits have been identified in macrovasculature and microvasculature endothelial cells giving rise to nAChR7 and nAChR3. No detectable Ca^{2+} response has been demonstrated for nAChR3, however, nAChR7 activation elicits pro-angiogenic Ca^{2+} entry [24, 25].

Ionotropic P2X receptors elicit membrane depolarization and Ca^{2+} influx. There are seven P2X receptor subtypes (P2X1–7). Vascular endothelial cells express all seven subtypes in both arteries and veins [26], with high expression of P2X4 observed in veins and also in balloon injured arteries [27], indicating that it may be involved in repair. The majority of functional responses elicited by ATP are largely mediated by P2X4 and P2X7. P2X4 has also been demonstrated to evoke shear stress induced Ca^{2+} influx, nitric oxide generation and flow mediated vasodilatation [28, 29].

Receptors activated by intracellular second messengers including DAG, arachidonic acid and cyclic nucleotides can also facilitate Ca^{2+} influx in endothelial cells. Cyclic nucleotide-gated Ca^{2+} channels (CNGCs) comprise of six family members, four α subunits (CNGC1-4) and two β subunits (CNGCB1 and CNGCB3), which are directly gated by cAMP or cGMP [30]. Endothelial cells express both CNGC1 and CNGC2 [31], which are activated following stimulation with adrenaline, ATP, adenosine or VEGF. Ca^{2+} influx evoked by ATP, adrenaline and adenosine via CNGCA2 result in nitric oxide production and vasorelaxation [32].

A number of mechanisms exist to remove Ca^{2+} from the cytosol to maintain low resting levels (Fig. 9.1). These include the plasma membrane calcium ATPase (PMCA), which extrudes Ca^{2+} across the plasma membrane, the sarcoendoplasmic reticulum (SERCA) which sequesters Ca^{2+} back into stores and the Na^+/Ca^{2+} exchanger, which exploits the Na^+ gradient across the plasma membrane and expels one Ca^{2+} ion in exchange for three Na^+ ions [4]. The expression and contribution of these regulating mechanisms varies between vessel type, vessel size and species.

An appropriately functioning endothelium is essential for a healthy cardiovascular system. Disruption in the endothelium's ability to detect or translate physical or humoral stimuli may contribute to a number of diseases including hypertension, diabetes, myocardial infarction and stroke. It is therefore important that we understand the molecular mechanisms that underpin the diverse physiological functions of endothelial cells in order to prevent or reduce the effects of endothelial dysfunction. This chapter will describe a method for measuring $[Ca^{2+}]_i$ in endothelial cells in a 96-well plate format, which can be used to investigate the role of specific GPCRs, RTKs or Ca^{2+} permeable ion channels in endothelial cell activation. It can also be used to examine potential inhibitors of endothelial cell activation or investigate the contribution of Ca^{2+} removal systems.

9.2 Materials

9.2.1 Equipment

1. Flexstation® 3 System (Molecular Devices, Wokingham, UK) (*see* **note 1**)
2. SoftMax Pro Software (Molecular Devices, Wokingham, UK)
3. 96-well tips (#9000-0911 black; #9000-0912 clear; Molecular Devices, Wokingham, UK) (*see* **note 2**)

9.2.2 Cell Culture

1. Endothelial cells – Human placenta artery endothelial cells (HPAECs), isolated from the chorionic vessels of human placentas [33]
2. Endothelial Growth Medium (EGM2; Lonza, UK)
3. VEGF-165 (Peprotech, UK)
4. FGF (R&D Systems, UK)
5. Trypsin (Lonza, UK)
6. HBSS (Ca^{2+} and Mg^{2+} free) (Lonza, UK)
7. Black clear flat bottom 96-well plates (Costar™, Fisher Scientific, UK)

9.2.3 Ca^{2+} Measurement

1. Fura-2-AM (Life Technologies Ltd., Paisley, UK)
2. Pluronic F-127 (Life Technologies Ltd., Paisley, UK)
3. Normal Physiological Saline (NPS): 145 mM NaCl, 5 mM KCl, 1 mM MgCl, 10 mM HEPES, 10 mM glucose, 2 mM CaCl pH 7.6)
4. Ca^{2+} loading buffer: Fura2-AM (4 μM), pluronic F-127 (0.02 %) and BSA (1 % w/v) in NPS

9.3 Methods

9.3.1 Cell Culture

This protocol describes the measurement of Ca^{2+} in HPAECs; however, it can be used as the basic guidelines for measuring Ca^{2+} in any type of endothelial cell.

1. Culture the freshly isolated HPAECs for 6–8 weeks to allow enough cells to grow. Maintain the cells in EGM-2 supplemented with VEGF-165 (6.66 ng/ml) and FGF (10 ng/ml) at 37 °C in a humidified atmosphere containing 5 % CO_2 and change the media every 2–3 days. (*see* **note 3**).
2. Prior to the experiment (24–48 h), wash the cells with HBSS (Ca^{2+} and Mg^{2+} free) and trypsinize for counting.
3. Count the cells and resuspend to a density of 4×10^5 cell/ml in VEGF/FGF supplemented EGM-2
4. Pipette 100 μl cells into each well of a black, clear bottom, 96-well plate a minimum of 24 h prior to the experiment.

9.3.2 Preparation of Fura-2-Loaded Endothelial Cells

1. When the cells are confluent and you are ready to perform the experiment, prepare your loading buffer (*see* **note 4 and note 5**). Ensure that you make enough buffer for 50 μl in each well and protect from the light.
2. Use a multichannel pipette to remove the media from the cells and gently wash three times with 200 μl NPS, once you have removed the final wash, add 50 μl of loading buffer to each well and incubate at 37 °C for 30 min (*see* **note 6**) in the dark. In the meantime prepare the agonist plate (see Sect. 9.3.3 Agonist plate preparation)
3. Following incubation, gently wash the cells three times with 200 μl NPS to remove any excess fura-2 and add 200 μl fresh NPS to the cells to perform the experiment.

9.3.3 Agonist Plate Preparation

1. Prepare a range of concentrations of the agonist to be tested.
2. The concentrations should be five times more concentrated than the desired final concentration since 50 μl will be added to the 200 μl NPS on the cells (Fig 9.2).
3. Each agonist well should contain the amount required for the stimulation e.g. 50 μl plus an additional 50 μl, which is residual and enables the required amount of agonist to be drawn up accurately by the automated pipettes.

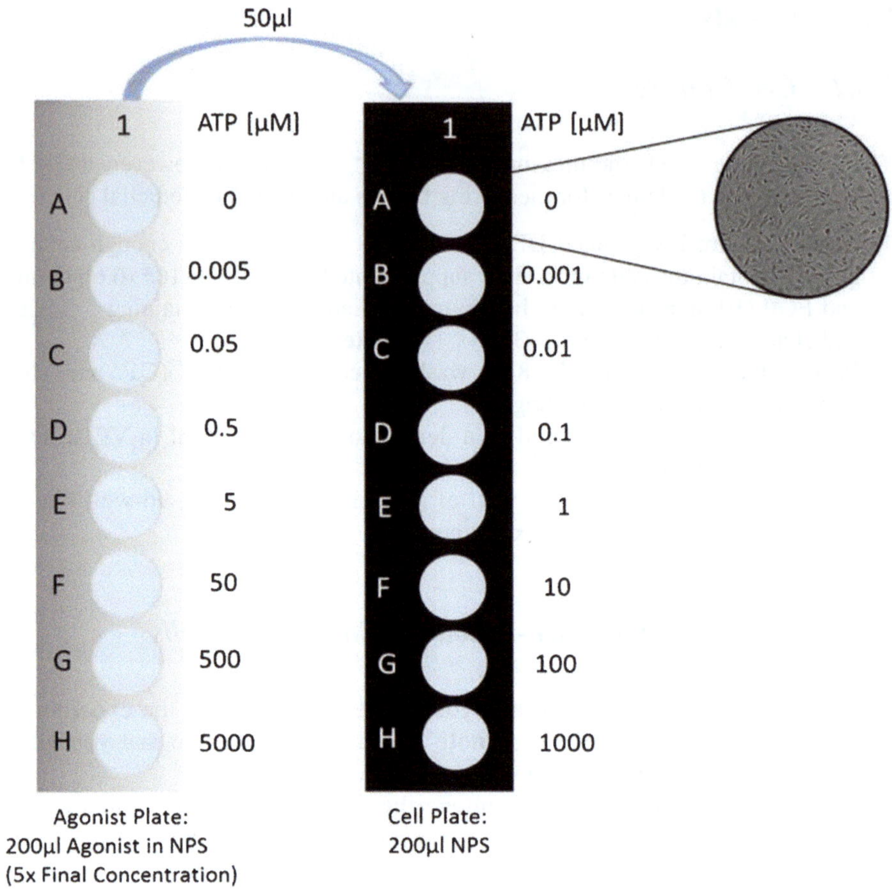

Fig. 9.2 Agonist and cell plate layout. The chosen agonist e.g. ATP is added to the wells of the clear 96-well agonist plate in ascending concentration starting with vehicle control in row A. The concentration of agonist in the agonist plate should be five times more concentrated than the desired final concentration. The *black walled*, clear bottom 96-well cell plate should contain a confluent monolayer of cells in each well, with 200 µl of normal physiological saline (NPS). At a user defined time point, 50 µl of agonist is automatically transferred from the agonist plate to the cell plate

9.3.4 Measurement of Fura-2-Signals

1. Turn the FlexStation on 15 min prior to use to allow it to warm up to 37 °C.
2. Insert the racked tips in the top drawer, the agonist plate into the middle drawer and the cell plate into the bottom drawer.
3. Open SoftMax Pro software and select Flex mode.

4. Open the settings window and apply the settings, which best suit your experiment. Below are the settings that we use to measure endothelial cell Ca^{2+} responses to ATP (Fig. 9.3):

   Flex

- Time: 200 s (length of experiment)
- Interval: 2 s (time between each reading)
- Reads: 100 (number of readings in the whole experiment, calculated automatically from the above information)

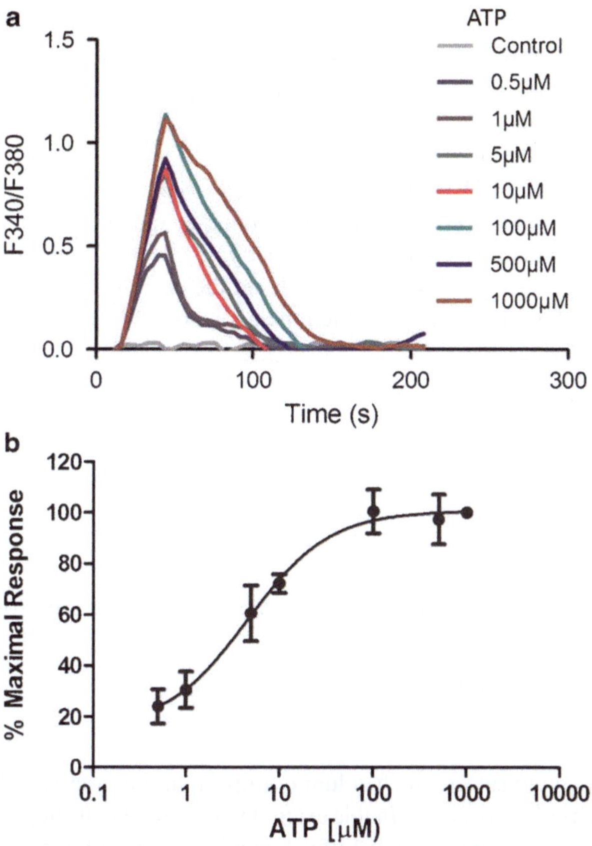

Fig. 9.3 ATP evoked Ca^{2+} responses in human placental artery endothelial cells (HPAECs). Cells loaded with fura-2 were stimulated with ATP (0.5 μM-1 mM) and intracellular Ca^{2+} measured as a ratio of 340 nm/380 nm (**a**). The peak Ca^{2+} response for each ATP concentration was calculated as a percentage of the maximal response, taken as 1 mM ATP (**b**). Data are presented as mean ± sem, $n=4$

Fluorescence bottom read

Ex	Em	Cutoff[a]
Lm1 340	510	495
Lm2 380	510	495

[a]Auto cutoff

- Automix: Off
- Calibrate: On
- PMT: High
- Settle time: Off

Assay Plate

- 96-well Costar black

Compound Transfer:

- Initial volume: 200 μl (initial volume of agonist in agonist plate)
- Transfers: 1 (more than one transfer can be made if you want to add one compound followed by another)
- Pipette Height: 205 μl
- Volume: 50 μl (Volume of agonist transferred to cell plate)
- Rate: 4 (~62 μl/s) (speed at which the agonist is added to the cells. A rate will appear in brackets next to the number that you select. If your cells are not very adherent, use a slower addition rate to avoid the cells lifting)
- Time point: 20 s (time point at which the agonist will be added to the cells)

Compound Source

- CoStar 96 flatbottom clear-3 ml

Titurate:

- T1: Not used

Pipette Tips layout:

- Tip area A1-H12 (The whole tip area can be selected or just the columns that you intend on using)

Compound & Tip Columns

- T: (1: 1/1)

This allows you to select which column of tips, to use to transfer which column of agonist to which column of cells. In this example, we have kept it simple and tip column 1, will transfer agonist from column 1 of the agonist plate to column 1 of the cell plate.

AutoCalibrate

- On

AutoRead

- Off

When you are happy with the settings that you have selected, click OK and run the experiment.

9.3.5 Data Analysis

The software measures emission at 340 nm and 380 nm and the 340/380 ratio is also recorded. The software can display the peak response, time to peak and area under the curve (AUC) as well as numerous other parameters. The 340/380 ratios (Fig 9.3a) can be exported to Microsoft Excel for data analysis. To construct a dose response curve the 340/380 ratio can be used directly or it can be used to calculate the percentage of the maximal response using the highest agonist concentration as 100 % (Fig. 9.3b).

9.3.6 Determination of Cytoplasmic Concentration

The 340/380 emission ratios can be converted to $[Ca^{2+}]_i$ using the following equation [34]

$$\left[Ca^{2+}\right]_i = K_d \times \left[\left(R - R_{min}\right) / \left(R_{max} - R\right)\right] \times \left(S_f / S_b\right)$$

Kd represents the fura-2 dissociation constant and R is the ratio of fluorescence intensities measured at 340 nM and 380 nM. R_{max} and R_{min} represent the maximum and minimum values of the 340/380 ratio under saturating Ca^{2+} concentrations and Ca^{2+} free conditions. S_f and S_b are the maximum and minimum fluorescence values at 380nM. The K_d of fura-2 for Ca^{2+} under the conditions used in this method (1 mM Mg^{2+}, at 37 °C and at physiological ionic strength) is 224 nM [34]. To obtain the minimum 380 nM value (S_f), measure Ca^{2+} saturation for fura-2 in the presence of 2 mM extracellular Ca^{2+} following cell lysis with Triton X-100 (0.1 %). The maximum 380 nm value (S_b) is then obtained by adding EGTA (8 mM). Prior to measurement of calibration constants, all 340 nm and 380 mm values should be corrected for autofluorescence.

9.4 Applications

A better understanding of the Ca^{2+}-dependent mechanisms that regulate endothelial cell function can be gained relatively quickly using this method. Simple stimulation experiments can be performed to screen for novel molecules, which stimulate a Ca^{2+} response in endothelial cells. Pharmacological inhibition or genetic manipulation to alter expression levels of receptors or signalling molecules can be used to dissect signalling pathways, which stimulate Ca^{2+} flux. The proportion of the Ca^{2+} response attributable to influx across the plasma membrane verses mobilisation from intracellular stores can be determined by altering the composition of the NPS buffer and performing experiments in Ca^{2+} free conditions. In addition to measuring the stimulation of Ca^{2+} responses, the real-time nature of this method allows the decay of Ca^{2+} to be measured and the contribution of the different Ca^{2+} removal mechanisms assessed.

The main limitation of the method is that it only measures global calcium from a population of cells, therefore oscillations are not observed and spatial and temporal aspects of Ca^{2+} signalling cannot be investigated. Also as with any experiments using cultured cells in vitro, expression levels of receptors and signalling molecules may be altered over time in culture. It is important to bear this in mind when interpreting the data from these studies and where possible compare expression levels to those observed in vivo.

9.5 Notes

1. Other multi-mode plate readers can be used e.g. FLUOstar (BMG Labtech, Germany). For the best data however an injection system is required to enable the addition of compounds whilst reading fluorescence.
2. Black tips can be used for the addition of compounds to the cells. In our experiments we have always used clear tips and never experienced and interference with the fluorescence.
3. The media and supplements used in the method are the standard media that we use for culture of HPAECs. For the best results, culture your endothelial cells under the same conditions that you usually maintain them in, with the same culture medium and supplements that you usually use.
4. Ca^{2+} sensitive fluorescent indicators other than fura-2 can be used, depending on the filters present in the multi-mode platelet reader that you are using. Another ratiometric indicator commonly used is indo-1. Alternatively single-wavelength indicators Fluo-3 or Fluo-4 can be used, and data expressed as change in fluorescence intensity relative to baseline (F/F0).
5. Pluronic F-127 is a non-ionic surfactant polyol, which facilitates the loading of cells with the fluorescent indicators by helping to disperse the acetoxymethyl (AM) ester group. If your cells appear to by leaky and have difficulty retaining the fluorescent indicator, probenecid, an inhibitor of organic-anion transporters can be used [35]. Caution however must be taken as probenecid can have many non-specific actions.
6. Cells can be loaded at room temperature for 45 min if preferred. It has been reported that loading at room temperature reduces the amount of dye taken up into intracellular compartments and reduces leakage of the indicator from cells.
7. When using inhibitors, they can be diluted in the NPS that is on the cells for the duration of the experiment or for more precise timings, multiple automated additions can pre-programmed so that inhibitors can be added to the cells at a set

point prior to agonist addition. If using the latter method ensure that you take in to account dilution factors for the first addition and second addition when calculating agonist/inhibitor concentrations.

References

1. Furchgott RF, Zawadzki JV (1980) The obligatory role of endothelial cells in the relaxation of arterial smooth muscle by acetylcholine. Nature 288(5789):373–376
2. Tran QK, Ohashi K, Watanabe H (2000) Calcium signalling in endothelial cells. Cardiovasc Res 48(1):13–22
3. Berridge MJ (1993) Inositol trisphosphate and calcium signalling. Nature 361(6410):315–325
4. Moccia F, Berra-Romani R, Tanzi F (2012) Update on vascular endothelial Ca(2+) signalling: a tale of ion channels, pumps and transporters. World J Biol Chem 3(7):127–158
5. Ridefelt P et al (1995) PDGF-BB triggered cytoplasmic calcium responses in cells with endogenous or stably transfected PDGF beta-receptors. Growth Factors 12(3):191–201
6. McLaughlin AP, De Vries GW (2001) Role of PLCgamma and Ca(2+) in VEGF- and FGF-induced choroidal endothelial cell proliferation. Am J Physiol Cell Physiol 281(5):C1448–C1456
7. Meyer RD, Latz C, Rahimi N (2003) Recruitment and activation of phospholipase Cgamma1 by vascular endothelial growth factor receptor-2 are required for tubulogenesis and differentiation of endothelial cells. J Biol Chem 278(18):16347–16355
8. Wood PG, Gillespie JI (1998) Evidence for mitochondrial Ca(2+)-induced Ca^2+ release in permeabilised endothelial cells. Biochem Biophys Res Commun 246(2):543–548
9. Falcke M et al (1999) Impact of mitochondrial Ca^2+ cycling on pattern formation and stability. Biophys J 77(1):37–44
10. Takemura H et al (1989) Activation of calcium entry by the tumor promoter thapsigargin in parotid acinar cells. Evidence that an intracellular calcium pool and not an inositol phosphate regulates calcium fluxes at the plasma membrane. J Biol Chem 264(21):12266–12271
11. Putney JW (2009) Capacitative calcium entry: from concept to molecules. Immunol Rev 231(1):10–22
12. Zitt C et al (1997) Expression of TRPC3 in Chinese hamster ovary cells results in calcium-activated cation currents not related to store depletion. J Cell Biol 138(6):1333–1341
13. Schaefer M et al (2000) Receptor-mediated regulation of the nonselective cation channels TRPC4 and TRPC5. J Biol Chem 275(23):17517–17526
14. Dietrich A et al (2007) Pressure-induced and store-operated cation influx in vascular smooth muscle cells is independent of TRPC1. Pflugers Arch 455(3):465–477
15. Varga-Szabo D et al (2008) Store-operated Ca(2+) entry in platelets occurs independently of transient receptor potential (TRP) C1. Pflugers Arch 457(2):377–387
16. Roos J et al (2005) STIM1, an essential and conserved component of store-operated Ca^2+ channel function. J Cell Biol 169(3):435–445
17. Liou J et al (2005) STIM is a Ca^2+ sensor essential for Ca^2+–store-depletion-triggered Ca^2+ influx. Curr Biol 15(13):1235–1241
18. Vig M et al (2006) CRACM1 is a plasma membrane protein essential for store-operated Ca^2+ entry. Science 312(5777):1220–1223
19. Feske S et al (2006) A mutation in Orai1 causes immune deficiency by abrogating CRAC channel function. Nature 441(7090):179–185
20. Zhang SL et al (2006) Genome-wide RNAi screen of Ca(2+) influx identifies genes that regulate Ca(2+) release-activated Ca(2+) channel activity. Proc Natl Acad Sci U S A 103(24):9357–9362
21. Bossu JL et al (1989) Voltage-dependent transient calcium currents in freshly dissociated capillary endothelial cells. FEBS Lett 255(2):377–380

22. Vinet R, Vargas FF (1999) L- and T-type voltage-gated Ca^2+ currents in adrenal medulla endothelial cells. Am J Physiol 276(4 Pt 2):H1313–H1322
23. Wei Z et al (2004) Ca^2+ flux through voltage-gated channels with flow cessation in pulmonary microvascular endothelial cells. Microcirculation 11(6):517–526
24. Gahring LC et al (2005) Pro-inflammatory cytokines modify neuronal nicotinic acetylcholine receptor assembly. J Neuroimmunol 166(1–2):88–101
25. Arias HR et al (2009) Role of non-neuronal nicotinic acetylcholine receptors in angiogenesis. Int J Biochem Cell Biol 41(7):1441–1451
26. Ray FR et al (2002) Purinergic receptor distribution in endothelial cells in blood vessels: a basis for selection of coronary artery grafts. Atherosclerosis 162(1):55–61
27. Pulvirenti TJ et al (2000) P2X (purinergic) receptor redistribution in rabbit aorta following injury to endothelial cells and cholesterol feeding. J Neurocytol 29(9):623–631
28. Yamamoto K et al (2000) Fluid shear stress activates Ca(2+) influx into human endothelial cells via P2X4 purinoceptors. Circ Res 87(5):385–391
29. Yamamoto K et al (2003) Endogenously released ATP mediates shear stress-induced Ca^2+ influx into pulmonary artery endothelial cells. Am J Physiol Heart Circ Physiol 285(2):H793–H803
30. Bradley J, Reisert J, Frings S (2005) Regulation of cyclic nucleotide-gated channels. Curr Opin Neurobiol 15(3):343–349
31. Kwan HY et al (2009) Role of cyclic nucleotides in the control of cytosolic Ca^2+ levels in vascular endothelial cells. Clin Exp Pharmacol Physiol 36(9):857–866
32. Cheng KT et al (2008) CNGA2 channels mediate adenosine-induced Ca^2+ influx in vascular endothelial cells. Arterioscler Thromb Vasc Biol 28(5):913–918
33. Lang I et al (2008) Human fetal placental endothelial cells have a mature arterial and a juvenile venous phenotype with adipogenic and osteogenic differentiation potential. Differentiation 76(10):1031–1043
34. Grynkiewicz G, Poenie M, Tsien RY (1985) A new generation of Ca^2+ indicators with greatly improved fluorescence properties. J Biol Chem 260(6):3440–3450
35. Di Virgilio F, Steinberg TH, Silverstein SC (1990) Inhibition of Fura-2 sequestration and secretion with organic anion transport blockers. Cell Calcium 11(2–3):57–62

Part II
In Vivo and Ex Vivo Manipulations

Chapter 10
Evaluation of Angiogenesis and Arteriogenesis in a Mouse Model of Prolonged Cerebral Hypoperfusion

Takakuni Maki, Loc–Duyen D. Pham, Nobukazu Miyamoto, Masafumi Ihara, Eng H. Lo, and Ken Arai

10.1 Introduction

In vertebrate organogenesis, the blood vessels constitute the first organ system that arises and reaches a functional state [1]. Blood vessels carry oxygen and nutrients to distant organs, remove metabolic waste products, provide gateways for immune surveillance, and produce various trophic signals [2]. Since blood vessels nourish nearly every organ of the body, the imbalance of vessel growth contributes to numerous diseases [3]. Insufficient vessel growth or inadequate maintenance can lead to stroke, myocardial infarction, ulcerative disorders, obesity-associated disorders, and neurodegeneration. Abnormal vessel growth or remodeling fuels cancer, inflammatory disorders, pulmonary hypertension and blinding eye diseases [4, 5]. Therefore, the study of mechanisms of vascular growth and repair are of fundamental importance.

Growth of blood vessels occurs by three major processes: vasculogenesis, angiogenesis, and arteriogenesis. Genetically determined vasculogenesis produces a primary vascular plexus during ontogenesis. Angiogenesis is defined as the growth of new capillaries leading to an increase in capillary density and the preservation of local oxygen and nutrition supply in response to metabolic stress within hypoxic tissues. Arteriogenesis describes the adaptive growth and outward remodeling of pre-existing collateral anastomoses in response to hemodynamic changes [6–9].

T. Maki • L.-D.D. Pham • N. Miyamoto • E.H. Lo • K. Arai (✉)
Neuroprotection Research Laboratory, Department of Radiology and Neurology, Massachusetts General Hospital and Harvard Medical School, MGH East 149-2401, Charlestown, MA 02129, USA
e-mail: karai@partners.org

M. Ihara
Department of Stroke and Cerebrovascular Diseases, National Cerebral and Cardiovascular Center, Osaka, Japan

M. Slevin, G. McDowell (eds.), *Handbook of Vascular Biology Techniques*,
DOI 10.1007/978-94-017-9716-0_10

The major determinants of vascular growth-genetic predisposition, metabolic factors (hypoxia), and hemodynamics- cannot be assigned in a mutually exclusive fashion to vasculogenesis, angiogenesis, and arteriogenesis, respectively. Vascular growth processes usually comprise more than one of these mechanisms and the interplay of genetics, hemodynamics, and oxygen tension determines vessel identity, morphology, and function during ontogenesis and in adult life [9]. Vascular growth in adult organisms occurs mainly by angiogenesis and arteriogenesis, although it has been debated whether vasculogenesis occurs in the adult.

The brain is a highly vascular organ and vascular cells are major cellular constituent of the brain. While the brain comprises ~2 % of total body mass, it receives up to 20 % of cardiac output and is responsible for ~20 % and ~25 % of the body's oxygen and glucose consumption, respectively [10]. The central nervous system (CNS) macro- and microvessels together occupy a significant 25–30 % of the total brain volume. The surface area of cerebral capillary endothelium is approximately 100 cm^2 per gram of brain tissue. In the adult human brain, the total surface area of microvasculature is 12 m^2. The total length of capillaries is 650 km. The diameter of a capillary lumen is approximately 6 μm. The capillaries are located 40 μm apart [11].

Vascular risk factors that impede adequate cerebral flow can substantially impair all aspects of cognitive function with aging [12]. Alzheimer's disease (AD) and vascular dementia (VaD) are the most common form of dementia in the elderly. In VaD, cerebrovascular insufficiency and ischemic injury are believed to cause the brain dysfunction that underlies the dementia [13]. Furthermore, accumulating evidence has suggested that AD and VaD belong to a continuous spectrum of diseases based on vascular pathologies [14–16]. Consistent with this concept, attenuations in cerebral blood flow (CBF) and cerebrovascular insufficiency have emerged as a powerful predictor of AD as well as that of VaD [17]. Importantly, ischemic white matter lesions, which are most likely caused by prolonged cerebral hypoperfusion following atherosclerosis and/or arteriosclerosis [18], are an established marker of risk for cognitive deterioration [19] in VaD [20] and AD [21].

A mouse model of prolonged cerebral hypoperfusion is useful in evaluating the substrates of VaD, exhibiting white matter lesions and working memory deficits specifically attributable to frontal-subcortical circuit damage [22, 23, 24]. The model is reproduced by bilateral common carotid artery stenosis (BCAS) using microcoils. In this BCAS model, CBF decreases to about 60 % just after the operation as measured by laser speckle blood flow imaging system, and then gradually recovers to about 80–85 % one month after the operation, mainly due to arteriogenesis, i.e., collateral growth at the leptomeningeal anastomoses and the circle of Willis. An angiogenic response also occurs at distal ischemic sites in the brain. Despite these attempts at repair, i.e., endogenous angiogenesis and arteriogenesis, white matter damage progress with BBB disruption, glial activation, oxidative stress, and oligodendrocyte loss, leading to cognitive decline [25, 26]. Thus, endogenous vascular growth after BCAS may be insufficient for complete functional recovery.

Therefore, therapeutic growth of functional blood vessels that can restore cerebral perfusion and vascular insufficiency may result in the preservation of white matter integrity and serve to maintain cognitive function in a mouse model of prolonged cerebral hypoperfusion as well as human subjects at risk of developing AD and VaD. However, no such treatment has been made available. Further research will be necessary to elucidate the related pathophysiological mechanisms, including insufficient vascular repair in these disorders while special attention should be paid to the techniques, which can evaluate vascular growth appropriately [9].

10.2 Methodology

Here, we will describe the protocol for a murine model of prolonged cerebral hypoperfusion (BCAS model), wherein surgical narrowing of bilateral common carotid arteries leads to arteriogenesis in the leptomenigeal anastomoses and the circle of Willis, and angiogenesis at ischemic areas in the brain.

The protocol comprises mouse preparation and surgery, evaluation of prolonged cerebral hypoperfusion by serial laser speckle blood flow imaging system, as well as analysis of angiogenesis and arteriogenesis by means of immunohistochemistry, fluorescence-perfusion, and latex-perfusion method.

10.3 Materials

10.3.1 Reagents

1. Experimental animals: male C57BL/6 mice, 10–12 weeks old weighing 20–30 g (Charles River Laboratory)
2. Chemical depilatory
3. Surgical scrub solutions: povidone-iodine scrub
4. Anesthetics: isoflurane
5. Silk suture
6. Gauze
7. Stainless steel wound clips
8. Microcoils: manufactured in the Sawane Spring Co., Ltd. (Hamamatsu, Japan)
9. Gel (Aquasonic, Parker Laboratories, Inc.)
10. Phosphate buffer saline (PBS) (GIBCO)
11. Citrate buffer (pH6.0, 0.01 mol/L): 120.6 g sodium citrate and 18.8 g citric acid (Sigma) are dissolved in 500 mL of deionised water.
12. Tween 20
13. Triton-X100
14. 4 % paraformaldehyde (PFA) (Wako)
15. Sucrose
16. TISSUE TEK O.C.T compound
17. Microscope slide (Fisher Scientific)
18. Goat/Bovine serum (Sigma)
19. Streptavidin-biotin complex (DAKO)
20. ABC Staining Kit (Pierce Biotechnology)
21. Diaminobenzidine tetrahydrochloride (Sigma)
22. H_2O_2 (Wako)
23. Tris HCl (Sigma)

24. Fluorescein-isothiocyanate (FITC)-dextran (Sigma)
25. VECTASHIELD mounting medium with DAPI (H-1200, Vector Laboratories)
26. Papaverine hydrochloride (Sigma)
27. Latex compound (Chicago Latex Products)
28. Carbon black (Fueki)
29. Ethanol
30. Xylene
31. Paraffin
32. Anti-CD31 antibody (BD Biosciences)
33. Anti-Ki-67 antibody (Santa Cruz)
34. Anti-F4/80 antibody (Abcam)
35. Anti-α-smooth muscle actin (SMA) antibody (Neomarkers)
36. Biotinylated secondary antibody (Vector Laboratories)
37. FITC-conjugated secondary antibody (Santa Cruz)
38. TRITC-conjugated secondary antibody (Dako Cytomation)
39. Donkey anti-mouse IgG (1:200, Jackson Immunoresearch Laboratories)

10.3.2 Equipment

1. Heating pad (CWE Inc.)
2. Rectal thermometer (CWE Inc.)
3. Surgical knife
4. Forceps: fine-tip and blunt-tip forceps
5. Digital microscope (DinoLite, AnMo Electronics Corp.)
6. Image analysis software (DinoCapture, AnMo Electronics Corp.).
7. Laser speckle blood flow imager (Omegawave, Inc.)
8. A calibration reference device (Calibrator S/N 080715–5, Omegawave, Inc.)
9. Cryostat (Leica)
10. Operating stereotactic microscope
11. Fluorescent microscope (BZ-9000, Keyence)
12. Confocal laser-scanning device (FV1000, Olympus)

10.4 Basic Protocol

Experiments must comply with national and institutional regulations concerning the use of animals for research purposes and permissions to carry out experiments have to be obtained.

10.4.1 Induction of Prolonged Cerebral Hypoperfusion

1. The C57BL/6 J mouse is anesthetized with gas anesthesia. Anesthesia is induced with 4 % isoflurane and maintained with 1.5 % isoflurane, via a face-mask. (**Note 1, 2**)

2. The anesthetized mouse is placed in dorsal recumbency position on the operating board with the tail toward the surgeon.
3. The ventral neck area is shaved with a chemical depilatory and swabbed with surgical scrub.
4. A 1.0–1.5 cm midline skin incision is made from the base of the neck to the point below the lower jaw.
5. The underlying fat is removed and the salivary glands are moved laterally or upwards using forceps to maximize the operating field.
6. Both common carotid arteries (CCAs) are exposed and freed from their sheaths under an operating microscope.
7. Two 4–0 silk sutures are placed around the distal and proximal parts of the CCA.
8. The artery is gently lifted by the sutures and placed between the loops of the microcoil just proximal to the carotid bifurcation. (**Note 3**)
9. Rotating the mircocoil around the CCA twines the microcoil.
10. Another microcoil of the same size is twined around the other CCA. (**Note 4**)
11. The incision is closed with stainless steel wound clips or fine sutures.
12. Return the mouse to the animal holding area after they appear normal. (**Note 5**)

10.4.2 CBF Monitoring

1. Anesthetized mouse is placed in ventral recumbency with the tail towards the surgeon.
2. The scalp is surgically removed to expose the skull.
3. The periosteum, which adheres to the skull, is widely removed with fine-tip forceps.
4. The skull surface is wiped with saline-soaked gauze and then covered with a thin layer of gel to prevent drying.
5. Calibration is carried out with a calibration reference device.
6. With the laser speckle blood flow imager, CBF recordings are performed through the skull. Circular regions of interest are defined on the image for quantitative measurement. (**Note 1–3**)

10.4.3 Evaluation of Angiogenesis

The cerebrovascular tree originates from large, interconnected arteries forming the circle of Willis at the base of the brain. These arteries sequentially divide and once inside the brain parenchyma, give rise to pial arteries, penetrating intracerebral arteries, arterioles and a vast capillary network.

Angiogenesis denotes capillary growth either by sprouting from or by splitting of pre-existing vascular structures (intussusception) [27]. Sprouting angiogenesis is initiated by the proteolytic degradation of the extracellular matrix (ECM) in response to local hypoxia. Endothelial cells proliferate and migrate toward the gradients of proangiogenic factors and form lumina, wherein vessels are stabilized, mature and acquire tissue-specific properties [6, 27]. While sprouting angiogenesis relies on endothelial cell mitosis and migration, intussusceptive growth multiplies vessels by splitting their lumina by the insertion of tissue pillars [28]. This chapter mainly focuses on sprouting angiogenesis.

10.4.3.1 Histological Quantification of Angiogenesis

Histologic and Immunohistochemical Evaluation

1. Mice are deeply anesthetized with isoflurane and perfused transcardially with 0.01 mol/L PBS, then with 4 % PFA.
2. The brains are removed, and immersed in 4 % PFA overnight at 4 °C.
3. 4 % PFA is subsequently replaced by 20 % sucrose in PBS and the brains are stored at 4 °C.
4. The coronal blocks are cut into serial sections (20 μm thick) in a cryostat at −20 °C and mounted onto microscope slides.
5. Non-specific antibody binding is blocked with 3 % goat serum in PBS for 30 min.
6. The sections are incubated with primary antibody (anti-CD31 antibody, 1:100) in PBS with 0.1 % Tween20 overnight at 4 °C.
7. The sections are washed with PBS and incubated with biotinylated secondary antibody (1:100) for 1 h at room temperature.
8. After washing with PBS, the streptoavidin-biotin complex (1:200) is applied for 30 min at room temperature.
9. The sections are visualized with 0.01 % diaminobenzidine tetrahydrochloride and 0.005 % H_2O_2 in 50 mmol/L Tris HCl (pH 7.6).
10. Five coronal sections (20 μm thick) from each animal (−1.0, −0.5, 0, +0.5, +1.0 mm posterior to bregma) are obtained. (Note1)
11. In each coronal brain slice, ROIs were set in the cortex (1×1 mm^2; 1.5–3.5 mm lateral from midline), in the caudoputamen (1×1 mm^2; 1.0–3.0 mm lateral from midline), and in the corpus callosum (0.5×0.2 mm^2; 1.0–1.5 mm lateral from midline).
12. A CD31-positive vessel with a diameter between 3 and 8 μm that is separated from adjacent vessels is regarded as a single capillary and counted in number. This number was added to the number of vascular branch points (number of vessel bifurcations) to yield 'the number of capillaries'. Capillary density is expressed as the number of capillaries per square millimeter.

10.4.3.2 Fluorescent Perfusion

1. Mice are transcardially perfused with 5 mL FITC-dextran (2×10^6 molecular weight, Sigma; 15 mg/mL) at a perfusion pressure of 120–140 mmHg.
2. The brains are removed, and immersed in 4 % PFA overnight at 4 °C.
3. 4 % PFA is subsequently replaced by 20 % sucrose in PBS.
4. Coronal sections are mounted on glass slides, which are covered with VECTASHIELD mounting medium with DAPI.

5. Fluorescence images are obtained using a fluorescent microscope (BZ-9000; Keyence).

10.4.4 Evaluation of Arteriogenesis

In the cerebrovascular system, a multitude of extra- and intracranial collateral systems, such as leptomeningeal anastomoses of Heubner, the anastomotic pathway via the ophthalmic artery, or the circle of Willis are present. Therefore, stimulation of such collateral circulation via arteriogenesis is a promising approach for the prevention of cerebrovascular disorders.

Arteriogenesis is the outgrowth of a collateral arteriole upon stenosis or occlusion of a major conductance artery, mainly driven by hemodynamic factors, such as an increase in stretch and/or fluid shear stress (FSS) on endothelial cells [9]. During the development of the arterial system, pericytes/smooth muscle cells grow alongside the nascent vessel or differentiate in situ from mesenchymal precursors [6]. Some authors refer to this process "arterialization" as "arteriogenesis". However, in this chapter, arteriogenesis will only be used in the former sense.

10.4.4.1 Latex Perfusion Method (Note 1)

1. The root of the ascending aorta is cannulated with flexible plastic tubing (0.65 mm external diameter).
2. The tubing is connected to a 5 mL syringe, the cannulated aorta, and a mercury manometer, establishing a closed circuit to monitor perfusion pressure.
3. A lethal dose of papaverine hydrochloride (40–50 mg/kg) is injected to produce maximal vasodilation.
4. Immediately after 2 mL saline injection, 4 mL white latex compound mixed with 50 μL/mL carbon black diluted 2:1 with saline is injected at a perfusion pressure of 150 mm Hg over a 5-min period.
5. After the initiation of infusion, the right atrium of the heart is incised to allow for venous outflow.
6. In order to harden the latex completely for the brain removal procedure, the dead animal is soaked in ice-cold water 20 min after the end of infusion, and the brain is subsequently removed 20 min later. (**Note 2**)
7. Photographs of the dorsal and ventral surface of the brain are taken using a digital microscope (DinoLite, AnMo Electronics Corp.) at ×80 magnification. The vessel diameters of the leptomeningeal anastomoses and the circle of Willis are measured using image analysis software (DinoCapture, AnMo Electronics Corp.).
8. The maximal diameter of the leptomeningeal anastomoses is measured at the point of confluence between the distal MCA and the distal ACA (Heubner's anastomoses) or between the distal MCA and the distal PCA. The diameter of the internal carotid artery (ICA), ACA, MCA and the posterior communicating artery (Pcom) are averaged across both sides. (**Note 3**)

10.4.4.2 Analysis of Monocyte Recruitments and Proliferation of Smooth Muscle Cells (Note 4)

1. Mice are euthanized by intracardiac perfusion with 4 % PFA.
2. The brains are post-fixed in 4 % PFA, and were stored in 20 % sucrose in PBS at 4 °C.
3. The brains are embedded in paraffin and sliced into 6 μm-thick coronal sections and then subjected to the immunofluorescent analysis for Ki-67 and F4/80 together with α-smooth muscle actin (α-SMA).
4. After deparaffinization with xylene and rehydration with ethanol, antigen retrieval for the Ki-67 antigen is performed using 0.01 mol/L citrate buffer (pH 6.0) at 95 °C for 20 min.
5. Sections are rinsed briefly with 0.1 mol/L PBS, and then blocked for 30 min at room temperature in a solution of PBS containing 3 % preimmune serum and 0.3 % Triton-X100 (PBS+).
6. Sections are subsequently incubated at 4 °C for 24 h with primary antibodies diluted in PBS+ and then washed extensively with PBS at room temperature.
7. Sections are incubated with secondary fluorochrome-conjugated antibodies in PBS for 90 min at room temperature and again washed extensively.
8. Stained tissue sections are evaluated by using a confocal laser-scanning device (FV1000, Olympus).
9. The number of the Ki-67-immunoreactive smooth muscle cells (SMCs) per total SMCs is analyzed on the dorsal surface of the brain in the coronally-cut sections at 0.5–1 mm posterior from the bregma.
10. The number of SMCs surrounded by F4/80-immunoreactive monocyte/macrophages is calculated per total SMCs at the same area of another section.

10.4.5 Evaluation of Blood Brain Barrier (BBB)

10.4.5.1 IgG Staining

1. Mice are deeply anesthetized with isoflurane and perfused transcardially with 0.01 mol/L PBS, then with 4 % PFA.
2. The brains are removed, and immersed in 4 % PFA overnight at 4 °C.
3. 4 % PFA is subsequently replaced by 20 % sucrose in PBS and the brains are stored at 4 °C.
4. The coronal blocks are cut into serial sections (20 μm thick) in a cryostat at −20 °C and mounted onto microscope slides.
5. After incubation in 3 % H_2O_2 followed by blocking in 10 % Bovine serum albumin (Sigma) in PBS, the sections are incubated overnight at 4 °C with antibody against Donkey anti-mouse IgG (1:200).
6. Immunoreactivity visualized using the avidin–biotin complex method (ABC Staining Kits).

10.5 Sample Results

10.5.1 Induction of Prolonged Cerebral Hypoperfusion [29]

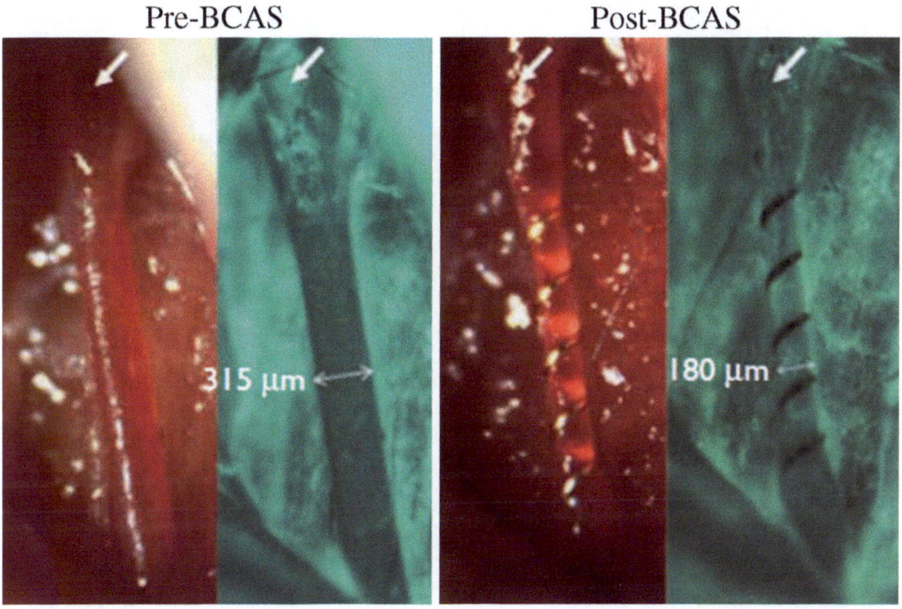

Representative photographs of common carotid arteries (CCAs; *left panel*) and fluorescein isothiocyanate-perfusion CCAs (*right panel*) of pre-bilateral common carotid artery stenosis (BCAS) and post-BCAS mouse (immediately after BCAS). *Arrows indicate* carotid bifurcation. The percentage of internal diameter reduction is 45.5 % ± 2.9 % (n = 4)

10.5.2 CBF Monitoring [26]

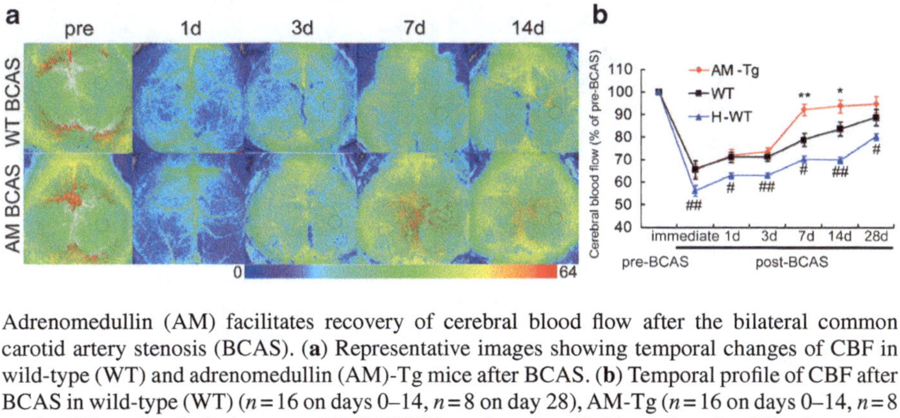

Adrenomedullin (AM) facilitates recovery of cerebral blood flow after the bilateral common carotid artery stenosis (BCAS). (**a**) Representative images showing temporal changes of CBF in wild-type (WT) and adrenomedullin (AM)-Tg mice after BCAS. (**b**) Temporal profile of CBF after BCAS in wild-type (WT) (n = 16 on days 0–14, n = 8 on day 28), AM-Tg (n = 16 on days 0–14, n = 8 on day 28), and hydralazine-treated WT (H-WT; n = 8 on days 0–14, n = 4 on day 28) mice, where the pre-BCAS value of CBF is adjusted to 100. Values are expressed as means ± SEM. *P < 0.05, **P < 0.01 in AM-Tg vs. WT; #P < 0.05, ##P < 0.01 in WT vs. H-WT

10.5.3 Evaluation of Angiogenesis [26]

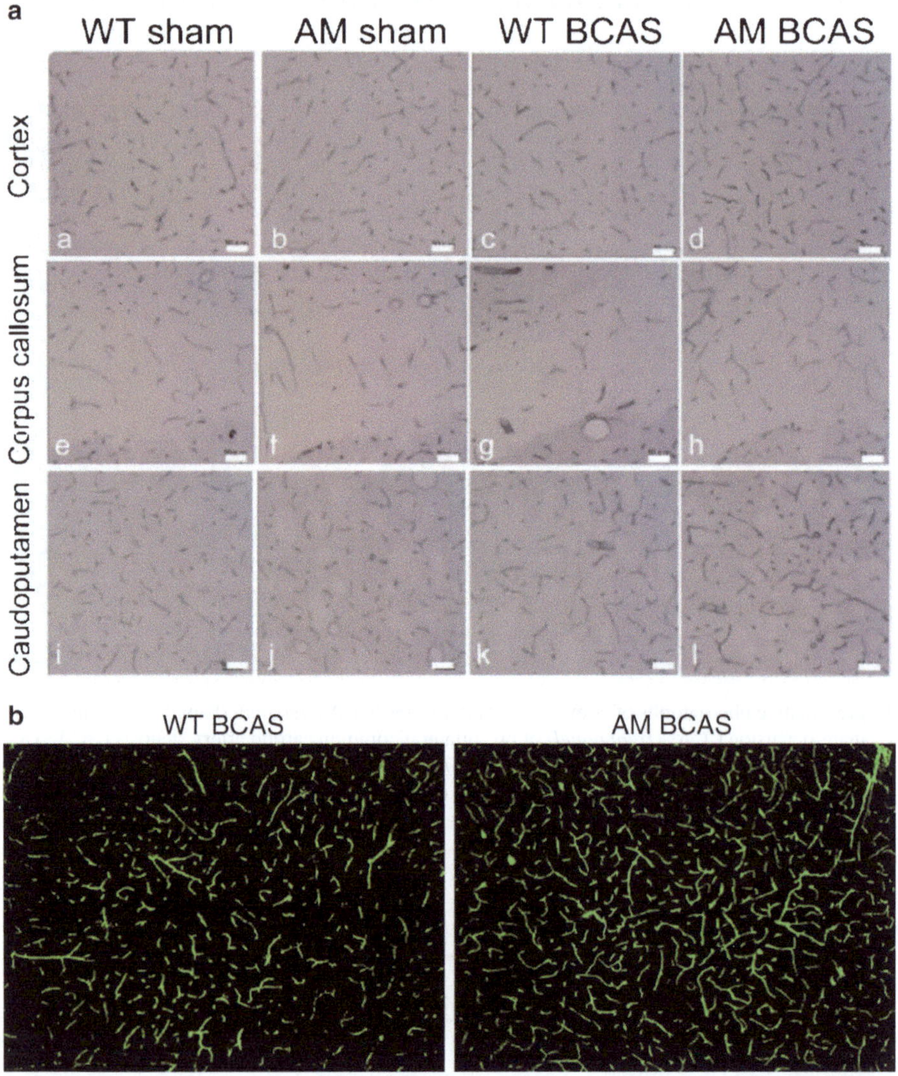

Adrenomedullin (AM) and bone marrow mononuclear cells (BMMNCs) enhance angiogenesis after the bilateral common carotid artery stenosis (BCAS). (**a**) Representative images of the CD31-positive capillaries in sections from the cortex (*a–d*), the corpus callosum (*e–h*) and the caudoputamen (*i–l*) of WT or AM-Tg mouse that is subjected to sham (WT sham or AM sham) or BCAS operation (WT BCAS or AM BCAS) on day 7. Scale bar, 50 μm. (**b**) Representative images of fluorescein-isothiocyanate (FITC)-dextran-perfused vessels in the cortex of WT BCAS and AM BCAS on day 7

10.5.4 Evaluation of Arteriogenesis [26]

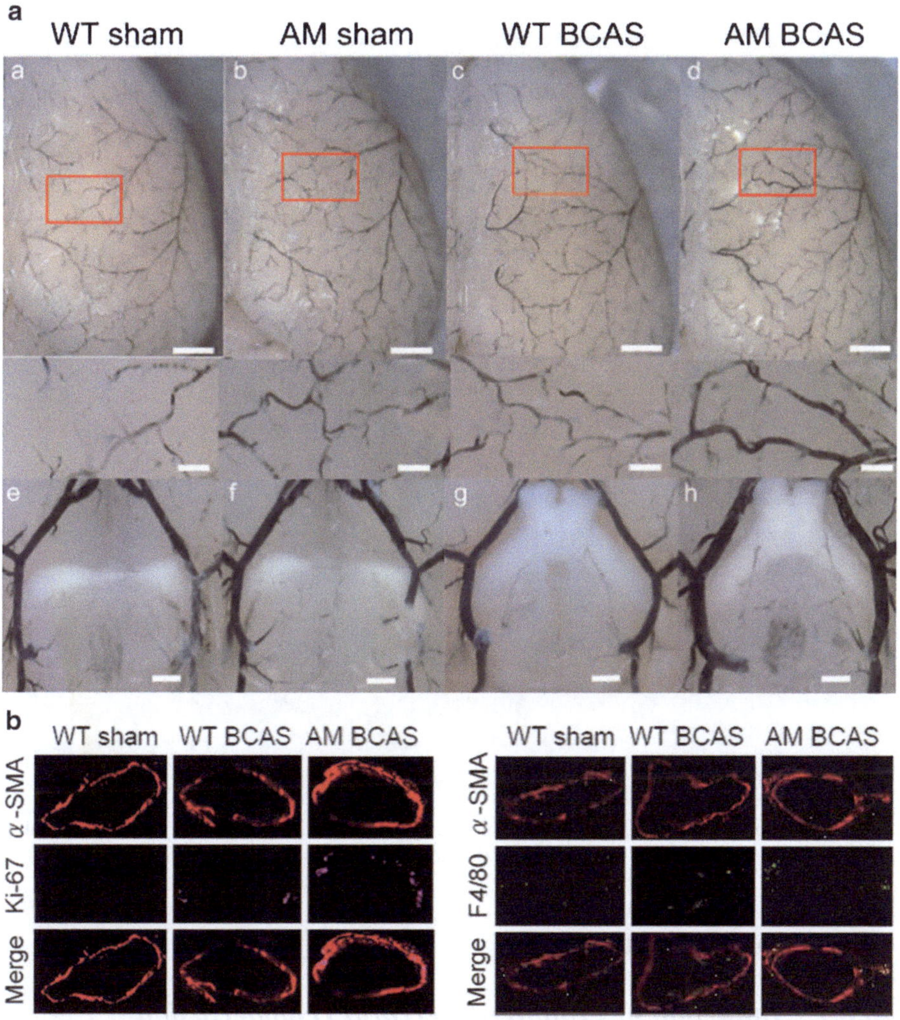

Adrenomedullin (AM) enhances arteriogenesis after the bilateral common carotid artery stenosis (BCAS). (**a**) Representative images of the dorsal (*a–d*) and ventral (*e–h*) cerebral angioarchitecture by postmortem latex perfusion method of WT or AM-Tg mouse that is subjected to sham (WT sham or AM sham) or BCAS operation (WT BCAS or AM BCAS) on day 7. *Scale bars*, 1 mm (*upper panels* of *a–d*), 250 μm (*lower panels* of *a–d*), and 500 μm (*e–h*). (**b**) (*left panel*) Representative double immunofluorescence images for Ki-67 (*green*)/α-SMA (*red*) of WT sham, WT BCAS, and AM BCAS on day 7. *Scale bar*, 10 μm. (*right panel*) Representative double immunofluorescence images for F4/80 (*green*)/α-SMA (*red*) of WT sham, WT BCAS, and AM BCAS on day 7. Scale bar, 10 μm

10.5.5 Evaluation of BBB [30]

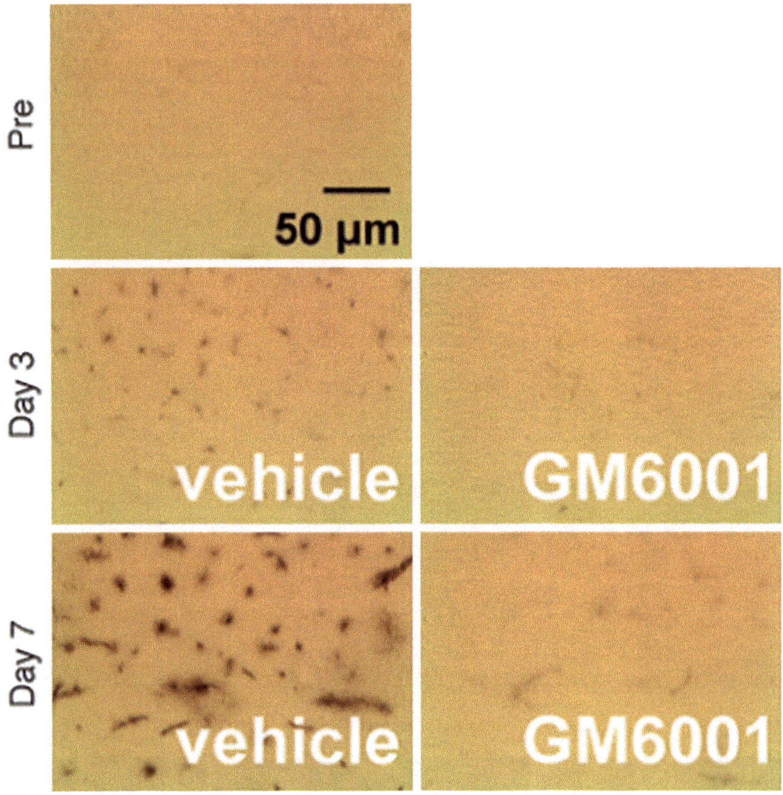

Treatment with GM6001 (MMP inhibitor) blocks IgG leakage at days 3 and 7 after white matter injury by prolonged cerebral hypoperfusion (bilateral common carotid artery stenosis)

10.6 Troubleshooting

10.6.1 Induction of Prolonged Cerebral Hypoperfusion

1. Separate the carotid from the vagus nerve, which is a white, string-like object directly lateral to the carotid artery. Take particular care to avoid damaging the vagus nerve.
2. Cessation of CBF for >1 min should be avoided.
3. One should avoid piercing the artery with the ends of the microcoil. Instead of rotating the microcoil around the CCA, it would be better to wind the artery along the groove of the coil.
4. As isoflurane anesthesia has been reported to abolish the ability to maintain constant CBF with changes in mean arterial blood pressure (autoregulation) [30], blood pressure should be measured and confirmed to keep constant CBF.

10.6.2 CBF Monitoring

1. It is important to remove the periosteum, which adheres tightly to the skull, with fine-tip forceps to minimize fibrous scar tissue buildup without significant changes in flow signals in sham-operated mice [19]. Swabbing the skull surface should be done gently. Rough or inappropriate treatment leads to scar tissue build up along the cranial sutures. It is difficult to reverse fibrous scar tissue build up and bone opacification. If they do occur, another animal should be used for the experiment. Although most bleeding stops spontaneously, the wound should be checked postoperatively for bloodstains on the skull, which, if present, must be removed gently with saline-soaked gauze.
2. For each recording, the skull surface was wiped with saline-soaked gauze, covered with a thin layer of gel (Aquasonic, Parker Laboratories, Inc.), and immersed for 5–10 min. Care should be taken to ensure the surface is fully wet as indicated by a semi-transparent appearance. A dry surface, as indicated by a white skull, or a partially wet surface results in a reduction in flow signal intensity.
3. Laser speckle imaging requires a baseline to anchor for repetitive measurement or comparison between different subjects. We use a calibration reference device (Calibrator S/N 080715–5, Omegawave, Inc.), and assign a value to this reference material (arbitrarily assigned value; 25.0). The value is attributed to the Brownian motion of red colored particles (0.35 μm, 24 °C). Calibration with this device before each test provides standardized values for comparison.
4. CBF should be measured after CBF is stabilized after anesthesia. During the CBF recordings, blood pressure should be measured by the tail cuff method and confirmed for consistency and the rectal temperature should be maintained between 36.5 °C and 37.5 °C. In addition, the stability of the level of anesthesia should be checked by testing corneal reflexes and monitor responses to tail pinch.
5. One can successfully image through the skull repeatedly up to 30 days after the operation [17, 19, 27]. Success is mostly dependent on the degree of removal of the periosteum.

10.7 Method Variations/Alternative

10.7.1 CBF Monitoring

Although, the laser speckle blood flow imaging system assesses only superficial areas of the brain, the following method uses small animal positron emission tomography (PET) imaging, which can evaluate and monitor the CBF of whole brain.

10.7.1.1 PET Acquisition and Image Analysis

Method

Dynamic PET studies were performed for 90 min after the intravenous application of 18.5–30.0 MBq 18 F-fluorodeoxyglucose (^{18}F-FDG) via a tail vein using a 23-frame protocol. (See detailed method in reference [31]) The PET studies were performed a week before and 2 h after BCAS in a mouse, 2 months after BCAS in a mouse, and 6 months after BCAS in 3 mice. First 5 min scans of ^{18}F-FDG uptake were used as an indicator of cerebral blood, which is suggested in tumor studies [32]. ^{18}F-FDG scans obtained between 45 and 90 min after injection were used as an indicator of cerebral glucose metabolism. Volumes of interest (VOIs) were defined on a representative magnetic resonance (MR) image obtained by male C57BL/6 J mice using a 7 T system (Bruker BioSpec 70/20 USR, Bruker, Karlsruhe, Germany): cortex (frontal area around from −1.5 mm posterior to the bregma), striatum (caudate-putamen), hippocampus, and cerebellum. Each reconstructed PET image during 45–90 min was rigidly coregistered to the MR image using the mutual information maximization algorithm [33] implemented in Statistical Parametric Mapping (SPM) version 2 (Welcome Department of Cognitive Neuroscience, London, UK). Each PET image during first 5 min was transformed by the same parameters obtained by a 45–90 min image. Average counts under each VOI were calculated. The PET images and counts under the three VOIs (cortex, striatum and hippocampus) were normalized to an average count of the cerebellum as a reference because ischemic changes were not observed in the cerebellum in mice with BCAS [22, 23].

Results

The first 5-min uptake of ^{18}F-FDG in the cerebral cortex decreased to ~70 % of the pre-BCAS level at 2 h after BCAS and recovered to 88 % of the pre-BCAS level at 2 months after BCAS (Figure, left panel). The early ^{18}F-FDG uptake scan in the striatum showed a similar temporal profile to that of the cerebral cortex. By contrast, the first 5-min uptake of ^{18}F-FDG in the hippocampus did not decrease at 2 h or 2 months after BCAS but decreased at 6 months. The late ^{18}F-FDG uptake scans showed that the glucose uptake of the hippocampus did not decrease by 2 months after BCAS but decreased by 20 % at 6 months after BCAS (Figure, right panel).

A limitation of this study is the effectiveness of the first 5-min ^{18}F-FDG blood flow analysis needs to be further confirmed. In tumor diagnosis, blood flow estimated from the early uptake of ^{18}F-FDG is linearly correlated with ^{15}O-measured blood flow25 (the current gold standard method for measuring blood flow in humans). However, the mean positron range of ^{15}O is far larger than that of ^{18}F (2.5 mm vs. 0.6 mm), ^{15}O-water is insufficient to apply to a small animal positron emission tomography scanner in terms of the spatial resolution. In addition, ^{15}O-water is a short-lived tracer with a 2-min half-life, necessitating an on-site cyclotron [32]. Therefore, ^{18}F-FDG imaging may be practical and feasible to estimate CBF noninvasively in rodent models for the simple first 5-min, despite the incomplete extraction of ^{18}F-FDG compared to ^{15}O-water. Although appropriate kinetic models of ^{18}F-FDG will be required to separate the flow component and K1 from the metabolic component of uptake, a method of measuring blood flow and metabolism from a single injection of ^{18}F-FDG uptake may be an important addition to functional imaging of rodent disease models with ^{18}F-FDG positron emission tomography [31].

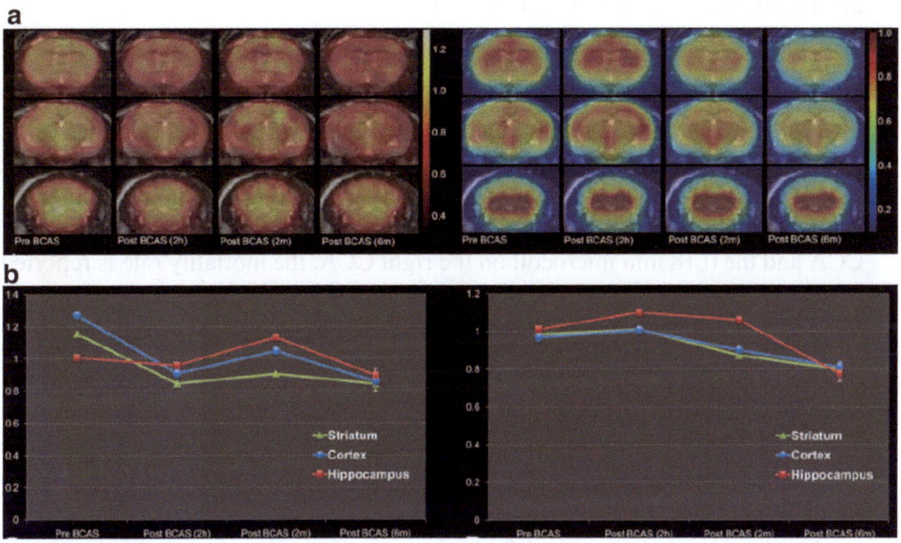

(**a**) Representative ^{18}F-FDG positron emission tomography coronal images during the first 5 min (*left panel*) and between 45 and 90 min (*right panel*) superimposed on MR image. (**b**) Temporal profiles of the mean normalized ^{18}F-FDG count during the first 5 min (*left panel*) and between 45 and 90 min (*right panel*) in the cortex, striatum, and hippocampus of a mouse 1 week before BCAS and 2 h after BCAS, a mouse 2 months after BCAS, and 3 mice 6 months after BCAS. All values were normalized to the average count in the cerebellum. Error bars represent standard deviation

10.8 Additional Note

10.8.1 Induction of Prolonged Cerebral Hypoperfusion

1. The animal must be maintained in a surgical plane of anesthesia, and its vital signs monitored and regulated throughout the procedure.
2. This model should be applied exclusively to C57BL/6 J strain, as CBF in the other strains may have a greater variability after BCAS [4, 8, 23].

3. Four types of microcoils are made from piano wire with varying inner diameters from 0.16 to 0.22 mm. Researchers in Japan may obtain the microcoils directly from the manufacturer (Sawane Spring Co., Ltd, Hamamatsu, Japan) but those outside Japan may purchase the microcoils from Invitrotech Co., Ltd. (Kyoto, Japan). The microcoils should be thoroughly disinfected with alcohol and air dried just before use.
4. In our original report, we used 30-min intervals between manipulations on the left and right CCAs to avoid early mortality [4]. However, no intervals may be required to generate this model. During the surgery, rectal temperature should be maintained between 36.5 °C and 37.5 °C.
5. The mortality rates range from 3 % to 5 % in mice with microcoils of 0.18 mm in diameter (unpublished data), although earlier studies suggested higher mortality rates: 13 % in mice with microcoils of 0.22 mm in diameter, 17 % in those of 0.20 mm, and 15–19 % in those of 0.18 mm [4, 25]. In contrast, 75 % (15/20) of mice with microcoils of 0.16 mm diameter administered died within 14 days after the surgery, most of whom were found to have cerebral infarctions [4]. In another study of a modified model with a 0.16 mm microcoil placed on the left CCA and the 0.18 mm microcoil on the right CCA, the mortality rate is reported to be 18.8 % [26].

10.8.2 CBF Monitoring

1. The mean CBF is usually measured in identically sized regions of interest (900 pixels) located 1 mm posterior and 2 mm lateral from the bregma.
2. During the measurement of CBF, the intact skull surface was diffusely illuminated by 780 nm laser light. The scattered light was filtered and detected by a CCD camera positioned over the head. The filter detected only scattered light that had a perpendicular polarization to the incident laser light. The raw speckle images were used to compute speckle contrast, which corresponds to the number and velocity of moving red blood cells, approximating CBF. Signal processing was performed by the algorithm developed by Forrester et al. [34]. Color-coded blood flow images were obtained in high-resolution mode (639×480 pixels; 1 image/s). The sample frequency was 60 Hz. One-blood flow image was generated by averaging the numbers obtained from 20 consecutive raw speckle images. The recordings were initiated after the examiner confirmed that CBF did not change over 1 min, and the five recordings of blood flow image were averaged.
3. Relative CBF determined by the laser speckle imager has been reported to have a linear relationship with absolute CBF values determined by [^{14}C]iodoamphetamine technique [35].

10.8.3 Evaluation of Angiogenesis

1. The evaluated regions are based on the mouse brain atlas (Paxinos G, Franklin KBJ. The mouse brain in stereotaxic coordinates. San Diego, Calif: Academic Press; 2001).

10.8.4 Evaluation of Arteriogenesis

1. The cerebral angioarchitecture is studied by a modification of the previous post-mortem latex perfusion technique [7, 36–40].
2. The injection volume and time of latex is much greater than those of previous reports (0.4–2 mL, 2 min) to harden the latex sufficiently [7, 36–40].
3. The distal middle cerebral artery (MCA) is identified from its branch angle and distinguished from the distal anterior cerebral artery (ACA) or posterior cerebral artery (PCA).

   The point of confluence is defined as the narrowest part of the vessel or half way between the nearest branching points of the ACA, the MCA, and the PCA branches, respectively.
4. The monocyte recruitments and proliferation of smooth muscle cells are essential in arteriogenesis [9].

10.9 Applications and Discussion

The various tools for assessment of angiogenesis and arteriogenesis discussed here provide a comprehensive picture of the state of cerebral vasculature. In addition, the BCAS mouse model provides a unique platform to investigate the mechanisms of angiogenesis and arteriogenesis following prolonged cerebral hypoperfusion and to explore potential drugs or cell therapies designed to enhance vascular growth as a preclinical step toward developing novel treatments for dementia of vascular origin.

In the BCAS mice, increased levels of circulating adrenomedullin, a vasoactive peptide, have been shown to restore cerebral hemodynamics, promote arteriogenesis, as well as angiogenesis, alleviate oxidative damage in cerebral microvessels, and preserve white matter integrity, leading to the prevention of cognitive dysfunction [26]. Furthermore, bone marrow mononuclear cell treatment has been shown to provide strong protection against white matter damage, dependent primarily on CBF recovery beginning from the early phase, and the subsequent endogenous restorative response, including angiogenesis in a later phase in this model mice [41].

Further cellular and molecular studies are warranted to elucidate how enhanced vascular growth can alleviate the tissue damage and functional impairment. Additionally, whether the growing/remodeling vasculature is functional or not should be carefully evaluated by the assessment of BBB/vascular integrity (such as interendothelial junctions, mural cells recruitment, remodeling of the surrounding matrix and elastic laminae), as well as vascular reactivity (such as functional hyperemia).

Acknowledgments and Funding Supported in part by the National Institutes of Health. Materials including figures in this chapter have been extensively drawn from our previously published papers including: Maki et al., Stroke, 2012; Maki et al., Neuroreport, 2012; Seo et al., J Clin Invest, 2013; Nishiko et al., Stroke, 2010.

References

1. Risau W, Flamme I (1995) Vasculogenesis. Annu Rev Cell Dev Biol 11:73–91
2. Potente M, Gerhardt H, Carmeliet P (2011) Basic and therapeutic aspects of angiogenesis. Cell 146:873–887
3. Carmeliet P, Jain RK (2011) Molecular mechanisms and clinical applications of angiogenesis. Nature 473:298–307
4. Folkman J (2007) Angiogenesis: an organizing principle for drug discovery? Nat Rev Drug Discov 6:273–286
5. Carmeliet P (2003) Angiogenesis in health and disease. Nat Med 9:653–660
6. Carmeliet P (2000) Mechanisms of angiogenesis and arteriogenesis. Nat Med 6:389–395
7. Buschmann IR, Busch HJ, Mies G, Hossmann KA (2003) Therapeutic induction of arteriogenesis in hypoperfused rat brain via granulocyte-macrophage colony-stimulating factor. Circulation 108:610–615
8. Schaper W (2009) Collateral circulation: past and present. Basic Res Cardiol 104:5–21
9. Persson AB, Buschmann IR (2011) Vascular growth in health and disease. Front Mol Neurosci 4:14
10. Zlokovic BV (2008) The blood–brain barrier in health and chronic neurodegenerative disorders. Neuron 57:178–201
11. Zlokovic BV, Apuzzo ML (1998) Strategies to circumvent vascular barriers of the central nervous system. Neurosurgery 43:877–878
12. Fotuhi M, Hachinski V, Whitehouse PJ (2009) Changing perspectives regarding late-life dementia. Nat Rev Neurol 5:649–658
13. Meguro K, Hatazawa J, Yamaguchi T, Itoh M, Matsuzawa T, Ono S, Miyazawa H, Hishinuma T, Yanai K, Sekita Y et al (1990) Cerebral circulation and oxygen metabolism associated with subclinical periventricular hyperintensity as shown by magnetic resonance imaging. Ann Neurol 28:378–383
14. Kalaria R (2002) Similarities between Alzheimer's disease and vascular dementia. J Neurol Sci 203–204:29–34
15. Viswanathan A, Rocca WA, Tzourio C (2009) Vascular risk factors and dementia: how to move forward? Neurology 72:368–374
16. Hachinski V, Iadecola C, Petersen RC, Breteler MM, Nyenhuis DL, Black SE, Powers WJ, DeCarli C, Merino JG, Kalaria RN et al (2006) National Institute of Neurological Disorders and Stroke-Canadian Stroke Network vascular cognitive impairment harmonization standards. Stroke 37:2220–2241
17. Iadecola C (2004) Neurovascular regulation in the normal brain and in Alzheimer's disease. Nat Rev Neurosci 5:347–360
18. Pantoni L, Garcia JH (1997) Pathogenesis of leukoaraiosis: a review. Stroke 28:652–659
19. De Groot JC, De Leeuw FE, Oudkerk M, Van Gijn J, Hofman A, Jolles J, Breteler MM (2002) Periventricular cerebral white matter lesions predict rate of cognitive decline. Ann Neurol 52:335–341
20. Roman GC, Erkinjuntti T, Wallin A, Pantoni L, Chui HC (2002) Subcortical ischaemic vascular dementia. Lancet Neurol 1:426–436

21. de Leeuw FE, Barkhof F, Scheltens P (2005) Progression of cerebral white matter lesions in Alzheimer's disease: a new window for therapy? J Neurol Neurosurg Psychiatry 76:1286–1288
22. Shibata M, Ohtani R, Ihara M, Tomimoto H (2004) White matter lesions and glial activation in a novel mouse model of chronic cerebral hypoperfusion. Stroke 35:2598–2603
23. Shibata M, Yamasaki N, Miyakawa T, Kalaria RN, Fujita Y, Ohtani R, Ihara M, Takahashi R, Tomimoto H (2007) Selective impairment of working memory in a mouse model of chronic cerebral hypoperfusion. Stroke 38:2826–2832
24. Ihara M, Taguchi A, Maki T, Washida K, Tomimoto H (2014) A mouse model of chronic cerebral hypoperfusion characterizing features of vascular cognitive impairment. Methods Mol Biol 1135:95–102 (PMID: 24510857)
25. Ihara M, Tomimoto H (2011) Lessons from a mouse model characterizing features of vascular cognitive impairment with white matter changes. J Aging Res 2011:978761
26. Maki T, Ihara M, Fujita Y, Nambu T, Miyashita K, Yamada M, Washida K, Nishio K, Ito H, Harada H et al (2011) Angiogenic and vasoprotective effects of adrenomedullin on prevention of cognitive decline after chronic cerebral hypoperfusion in mice. Stroke 42:1122–1128
27. Risau W (1997) Mechanisms of angiogenesis. Nature 386:671–674
28. Patan S (2000) Vasculogenesis and angiogenesis as mechanisms of vascular network formation, growth and remodeling. J Neurooncol 50:1–15
29. Maki T, Ihara M, Fujita Y, Nambu T, Harada H, Ito H, Nakao K, Tomimoto H, Takahashi R (2011) Angiogenic roles of adrenomedullin through vascular endothelial growth factor induction. Neuroreport 22:442–447
30. Seo JH, Miyamoto N, Hayakawa K, Pham LD, Maki T, Ayata C, Kim KW, Lo EH, Arai K (2013) Oligodendrocyte precursors induce early blood–brain barrier opening after white matter injury. J Clin Invest 123:782–786
31. Nishio K, Ihara M, Yamasaki N, Kalaria RN, Maki T, Fujita Y, Ito H, Oishi N, Fukuyama H, Miyakawa T et al (2010) A mouse model characterizing features of vascular dementia with hippocampal atrophy. Stroke 41:1278–1284
32. Mullani NA, Herbst RS, O'Neil RG, Gould KL, Barron BJ, Abbruzzese JL (2008) Tumor blood flow measured by PET dynamic imaging of first-pass 18 F-FDG uptake: a comparison with 15O-labeled water-measured blood flow. J Nucl Med 49:517–523
33. Maes F, Collignon A, Vandermeulen D, Marchal G, Suetens P (1997) Multimodality image registration by maximization of mutual information. IEEE Trans Med Imaging 16:187–198
34. Forrester KR, Stewart C, Tulip J, Leonard C, Bray RC (2002) Comparison of laser speckle and laser Doppler perfusion imaging: measurement in human skin and rabbit articular tissue. Med Biol Eng Comput 40:687–697
35. Ayata C, Dunn AK, Gursoy OY, Huang Z, Boas DA, Moskowitz MA (2004) Laser speckle flowmetry for the study of cerebrovascular physiology in normal and ischemic mouse cortex. J Cereb Blood Flow Metab 24:744–755
36. Maeda K, Hata R, Hossmann KA (1998) Differences in the cerebrovascular anatomy of C57black/6 and SV129 mice. Neuroreport 9:1317–1319
37. Todo K, Kitagawa K, Sasaki T, Omura-Matsuoka E, Terasaki Y, Oyama N, Yagita Y, Hori M (2008) Granulocyte-macrophage colony-stimulating factor enhances leptomeningeal collateral growth induced by common carotid artery occlusion. Stroke 39:1875–1882
38. Woitzik J, Hecht N, Schneider UC, Pena-Tapia PG, Vajkoczy P (2006) Increased vessel diameter of leptomeningeal anastomoses after hypoxic preconditioning. Brain Res 1115:209–212
39. Schneider UC, Schilling L, Schroeck H, Nebe CT, Vajkoczy P, Woitzik J (2007) Granulocyte-macrophage colony-stimulating factor-induced vessel growth restores cerebral blood supply after bilateral carotid artery occlusion. Stroke 38:1320–1328
40. Ergul A, Elgebaly MM, Middlemore ML, Li W, Elewa H, Switzer JA, Hall C, Kozak A, Fagan SC (2007) Increased hemorrhagic transformation and altered infarct size and localization after experimental stroke in a rat model type 2 diabetes. BMC Neurol 7:33
41. Fujita Y, Ihara M, Ushiki T, Hirai H, Kizaka-Kondoh S, Hiraoka M, Ito H, Takahashi R (2010) Early protective effect of bone marrow mononuclear cells against ischemic white matter damage through augmentation of cerebral blood flow. Stroke 41:2938–2943

Chapter 11
Sponge Implant Model of Inflammatory Angiogenesis

Silvia Passos Andrade, Paula Peixoto Campos, and Mônica A.N.D. Ferreira

11.1 Introduction

The realization that advances in angiogenesis research depend on making the assays more quantitative and reproducible in vitro and in vivo, has led to the development of new techniques and improvement of old or current models to comply with such requirements. The implantation technique for assessment of the inflammatory process is an old surgical procedure. Grindlay and Waugh [1], Woessner and Boucek [2], and Edwards [3] were the first to use polyvinyl sponge implants in dogs, rats, and rabbits as a framework for the ingrowth of vascularized connective tissue and measurement of enzyme activities in the newly formed fibrovascular tissue. This technique has further been developed to determine other biochemical variables of the fibrovascular tissue, including collagen metabolism [4], fibronectin deposition [5], and proteoglycan turnover [6]. In addition, the technique has been employed to characterize the sequence of histological changes in granulation tissue formation [7] and to monitor the kinetics of cellular proliferation [8]. The extent of neutrophil and macrophage accumulation in the sponge compartment has also been possible by assaying the inflammatory enzymes myeloperoxidase (MPO) and *N*-acetyl-*β*-*D*-glucosaminidase (NAG) and cytokines [9–11]. The effects of various agents on the modulation of cell migration/

S.P. Andrade (✉)
Departments of Physiology and Biophysics, Federal University of Minas Gerais, Belo Horizonte, MG, Brazil
e-mail: andrades@icb.ufmg.br

P.P. Campos • M.A.N.D. Ferreira
Department of General Pathology, Institute of Biological Sciences, Federal University of Minas Gerais, Belo Horizonte, MG, Brazil
e-mail: paulapc@icb.ufmg.br; monicadf@icb.ufmg.br

M. Slevin, G. McDowell (eds.), *Handbook of Vascular Biology Techniques*,
DOI 10.1007/978-94-017-9716-0_11

proliferation, inflammation, angiogenesis and fibrogenesis have been also evaluated in sponge implants [12–15]. The model of acute inflammation was particularly useful, allowing the collection and examination of both cellular and fluid phases of the exudates formed within the sponge [12]. Sponge implantation has also been used as framework to host rodent cell lines [13–15]. The advantage of the implantation technique for investigating tumor-induced angiogenesis is that the assessment of the relative contributions of the tumor cells to early changes in the implant blood flow can be detected even before visible growth of the tumor mass is evident. Using a 133Xe clearance technique or fluorescein diffusion method, the development of Colon 26, melanoma B16, or Ehrlich tumor has been evaluated regarding solid tumor hemodynamic features [15, 16]. In addition, the inflammatory microenvironment of the implant compartment has been exploited to study the role of inflammation on tumor development [10, 15]. The implantation technique has been also employed to study the adverse healing response (foreign body reaction) to medical devices, a critical issue that impairs the in vivo functionality and durability of implantable biomaterials [17]. Thus, the type of injury induced by a synthetic scaffold can be instrumental in the assessment of various components of the processes that result in blood vessel formation in a number of physiological or pathological conditions.

11.2 Materials

11.2.1 Sponge Matrix

A number of different sponge matrices have been used for inducing fibrovascular growth and for hosting tumor cells. The synthetic materials are mainly polyvinyl alcohol, cellulose acetate, polyester, polyether, and polyurethane alone or in combination. In our laboratory, we use sponge discs made of polyether polyurethane. This type of material possesses the following characteristics: uniform pore size and intercommunicating pore structure, ability to resist chemical treatment, and biocompatibility. **Note 1**.

11.2.2 Anaesthesia

Mice are anaesthetized with subcutaneous injection of xylazine (10 mg/kg) and ketamin (60 mg/kg) solution. Post-operatively, the animals are monitored for any sign of infection at the operative site, discomfort or distress.

11.3 Methods

11.3.1 Preparation of Sponge Implants

Circular sponge discs are cut from a sheet of sponge using a cork borer. The recommended dimensions of the implants for mice are 8 × 4 mm (diameter/thickness). The sponge allows accurate injection of tracers and test substances, and withdrawal of fluid into and from the interior of the implant. **Note 2**.

11.3.2 Surgical Procedure for Sponge Disc Implantation

- Implantation of sponge discs is performed with aseptic techniques following induction of anesthesia.
- The hair on the dorsal side is shaved and the skin wiped with 70 % v/v ethanol in distillated water.
- A 1-cm midline incision is made, and through it one subcutaneous pocket is prepared by blunt dissection using a pair of curved scissors.
- A sterilized sponge implant is then inserted into the pocket.
- The midline incision is closed by three interrupted 5–0 silk stitches, and the animals are kept singly with free access to food and water after recovery from anesthesia Fig. 11.1.

11.3.3 Procedure for Estimating Blood Flow Development By Diffusion Rate of Sodium Fluorescein Applied Intraimplant

The sequential development of blood flow in the implanted sponges, originally acellular and avascular, can be determined by the diffusion rate of fluoregenic dyes injected into implants. This technique was developed to measure blood flow and thus monitor the vascular changes indirectly. Compared with radioactive isotope compounds, the advantages of fluorescent dyes are clear. Fluorescence is relatively atoxic, nonradioactive, and inexpensive [18]. The measurement of fluorochrome-generated emission in bloodstream following its application in the sponge implant compartment at various intervals postimplantation reflects the degree of local blood flow development and the interaction of the angiogenic site with the systemic circulation [19]. This approach can be used to study sponge-induced angiogenesis quantitatively and to investigate the pharmacological reactivity of the neovasculature. **Note 3**.

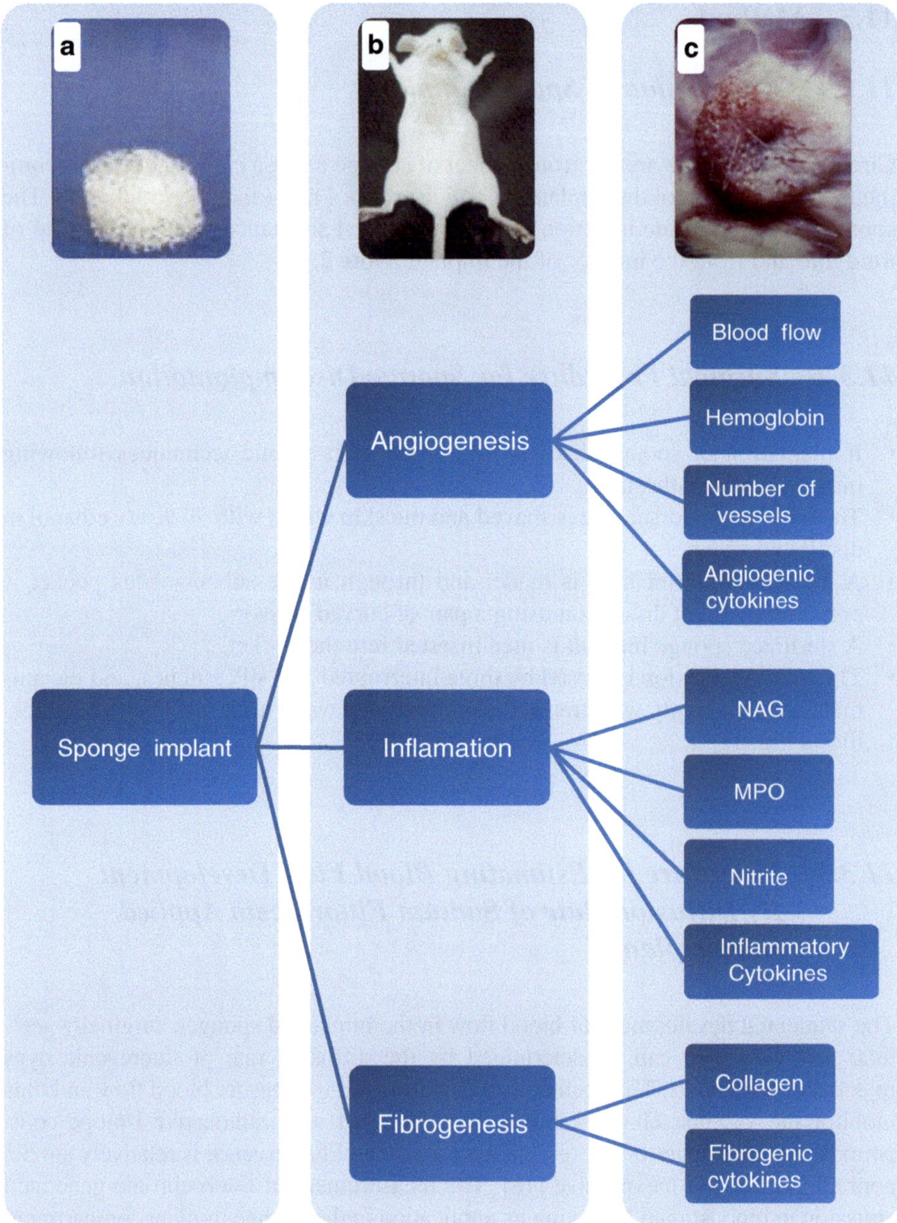

Fig. 11.1 The sponge implant system and parameters evaluated. An implant sponge disc (**a**) the subcutaneous arrangement of the implant in the mouse (**b**) and macroscopic view of the fibrovascular tissue (**c**). The sponge discs can be left in situ for periods ranging from days to weeks

Measurements of the extent of vascularization of sponge implants are made by estimating *t* ½ (min) of the fluorescence peak in the systemic circulation following intraimplant injection of sodium fluorescein (10 μL of a sterile solution of 10 % sodium fluorescein) at fixed time intervals (e.g., day 1, 4, 7, 10 and 14) postimplantation.

- Anaesthetize the animal.
- Determine blood background fluorescence by piercing the extremity of the tail vein and collecting 5 μL of blood with a heparinized yellow tip. Transfer the blood sample to a centrifuge tube containing 1 mL of isotonic saline (0.9 %).
- "At time 0, inject intra-implant sodium fluorescein (10 μL of a sterile solution of 10 % sodium fluorescein) to anesthetized animals".
- After 1 min, collect the first blood sample following dye injection as in step 2.
- At 3 min, collect a second blood sample. Repeat this procedure every 2–3 min for 25–30 min.
- Centrifuge the blood samples for 10 min at 1,400 *g*. Keep the supernatant for fluorescence determination (excitation 485 nm/emission 519 nm).
- From the fluorescence values, estimate the time for the fluorescence to peak in the bloodstream (absorption) and the time required for the elimination of the dye from the systemic circulation (elimination). These parameters are expressed in terms of half-time (*t* ½; time taken for the fluorescence to reach 50 % of the peak value in the systemic circulation).

11.3.4 Procedure for Estimating Biochemical Parameters in the Implants

Quantification of various biochemical parameters further supports the functional characterization of the fibrovascular tissue that infiltrates the implants and have been used to corroborate assessment of angiogenesis [12, 20–24].

- Remove the implants at any time post-implantation as required. **Note 4**
- Immediately after, weigh and homogenize the implant (Tekmar TR-10, OH, U.S.A.) in 2 mL of Drabkin reagent (Labtest, São Paulo, Brazil)
- Store the pellet at −20 °C for determination of the proinflammatory enzymes MPO (neutrophil influx) and NAG (macrophage recruitment), collagen, nitrite, cytokines measurement.

11.3.4.1 Hemoglobin Determination

The vascularization of the implant can be assessed by measuring the amount of hemoglobin contained in the tissue using the Drabkin method.

- After centrifuging the homogenate (implant plus 2 mL Drabkin reagent) at 10,000 g for 40 min, filter it in a 0.22 μm Millipore filter.

- Determine hemoglobin concentration by measuring absorbance at 540 nm using an enzyme-linked immunosorbent assay (ELISA) plate reader and compare against a standard curve of hemoglobin.

11.3.4.2 Cytokine Determination

- Take 50 μL of the supernatant previously homogenized in Drabkin reagent (to remove hemoglobin) and centrifuged (12,000 g, 20 min at 4 °C) and add 500 μL of PBS at pH 7.4 containing 0.05 % Tween-20 (Difco) and centrifuge at 12,000 g, at 4 °C for 30 min.
- The amount of the cytokines and chemokines in each sample is determined using immunoassay kit (R&D Systems, USA) following the manufacture's protocol.
- Briefly, cell-free supernatants are added in duplicate to ELISA plates coated with specific murine polyclonal antibody against the cytokine/chemokine, followed by the addition of a second polyclonal antibody against the cytokine/chemokine. After washing to remove any unbound antibody-enzyme reagent, a substrate solution (a 1:1 solution of hydrogen peroxide and tetramethylbenzidine) is added. To stop the reaction add H_2SO_4. Plates are read at 492 nm in a spectrophotometer. Standards of recombinant murine chemokines are diluted in a range from 7.5 to 1,000 pg/mL. Express the results as pictograms/milligram wet tissue.

11.3.4.3 Measurement of Myeloperoxidase (MPO) and *N*-acetyl-*β*-*D*-glucosaminidase (NAG) Activities

The Extent of Neutrophil Accumulation in the Implants Is Measured by Assaying MPO Activity in Whole Tissue

- After processing the supernatant of the tissue for hemoglobin determination (see Sect. 11.3.4.1), a part of the corresponding pellet is weighed, homogenized in pH 4.7 buffer (0.1 M NaCl, 0.02 M Na_3PO_4, 0.015 M Na_2EDTA -disodium ethylenediamine tetraacetate). Centrifuges at 12,000 *g* for 10 min at 4 ° C.
- The pellets are then resuspended in 0.05 M sodium phosphate buffer (pH 5.4) containing 0.5 % hexa-1,6-bis-decyltrimethylammonium bromide (HTAB; Sigma). The suspensions are freeze-thawed three times using liquid nitrogen and finally centrifuged at 10,000 g for 20 min at 4 ° C.
- MPO activity in the resulting supernatant is assayed by mixing 25 μL of 3,3′-5,5′- tetramethylbenzidine (TMB, Sigma), prepared in dimethylsulphoxide (DMSO; Merck) in a final concentration of 1.6 mM; add 100 μL H_2O_2 in a final concentration of 0.003 % v/v, dissolved in sodium phosphate buffer (pH 5.4) and 25 μL of the supernatant from the tissue sample.
- The assay is carried out in a 96-well microplate and is started by adding the supernatant sample to the H_2O_2 and TMB solution and incubated for 1 min at 37 °C.

- The reaction is terminated by adding 100 μL 4 M H_2SO_4 at 4 ° C and is quantified colorimetrically at 450 nm in a spectrophotometer. MPO activity is finally expressed as optical density (OD/mg wet tissue). Results can be expressed as change in optical density per gram of wet tissue.

Recruitment/Activation of Monocyte/Macrophages Can be Evaluated by Measuring the Activity of *N*-acetyl-*β*-*D*-glucosaminidase [9, 24, 25]

- Part of the pellet remaining after the hemoglobin measurement is kept for this assay. These pellets are weighed, homogenized in NaCl solution (0.9 % w/v) containing 0.1 % v/v Triton X-100 (promega), and centrifuged (3,000 *g* for 10 min at 4 ° C).
- Samples of the resulting supernatant (100 μL) are incubated for 30 min with 100 μL *p*-nitrophenyl-*N*-acetyl-*β*-*D*-glucosaminide (Sigma) prepared in 0.1 M citrate/sodium phosphate buffer (pH 4.5) in a final concentration of 2.24 mM.
- The reaction is terminated by addition of 100 μL 0.2 M glycine buffer, pH 10.6. Hydrolysis of the substrate is determined by measuring the color absorption at 405 nm. NAG activity is finally expressed as optical density (OD/mg wet tissue).

11.3.4.4 Measurement of Nitric Oxide (NO) Production

NO release was evaluated by measuring nitrite levels according to the method described by Green et al. [26].

- Sponge removed days post-cell implantation must be weighed and incubated for 15 min at 37 °C with PBS (500 μl).
- The incubation medium (100 μl) is mixed with 10 μl of Griess reagent (0.1 % *N-1-naphthylediamine*, 1 % sulfanilamide in 5 % H_3PO_4 and optical density is measured at 540 nm).
- The amount of nitrite in the incubation media is calculated using sodium nitrite (Sigma-Aldrich, St Louis, MO, US) as standard.

11.3.4.5 Collagen Measurement

Total soluble collagen is measured in whole implant homogenates by the Sirius Red reagent based-assay [27, 28].

- Homogenize the implants in 1 ml of PBS and centrifuge (5,000 g).
- Take 50 μL of the supernatant and mix it with 50 μl of Sirius Red reagent. Gently, mix the samples by inversion.
- After centrifuging the homogenate (50 μl of Sirius Red reagent plus 50 μl of sample) at 5,000 g for 10 min, drain off the supernatant.

- Wash the resulting pellet with 500 μl of ethanol (99 % pure and methanol free).
- Add 1 ml of a 0.5 M NaOH solution to the remaining pellet of collagen-bound dye.
- Following solubilization, transfer the samples to a 96-well plate and read at 540 nm. The calibration curve is set up on the basis of gelatin standard (Merck). The results are expressed as μg collagen/mg wet tissue.

11.3.5 Histological Analysis of the Implants

To establish the sequential development of granulation tissue and blood vessels in the implants, several histologic techniques have been employed.

- Kill the animals bearing the implants.
- Dissect the implants free of adherent tissue, fix in formalin (10 % w/v in isotonic saline), and embed them in paraffin.
- Cut the sections (5–8 μm) from halfway through the sponge's thickness.
- Stain and process for light microscopy studies.

11.3.5.1 Routine Histological Stainings

Hematoxylin and eosin (HE), Masson's and Gomori's trichrome, Dominici blue, and picrossirius red staining can be used for determining the implant fibrovascular tissue main features (area, vessels, inflammatory cells and collagen content) Fig. 11.2.

11.3.5.2 Specific Histological Stainings

Terminal Deoxynucleotidyl Transferase (TdT) Mediated dUTP-biotin Nick End Labeling (TUNEL) Assay

Apoptotic cells can be detected by the terminal deoxynucleotidyl transferase-mediated deoxyuridine triphosphate fluorescence nick end labeling (TUNEL) method Gavrieli et al. [29], modified by Campos et al. [30] in paraffin-embedded sponge implant sections mounted on glass slides. "The TUNEL positive cells estimate the apoptotic event (commercial kit used TdT-FragEL™ DNA Fragmentation Detection Kit; Calbiochem, EMD Biosciences, Inc., Darmstadt, Germany)."

- Tissue sections are deparaffinized and rehydrated by transferring the slides through the following solutions: xylene three times for 5 min, 100 % ethanol two times for 3 min, 95 % ethanol two times for 3 min, 80 % ethanol for 3 min, 70 % ethanol for 3 min, and finally PBS for 5 min;
- Incubated with 0.3 % H_2O_2 in PBS to quench endogenous peroxidase activity;

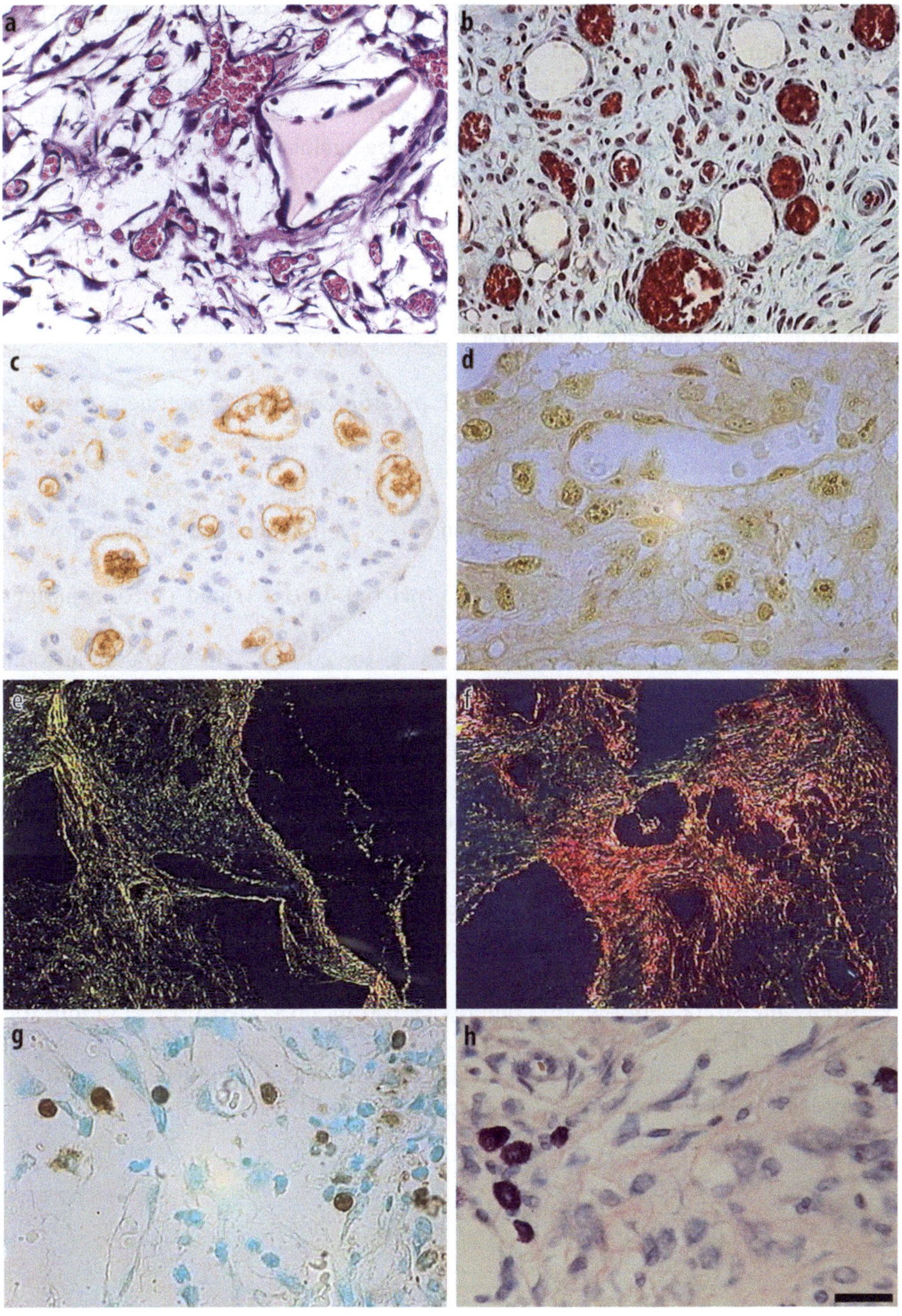

Fig. 11.2 Representative histological sections of a range of stainings used to identify various components of the fibrovascular tissue induced by the synthetic matrix in mice, (**a**) hematoxylin and eosin; (**b**) Giemsa; (**c**) CD31; (**d**) AgNOR; (**e**) and (**f**) Picrossirius *red*; (**g**) TUNEL; (**h**) Dominicci

- Nuclei are stripped of proteins by incubation with proteinase K (kit Calbioquem) for 15 min.
- DNA fragments are labeled with TdT and digoxigenin-dUTP at 37 °C for 1 h.
- The enzymatic labeling of the DNA fragments with digoxigenin-dUTP is detected with peroxidase-conjugated antibody against digoxigenin.
- Diaminobenzidine (DAB) is used as the substrate for peroxidase.
- Cover the entire specimen with methyl green counterstain solution.
- Sections must be dehydrated in 2× ethanol 100 % for 3 min and xylene to mount.

Immunohistochemistry Assay for Blood Vessel Counting

- Cut 5 μm sections from formalin fixed, paraffin embedded specimen of sponge and mount of gelatin coated microscope slides.
- Dry the slides overnight at 37 °C.
- Dewax the sections using xylene for 15 min 2× followed by four changes of graded ethanol (100 %; 90 %; 70 % and 50 %) 3 min each and wash with water prior to staining.
- Place tissue sections in citrate buffer (pH 6.0, 0.01 M) and perform antigen retrieval.
- Place tissue sections in water bath at 98 °C for 20 min. Allow to cool at room temperature and wash in PBS/TBS (buffer Tris–HCl) prior to ethan-staining (CD 31, factor VIII).
- Block endogenous peroxidase activity by incubating sections with 3 % hydrogen peroxide in ethanol for 10 min.
- Incubate serial sections with primary antibody (monoclonal antibody to CD31 or factor VIII; 1:100, Fitzgerald).
- Wash slides with TBS for 5 min.
- Incubate biotinylated secondary antibody 1:100 polymer (ADVANCE HRP DAKO) in 1 % goat serum in TBS for 30 min followed by washing in TBS for 5 min.
- Apply the streptavidin-biotin complex (1:100 in TBS; DAKO) for 30 min followed by 3× rinses in TBS.
- Apply chromogenic substrate 0.08 % diaminobenzidine (Sigma) and hydrogen peroxide for 5–10 min and rinse in tap water.
- Counterstain as required and mount.

11.4 Additional Notes

1. Depending on the material, the inflammatory response can cause excessive matrix deposition and unwanted fibrosis. Because of the variety of materials used (size, structure, composition, porosity), the pattern of the response varies widely.

2. It is advisable that administration of drugs intra-implant be performed 24 h postimplantation to allow the sponges to be soaked in the inflammatory exudate.
3. To eliminate acute effects of the vasoactive substances tested (vasodilation or vasoconstriction), they should be given 6–8 h prior to blood flow measurement.
4. To avoid possible “contamination” of blood spilled during and after the surgical procedure or with surrounding preexisting vessels, removal of the implants must be done 30 min after death of the animals.

Acknowledgement This work was supported by grant from FAPEMIG and CNPq-Brazil.

References

1. Grindlay JH, Waugh JM (1951) Plastic sponge which acts as a framework for living tissue; experimental studies and preliminary report of use to reinforce abdominal aneurysms. AMA Arch Surg 63:288–297
2. Woessner JF, Boucek RJ (1959) Enzyme activities of rat connective tissue obtained from subcutaneously implanted polyvinyl sponge. J Biol Chem 234:3296–3300
3. Edwards RH, Sarmenta SS, Hass GM (1960) Stimulation of granulation tissue growth by tissue extracts. Study in intramuscular wounds in rabbits. Arch Pathol 69:286–302
4. Paulini K, Körner B, Beneke G et al (1974) A quantitative study of the growth of connective tissue: investigations on implanted polyester-polyurethane sponges. Connect Tissue Res 2:257–264
5. Hølund B, Clemmensen I, Junker P et al (1982) Fibronectin in experimental granulation tissue. Acta Pathol Microbiol Immunol Scand A 90:159–165
6. Bollet AJ, Goodwin JF, Simpson WF et al (1958) Mucopolysaccharide, protein and desoxyribosenucleic acid concentration of granulation tissue induced by polyvinyl sponges. Proc Soc Exp Biol Med 99:418–421
7. Hølund B, Junker P, Garbarsch C et al (1979) Formation of granulation tissue in subcutaneously implanted sponges in rats. A comparison between granulation tissue developed in viscose cellulose sponges (Visella) and in polyvinyl alcohol sponges (Ivalon). Acta Pathol Microbiol Scand A 87A:367–374
8. Davidson JM, Klagsbrun M, Hill KE et al (1985) Accelerated wound repair, cell proliferation, and collagen accumulation are produced by a cartilage-derived growth factor. J Cell Biol 100:1219–1227
9. Bailey PJ (1988) Sponge implants as models. Methods Enzymol 162:327–334
10. Belo AV, Barcelos LS, Ferreira MA et al (2004) Inhibition of inflammatory angiogenesis by distant subcutaneous tumor in mice. Life Sci 74:2827–2837
11. Ferreira MA, Barcelos LS, Campos PP et al (2004) Sponge-induced angiogenesis and inflammation in PAF receptor-deficient mice (PAFR-KO). Br J Pharmacol 141:1185–1192
12. Ford-Hutchinson AW, Walker JA, Smith JA (1977) Assessment of anti-inflammatory activity by sponge implantation techniques. J Pharmacol Methods 1:3–7
13. Mahadevan V, Hart IR, Lewis GP (1989) Factors influencing blood supply in wound granuloma quantitated by a new in vivo technique. Cancer Res 49:415–419
14. Andrade SP, Bakhle YS, Hart I et al (1992) Effects of tumour cells on angiogenesis and vasoconstrictor responses in sponge implants in mice. Br J Cancer 66:821–826
15. Lage AP, Andrade SP (2000) Assessment of angiogenesis and tumor growth in conscious mice by a fluorimetric method. Microvasc Res 59:278–285
16. Andrade SP, Hart IR, Piper PJ (1992) Inhibitors of nitric oxide synthase selectively reduce flow in tumor-associated neovasculature. Br J Pharmacol 107:1092–1095

17. Mendes JB, Campos PP, Ferreira MA et al (2007) Host response to sponge implants differs between subcutaneous and intraperitoneal sites in mice. J Biomed Mater Res B Appl Biomater 83:408–415
18. McGrath JC, Arribas S, Daly CJ (1996) Fluorescent ligands for the study of receptors. Trends Pharmacol Sci 17:393–399
19. Andrade SP, Machado RD, Teixeira AS et al (1997) Sponge-induced angiogenesis in mice and the pharmacological reactivity of the neovasculature quantitated by a fluorimetric method. Microvasc Res 54:253–261
20. Andrade SP, Vieira LB, Bakhle YS et al (1992) Effects of platelet activating factor (PAF) and other vasoconstrictors on a model of angiogenesis in the mouse. Int J Exp Pathol 73:503–513
21. Andrade SP, Cardoso CC, Machado RD et al (1996) Angiotensin-II-induced angiogenesis in sponge implants in mice. Int J Microcirc Clin Exp 16:302–307
22. Hu DE, Hiley CR, Smither RL et al (1995) Correlation of 133Xe clearance, blood flow and histology in the rat sponge model for angiogenesis. Further studies with angiogenic modifiers. Lab Invest 72:601–610
23. Buckley A, Davidson JM, Kamerath CD et al (1985) Sustained release of epidermal growth factor accelerates wound repair. Proc Natl Acad Sci U S A 82:7340–7344
24. Plunkett ML, Hailey JA (1990) An in vivo quantitative angiogenesis model using tumor cells entrapped in alginate. Lab Invest 62:510–517
25. Belo AV, Barcelos LS, Teixeira MM et al (2004) Differential effects of antiangiogenic compounds in neovascularization, leukocyte recruitment, VEGF production, and tumor growth in mice. Cancer Invest 22:723–729
26. Green LC, Wagner DA, Glogowski J et al (1982) Analysis of nitrate, nitrite and [15 N] nitrate in biological fluids. Anal Biochem 126:131–135
27. Campos PP, Bakhle YS, Andrade SP (2008) Mechanisms of wound healing responses in lupus-prone New Zealand White mouse strain. Wound Repair Regen 16:416–424
28. Marques SM, Campos PP, Castro PR et al (2011) Genetic background determines mouse strain differences in inflammatory angiogenesis. Microvasc Res 82:246–252
29. Gavrieli Y, Sherman Y, Ben-Basson SA (1992) Identification of programed cell death in situ via specific labeling of nuclear DNA fragmentation. J Cell Biol 119:493–501
30. Campos PP, Vasconcelos AC, Ferreira MA et al (2011) Alterations in the dynamics of inflammation, proliferation and apoptosis in subcutaneous implants of lupus-prone mice. Histol Histopathol 26:433–442

Chapter 12
The Chick Embryo Chorioallantoic Membrane Assay

Domenico Ribatti

12.1 Introduction

The chick embryo chorioallantoic membrane (CAM) is an extraembryonic membrane involved in gas exchange through a dense capillary network [1]. Due to its extensive vascularization, the CAM has been used to study the morpho-functional aspects of angiogenesis *in vivo* and to investigate the activity of pro-angiogenic and anti-angiogenic molecules [1]. Moreover, due to the lack of a fully developed immunocompetence system in the chick embryo, the CAM represents an ideal host tissue for tumor engrafting [2].

The allantoic vesicle enlarges very rapidly from days 4 to 10: the CAM surface increases from 6 cm^2 at day 6 to 65 cm^2 at day 14. In this process the mesodermal layer of the allantois fuses with the adjacent mesodermal layer of the chorion to form the CAM. An extremely rich vascular network connected to embryonic circulation by the allantoic arteries and veins develops between the two layers. Immature blood vessels scattered in the mesoderm grow very rapidly until day 8, giving rise to a capillary plexus, associated with the overlaying chorionic epithelium and mediating gas exchange with the outer environment. Capillary proliferation continues until day 10; then, the endothelial cell mitotic index declines and the vascular system attains its final arrangement on day 12 [3]. Besides sprouting angiogenesis that characterizes the early phases of CAM development, late CAM vascularization is supported by intussusceptive microvascular growth in which capillary network increases its complexity and vascular surface by insertion of transcapillary pillars [4].

D. Ribatti (✉)
Department of Basic Medical Sciences, Neurosciences and Sensory Organs, University of Bari Medical School, Piazza Giulio Cesare, 11, Policlinico, 70124 Bari, Italy

National Cancer Institute "Giovanni Paolo II", 70124 Bari, Italy
e-mail: domenico.ribatti@uniba.it

M. Slevin, G. McDowell (eds.), *Handbook of Vascular Biology Techniques*,
DOI 10.1007/978-94-017-9716-0_12

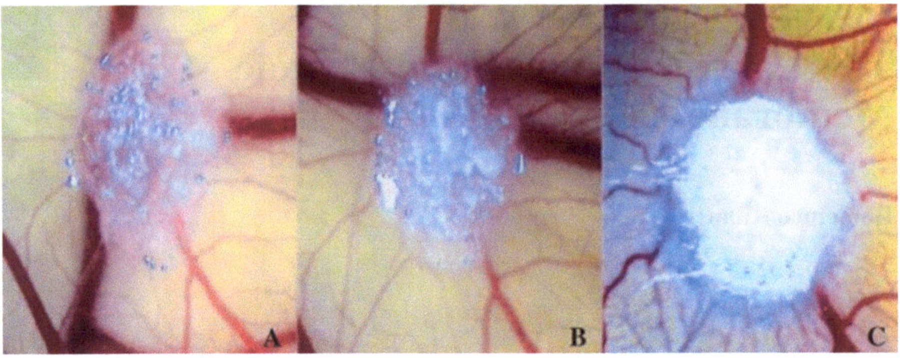

Fig. 12.1 Time-course of the macroscopic appearance of a chorioallantoic membrane implanted at day 8 (**a**) with a sponge loaded with 18,000 plasma cells of an active multiple myeloma patient. Note that, whereas on day 9 (**b**) no vascular reaction is detectable, in day 12 (**c**) numerous allantoids vessels develop radially towards the implant in a "spoked-wheel" pattern (Reproduced from Ribatti et al. [8])

Many protocols have been proposed to deliver macromolecules and low molecular weight compounds onto the CAM by using as vehicles silostatic rings, methylcellulose discs, silicon rings, filters, and plastic rings collagen and gelatin sponges. The gelatin sponge is also suitable for the delivery of cell suspensions onto the CAM surface and the evaluation of their angiogenic potential [5]. As compared with the application on the CAM of large amounts of a pure recombinant angiogenic cytokine in a single bolus, implants of cells overexpressing angiogenic cytokines enables a more physiological continuous delivery of growth factors.

Besides *in ovo* experimentation, a number of shell-less culture techniques have been devised, involving cultures of avian embryos with associated yolk and albumin outside of the eggshell. Shell-less cultures facilitate experimental access and continuous observation of the growing embryo.

An angiogenic response occurs 72–96 h after stimulation in the form of increased vessel density around the implant, with the vessels radially converging toward the center like spokes in a wheel (Fig. 12.1). Conversely, when an angiostatic compound is tested, the vessels become less dense around the implant and eventually disappear (Fig. 12.2).

Several semiquantitative and quantitative methods are used to evaluate the extent of vasoproliferative response or angiostatic activity at macroscopic and microscopic levels.

Many techniques can be applied within the constraints of paraffin and plastic embedding, including histochemistry and immunohistochemistry. Electron microscopy can also be used in combination with light microscopy. Moreover, unfixed CAM can be utilized for biochemical studies, such as the determination of DNA, protein and collagen content, and for reversal-transcriptase polymerase chain reaction (RT-PCR) analysis of gene expression. Finally, the study of intracellular signaling pathways mediating the angiogenic response to growth factors and cytokines have been successfully performed.

The main advantages of the CAM *in vivo* assay are the high embryonic survival rate, its low cost, simplicity, reproducibility and reliability. The major disadvantage

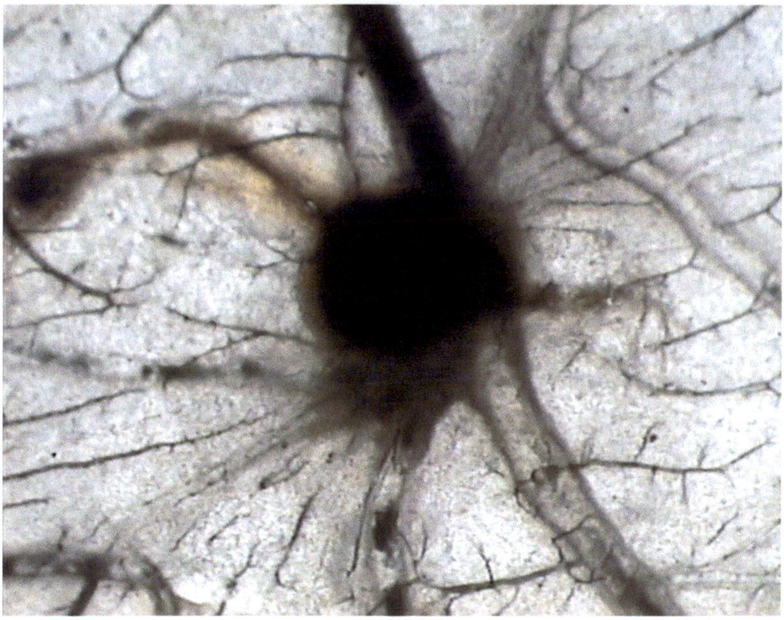

Fig. 12.2 Inhibitory effect of the urotensin receptor antagonist Palosuran on the angiogenic effect induced by urotensin II on the chorioallantoic membrane (Reproduced from Spinazzi et al. [9])

is that the CAM already contains a well-developed vascular network and the vasodilatation that follows its manipulation may be hard to distinguish from the effects of the test substance. Moreover, real neovascularization can hardly be distinguished from a falsely vascular density due to the rearrangement of existing vessels that follows contraction of the membrane [6]. Another limitation is nonspecific inflammatory reactions that may develop as a result of grafting, inducing a secondary vasoproliferative response [7]. Species-specific differences and the lack of avian-specific reagents (as well as limited genomic information) may represent other disadvantages. However, in the last years retroviral, lentiviral, and adenoviral vectors have been used to infect the CAM (as well as the whole chick embryo), leading to the expression of the viral transgene. This allows the long-lasting presence of the gene product that is expressed directly by CAM cells, and makes feasible the study of the effects of intracellular or membrane-bound proteins as well as of dominant-negative gene products.

12.2 Methodology

12.2.1 Reagents

1. White chicken egg obtained at day 1–2 postlaying (see Note 1).
2. 70 % (vol/vol) ethanol in dH_2O.
3. Sterile routine tissue culture medium (MEM Amino Acid Solution, Sigma).
4. Sterilized gelatin sponges (Gelfoam, Upjohn).

5. Angiogenic molecule (dissolved in 1–3 μl of sterile routine tissue culture medium at doses ranging between 10 and 500 ng per implant) or tumor cell suspension (from 1×10^4 to 6×10^6 cells per sponge resuspended in 3–4 μl of sterile routine tissue culture medium).
6. Bouin's fluid solution.
7. Standard solutions for paraffin embedding.
8. Toluene.
9. Embedding paraffin (Tissue-Tek VIP, electron Microscopy Science).
10. Toluidine blue 0 (Sigma; use a 0.5 % (vol/vol) aqueous solution).

12.2.2 Equipment

1. Egg incubator (Kemps Koops), 37 °C and 60 % humidity.
2. 25- or 26-G hypodermic needles and 1–ml syringes.
3. Curved-tip forceps.
4. Small dissecting scissors.
5. Transparent tape or glass coverslip, approximately 10×10 mm.
6. Microtome.
7. Stereomicroscope (Olympus, Italia).
8. Double-headed light microscope (BX51, Olympus, Italia), including a square mesh insert consisting of 12 lines per side for each eyepiece.

12.3 Methods

12.3.1 CAM Samples Preparation

1. Sterilize all instruments in 70 % ethanol before use.
2. Clean the fertilized white chicken eggs with 70 % ethanol and incubated at 37 °C and 60 % humidity in an egg incubator for 48 h (Note 2).
3. Aspirate 2–3 ml albumen from the egg using a 25- or 26-G hypodermic needle and 1-ml syringe at the acute pole of the egg on day 3 of incubation (Note 3).
4. After albumen removal, cut a square window into the shell with the aid of small dissecting scissors (Note 4).
5. On day 8 of incubation, open the window under sterile conditions with a laminar-flow hood and implant a 1 mm^3 sterilized gelatin sponge onto the CAM (Note 5).
6. Pipe the angiogenic molecule onto the sponge. Use a sponge containing vehicle alone as the negative control. Similarly, the gelatin sponge is suitable for the delivery of tumor cell suspensions onto the CAM surface and for the evaluation of their angiogenic potential. The sample may be a mixture of an angiogenic molecule and potential anti-angiogenic compounds (Note 6).

7. On day 12, fix the embryos and their membranes *in ovo* by pipetting 5 μl of Bouin's fluid solution onto the CAM surface and allowing the embryos to fix for 3 h at room temperature.
8. Cut the sponges and the underlying and immediately adjacent CAM portions with curved-tip forceps and transfer each specimen to a culture tube.
9. Dehydrate tissue samples in an ethanol series, clear them in toluene and immerse them in embedding paraffin for 2 h, according to manufacturer's instructions.
10. Using a microtome, cut 8-μm serial sections from each sample of paraffin-embedded CAMs in parallel to the surface of the membrane and stain the sections with a 0.5 % aqueous solution of toluidine blue for 1 min at room temperature.

12.3.2 Macroscopic Evaluation of the Vasoproliferative Response

On incubation day 12, macroscopic observation shows that the gelatin sponge treated with an angiogenesis stimulator is surrounded by allantoic vessels that develop radially towards the implant in a 'spoked-wheel' pattern (Note 7).

1. Analyzing the convergence of blood vessels toward the graft. For each egg, count the total number of macroscopic vessels that converge toward the graft under the stereomicroscope at 10 × magnification at different time points after implantation from day 8 to day 12. Express the data for each experimental group as the mean +/−1 standard deviation and obtain kinetics curves for proangiogenic or antiangiogenic stimuli compared with controls.
2. Analyzing variations in the distribution and density of CAM vessels next to the site of grafting. The intensity of the angiogenic response is scored under a stereomicroscope at regular intervals following the grafting procedure from day 8 to day 12 by means of a 0–5 scale of arbitrary values: 0 describes a condition of the vascular network that is unchanged with respect to the time of grafting; 1 marks a slight increment in vessel density associated to occasional changes in the course of vessels converging towards the grafting site; 2, 3, 4, and 5 correspond to a gradual increase in vessel density associated with increased irregularity in their course; a 5 rating also highlights strong hyperemia. A coefficient describing the degree of angiogenesis can also be derived from the ratio of the calculated value to the highest attainable value; thus, the coefficient's lowest value is equal to 0 and the highest value is 1.
3. Analyzing blood-vessel branching. A vascular index based on blood vessel branching may represent an alternative semiquantitative method to assess the vasoproliferative response. According to this procedure, all the vessels converging toward the implant and contained inside a 1 mm in diameter ring superposed to the CAM are enumerated: the ring is drawn around the implant in such a way that it will form an angle of less than 45° with a straight line drawn starting from the implant's center. Vessels branching outside the ring are scored as two, while those branching inside the ring are scored as 1.

12.3.3 Microscopic Evaluation of the Angiogenic Response

Quantitative evaluation of the angiogenic response, expressed as microvessel density, can be obtained by applying a morphometric method of 'point counting' on histological CAM sections. Briefly, two investigators simultaneously identify transversally cut microvessels (diameter ranging from 3 to 10 μm) among the gelatin sponge trabeculae with a double headed photomicroscope at × 250 magnification. A square mesh consisting of 12 lines per side, giving 144 intersection points, is inserted in the eyepiece. Ten sections are analyzed for each CAM specimen by observing every third section within 30 serial slides. For each section, six randomly chosen microscopic fields are evaluated for the number of intersection points occupied by microvessels. Then, mean values ± 1 standard deviation are determined for each CAM specimen. The microvessel density is expressed as percentage of intersection points occupied by microvessels. Statistically significant differences between the mean values of the intersection points in the experimental CAMs and control ones are determined by Student's t test for unpaired data.

At microscopic level, a highly vascularized tissue is recognizable among the trabeculae of the sponges treated with the angiogenesis stimulator. The tissue consists of newly formed blood vessels growing perpendicularly to the plane of the CAM, mainly capillaries with a diameter ranging from 3 to 10 μm within an abundant network of collagen fibers. In contrast, no blood vessels are present among trabeculae of the implants treated with vehicle alone. Otherwise, in the specimens treated with an angiogenesis inhibitor or with vehicle alone, few blood vessels are detectable around the sponge. Also, angiogenesis inhibitor causes the progressive regression of blood vessels distributed at the boundary between sponge and CAM mesenchyme, while they are still detectable in embryos treated with vehicle alone.

12.4 Notes

1. Chicken embryos from different vendors can vary significantly in their degree of vascularization and developmental status. Thus, consistent use of the same vender can decrease experimental variability.
2. Cleaning the egg shell before incubation will remove any debris associated with the outer surface and decrease the risk of infection. The eggs are very susceptible to mixed bacterial infections from feces-derived shell organisms, including *Aspergillus fumigatus*, *Staphylococcus aureus*, and *Pseudomonas aeruginosa*. Furthermore, specific incubation conditions, including constant temperature and humidity, are of critical importance for proper vascularization and embryo survival.
3. This procedure creates a false air sac directly over the CAM, allowing its dissociation form the egg shell membrane. Make sure the needle opening is pointing away from the embryo during albumen aspiration. In addition, regular changes of needle and syringe limit the carryover of infection from egg to egg.

4. This window can be enlarged to approximately 10×10 mm. The underlying embryo and CAM vessels are revealed. Seal the window with transparent tape or a glass coverslip of the same dimension, and return the egg to the incubator. This step should be done in an enclosed area such as a laminar-flow hood to minimize the risk of infection. If large pieces of shell fall onto the CAM it may possible to remove these using fine forceps. It is probably best to discard eggs where pieces of shell have fallen onto the CAM and have not been removed very easily, as these eggs may develop false positive response due to inflammation.
5. For this purpose, the sponge is cut by hand with a blade and gently placed on top of the growing CAM. Use sterile gloves to minimize the risk of infection. Reject eggs with an excessively humid CAM; otherwise, sponges may float off during the incubation period. The CAM is an expanding membrane with vessels developing over its entire surface. It is preferable place the sponges between two large vessels and not place outer edges of the CAM.
6. Although most substances are soluble in water, small organic molecules, such as synthetic angiogenesis inhibitors, may require organic solvents. Ethanol, but not DMSO, can be used without adverse effects by soaking the gelatin sponge in the dissolved compound of interest and allowing the solvent to evaporate before implantation onto the CAM.
7. Two operators, preferably blinded to the sample identity should grade the angiogenic/antiangiogenic response. Different test substances can produce a range of different types of angiogenic response, such as a mixed response of microvascular growth and large vessel deformation/growth towards the point of application. Samples may also induce local bleeding and the presence and severity of these reactions should be noted, as the response may be secondary to the bleeding or inflammation.

Acknowledgements The research leading to these results has received funding from the European Union Seventh Framework Programme (FP7/2007–2013) under grant agreement n°278570.

References

1. Ribatti D (2010) The chick embryo chorioallantoic membrane in the study of angiogenesis and metastasis. Springer, Dordrecht
2. Leene W, Duyzings MJM, Von Steeg C (1973) Lymphoid stem cell identification in the developing thymus and bursa of Fabricius of the chick. Z Zellforsch 136:521–533
3. Ausprunk DH, Knighton DR, Folkman J (1974) Differentiation of vascular endothelium in the chick chorioallantois: a structural and autoradiographic study. Dev Biol 38:237–248
4. Patan S, Haenni B, Burri PH (1993) Evidence for intussusceptive microvascular growth in the chicken chorio-allantoic membrane (CAM). Anat Embryol 187:121–130
5. Ribatti D, Nico B, Vacca A et al (2006) The gelatin sponge-chorioallantoic membrane assay. Nature Protoc 1:85–91
6. Knighton DR, Fiegel VD, Phillips GD (1991) The assay of angiogenesis. Prog Clin Biol Res 365:291–299

7. Jakob W, Jentzsch KD, Mauersberger B et al (1978) The chick embryo choriallantoic membrane as a bioassay for angiogenesis factors: reactions induced by carrier materials. Exp Pathol (Jena) 15:241–249
8. Ribatti D, De Falco G, Nico B et al (2003) In vivo time-course of the angiogenic response induced by multiple myeloma plasma cells in the chick embryo chorioallantoic membrane. J Anat 203:323–328
9. Spinazzi R, Albertin G, Nico B et al (2006) Urotensin-II and its receptor (UT-R) are expressed in rat brain endothelial cells, and urotensin-II via UT-R stimulates angiogenesis *in vivo* and *in vitro*. Int J Mol Med 18:1107–1112

Chapter 13
Dorsal Air Sac Assay

Ben K. Seon

Abbreviations Used

DASA	Dorsal air sac assay
VEGF	Vascular endothelial growth factor
mAb	Monoclonal antibody

13.1 Introduction

The DASA was developed by researchers in Japan [1]. Subsequently it has been used by many Japanese researchers [2–9] and others [10–12] to measure antiangiogenic activities of various substances that include a bacterial polysaccharide, an anti-human VEGF monoclonal antibody (mAb), irsogladine (an anti-gastric ulcer agent), cytogenin (a microbial product), TNP-470 (a semisynthetic analogue of fumagillin), 5-fluorouracil (5FU), CS-706 (a cyclooxygenase-2 inhibitor), an anti-human endoglin mAb and its conjugate with deglycosylated ricin A chain, and a tissue inhibitor of metalloproteinase 2. In addition, many other substances were tested for their antiangiogenic activities in the DASA but not mentioned in this review. In some cases, selected substances were measured for their angiogenesis-inducing activities in the DASA [13, 14].

B.K. Seon (✉)
Department of Immunology, Roswell Park Cancer Institute, Buffalo, NY, USA
e-mail: Ben.seon@roswellpark.org

M. Slevin, G. McDowell (eds.), *Handbook of Vascular Biology Techniques*,
DOI 10.1007/978-94-017-9716-0_13

13.2 Methodology

An example of the DASA is described to measure tumor cell-induced angiogenesis and to measure antiangiogenic activity of selected substances.

Tumor cells (e.g., 1×10^7 HT1080 cells in 0.2 ml PBS) or control media (PBS) is placed in a diffusion chamber (e.g., a Millipore chamber, 14 mm in diameter) that is sealed at each chamber end with a cellulose membrane filter (0.45-μm pore size). A chamber, so prepared, is implanted into a dorsal air sac of each of several BALB/c mice (either female or male mice). The dorsal air sac is generated in anesthetized mice by lifting the dorsal skin and injecting air (8–10 ml) under the skin. The chamber is implanted through a transverse section cut on the back followed by suturing. These mice are divided into several groups. One group (three mice per group) is a positive control for angiogenesis while another group is a negative control with a chamber containing control media. Additional groups of mice bearing a chamber containing tumor cells are treated by systemic (e.g., i.v. or p.o.) administration of test materials (e.g., anti-endoglin mAb) 24, 48 and 72 h post chamber implantation. The mice are sacrificed on day 4 or 7 and blood vessels in the excised skin are examined under a stereoscopic microscope and photographed. The newly formed blood vessels on the photograph can be counted or quantified with an image processor (e.g., Nexus Inc., Tokyo).

13.3 Troubleshooting

Tumor cells that release angiogenic factors may be used to induce angiogenesis in the DASA. We used HT1080 human fibrosarcoma cell line [10] and MCF-7 human breast cancer cell line [11] for this purpose. Other investigators used different tumor cells that include HepG2 human hepatic cancer cell line [3], RENCA murine renal cell carcinoma cell line [5] and M5076 murine ovarian ascites tumor cells [1].

Some of the newly generated blood vessels in the DASA are tortuous and can be readily distinguished from the preexisting straight blood vessels [3, 5, 10, 11].

13.4 Discussion and Conclusion

The DASA is a relatively simple assay. Nevertheless, the first time users need to perform ample practice tests and to define optimal conditions for their particular need to obtain reliable data. Once these requirements are met, the DASA is a convenient assay to measure angiogenic or antiangiogenic activities of substances of interest.

References

1. Tanaka NG, Sakamoto N, Inoue K et al (1989) Antitumor effects of an antiangiogenic polysaccharide from an Arthrobacter species with or without a steroid. Cancer Res 49(23):6727–6730
2. Asano M, Yukita A, Matsumoto T, Kondo S, Suzuki H (1995) Inhibition of tumor growth and metastasis by an immunoneutralizing monoclonal antibody to human vascular endothelial growth factor/vascular permeability factor121. Cancer Res 55(22):5296–5301
3. Ono M, Kawahara N, Goto D et al (1996) Inhibition of tumor growth and neovascularization by an anti-gastric ulcer agent, irsogladine. Cancer Res 56(7):1512–1516
4. Oikawa T, Sasaki M, Inose M et al (1997) Effects of cytogenin, a novel microbial product, on embryonic and tumor cell-induced angiogenic responses in vivo. Anticancer Res 17(3C):1881–1886
5. Yonekura K, Basaki Y, Chikahisa L et al (1999) UFT and its metabolites inhibit the angiogenesis induced by murine renal cell carcinoma, as determined by a dorsal air sac assay in mice. Clin Cancer Res 5(8):2185–2191
6. Ogawa H, Sato Y, Kondo M et al (2000) Combined treatment with TNP-470 and 5-fluorouracil effectively inhibits growth of murine colon cancer cells in vitro and liver metastasis in vivo. Oncol Rep 7(3):467–472
7. Senzaki M, Ishida S, Yada A et al (2008) CS-706, a novel cyclooxygenase-2 selective inhibitor, prolonged the survival of tumor-bearing mice when treated alone or in combination with anti-tumor chemotherapeutic agents. Int J Cancer 122(6):1384–1390
8. Matsumoto K, Obara N, Ema M et al (2009) Antitumor effects of 2-oxoglutarate through inhibition of angiogenesis in a murine tumor model. Cancer Sci 100(9):1639–1647
9. Murakami Y, Watari K, Shibata T et al (2013) N-myc downstream-regulated gene 1 promotes tumor inflammatory angiogenesis through JNK activation and autocrine loop of interleukin-1alpha by human gastric cancer cells. J Biol Chem 288(35):25025–25037
10. Seon BK, Matsuno F, Haruta Y, Kondo M, Barcos M (1997) Long-lasting complete inhibition of human solid tumors in SCID mice by targeting endothelial cells of tumor vasculature with antihuman endoglin immunotoxin. Clin Cancer Res 3(7):1031–1044
11. Matsuno F, Haruta Y, Kondo M, Tsai H, Barcos M, Seon BK (1999) Induction of lasting complete regression of preformed distinct solid tumors by targeting the tumor vasculature using two new anti-endoglin monoclonal antibodies. Clin Cancer Res 5(2):371–382
12. Chetty C, Lakka SS, Bhoopathi P, Kunigal S, Geiss R, Rao JS (2008) Tissue inhibitor of metalloproteinase 3 suppresses tumor angiogenesis in matrix metalloproteinase 2-down-regulated lung cancer. Cancer Res 68(12):4736–4745
13. Takahashi F, Akutagawa S, Fukumoto H et al (2002) Osteopontin induces angiogenesis of murine neuroblastoma cells in mice. Int J Cancer 98(5):707–712
14. Hamada Y, Yuki K, Okazaki M et al (2004) Osteopontin-derived peptide SVVYGLR induces angiogenesis in vivo. Dent Mater J 23(4):650–655

Chapter 14
Scanning Electron Microscopy of Blood Vascular Corrosion Casts in Mammals

Guido Macchiarelli, Maria Grazia Palmerini, and Stefania Annarita Nottola

14.1 Introduction

Scanning Electron Microscopy applied to observe Vascular Corrosion Casts (SEM of VCC) is a technique introduced in 1971 by Takuro Murakami, who firstly successfully injected prepolymerized methyl methacrylate resin to cast small vessels, including capillaries, thank to a low resin viscosity [1]. SEM of VCC was primarily applied to perform qualitative studies on the fine distribution of blood and lymphatic vessels, since it provides *quasi* three-dimensional (3D) models of the vasculature bed (Fig. 14.1). Thank to progresses in morphometry, stereology and computer hardware/software, this technique has been continuously upgraded. Today, it is widely considered one of the best method to quantify important variables of the blood and lymphatic vessel system. Vessel diameter, length, circumference, interbranching and intervascular distance, branching and interbranching angle, surface, volume and geometry of cast structures may be precisely calculated [2–7].

Latest advances are represented by X-ray Micro-Computed Tomography (Micro-CT) methods, employed for 3D analyses of the vascular corrosion casts in physiological and pathological conditions [8]. Recently, to enhance micro-CT imaging of corrosion casts of fine vasculature, the osmium tetroxide coating of methyl methacrylate vascular corrosion casts was used to increase resolution [9].

A huge literature has been produced in the last 40 years since Murakami's paper has been published, indeed decreeing the undiscussed scientific success of this method.

G. Macchiarelli (✉) • M.G. Palmerini
Department of Life, Health and Environmental Sciences, University of L'Aquila, L'Aquila, Italy
e-mail: gmacchiarelli@univaq.it

S.A. Nottola
Department of Anatomy, Histology, Forensic Medicine and Orthopaedics, Laboratory of Electron Microscopy "Pietro M. Motta", University La Sapienza, Rome, Italy

M. Slevin, G. McDowell (eds.), *Handbook of Vascular Biology Techniques*, DOI 10.1007/978-94-017-9716-0_14

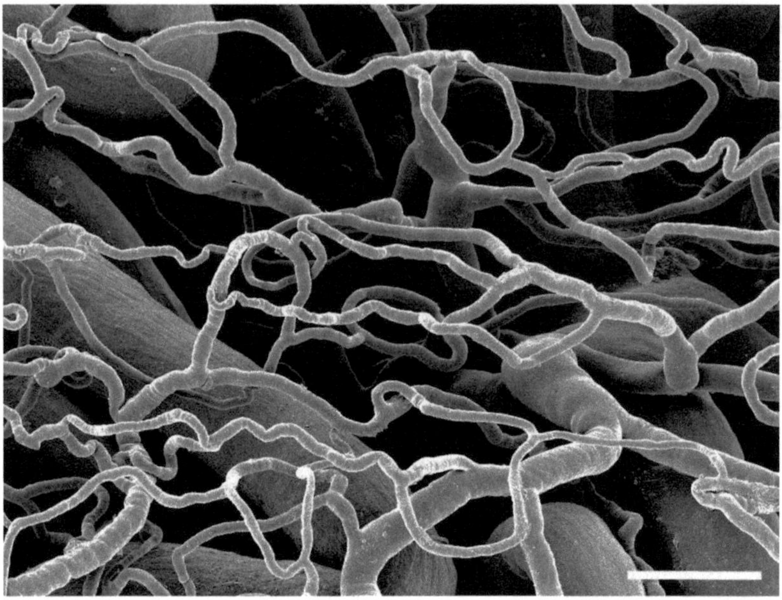

Fig. 14.1 Representative image obtained by SEM of VCC, showing the *quasi* tree-dimensional arrangement of a capillary network in physiological conditions (Mouse. Bar: 200 μm)

14.2 Methods

14.2.1 Pre-casting Procedures

Prior to casting laboratory animals, several procedures should be carefully applied in order to ensure an appropriate and regular anaesthesia and avoiding incomplete filling of the vascular network. It is fundamental avoiding blood clotting, vasoconstriction or excessive vasodilation, and allowing a smooth casting medium flow.

14.2.1.1 Premedication

In sensible animals, subjected to systemic vasoconstriction related to stress induced by animal handling, low dosage benzodiazepine may be used. Vasoactive drugs or spasmolytic agents are added to the solutions to prevent vasospasm and/or blood vessel compression by contraction of the surrounding skeletal muscles [10]. However, it should be considered that vasoactive drugs may change local hemodynamics. Animal heparinization may be performed prior to casting to prevent blood clotting. Pre-fixation with low concentration of fixatives, such as 0.5–1.0 % glutaraldehyde, is sometime attempted to replicate the endothelial surface structures [10].

14.2.1.2 Anaesthesia

Anaesthesia should be performed in authorized animal facilities, according to the Good Laboratory Practice (GLP). In animals, anaesthesia can occur by inhalation and injection. The former is becoming less frequent, and reserved to small animals. Differently, injection is the most frequently applied method [11]. In rats, it was used pentobarbital sodium (Abbot Laboratories, Illinois). In rabbit, we used intravenous injection of Nembutal (Abbot U.S.A.) (50 mg/Kg body weight) in the external ear vein [12].

14.2.1.3 Cannulation

Complete wash out of blood is necessary to inject the casting medium into the blood vascular system as well as to obtain endothelial cell impressions onto the cast surface [1, 10]. Cannulation can occur close or far to the target organ or tissue. For practical purposes, the rule is to inject through an artery, as large as possible, and close to the target. However, since cannulation close to the target organ can result difficult or troublesome, the casting medium is generally injected via the left ventricle, the aorta, or large arterial branches [10, 13]. In any case, the preparation of the cannulating vessels deserves a special care, paying attention to avoid any stretching and rupture of the surrounding vascular bed. Heparin may be added to the rinsing solutions to prevent blood clotting [14].

Cannulation Site in Small Animals

Small sized animals, such as mice and rats are often cannulated directly through the left ventricle [15] or the ascending aorta, together with the right atrium opened to enable an appropriate drainage [11].

Cannulation Site in Medium Animals

Medium sized animals are firstly subjected to laparotomy and median sternotomy, by opening the body along the thoracic-abdominal plane. This procedure allow an easy cannulation of major vessels and at the some tome make easier the blood flow through all body systems. Then, cannulation preferably occurs through the thoracic aorta, as made in rabbits [14].

Cannulation Site in Large Animals

In large animals, cannulation occurs into a large artery close the target. This is mainly consequence of the difficulties in reaching the location of specific tissues or organs and also to avoid casting media waste. In any case, it should be taken into account that the presence of vascular anastomosis can flow away perfusates, thus impairing the cast outcome. It is, indeed, recommended to cannulate upstream the anastomotic site/s. In pig, cow and sheep, ovaries were perfused with heparinised saline solution through the ovarian artery [16–18].

14.2.2 Rinsing/Perfusion

Perfusion serves to flush the blood circulatory system with rinsing solution (such as sodium chloride solution, Ringer solution, Tyrode solution, phosphate buffered saline (PBS), Dulbecco phosphate buffered saline) in order to remove any obstacles that might obstruct the casting media flow. After having placed the injection cannula securely, a first rinse is done, closely watching the cut edges of the tissue samples to detect immediate leakage of injected solution. By ligating successively all recognizable cut end vessels, possibly under binocular control, with suture silk leakage, the pressure needed for perfusion will rise. If all major cut arterial vessels are ligated, rinsing and casting can be done properly [11].

14.2.2.1 Perfusion Pressure

Perfusion pressure should be equal to the mean arterial blood pressure at the site of injection; in casting the peripheral vascular territories, the pressure in the target organ is more important than that recorded at the injection site. The law of Hegen-Poiseuille states that a high-viscosity medium flows slowly in a long and wide pathway. Thus, the use of a short, narrow and thin-walled cannula, as well as a low viscosity casting medium, is recommended in each casting [11].

14.2.3 Casting Media

Casting media are supplemented with catalysts to accelerate their polymerization in vessels prior to injection. The casting medium have to be non toxic to the investigator; to cause neither morphological nor physiological changes to the tissues or their vessels; to be of sufficient low viscosity or particle size to pass through capillaries; to show no marked leakage into the tissues or their spaces; to polymerize within 3–15 min; to replicate fine vascular connections as well as delicate luminal and endothelial structures; to show no marked shrinkage during curing (solidifying) or hardening; to permit microdissection with intact surrounding tissue; to be resistant to corrosion procedures; to be visible in the dissection light microscope; to retain structural configuration during drying; to be suitably for microdissection after drying; to be electron conductive or show no marked structural changes during conductive treatment; to be resistant to electron beam bombardment; to produce well-highlighted and well-contrasted SEM images; and to be useful for quantitative analysis [10, 13]. There is a broad spectrum of acrylic resins (methyl methacrylate mixture introduced by [1]), laboratory-prepared methacrylate mixtures (Murakami's mixture) or commercially available resins, such as Mercox® CL (Okenshoji, Tokyo, Japan and Dainippon Ink and Chemicals, Tokyo, Japan), Batson's No. 17 plastic, araldite, CY 223 and tardoplast [19]. More recently, a new polyurethane-based casting resin (PU4ii) has been successfully used to produce form- and shape-retaining, resilient yet elastic casts and molds of any size, with minimal shrinkage and almost unlimited possibilities for additional processing, analysis, and storage [20]. Mercox®-CL is a blue (2B), red (2R) or yellow (2Y) colored prepolymerized casting medium. It is supplied in sets comprising 500 ml prepolymerized Mercox®-CL and a vial containing 100 g of wetted paste MA (initiator, catalyst). Because of its toxicity any handling of Mercox® has to be done under the fume hood wearing gloves to prevent contact with the skin. As soon as both components are mixed by gentle stirring, the polymerization starts immediately [11, 13, 21].

14.2.4 Injection of Casting Medium

The ready-to-use mixture is then sucked up either directly via the syringe conus or via a short plastic tubing from the beaker glass into a hand-controlled disposable syringe or a syringe connected to a perfusion apparatus with a flow meter or manometer, until the resin leaking out from the effluent vein(s) become virtually water-free [11, 22]. The use of the injection apparatus allows monitoring of the pressure of compressed air that drives the resin into the vasculature. Depending on the medium, the working life ranges from 2 to 60 min. Cast medium coloration indicates filling conditions in the target organs, although it does not guarantee complete filling [10]. It may be useful, just prior to inject casting medium, to add to the cleaning solution some drops of blu metilene or ink solutions, to monitor if the district under investigation is actually well perfused.

14.2.4.1 Vascular Corrosion Casts of Tissues and Organs

Isolated organs and surgically excised tissues (biopsies) should be immediately submerged into warm heparinised physiological solution and be transferred to the laboratory for further washing and other processing. When isolating organs, arterial feeders can be marked by ligatures thus making subsequent cannulation much easier. In excised tissues these feeders have to be dissected painstakingly under the dissecting microscope [11]. Normally, blood drainage occurs via cut vessels but in any case vessel patency should be carefully checked.

14.2.4.2 Vascular Corrosion Casts of Whole Animals

Casting of the entire circulatory system of animals for subsequent scanning electron microscopy evaluation is normally performed in small animals. Precasting procedures are similar to what above described for organs and tissues. To allow drainage of blood, surplus perfusates and injection medium either the caudal vena cava or the left atrium are opened simultaneously with the onset of flushing [11]. A wide exposure of the entire visceral organs is recommended in order to facilitate the perfusion of all districts.

14.2.5 Polymerization

After casting, the injected specimen has to be left undisturbed for a certain minimal period of time in order to allow the resin to polymerize. With Mercox®, the injected specimen is left untouched for a minimum of 30 min at room temperature. Once the viscosity of the injected casting medium has increased, the specimen is generally placed into hot water for tempering (60 °C for up to 12 h) [10]. In our experience, after Mercox® resin injection, casted ovaries were placed in warm tap water for 3–4 h until polymerization was completed [12, 13, 23–28].

14.2.6 Corrosion

Maceration (corrosion) removes tissues around cast systems and allows the cast to be exposed. Maceration of the tissue can be accomplished by immersion the resin-injected tissues into solutions of alkali or acid (i.e. 10–20 % NaOH, KOH or HCl). Often, the resin-injected tissues are placed in warm distilled water (60 °C) for several hours to 1–2 days before corrosive maceration. This starts tissue disruption, increase the efficacy of corrosive agents, and reduces the corrosion time considerably. The time required to macerate tissues varies from several hours to several days, depending on the size of the resin-injected tissues and the corrosive agents used [11, 13]. Casted ovaries were corroded in 10 % NaOH at 60 °C per 24–48 h [12, 23–28].

14.2.7 Cleaning

Between and following maceration, the cast is washed in gently running warm tap water (40–60 °C) for several hours for a day or two to dissolve saponified materials caused by maceration [13]. Finally, deionated or even distilled water should be used to prevent any specimen contamination [11]. To obtain high quality micrographs it is, in fact, very important to keep clean the specimen before conductive coating for SEM observation. Often, after cleaning, a further removal of debris is performed manually, with appropriate instruments and under stereoscopic magnification.

14.2.8 Drying

After cleaning, the cast should be totally dried, before conductive coating for SEM observations. Air drying may be performed in a oven (60 °C) for several hours or overnight. Critical point-drying and freeze-drying are also performed, the latter yields better results, because the deep-frozen in distilled water better preserve the three-dimensional structure of the microvascular network [26]. An example of ovarian cast after drying is shown in Fig. 14.2.

14.2.9 Specimen Dissection, Mounting and SEM Observation

To study the microvascular network, microdissection could be necessary to expose the structures of interest. Microdissection could be performed before or after mounting with fine tweezers and needles under a binocular microscope. Specimens may

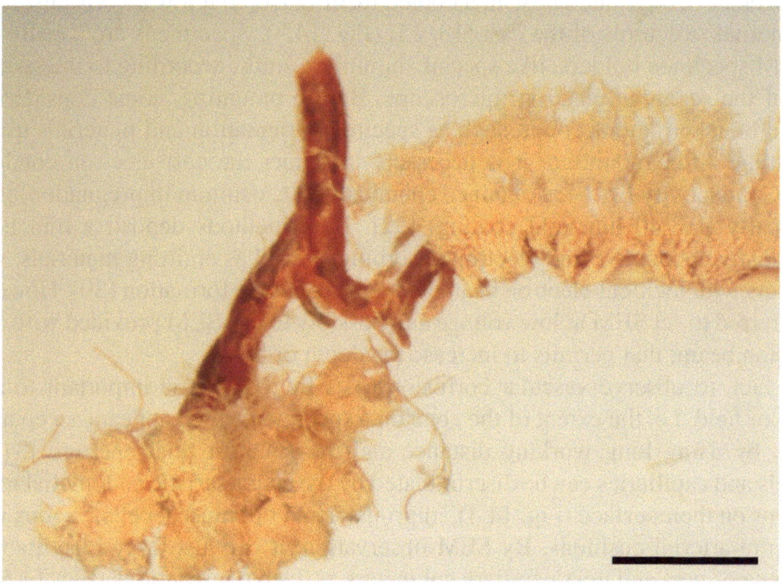

Fig. 14.2 Representative micrographs of mouse ovary and fallopian tube casted with Mercox® CL resin and observed before mounting by stereomicroscopy (Bar: 2 mm)

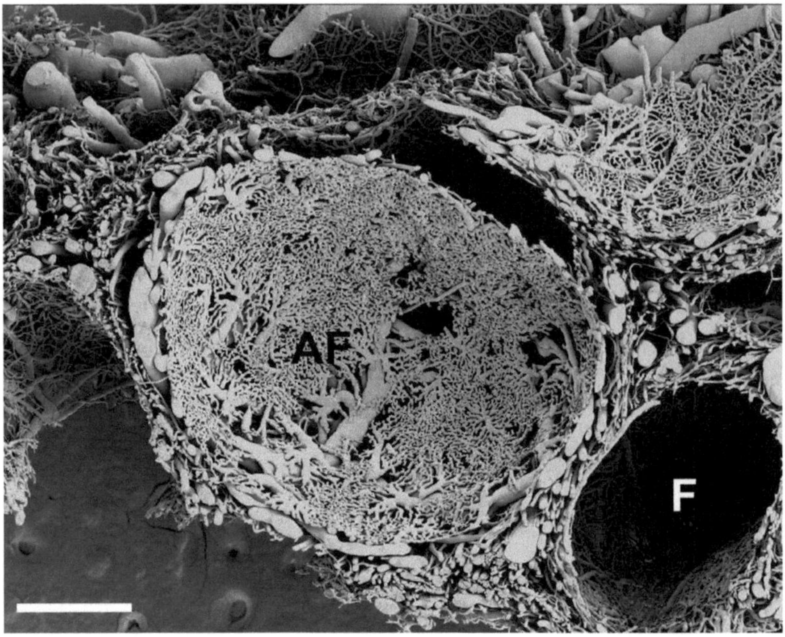

Fig. 14.3 View of a freeze-fractured vascular corrosion cast of a pig ovary, showing antral (*AF*) and preantral follicles (*F*). SEM of VCC (Modified from [17]. Bar: 700 μm)

be frozen at −18 °C and cut with a cooled razor blade, to allow the visualization of the internal structures of the cast [13, 25] (Fig. 14.3). Specimens are usually glued to SEM specimen holders, like special aluminium stubs, according to the specification of the scanning electron microscope. Before mounting, some considerations should be taken into account, such as specimen orientation and materials used for mounting. After mounting, it is necessary to render the cast electron conductive [10]. Conductivity has been obtained chemically (i.e. osmium impregnation) [29] or physically (i.e. 40 nm gold coating) [22]. Both methods deposit a thin film of electrically conductive and electron absorbing as well as emitting materials, which interacts with incident electron beam and enables image formation [30]. However it is preferred to set SEM at low voltage and possibly to use SEM provided with a field emission beam, that permits to increase the depth of field.

In fact, to observe vascular corrosion casts by SEM, it is important to adjust depth of field, i.e. the extent of the zone on a specimen which appears acceptably in focus, by using long working distance and the smallest final aperture [31, 32]. Vessels and capillaries can be discriminated by the characteristic endothelial imprint patterns on their surface (Fig. 14.4), imprints of smooth muscle cells, venous valves and intra-arterial cushions. By SEM observation, it can also easy identifiable the presence of pre- and post-capillary sphincters as deep constrictions (Fig. 14.5).

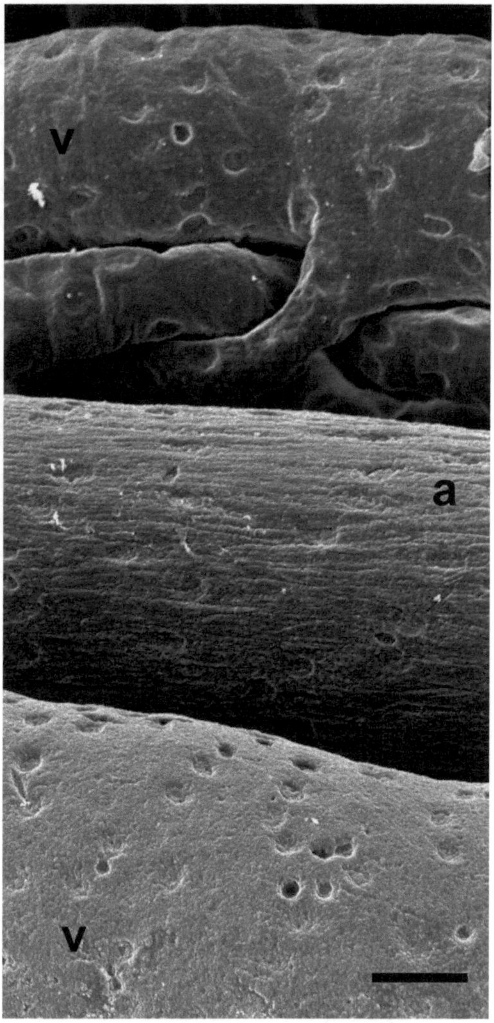

Fig. 14.4 Endothelial nuclear imprints left on the vessel surface allow to distinguish between arterial and venous vessels. Veins (*v*) present circular imprints of endothelial cell nuclei, in contrast to the longish arterial ones (*a*). SEM of VCC (Bar: 40 μm)

14.3 Recent Advances of Vascular Corrosion Casts in Major Mammalians Organs and Systems

14.3.1 Heart and Great Vessels

SEM of vascular corrosion casts have been applied to study the microvascular architecture of great vessels in several mammals. In mouse, colored plastic solutions were injected directly into the heart to visualize the fetal vasculature [33]. Fetoplacental vascularization has also been studied in murine model of eNOS deficiency to study fetal growth restriction [7]. In rabbit, the *vasa vasorum* of carotid

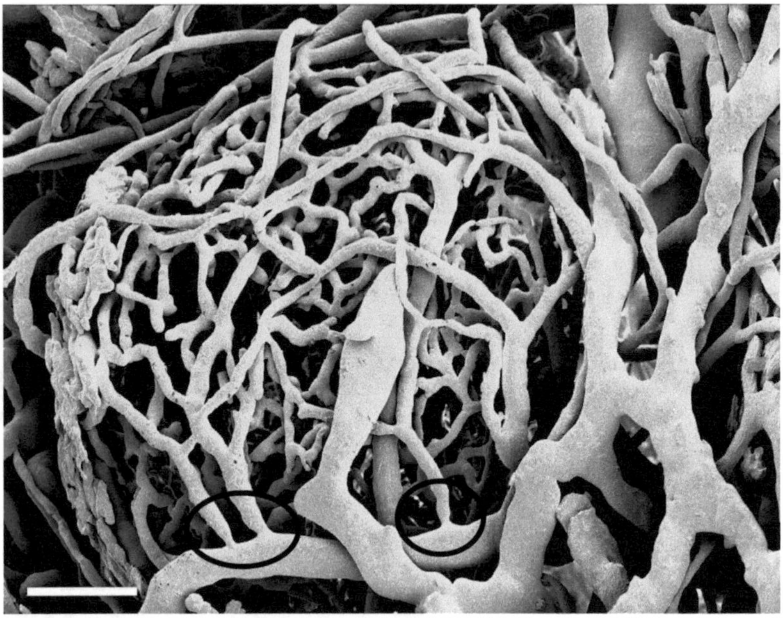

Fig. 14.5 Several roundish cortical vascular plexuses from a ovary showing common arteriolar (*a*) supply and venular (*v*) drainage (SEM of VCC. Bar: 20 μm)

artery were casted with a methylmethacrylate casting compound to give an overview to of the amount of neovascularization [34], while the entire human *vasa vasorum* circulatory bed was casted with polymerizing resin Mercox®-Cl-2B from the feeding arterial stem vessels throughout the capillaries to the draining venous *vasa vasorum* [35]. VCC, other than ensure a morphological evaluation of animal vascular network models in physiological and pathological conditions, can also be applied to define the biological response of gross vein wall to hemodynamic changes, as studied by [36] in the portal vein in a model of portal hypertensive and cirrhotic rodents.

14.3.2 Respiratory System

Changes in the number of pulmonary vessels with postnatal growth was studied in rats by means of a technique involving arterial casting with Microfil silicone polymer, high-resolution micro-computed tomography (micro-CT) and image data analysis [6]. SEM of VCC of bronchial artery casts, demonstrated in a murine model of lung inflammation, the presence of the peripheral bronchial circulation, interconnections of the two systems in the distal bronchial arteries and at the level of alveolar capillaries, and functional evidence of increased bronchial perfusion of alveolar capillaries during mononuclear inflammation. The AA suggested an important adaptive role of the bronchial circulation in pulmonary inflammation [37]. In Baird's beaked whales the alveolar microvasculature was successfully visualized by SEM of VCC, giving information on the structure of alveoli septa, made by partially interconnected capillary networks [38]. Casts of the entire tracheal-bronchial tree and casts of vascular kidney from different animals were prepared by injection of polyurethane, retaining sufficiently fine details [39].

14.3.3 Digestive System

In a chemically-induced model of murine colitis, morphometric changes -such as vessel diameter, interbranch distance, intervascular distance- were studied after VCC, concluding that inflammation-associated intussusceptive angiogenesis was associated with vessel angle remodelling and suggesting the influence of blood flow on the location and orientation of remodeled vessels [40]. Marked microvascular changes were observed in a rat model of steatosis-related disease, as demonstrated by a dense and markedly disorganized pattern with loss of the classical trabecular ordering in the liver, also associated to the presence of blebs [41]. The spatial arrangement of the colon microcirculatory bed was described in humans cadavers or biopsies, after methylmetacrylate Mercox® resin for microdissection and corrosion casting for scanning electron microscopy [42]. The gastrocolic trunk of Henle was casted trough the superior mesenteric vein and artery in cadavers with cold polymerized methylacrylate, thus allowing to define unknown parameters as caliber, length and three-dimensional position [43].

14.3.4 Urinary System

Rat kidneys were analyzed in a multidisciplinary approach involving three distinct SEM methods: coronal sections, cryofracturing and vascular corrosion casts/angioarchitecture. Among the specificity of each technique, vascular corrosion casts with Mercox® CI-2 Red showed the organization and arrangement of the angioarchitecture. Capillaries organized throughout the cortical and medullary regions of the kidney formed a dense network containing various types of anastomoses [44]. In mouse, the urinary bladder was casted to study its special features, such as vessel coiling, abundant collateral circulation, arterial sphincters, and a dense mucosal capillary plexus [45]. An entire mouse kidney was casted and scanned with a modular multiresolution X-ray nanotomography system, that ensured a high-resolving power of this system, a large field of view and the three-dimensional nature of the resulting image [46].

14.3.5 Locomotor System

The arterial arrangement of the cervical spinal cord was studied in detail in rabbit after injection of Batson no. 17 resin, to examine the pathophysiology of spinal cord injury [47]. Others investigated the microvascular system of anterior cruciate ligament (ACL) using dogs, to study the microvascular architecture and the status of the barrier function of the capillary wall [48].

14.3.6 Nervous System

Microvascular changes that occur during the first 12 h after traumatic brain injury (TBI) were analyzed using SEM of VCC [49]. Results obtained confirmed how this was a reliable method for studying the pathophysiology of the vascular alterations occurring at acute and subacute stages after brain injury. It was also possible to obtain a topographical localization of the vascular and cellular events that usually lead to hyperemia, edema, and brain swelling. Moreover, by applying informatic software to anatomical images, quantification and statistical analysis of the observed events were performed [48]. SEM of VCC allowed to observe a higher density of vessels in the lumbar dorsal root ganglion than the nerve root in rats and the placement of blood flow control structures (ring-shaped constrictions) in both the arterial supply and the capillary network [50].

14.3.7 Sensory Organs and Lymphatic System

The vasculature of the hemal node (HN) from the bovine cervical region was investigated using a combination of vascular corrosion casting and scanning electron microscopy [51]. A dense vessel network of the capsule was found surrounding the cast of HN parenchyma and had no connection with the subcapsular sinus; these vessels converged and exited the HN via the hilar vein. In rats, the angioarchitecture of the submandibular and sublingual glands and lymph nodes was studied by means of corrosion casts. Scanning electron microscopy showed that the three structures have distinct vascular patterns [52]. The general structure of the lymphatic vessels and the organization of the lymphatic system was systematically studied also by SEM of VCC [53]. The ventral surface of the rat tongue presented simple, even and abundant vasculature made by a vascular plexus consisting of a superficial vascular network and by the ranine veins. The superficial vascular network, made up of the ascending and descending branches, presents as a loose network, with little morphological variation between the capillary loops [54]. In rabbit, SEM of VCC evidenced an elaborate microvasculature of the conjunctiva, probably prerequisite for the exchange of nutrients and gasses between the cornea and the vessels across the conjunctival epithelium when the eyelids are shut during sleep, and possibly for the dynamics of eye drop delivery [38].

14.3.8 Endocrine System

Morphological and microvascular alterations of adrenal glands were observed in streptozotocin (STZ)-induced long-term diabetic rats, to show dilated luminal diameters of capillaries in the diabetic group, that were also arranged in tortuous course [55].

14.3.9 Reproductive System

The entire vascular system of the human cervical wall was casted in human uteri collected at autopsy, especially the vasculature of the endocervical mucosa [56]. Authors found an extremely high density of endocervical mucosal capillaries and the presence of small veins located in the direct vicinity of the endocervical canal, thus suggesting bidirectional draining of endocervical mucosal capillary plexus: centrifugal, by the venules/veins located in the middle and peripheral zones of the cervical wall and centripetal, by the subepithelial veins. In another study, the vascular architecture of the vaginal and supravaginal parts of the human uterine cervix was studied by SEM of VCC. This study introduced the idea of two systems responsible for draining blood from the mucosal capillaries. It also suggested the possible existence of a countercurrent transport between adjoining veins and arteries [57]. Ovaries were extensively studied by our group in different mammals by SEM of VCC and correlated techniques (reviewed in [26]). In rats, rabbits, pigs and cows specie-specific differences in vascular density, presence and type of vascular plexuses (Fig. 14.6), numbers and disposition of vascular layers, vessel caliber, were found. Capillary remodelling was demonstrated during both follicle and corpus luteum development, with phases of angiogenesis (Fig. 14.7a) followed by vessel maturation and a period of physiological angioregression (Fig. 14.7b) during luteolysis [12, 23–28].

14.4 Angiogenesis

Angiogenesis, the growth of new blood vessels, occurs (i) physiologically, during tissue development, growth and repair and (ii) pathologically, when new vessels grow excessively (such as in malignant or benign tumors, liver cirrhosis, fibroses, rheumatoid arthritis, psoriasis, or diabetic retinopathies) or insufficiently (coronary artery disease, stroke, delayed tissue healing and varicosity) ([25] and references herein cited).

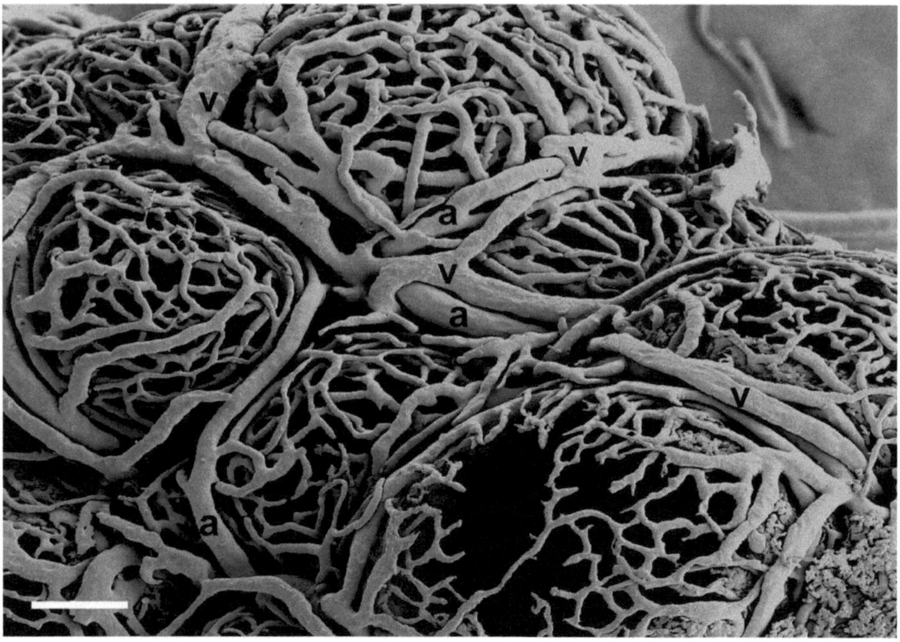

Fig. 14.6 Post-capillary sphincters in capillaries connecting to venular vessels are a sign of vasoconstriction. SEM of VCC (Modified from [17]. Bar: 50 μm)

Since angiogenesis is a morphologically dynamic phenomenon, evolving in a three-dimensional pattern, SEM of VCC was considered a technique of choice to study subtle ultrastructural details, and to describe different structural conformations [18].

In adults, only the female genital tracts have organs able to display a cyclic, exuberant and physiological angiogenesis. The uterus, placenta, and ovary, in fact, shown a rapid formation of new vessels under physiological stimuli throughout the reproductive life. An equivalent rapid regression and disruption of the newly formed vessels regularly follows such a considerable angiogenesis. These aspects made the female genital tract a unique experimental model for the study of formation and regression of blood vessels ([25] and references herein cited) [58, 59] (Fig. 14.7b). Two main types of angiogenesis are known: sprouting and non-sprouting. In sprouting angiogenesis, new capillaries generally form by outward growth of endothelial cells from pre-existing vessels. Sprouting is the most represented process of vascular growth in adults (Fig. 14.7a). Non-sprouting angiogenesis is an alternative method of vessel formation from pre-existing vessels (reviewed in [60]) (Fig. 14.7a). It consists in splitting of capillaries inward from their vessels of origin (intussusception i.e. "growth within itself"). Non-sprouting angiogenesis occurs with low endothelial cell proliferation, is achieved at low vascular permeability levels, requires only 4–5 h for completion and is the optimal physiological solution to sustain a rapid angiogenesis during the development of ovarian endocrine functions [60].

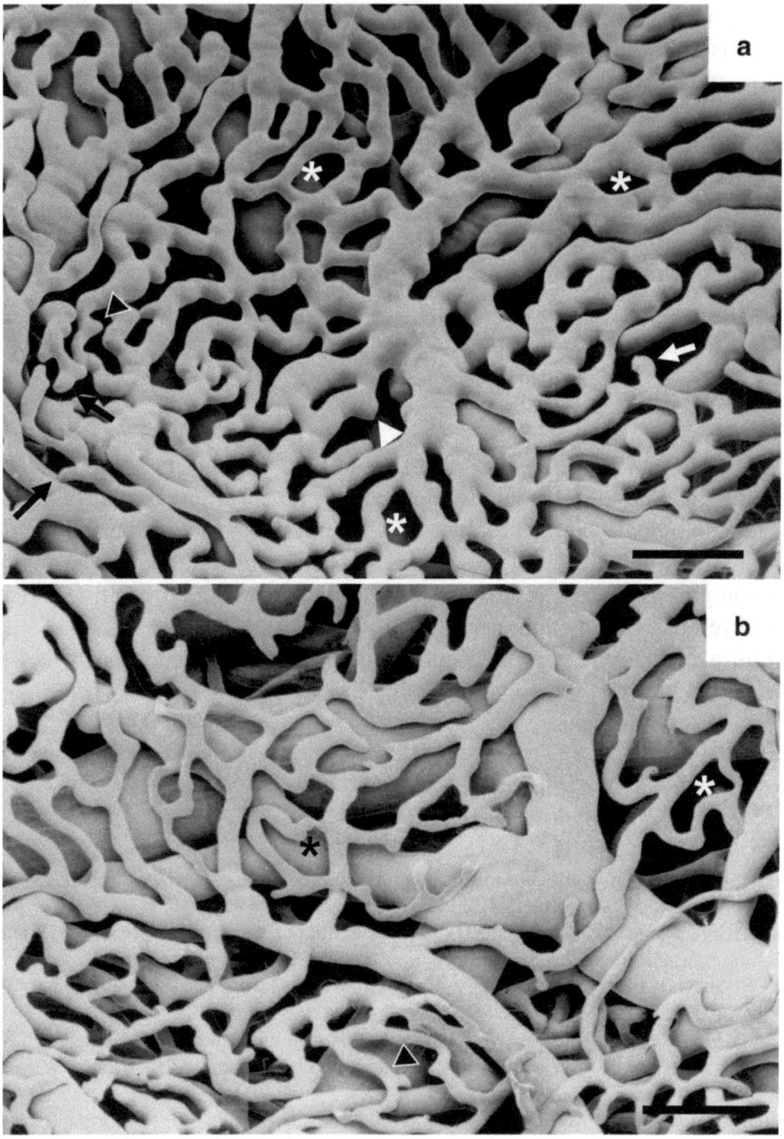

Fig. 14.7 Angiogenic (**a**) and angioregressive (**b**) vascular networks from rabbit preovulatory follicles. In (**b**), angioregressive, degenerative characteristics are represented by thin or incompletely filled capillaries. *Asterisks* intussusception, *arrowheads* budding, *arrows* sprouting (SEM of VCC. Bar: 20 μm)

In the ovary, the initial rapid proliferation of granulosa and thecal cells asks for the cooptation of interstitial vessel and a quick subsequent sprouting angiogenesis for providing nutrients and extraovarian endocrine stimuli. As soon as maturing follicles differentiate towards endocrine function, by expressing the proper steroidogenic activity, intussusceptive angiogenesis is required in order to create a sinusoidal network that is able to sustain both nutritional and functional microcirculation. After ovulation, the growth of corpus luteum requires an intense angiogenesis, cyclically followed by angioregression at luteolysis. These processes are not only "structural", but also functional. In fact, the follicular and luteal microcirculation shows hemodynamic control activation at capillary and postcapillary level, as demonstrated by capillary activation, enlargement and sinusoid formation; pericyte proliferation, activation of venular sphincters (Fig. 14.5) [26]. All of these data clearly indicate the existence of mechanisms of up- and down-regulation of angiogenesis.

14.5 Conclusions

SEM of VCC is a technique of choice to study the vascular beds in physiological and pathological conditions, and to highlight organ-specific vascular patterns. The recent technological approaches are allowing not only to describe the microanatomy of capillary networks, but also to give important information on the angiogenic and angioregressive dynamics that sustain vascular growth, remodelling and regression in different experimental models.

A future interdisciplinary research, with a translational approach, is warmly encouraged to improve the potentiality of SEM of VCC.

References

1. Murakami T (1971) Application of the scanning electron microscope to the study of the fine distribution of the blood vessels. Arch Histol Jpn 32(5):445–454
2. Lametschwandtner A, Aharinejad SH (1997) Scanning electron microscopy/corrosion casting technique in biological and medical research. State of the art and perspectives. In: Motta PM (ed) Recent advances in microscopy of cells, tissues and organs. Antonio Delfino Editore, Rome, pp 51–58
3. Hossler FE, Douglas JE (2001) Vascular corrosion casting: review of advantages and limitations in the application of some simple quantitative methods. Microsc Microanal 7(3):253–264
4. Lametschwandtner A, Minnich B et al (2005) Analysis of microvascular trees by means of scanning electron microscopy of vascular casts and 3D-morphometry. Ital J Anat Embryol 110(2 Suppl 1):87–95
5. Mondy WL, Cameron D et al (2009) Micro-CT of corrosion casts for use in the computer-aided design of microvasculature. Tissue Eng Part C Methods 15(4):729–738
6. Razavi H, Dusch MN et al (2012) A method for quantitative characterization of growth in the 3-D structure of rat pulmonary arteries. Microvasc Res 83(2):146–153
7. Kulandavelu S, Whiteley KJ et al (2013) Endothelial NO synthase augments fetoplacental blood flow, placental vascularization, and fetal growth in mice. Hypertension 61(1):259–266
8. Folarin AA, Konerding MA et al (2010) Three-dimensional analysis of tumour vascular corrosion casts using stereoimaging and micro-computed tomography. Microvasc Res 80(1):89–98
9. Mondy WL, Casteleyn C et al (2013) Osmium tetroxide labeling of (poly)methyl methacrylate corrosion casts for enhancement of micro-CT microvascular imaging. Microsc Microanal 19(6):1416–1427

10. Lametschwandtner A, Lametschwandtner U (1992) Historical review and technical survey of vascular casting and scanning electron microscopy. In: Motta PM, Murakami T, Fujita H (eds) Scanning electron microscopy of vascular casts: methods and applications. Kluwer Academic, Boston, pp 1–11
11. Aharinejad SH, Lametschwandtner A (1992) Microvascular corrosion casting in scanning electron microscopy: techniques and applications. Springer, Wien
12. Macchiarelli G, Nottola SA et al (1993) Microvasculature of growing and atretic follicles in the rabbit ovary: a SEM study of corrosion casts. Arch Histol Cytol 56(1):1–12
13. Ohtani O, Murakami T (1992) Routine methods for vascular casting and SEM. In: Motta PM, Murakami T, Fujita H (eds) Scanning electron microscopy of vascular casts: methods and applications. Kluwer Academic, Boston, pp 13–25
14. Macchiarelli G, Nottola SA et al (1998) The microvasculature of the corpus luteum in pregnant rabbit. A scanning electron microscopy study of corrosion casts. Ital J Anat Embryol 103(4 Suppl 1):191–202
15. Jiang JY, Miyabayashi K et al (2008) Thyroxine treatment stimulated ovarian follicular angiogenesis in immature hypothyroid rats. Histol Histopathol 23(11):1387–1398
16. Jiang JY, Macchiarelli G et al (2002) Follicular microvasculature in the porcine ovary. Cell Tissue Res 310(1):93–101
17. Jiang JY, Macchiarelli G et al (2003) Capillary angiogenesis and degeneration in bovine ovarian antral follicles. Reproduction 125(2):211–223
18. Martelli A, Palmerini MG et al (2009) Blood vessel remodeling in pig ovarian follicles during the periovulatory period: an immunohistochemistry and SEM-corrosion casting study. Reprod Biol Endocrinol 7:72
19. Martín-Orti R, Stefanov M et al (1999) Effect of anticoagulation and lavage prior to casting of postmortem material with Mercox® and Batson 17. J Microsc 195(Pt 2):150–160
20. Krucker T, Lang A et al (2006) New polyurethane-based material for vascular corrosion casting with improved physical and imaging characteristics. Microsc Res Tech 69(2):138–147
21. Lametschwandtner A, Lametschwandtner U et al (1984) Scanning electron microscopy of vascular corrosion casts–technique and applications. Scan Electron Microsc 2:663–695
22. Christofferson RH, Nilsson BO (1988) Morphology of the endometrial microvasculature during early placentation in the rat. Cell Tissue Res 253(1):209–220
23. Macchiarelli G, Nottola SA et al (1995) Changes of ovarian microvasculature in hCG stimulated rabbits. A scanning electron microscopic study of corrosion casts. Ital J Anat Embryol 100(Suppl 1):469–477
24. Macchiarelli G (2000) The microvasculature of the ovary: a review by SEM of vascular corrosion casts. J Reprod Dev 46:207–225
25. Macchiarelli G, Jiang JY et al (2006) Morphological patterns of angiogenesis in ovarian follicle capillary networks. A scanning electron microscopy study of corrosion cast. Microsc Res Tech 69(6):459–468
26. Macchiarelli G, Nottola SA et al (2010) Morphological expression of angiogenesis in the mammalian ovary as seen by SEM of corrosion casts. Ital J Anat Embryol 115(1–2):109–114
27. Macchiarelli G, Palmerini MG et al (2013) Restoration of corpus luteum angiogenesis in immature hypothyroid rdw rats after thyroxine treatment: morphologic and molecular evidence. Theriogenology 79(1):116–126
28. Nottola SA, Macchiarelli G et al (1997) The angioarchitecture of estrous, pseudopregnant and pregnant rabbit ovary as seen by scanning electron microscopy of vascular corrosion casts. Cell Tissue Res 288(2):353–363
29. Murakami T (1975) Scanning electron microscopy of tannin-osmium treated and resin-embedded specimens. A demonstration of transmural passage of blood cells in the bone marrow. Arch Histol Jpn 38(3):229–235
30. Murphy JA (1982) Considerations, materials, and procedures for specimen mounting prior to scanning electron microscopic examination. Scan Electron Microsc 2:657–696

31. Hodde KC, Steeber DA et al (1990) Advances in corrosion casting methods. Scanning Microsc 4(3):693–704
32. Postek MT, Howard KS et al (1980) Scanning electron microscopy a student's handbook. Ladd Research Industries, Burlington
33. Liu C, Martin J (2010) Visualizing fetal mouse great blood vessels by plastic casting. Cold Spring Harb Protoc 2010(5):pdb-prot5433
34. Bayer IM, Caniggia I et al (2002) Experimental angiogenesis of arterial vasa vasorum. Cell Tissue Res 307(3):303–313
35. Lametschwandtner A, Minnich B et al (2004) Three-dimensional arrangement of the vasa vasorum in explanted segments of the aged human great saphenous vein: scanning electron microscopy and three-dimensional morphometry of vascular corrosion casts. Anat Rec A Discov Mol Cell Evol Biol 281(2):1372–1382
36. Van Steenkiste C, Trachet B et al (2010) Vascular corrosion casting: analyzing wall shear stress in the portal vein and vascular abnormalities in portal hypertensive and cirrhotic rodents. Lab Invest 90(11):1558–1572
37. Ravnic DJ, Konerding MA et al (2007) The murine bronchopulmonary microcirculation in hapten-induced inflammation. J Thorac Cardiovasc Surg 133(1):97–103
38. Ninomiya H, Inomata T et al (2008) Microvascular architecture of the rabbit eye: a scanning electron microscopic study of vascular corrosion casts. J Vet Med Sci 70(9):887–892
39. Viggiano D, Sangiorgi S, Reguzzoni M, Manelli A, Marano L, Dell'Orbo C, Passiatore C (2003) A new method to make vascular and bronchial casts of voluminous organs. Eur J Morphol 41(5):161–165
40. Ackermann M, Tsuda A et al (2013) Intussusceptive remodeling of vascular branch angles in chemically-induced murine colitis. Microvasc Res 87:75–82
41. Francque S, Laleman W et al (2012) Increased intrahepatic resistance in severe steatosis: endothelial dysfunction, vasoconstrictor overproduction and altered microvascular architecture. Lab Invest 92(10):1428–1439
42. Kachlik D, Baca V et al (2010) The spatial arrangement of the human large intestinal wall blood circulation. J Anat 216(3):335–343
43. Ignjatovic D, Stimec B et al (2004) Venous anatomy of the right colon: three-dimensional topographic mapping of the gastrocolic trunk of Henle. Tech Coloproctol 8(1):19–21, discussion 21–2
44. Bolina Cde S, Bolina-Matos Rde S et al (2013) Three-dimensional aspects of the structural characteristics and kidney angioarchitecture of adult and aged Wistar rats: a scanning electron microscopy study. Microsc Res Tech 76(5):538–544
45. Hossler FE, Lametschwandtner A et al (2013) Microvascular architecture of mouse urinary bladder described with vascular corrosion casting, light microscopy, SEM, and TEM. Microsc Microanal 19(6):1428–1435
46. Wagner R, Van Loo D et al (2011) High-resolution imaging of kidney vascular corrosion casts with Nano-CT. Microsc Microanal 17(2):215–219
47. Mazensky D, Danko J et al (2012) Arterial arrangement of the cervical spinal cord in rabbit. Anat Sci Int 87(3):155–159
48. Kobayashi S, Baba H et al (2006) Microvascular system of anterior cruciate ligament in dogs. J Orthop Res 24(7):1509–1520
49. Sangiorgi S, De Benedictis A et al (2013) Early-stage microvascular alterations of a new model of controlled cortical traumatic brain injury: 3D morphological analysis using scanning electron microscopy and corrosion casting. J Neurosurg 118(4):763–774
50. Kobayashi S, Mwaka ES et al (2010) Microvascular system of the lumbar dorsal root ganglia in rats. Part I: a 3D analysis with scanning electron microscopy of vascular corrosion casts. J Neurosurg Spine 12(2):197–202
51. Zhang W, Yasuda M et al (2013) Scanning electron microscopic study of the vascular system in the hemal node of the bovine cervical region. J Vet Med Sci 75(1):79–83

52. Rossi-Schneider TR, Verli FD et al (2008) Contribution to the study of the vasculature of submandibular and sublingual glands and lymph nodes of rats by corrosion cast technique combined with scanning electron microscopy. Microsc Res Tech 71(10):737–741
53. Ohtani O, Ohtani Y (2008) Organization and developmental aspects of lymphatic vessels. Arch Histol Cytol 71(1):1–22
54. Verli FD, Marinho SA et al (2008) Angioarchitecture of the ventral surface of the tongue from Wistar rats. Scanning 30(5):414–418
55. Sricharoenvej S, Boonprasop S et al (2009) Morphological and microvascular changes of the adrenal glands in streptozotocin-induced long-term diabetic rats. Ital J Anat Embryol 114(1):1–10
56. Walocha JA, Litwin JA, Bereza T, Klimek-Piotrowska W, Miodoński AJ (2012) Vascular architecture of human uterine cervix visualized by corrosion casting and scanning electron microscopy. Hum Reprod 27(3):727–732
57. Bereza T, Tomaszewski KA et al (2012) The vascular architecture of the supravaginal and vaginal parts of the human uterine cervix: a study using corrosion casting and scanning electron microscopy. J Anat 221(4):352–357
58. Kikuta A, Macchiarelli G et al (1991) Microvasculature of the ovary. In: Motta PM (ed) Ultrastructure of the ovary. Kluwer Academic, Boston, pp 239–254
59. Macchiarelli G, Nottola SA, Kikuta A, Ohtani O, Murakami T (1992) The ovary: Three-dimensional morphodynamics of the luteo-follicular complex by SEM of corrosion casts and other EM techniques. In: Murakami M, Fujita H (eds) Scanning electron microscopy of vascular casts: methods and applications. Kluwer Academic, Boston/Dordrecht/London, pp 245–259
60. Burri PH, Hlushchuk R et al (2004) Intussusceptive angiogenesis: its emergence, its characteristics, and its significance. Dev Dyn 231(3):474–488

Chapter 15
Hypoxia-Induced Retinal Angiogenesis in Adult Zebrafish

Zaheer Ali and Lasse Dahl Jensen

15.1 Introduction

Hypoxia is perhaps the most important pathophysiological trigger of angiogenesis and is crucial for developmental angiogenesis in human embryos as well as pathological and/or regenerative angiogenesis in adults [1]. Hypoxia-induced angiogenesis is in turn a very important driving force of diseases such as cancer [2], retinopathies [3] and during regeneration/functional recovery of heart or brain tissue following myocardial infarction and stroke [1]. Because of its importance in disease, many mouse models have been developed in which tissue hypoxia can be induced, although in an uncontrollable fashion and in combination with ischemia (lack of blood flow), by ligation of large arteries [4]. However the lack of flow-mediated shear-stress stimulation of the endothelium in blood vessels is by itself a strong stimulus for degeneration of those unused vessels in vivo [5]. In addition, lack of blood flow will also lead to reduced delivery of blood-borne mediators including glucose, blood-borne inflammatory cells, and endocrine signal molecules while also reducing the clearance of toxic waste products such as acids, reactive oxygen species and dead or decaying cellular components. Zebrafish are in contrast to mice very tolerant to environmental hypoxia [6]. Developing zebrafish embryos or adult zebrafish can manage prolonged exposure to as little as 10 % of the normal oxygen levels in the water without suffering obvious adverse reactions [2, 6]. This exposure to environmental hypoxia leads to induction of hypoxia signaling in various tissues in the zebrafish including the retina. Importantly, this hypoxia-induced retinal hypoxia is also rapidly and robustly driving angiogenesis following only 1 week exposure to hypoxia [3]. This is so far the only method for studying hypoxia-induced angiogenesis in living, adult tissues as an isolated event, free from confounding

Z. Ali • L.D. Jensen (✉)
Department of Medicine and Health Sciences, Linköping University, Linköping, Sweden
e-mail: lasse.jensen@liu.se

M. Slevin, G. McDowell (eds.), *Handbook of Vascular Biology Techniques*,
DOI 10.1007/978-94-017-9716-0_15

factors associated with ischemia. Here we will provide detailed protocols necessary for building hypoxia chambers for adult zebrafish, how to use such chambers for long-term hypoxia exposure and how to dissect out the retina for histological analysis in order to study hypoxia-induced angiogenesis. This protocol furthermore offers the opportunity to regulate the level of hypoxia in the water – and therefore also in the fish – to any level the researcher deems appropriate; and for as long time as needed, giving complete control over the level and duration of hypoxia in the zebrafish.

15.2 Methodology

15.2.1 Materials and Equipment

Hypoxia setup including aquarium 1–25 l, lid, magnet stirrer and magnet, nitrogen gas source including pressure-regulator and tubing.

Oxygen regulator setup constructed following the instructions on http://www.mbl.ku.dk/JFSteffensen/OxygenTemperatureMonitorRegulator/, alternatively purchased from Loligo Systems A/S, Denmark.

Dissection tools including fine forcepts (i.e. Dumont #5), scissors (i.e. Vannas), probes, micro-spoon and scalpel blades (i.e. type #10) (AgnThos AB, Sweden).

Stereomicroscope with fluorescence for dissection and imaging of the samples (Nikon SMZ1500, BergmanLabora, Sweden).

Materials for histology including glass slides, cover slides (VWR international, Sweden), nail polish, markers and mounting medium (Vector Laboratories).

Software for image analysis (i.e. ImageJ or Photoshop).

15.3 Protocol for Setting Up the Hypoxia Chamber

Equip an aquarium with a lid and make sure there are no large gaps between the lid and the aquarium while the aquarium is also not sealed completely as that will increase pressure within the aquarium during the experiment.

The aquarium should be of an appropriate size for the number of fish used in the experiment. We recommend that larger aquaria be used (i.e. 3 l or more), as it is easier to accurately regulate oxygen concentration and maintain high water quality and therefore the well-being of the fish in larger volumes of water.

Fill the aquarium 2/3 with the same type of water that the fish are housed in (i.e. system water).

Put a clean magnet bar (1 in Fig. 15.1) inside the aquarium and put the aquarium on a magnet stirrer to ensure proper mixing during the experiment.

Install the oxygen electrode (2 in Fig. 15.1) such that the tip is fixed in close proximity to the magnet bar without touching it.

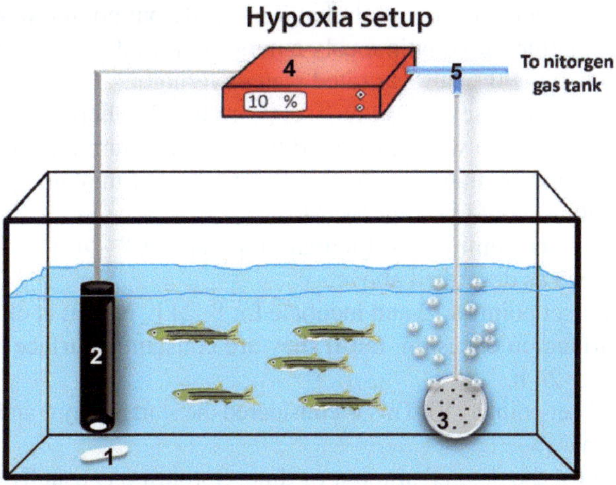

Fig. 15.1 Diagram of the hypoxia setup. A magnet bar (*1*), oxygen electrode (*2*) and an air-stone (*3*) are placed inside the hypoxia chamber, in which the oxygen level is controlled by an oxygen regulator (*4*) by controlling the perfusion of nitrogen into the water through a solenoid valve (*5*). See the text for more details

Install the air-stone (3 in Fig. 15.1), which is used for more effective delivery and dissolution of nitrogen gas in the water on the other side of the aquarium, so it is far from the tip of the oxygen electrode.

Connect the oxygen electrode to the oxygen regulator (4 in Fig. 15.1).

Connect a tube from the air-stone to the solenoid valve (5 in Fig. 15.1), which is connected to and thus controlled by the oxygen regulator, and a tube from the solenoid valve to the nitrogen source. The whole setup is shown in Fig. 15.1.

Turn on the oxygen regulator and calibrate the system using the CA.HI function such that it shows 100 on the display.

Add the fish by transferring them through a wetted net to the aquaria.

Put on the lid so it is tight, but not too tight allowing pressure equilibration between the environment inside and outside the aquarium.

15.4 Protocol for Exposing Adult Zebrafish to Hypoxia

If you use the Loligo system, make sure the system is calibrated for both high (100 % air saturated water) and low (0 % air saturated water) and is set up to display values between 0 and 100, the output should be set to potentiometer with an output between 4 and 20.

If the solenoid valve was connected to relay 1, make sure relay 2 is disabled, and set relay 1 to 30 (i.e. 30 % air saturated water). Relay 1 should be engaged on increased oxygen levels (act on inc).

Turn on the nitrogen gas flow such that it is steady but not too high – a pressure in which 10–20 bubbles are continuously emerging from the air-stone is optimal.

Incubate the fish in 30 % air-saturated water overnight.

On the next morning, set the relay 1 set point to 20, whereby the oxygen concentration in the aquarium will be lowered to 20 % air saturation. Incubate the fish for 8–24 h depending on their reaction to the hypoxia (if they go up to the surface gasping for air, they should be kept for 24 h).

Set the relay 1 set point to 15. Incubate the fish for 8–24 h depending on the reaction to hypoxia as described above.

Set the relay 1 set point to 12 and incubate for 8–24 h. Now it is ok that some fish go up to the surface on occasion, but if they are constantly surface breathing they should be kept for 24 h.

Set the relay 1 set point to 10. This is considered the start of day 0 and the beginning of the experiment.

Incubate fish for the desired amount of time. Sprouting is usually starting at day 3–5 after lowering to 10 % oxygen whereas branch point formation and stabilization is seen later – at day 8–12 following exposure to 10 % oxygen.

15.5 Reperfusion

The re-oxygenation of a hypoxic tissue, often referred to as reperfusion in ischemic diseases, is an important pathological event causing drastically increased production of reactive oxygen species and associated lipid peroxidation and DNA-damage in affected tissues. The protocol for exposing zebrafish to hypoxia, as described above, can be adapted to studies of reperfusion injury in zebrafish by shifting the gas source to normal air or pure oxygen in order to stimulate a hyperoxic environment in the previously hypoxia-exposed fish, following the following procedure.

After 6–8 days of exposure to 10 % relative air-saturated water, turn off the nitrogen gas flow on the nitrogen tank and remove the tube connecting the nitrogen gas tank to the valve. Make sure the valve is located at an elevated position relative to the aquarium as it may otherwise be "back-filled" with water from the aquarium.

Change the settings on the oxygen regulator such that relay 1 is now instead disabled.

Couple a normal air source to the valve – for example an air pump used in normal "home-type" aquaria (i.e. from RenaAir or similar companies, available in most pet stores).

Make sure that the oxygen electrode still measures 100 % at steady state after having perfused the aquarium with normal air for some time. If not, re-calibrate the electrode on the oxygen regulator (Cal. Hi).

Change the normal air source to a source of oxygen gas and continue to perfuse the water with oxygen until a desired oxygen level has been reached (i.e. 250 % air-saturation).

Change the Disp-Hi setting on the oxygen regulator to this value (i.e. 250) and re-calibrate high (Cal. Hi).

Change relay 1 to activate on decreased oxygen levels (Act on decrease) and set the set point to 250. The oxygen regulator now ensures that the oxygen tension never drops below 250 % air-saturation.

Incubate the fish for the desired time period under these hyperoxic circumstances for example 1–3 days.

Euthanize the fish by transferring them from the aquarium setup to a container with 0.04 % MS-222 (Sigma-Aldrich, St. Luis, US), use at least 40 mL per fish and incubate for at least 15 min after cessation of gill movements. Crush the brain with forceps taking care not to hurt the eyes.

Put the dead fish in 4 % freshly prepared PFA (Sigma-Aldrich, St. Luis, US) in PBS. At least 10 mL PFA per fish should be used. Incubate overnight in the refrigerator at 4°.

Wash the fixed fish 3× with PBS or salt-water. The tissues can now be dissected and used for histological examination.

15.6 Dissection, Flat-Mounting and Visualization of Blood Vessels in the Retina

The extent of hypoxia-induced angiogenesis can be easily determined in the retina of the fish. The zebrafish retinal blood vessels are highly similar to mammalian blood vessels including being covered with vascular mural cells and exhibiting low degree of leakage. They are furthermore exhibiting a relatively low vascular density and are organized in a vascular tree-like structure, which is so simple that any alteration of the vasculature can be easily and readily identified from histologic examinations (see Fig. 15.2). In order to obtain high quality images of the retinal vasculature, we recommend that transgenic zebrafish expressing EGFP in endothelial cells (i.e. kdrl: EGFP or fli1a: EGFP strains, both available from ZIRC, Oregon, USA) are used.

To isolate the retina, punch a small hole through the cornea using very sharp, fine forceps. Insert a pair of microsurgery scissors through the hole and cut all around the cornea to release it from the rest of the eye (B in Fig. 15.3).

Gently push out the lens using fine forceps (B in Fig. 15.3).

Cut off the gill lids and gills with fine scissors. This exposes the optic nerve, which can then be cut to release the eyes from their socket.

Use very sharp, fine forceps to create a small hole between the sclera and the retina entering from the optic nerve on the back of the eye. Expand the hole if necessary to create room for the scissors.

Cut the sclera along one side of the eye all the way to the rim. It should now be possible to gently peel of the sclera from the retina using double forceps (one for holding another for peeling). The result should look like photo C in Fig. 15.3.

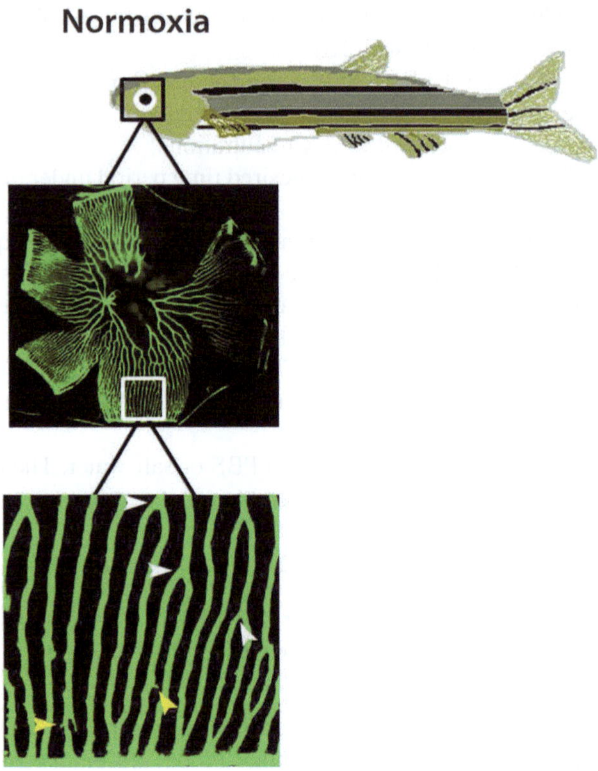

Fig. 15.2 Example of the retinal vasculature in the adult zebrafish. *Green* endothelial cells are revealed by fluorescence microscopy in transgenic fli1a: EGFP zebrafish. A minimal amount of sprouts (*yellow arrowheads*) and branch points (*white arrowheads*) are present under normoxic conditions

Cut the retina three to four times from the rim and almost to the center but not quite such that the tissue is still in one piece. This will enable the flattening of the retina in the characteristic flower-like shape (D in Fig. 15.3).

Prepare a viewing chamber by attaching two small cover slips (i.e. 22×22 mm) to either end of a glass slide using nail polish. Seal the sides facing the other cover slip with nail polish to avoid capillary forces to pull water or mounting medium under the slips.

Place the retina using a small spoon and forceps between the cover slips. Remove excess water and add just enough VectaShield to cover the retina.

Place a small drop of nail polish close to the inner edge of each cover slip "spacer" and carefully place a third cover slip on top of the retina such that it touches the nail polish on the two spacers. This allows fixation of the cover slip and ensure that the retina is not squeezed but retain its structure during visualization.

Put the mounted retina in the freezer at –20° for visualization later (it will keep in good condition for at least a week), or visualize it directly.

Using a fluorescent stereomicroscope (Nikon SMZ1500, BergmanLabora, Sweden) take an image of the entire retina, and smaller images at 5–10× magnification or higher as appropriate to visualize the finer details of the hypoxia-induced angiogenic response (i.e. emergence of small sprouts and tips etc.). A typical example

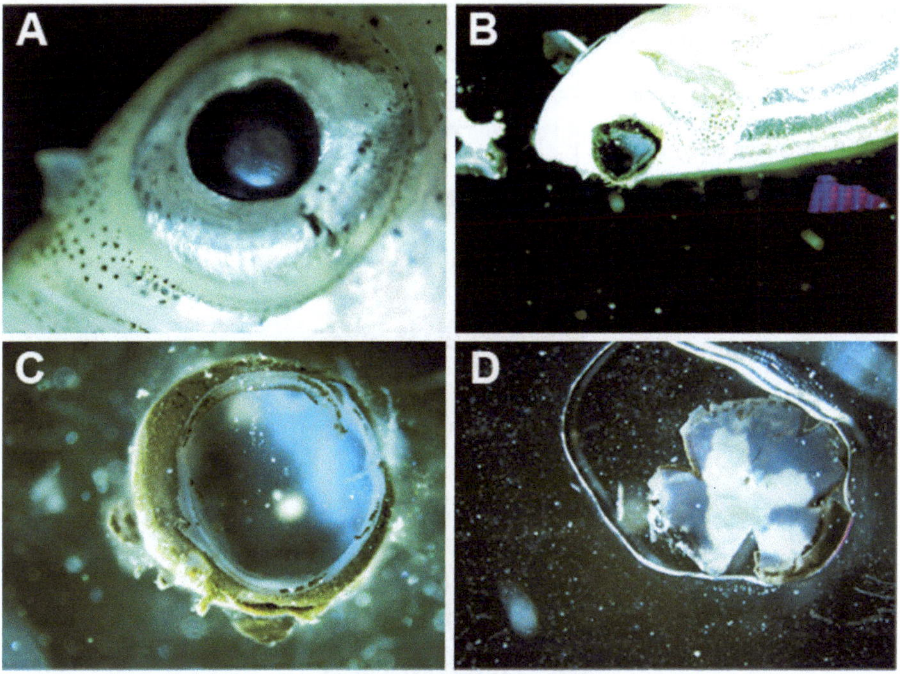

Fig. 15.3 Isolation and flat-mounting of the retina. The retina can be extracted from the whole zebrafish head. (**a**) By punching through and cutting around the cornea, releasing it and the lens from the retina. (**b**) By cutting the optic nerve bundle, the retina can be released from the head and the sclera can be gently peeled off. (**c**) Flat-mounting is accomplished by cutting 3–4 times from the periphery almost to the center, creating a flat, flower-like shape of the retina (**d**)

of the angiogenic response in the retina of zebrafish exposed to hypoxia for 8 days is shown in Fig. 15.4.

These images can, if saved in an appropriate format such as .tif or .jpg be analyzed in the available image-analysis software ImageJ, or other software that the user is familiar with.

15.7 Interpretation of the Results

The retinal vasculature originates from the central optic artery, which run alongside the optic nerve from the back of the eye, through the retina and emerge in the center of the optic disc. The central optic artery gives rise to 3–7 branches – called grade I retinal arteries, which run toward the periphery of the retina, splitting themselves in two a few times on the way giving rise to grade II, III and IV retinal arterioles. The arterioles feed into the retinal capillary network, which continue on the peripheral course before finally draining into the circumferential vein which lead the venous

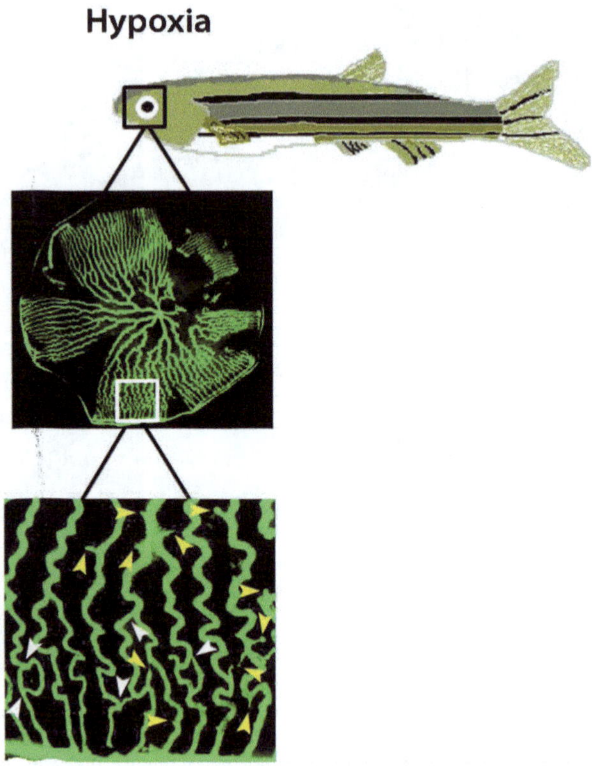

Fig. 15.4 Example of the retinal vasculature in adult transgenic fli1a: EGFP zebrafish exposed to hypoxia for 8 days as revealed by fluorescence microscopy of a flat mounted retina. Many sprouts (*yellow arrowheads*) and branch points (*white arrowheads*) are clearly discernible and quantifiable (See text for more details)

blood back into the general circulation (see Figs. 15.2 and 15.4). As a consequence of this vascular organization, all vessels in the center of the retina are arteries, whereas all vessels in the high-density, peripheral area are capillaries. This makes it easy to distinguish between arteriogenesis (the growth of arterial vessels) and angiogenesis (the growth of blood vessels in general – including capillaries and veins).

Commonly seen effects of hypoxia on the retinal vasculature are illustrated in Fig. 15.4. It is convenient to simply count the number of sprouts (yellow arrowheads in Figs. 15.2 and 15.4) and branch points (white arrowheads in Figs. 15.2 and 15.4) in regions of a defined size, to illustrate the extent of hypoxia-induced angiogenesis observed. More sophisticated analysis involve quantifying parameters such as vascular dilation (changes in diameter), inter-capillary distance or capillary density by using software such as ImageJ. ImageJ is freely available and is quite easy to use, and is therefore recommended for this type of analysis. These parameters can for example be obtained from ImageJ using the following protocol.

Open the picture pressing "o" or file + open and choose the file you wish to open.

To measure vessel diameter, select the line tool (fifth box from the left) and draw the shortest possible line from one end to the other of a vessel.

Press "m" and the length along with other parameters related to that line will be quantified and indicated in a new window.

Repeat this procedure to obtain the diameter of multiple vessels or vessels in different regions of the eye as needed.
Note the lengths in a data analysis program such as Excel or similar.
To measure the inter-capillary distance use the same method but draw the shortest possible line between the vessels instead.
To measure the capillary density go to image + adjust + brightness/contrast and increase the brightness and contrast to the maximum (slide the last two bars all the way to the right) and press apply. This will ensure that all vessels are selected during the following thresholding operation.
Select process + binary + make binary. The image is now converted to black and white and all the vessels but none of the background should appear as black on a white background.
Select an area of appropriate size i.e. 200 × 200 pixels by drawing a box of that size using the rectangle tool (first box from the left). The size of the box can be followed as its drawn in the data representation under the tool-images.
Measure the vessel area by selecting analyze + analyze particles. The parameters should be set to 0-infinity in size and 0.00–1.00 in circularity. The choice "summarize" should be the only selection in this operation.
The vascular area is now displayed in a new window. This value can be noted in an image analysis program such as Excel, where the density may be calculated by dividing this value by the total area (i.e. 200 × 200 = 40,000).

Repeat this operation in additional boxes as required to obtain more data from the same image.

15.8 Exploiting the Potential of the Model

The protocols described here will give basic insights into the angiogenic response caused by hypoxia and reperfusion injury. There are however many more sophisticated types of experiments/analysis that can be integrated with this model to gain information on more specific aspects of hypoxia-induced changes in tissues in vivo. For example, staining with antibodies can be done to investigate other cell types such as inflammatory cells or neuronal cells in the retina, or whether cells are proliferating or dying [7]. Dyes can be injected to look at the leakiness and/or perfusion of the vessels under hypoxia compared to normoxia [8]. Also, fresh tissues can be extracted and used to isolate protein or RNA for analyses such as ELISA, Western blot or qPCR [9]. The options are many, but such analyses are often performed in the same or very similar ways as when using tissues from other model systems such as mice or human tissues. Thus this system is highly compatible with other methods and will therefore be useful also to derive information on which genes are regulated by hypoxia in vivo, how does hypoxia influence perfusion in different tissues or how could hypoxia influence inflammation or cell proliferation/death.

15.9 Problems and Solutions

The hypoxia-exposure of the zebrafish is quite a strain, and therefore only the strongest, healthiest fish should be used for such experiments. This means that you will probably not be able to keep zebrafish in sufficiently low oxygen concentrations for sufficiently long time if you are using old, crooked, fat/very large, sick or stressed fish. Also the water quality has to be optimal. While the fish may survive in suboptimal water (such as tap-water in Scandinavian countries at least) in normoxia and appear completely fine, they cannot handle the low levels of oxygen as well under such conditions. Also it is important to make sure that the rotational speed of the mixing-magnet, the size of the magnet, the position of the electrode relative to the magnet and the air-stone as well as the amount of water and air in the aquaria is as described in this protocol. Too little mixing will result in the generation of oxygen gradients (especially at the top layers of the water, in close association with the air), whereas too rigorous mixing will force the fish to swim too fast, making them exhausted and less adept at handling the low oxygen levels.

When dissecting and flat-mounting the retina, take good care to not damage the tissue, which is very soft and brittle. We recommend that you practice on 10–20 retinas at least before working on experimental animals. The vessels are loosely associated with the luminal side of the retina, and can sometimes be removed together with the lens, if great care is not observed during this critical step. Also be careful neither to remove too much nor too little of the limbal region onto which the cornea is attached, as this will influence the visualization of the circumferential vein in the preparation. Please refer to [6] for more trouble shooting tips.

References

1. Jensen LD, Rouhi P, Cao Z, Lanne T, Wahlberg E, Cao Y (2011) Zebrafish models to study hypoxia-induced pathological angiogenesis in malignant and nonmalignant diseases. Birth Defects Res C Embryo Today 93:182–193
2. Rouhi P, Jensen LD, Cao Z, Hosaka K, Lanne T, Wahlberg E, Steffensen JF, Cao Y (2010) Hypoxia-induced metastasis model in embryonic zebrafish. Nat Protoc 5:1911–1918
3. Cao R, Jensen LD, Soll I, Hauptmann G, Cao Y (2008) Hypoxia-induced retinal angiogenesis in zebrafish as a model to study retinopathy. PLoS One 3:e2748
4. Cao R, Brakenhielm E, Pawliuk R, Wariaro D, Post MJ, Wahlberg E, Leboulch P, Cao Y (2003) Angiogenic synergism, vascular stability and improvement of hind-limb ischemia by a combination of PDGF-BB and FGF-2. Nat Med 9:604–613
5. dela Paz NG, Walshe TE, Leach LL, Saint-Geniez M, D'Amore PA (2012) Role of shear-stress-induced VEGF expression in endothelial cell survival. J Cell Sci 125:831–843
6. Cao Z, Jensen LD, Rouhi P, Hosaka K, Lanne T, Steffensen JF, Wahlberg E, Cao Y (2010) Hypoxia-induced retinopathy model in adult zebrafish. Nat Protoc 5:1903–1910
7. Brautigam L, Jensen LD, Poschmann G, Nystrom S, Bannenberg S, Dreij K, Lepka K, Prozorovski T, Montano SJ, Aktas O et al (2013) Glutaredoxin regulates vascular development by reversible glutathionylation of sirtuin 1. Proc Natl Acad Sci U S A 110:20057–20062

8. Dahl Ejby Jensen L, Cao R, Hedlund EM, Soll I, Lundberg JO, Hauptmann G, Steffensen JF, Cao Y (2009) Nitric oxide permits hypoxia-induced lymphatic perfusion by controlling arterial-lymphatic conduits in zebrafish and glass catfish. Proc Natl Acad Sci U S A 106:18408–18413
9. Jensen LD, Cao Z, Nakamura M, Yang Y, Brautigam L, Andersson P, Zhang Y, Wahlberg E, Lanne T, Hosaka K et al (2012) Opposing effects of circadian clock genes bmal1 and period2 in regulation of VEGF-dependent angiogenesis in developing zebrafish. Cell Rep 2:231–241

Chapter 16
Angiogenesis in the Regenerating Adult Zebrafish Tail Fin

Zaheer Ali and Lasse Dahl Jensen

16.1 Introduction

Angiogenesis is important for regeneration and wound healing both as a source of oxygen and nutrients to the re-growing tissue, but also as a supply route for inflammatory cells, bone-marrow derived progenitor cells and potentially as a signaling/differentiation hub as blood vessel endothelial and perivascular cells are important sources of many growth- and differentiation factors [1]. Following surgery or in diseases such as myocardial infarction and stroke, regeneration of the damaged tissue is crucial for recovery. Many attempts have been made to speed up regeneration, and in particular regenerative angiogenesis in ischemic tissues such as the heart and the brain [2], but so far the process have proven more complex and difficult to target therapeutically than first anticipated. As such there is a need for good animal models in which regeneration-induced angiogenesis can be studied.

Zebrafish are as other fish and amphibians much more adapt at regenerating their tissues compared to mammals such as mice, rats and humans. Therefore, zebrafish models are often superior to study signaling pathways and molecular and cellular factors important for the regeneration process. Whereas regeneration of the heart, brain and peripheral muscle tissues are largely non-existing in adult mammals, adult zebrafish can regenerate heart [3], brain [4], retina [5], muscle [6] and other tissues with remarkable speed and without scarring. For example, following resection of up to 1/3 of their heart (ventricle), adult zebrafish not only survive, the heart grows back in a matter of a few weeks [3]. Usually however, less invasive models are used, and the tail fin regeneration model is a popular way to study regenerative angiogenesis in vivo [7–10]. A major benefit of studying the regeneration of the fins is that these tissues remain transparent (especially in non-pigmented zebrafish strains such

Z. Ali • L.D. Jensen (✉)
Department of Medicine and Health Sciences, Linköping University, Linköping, Sweden
e-mail: lasse.jensen@liu.se

M. Slevin, G. McDowell (eds.), *Handbook of Vascular Biology Techniques*,
DOI 10.1007/978-94-017-9716-0_16

as Casper), during adulthood, and angiogenesis as well as the growth and patterning of other cell types can therefore be followed over time in the same animal by intravital imaging using regular microscopes. This protocol will outline the materials, equipment and steps involved in tail fin amputation as well as genetic manipulation of the regenerating tissues by microinjection and electroporation.

16.2 Methodology

16.2.1 Materials and Equipment

Fish aquaria (3 L), anaestisia tanks (200 mL), nets and spoons for fish transfer.

Scalpel blades with rounded cutting edge (i.e. type #10) and forceps for holding the fish in place.

Microinjection setup including micromanipulator (Narishige, Japan), Microinjection pump (MINJ-D, TriTech Research, USA), capillary needle puller (PC-10, Narishige, Japan), Borosillicate capillaries with filament and outer diameter of 1.0 mm (World Precision Instruments, USA), microloaders and pipettes (Eppendorf, Germany).

Square wave electroporator, including small planer forceps-like electrodes (ECM830, Harvard Instruments, USA with "tweezertrodes" 3 mm in diameter).

Dissection microscope with fluorescence (i.e. Nikon SMZ1500, BergmanLabora, Sweden).

16.3 Tail Fin Amputation

Prepare a solution of 0.02 % (0.2 mg/mL) MS-222 (Sigma-Aldrich, St Luis, USA) in fish facility system water. It is advised to prepare a buffered stock of MS-222 at 0.5 % (5 mg/mL) at pH 7.4 and keep frozen aliquots that may be diluted to the final concentration of 0.02 % by diluting the stock 25× with system water. MS222 is unstable in aquous solutions so therefore higher concentrations kept frozen is needed for long-term storage. Transfer one adult zebrafish of at least 3 cm in length (4–18 months old) into this solution using a fish transfer net.

Wait until the fish no longer respond to being poked by blunt-ended forceps, but still has slow and vague gill movements (this usually takes 2–4 min.) and move the fish with a spoon to a 10 cm bacterial culture petri dish filled with 0.01 % MS-222.

Under a dissection microscope, gently spread out the tail fin without damaging it using blunt ended forceps held in the non-dominant hand (i.e. left hand for most people). Identify the fin rays of the fish and cut the tail fin in a straight line just below the second-last bifurcation of the rays with a scalpel held in the dominating hand (i.e. right hand for most people), while still keeping the tail fin immobilized and spread out with the forceps. A nice straight cut can be achieved by putting the

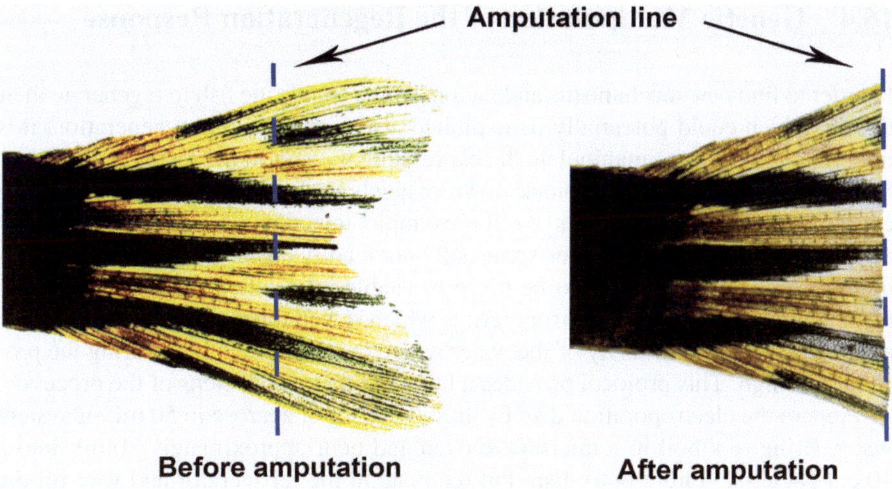

Fig. 16.1 Adult zebrafish with an amputated tail fin. Photograph is taken immediately before (*left image*) or after (*right image*) amputation. The *blue dashed line* indicate the amputation line relative to the structure of the fin

tip of the round-edged scalpel blade onto the plastic bottom of the petri dish just above the tail fin and "rolling" in backwards cutting the fin in one smooth motion. The amputation line should be just below where the two lobes of the fins merge such that both fin lobes are cut off and a triangle-like shape of the fin is created (see Fig. 16.1).

Put the fish somewhere else in the petri dish and take a photograph using a camera mounted on the dissecting microscope if available, to document the position of the amputation without other disturbing elements in the image such as the cut-off tail fin part and the line made in the plastic during the amputation procedure (see Fig. 16.1).

Move the fish back into clean system water without MS-222. The fish should wake up and start swimming normally again within 3–5 min depending on the size.

Follow the regeneration over time by using the same method for anaestisia as described above. Fish can be safely anaestitized like this multiple times and at least once per day without causing health problems. It is recommended to take photos documenting the regeneration process at least twice per week for at least 2 weeks. Full regeneration is usually observed at 4 weeks post amputation.

This protocol is often used also for obtaining adult fish tissues for genotyping of transgenic or mutant zebrafish strains, which do not have a phenotype that can be accurately picked up by visual screening during embryogenesis. As the fin grows back relatively quick, and this procedure is very well tolerated by the fish, it can be considered similar to cutting a persons hair and is therefore a transient, easy and safe procedure for obtaining tissues in adult zebrafish.

16.4 Genetic Manipulation of the Regeneration Response

In order to find new mechanisms and factors which enable the fish to regenerate their tail fin, which could potentially be exploited also in mammalian regeneration, it is important to be able to manipulate the expression of genes in the regenerating tissue by for example injection of knock down reagents such as morpholinos or forcing elevated production of factors by for example injection of expression vectors/ mRNA. However, as nucleotides are usually not readily taken up over the cell membrane additional steps needs to be taken to enable intracellular delivery. A convenient method often used is electroporation, which is well tolerated in the fish fins and effective, as the conductivity of the water where the fish is localized during the procedure, is high. This protocol provides a layout of the overall steps of the process.

Prepare the electroporation dish by dissolving 1 g of agarose in 50 mL of system water. Bring to a boil in a microwave oven and pour approximately 30 mL into a 10 cm bacterial culture petri dish. Put the plate in the refrigerator and wait till the agarose has solidified. Pour over the rest of the still warm agarose solution and gently place into the molten agarose a mold, which will create the desired shape of the holding cell for the fish. We recommend using a rectangular mold with a small triangular tip at one of the short ends, as this shape enable convenient holding of the fish during the electroporation step without damaging the skin or fins (see Fig. 16.2). When the agarose has solidified, gently remove the mold and pour in 0.01 % MS-222 in system water. The plate is now ready to use.

Pull glass capillaries in the needle puller using a program for making needles suitable for morpholino injection; the taper should be several mm long and fused at the tip. If using the PC-10 needle puller, this can be achieved through a one stage pull using full weights and the heater level set to 65.

Load the needle with 2 μL of the substance to be injected, for example a 0.6 mM solution of scrambled control morpholino dissolved in 33 % injection buffer (9 μM

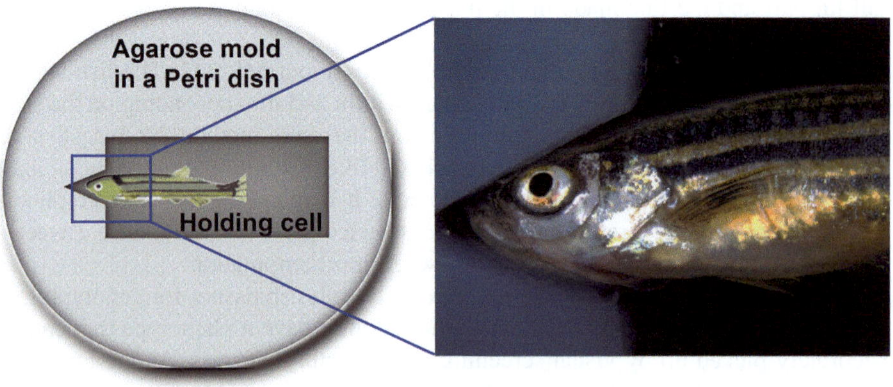

Fig. 16.2 Adult zebrafish placed in the holding cell. Note the triangular extension to the rectangular shape of the holding cell which enable covering the head in water

spermine, 0.21 mM spermidine, 0.3 % phenol red in PBS) and Nuclease-free, non-DOPC-treated water.

Break the needle at the very tip using sharp forceps, and insert the needle in the needle holder of the micromanipulator. Make sure the needle is pointed down in an angle of approximately 30° with the screws on the manipulator placed such that the needle can be advanced in this trajectory by turning the screws.

Turn on the gas supply and the microinjector. The micro-injection system is now ready to use.

Calibrate the needle by injecting into mineral oil placed on a microscope ruler. Adjust the injection time and pressure such that a drop of 6 nL is injected when the foot pedal is engaged. The injection volume can be calculated from the diameter of the drop produced in oil (hence the placement of the oil on a microscope ruler) using the formula: 1/6 × pi × d^3, where d is the diameter and pi can be approximated by 3. Thus a diameter of 0.23 mm will give a drop volume of approximately 6 nL.

Turn on the electroporator (for example ECM830, Harvard instruments, USA) and program the instrument to give 10 pulses of 15 V and 60 ms per pulse and a waiting time of 1 s between pulses.

Anaestitize adult zebrafish in which the tail fin has been amputated 2 days before, following the steps outlined above in Sect. 16.3.

Move the anaestitized fish with a spoon into the holding cell in the agarose plate, with the head in the triangle, lying on its side (see Fig. 16.2) such that the fin is spread out and touching the agarose bed inside the cell. Lift the plate a cm or two at the end where the tail fin is placed such that the fin is lying only covered by a thin film of water whereas the head is completely submerged.

Put the dish with the fish under the dissection microscope and advance the needle toward the blastema (regenerating tissue) at the very tip of the regenerating tail fin. Gently and carefully insert the needle into the blastema and inject approximately 60–90 nL by engaging the foot pedal 10–15 times (see Fig. 16.3).

Pull out the needle and repeat the injection procedure next to the site of the first injection.

Continue like this until 6–12 small injections have been done next to each other, covering one half of the blastema, corresponding to one lobe (see Fig. 16.3). The other half/lobe can be used as a control. Keep track of which lobe that has been injected – i.e. always inject either the dorsal or the ventral lobe.

Put the dish back in a horizontal position and gently turn the fish ventral side up. The tail fin should now be in a vertical position and completely covered in water (see Fig. 16.4).

Place the electrodes close to the fin without touching it, there should not be more than one or a few mm between the fin and the electrodes (see Fig. 16.4). Engage the electroporation by pressing the foot pedal of the electroporator. Bubbles are likely to arise from hydrolysis at each electrode by each pulse, but they should move away from the electrodes quickly. Make sure the fin is not pushed into the electrodes during the electroporation, which takes approximately 11 s in total.

Move the fish back into fresh system water to recover, using the spoon. The fish should be swimming normally again within 3–5 min.

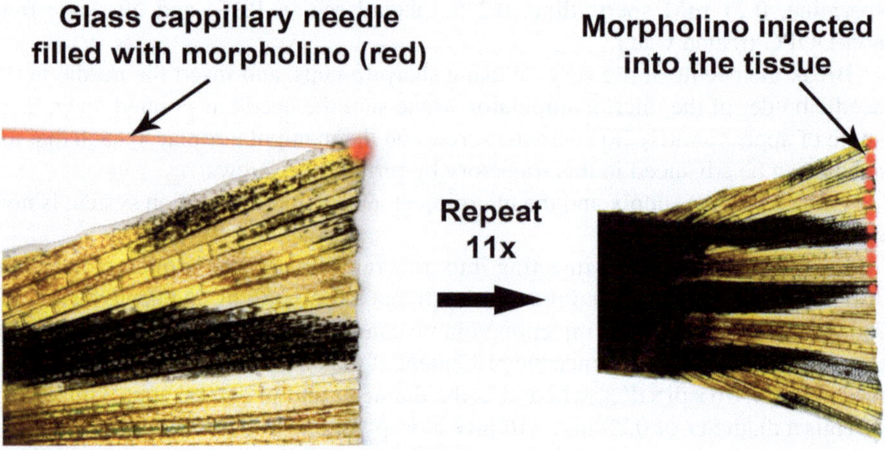

Fig. 16.3 Microinjection of morpholinos into the tail fin blastema of 2 days post amputation zebrafish. The glass capillary is shown loaded with the red morpholino solution, and the injected site is indicated with a *red dot*. In the image to the *left* the zebrafish is being injected for the first time, in the image to the *right* the zebrafish has been injected 11 times, each injection close to the previous one such that the entire lobe (half of the fin) is injected

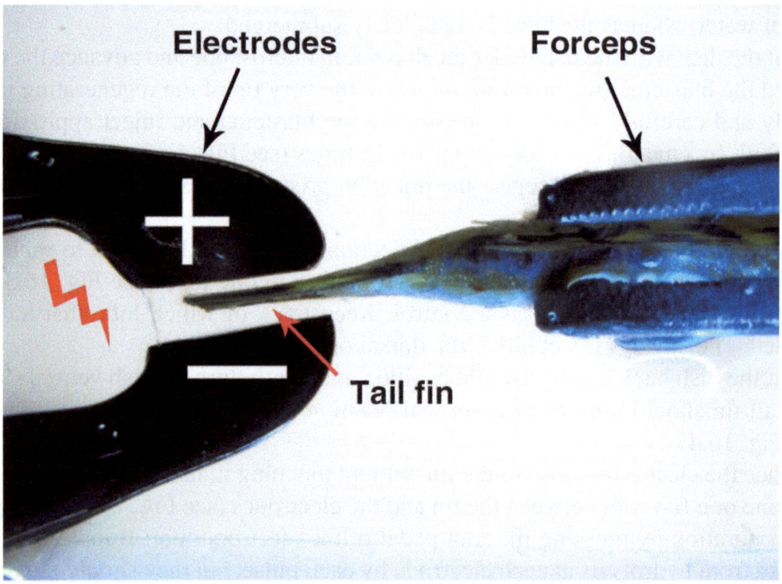

Fig. 16.4 Electroporation of the tail fin. Note that the tail fin is placed close to, but not touching the electrodes giving optimal transfection efficiency without causing electric burns on the skin. The zebrafish is kept immobilized by gently holding it with a pair of rounded forceps as indicated in this image

Track the regeneration of the fin by anaestitizing the fish and photographing the fin under the microscope as described above in Sect. 16.3, and compare the injected lobe to the non-injected lobe. If control morpholino has been used, there should be no difference in the regeneration but knock-down or over-expression of specific genes may affect this process.

16.5 Determining Regenerative Angiogenesis by Fluorescent Imaging

In order to determine the angiogenic response into the re-growing tissue during tail fin regeneration, it is convenient to use transgenic zebrafish such as the Tg(fli1a: EGFP) or Tg(kdrl: EGFP) lines. In these fish, the endothelial cells are labeled with enhanced green fluorescent protein, and can readily be observed under the fluorescent microscope. The following protocol can be used to follow the growth of blood vessels over time in the same fish.

Label the aquaria in which each fish is kept in isolation with the number of that fish.

Anaestitize the fish following the procedure from Sect. 16.3 and place it in a holding chamber similar to that used during microinjection and electroporation filled with approximately 10–20 mL 0.01 % MS-222.

Place the fish and the dish under the microscope and tilt the dish such that the tail fin is only covered with a thin film of water and is flattened and spread out.

Turn on the fluorescence light path, zoom in to an appropriate level (1–2× is often good for the whole fin whereas 6–12× is good for recording details of the sprouting process) and focus on the vessels. Take pictures following the instructions of the manufacturer of the microscope/camera. If using a stereomicrscope such as SMZ1500 (Bergman/Labora, Sweden), remember to note, for example when naming the images you save, how high magnification each image has been taken by, or if available include a bar indicating the length of a particular distance such as 500 μm in the picture. Also note which rays are being photographed by numerical identifiers related to their position such that the first ray from the dorsal side is denoted no. 1, the next one no. 2 and so on.

Move the fish back to a recovery aquarium and when it is swimming normally again, move it back to its home aquarium.

This procedure can be repeated for example every day in order to get a series of images showing the time course of regenerative angiogenesis. Remember to only compare the same rays in the same fish with each other as different rays may grow back at different speed (i.e. rays 3, 4 and 5 grow faster than any other ray and the rays closest to the center of the tail fin grow slower).

Use programs such as ImageJ to count the number of sprouts, measure the length of the newly formed blood vessels, the density of the vasculature at the regenerating front or the diameter of the blood vessels. More details on how to do this is given in the chapter on hypoxia-induced retinal angiogenesis.

16.6 Problems and Solutions

The tail fin amputation is a very simple procedure but it is important to choose the right site for the amputation. If too little or too much of the fin is amputated, it will either grow back very quickly (i.e. within a week if only the most distal 20 % or so is removed) or very slowly if at all (i.e. if more than half is amputated). This in turn will affect the results as changes induced by knock-down or drugs added to the water is less or not apparent if the fin regenerate too quickly or too slowly. Approximately 33 % is the optimal amount to remove, and as described the cut should be placed just below the second bifurcation of the rays.

Also the genetic background of the fish is important. Some strains such as TL have very long fins, which may exhibit different regenerative characteristics compared to the fins of AB strain. Therefore results should always only be compared within the same strain of fish.

Also the number of fish in the aquaria is important, as regeneration will be reduced due to fighting if 2–6 fish are kept together. The fish should either be isolated throughout the regeneration procedure (that would give the least variability) or kept in groups of preferably 10 or more.

Different electroporators work slightly differently and it is therefore important to evaluate which parameters give the best transfection efficiency for each electroporator. For example it may be required to increase the voltage to 20 or 25 V, reduce the pulse duration to 50 or 45 ms or do fewer pulses in order to get good efficiency with some electroporators. Also the size and shape of the electrodes are important (we use electrodes of 3 mm in diameter and only slightly rounded at the periphery, but of course the electric field would be different if larger or more round electrodes are used). The efficiency can for example be evaluated by injecting fluorescently labeled morpholinos, performing electroporation, waiting 24 h (during which time the extracellular morpholino should be largely gone), amputate the electroporated area and measure the fluorescence intensity in lysed tissue using a plate reader for example. This intensity can then be compared with that of non-electroporated fish or with electroporated with a different protocol to find out what works best for a particular electroporator.

References

1. Madeddu P (2005) Therapeutic angiogenesis and vasculogenesis for tissue regeneration. Exp Physiol 90(3):315–326, Epub 2005/03/22
2. Cao Y (2010) Therapeutic angiogenesis for ischemic disorders: what is missing for clinical benefits? Discov Med 9(46):179–184, Epub 2010/03/31
3. Poss KD, Wilson LG, Keating MT (2002) Heart regeneration in zebrafish. Science 298(5601):2188–2190, Epub 2002/12/14
4. Kroehne V, Freudenreich D, Hans S, Kaslin J, Brand M (2011) Regeneration of the adult zebrafish brain from neurogenic radial glia-type progenitors. Development 138(22):4831–4841, Epub 2011/10/19

5. Vihtelic TS, Hyde DR (2000) Light-induced rod and cone cell death and regeneration in the adult albino zebrafish (Danio rerio) retina. J Neurobiol 44(3):289–307, Epub 2000/08/16
6. Rodrigues AM, Christen B, Marti M, Izpisua Belmonte JC (2012) Skeletal muscle regeneration in Xenopus tadpoles and zebrafish larvae. BMC Dev Biol 12:9, Epub 2012/03/01
7. Poss KD, Shen J, Nechiporuk A, McMahon G, Thisse B, Thisse C et al (2000) Roles for Fgf signaling during zebrafish fin regeneration. Dev Biol 222(2):347–358, Epub 2000/06/06
8. Bayliss PE, Bellavance KL, Whitehead GG, Abrams JM, Aegerter S, Robbins HS et al (2006) Chemical modulation of receptor signaling inhibits regenerative angiogenesis in adult zebrafish. Nat Chem Biol 2(5):265–273, Epub 2006/03/28
9. Akimenko MA, Johnson SL, Westerfield M, Ekker M (1995) Differential induction of four msx homeobox genes during fin development and regeneration in zebrafish. Development 121(2):347–357, Epub 1995/02/01
10. Alvarez Y, Astudillo O, Jensen L, Reynolds AL, Waghorne N, Brazil DP et al (2009) Selective inhibition of retinal angiogenesis by targeting PI3 kinase. PLoS One 4(11):e7867, Epub 2009/11/20

Chapter 17
Methods for Studying Developmental Angiogenesis in Zebrafish

Zaheer Ali, Jian Wang, Yihai Cao, and Lasse Dahl Jensen

17.1 Introduction

Zebrafish have recently gained overwhelming attention as a model organism to study mechanisms of development and disease. Today, zebrafish are in many countries the second largest laboratory animal research platform, second only to mice, and most universities have their own zebrafish facility (or several facilities). The model is therefore available to most researchers, and as these protocols will indicate, they are very easy to use giving researchers ample opportunities to include zebrafish models in their research. The most celebrated benefits of the zebrafish models are transparency and extra-uterine (outside of the uterus) development of the embryos which enable continuous visualization and even video-recording of highly dynamic processes during development such as the growth, pathfinding, maturation, perfusion and regression of blood vessels [1]. They furthermore are produced in large numbers every breeding cycle; often a few hundred embryos per week per female. As adult zebrafish can be housed in high density, a small facility will therefore be able to produce thousands if not tens of thousands of embryos per day – enabling medium-to-high throughput screenings of the effects of biological compounds in a living animal system – studies that were previously only done in cells. As zebrafish

Z. Ali • L.D. Jensen (✉)
Department of Medicine and Health Sciences, Linköping University, Linköping, Sweden
e-mail: lasse.jensen@liu.se

J. Wang
Department of Microbiology, Tumor and Cell Biology, Karolinska Institute, Stockholm, Sweden

Y. Cao
Department of Medicine and Health Sciences, Linköping University, Linköping, Sweden

Department of Microbiology, Tumor and Cell Biology, Karolinska Institute, Stockholm, Sweden

M. Slevin, G. McDowell (eds.), *Handbook of Vascular Biology Techniques*,
DOI 10.1007/978-94-017-9716-0_17

are vertebrates, there are many aspects of nervous system, vascular and organ development in this animal that more closely resemble those in humans, compared to other developmental systems such as flies or worms. The zebrafish genome has been fully sequenced and annotated in 9 rounds providing a high-coverage and well-established platform for genetic studies in this animal. Today, following three decades of intensive zebrafish research, many tools have been developed for use in/with zebrafish which are often freely available to be exploited by everyone who have the use for them in their research. The large number and collaborative nature of the researchers involved using zebrafish furthermore ensure that novel, innovative tools are continuing to be put out at a rapid pace. Many of these tools are specific for zebrafish (i.e. specific genetically modified strains, transplantation techniques, regeneration models, disease models etc.), which gives zebrafish several unique advantages over other model systems. Finally, as discussed in more detail in two other chapters in this book, zebrafish are particularly good for studying mechanisms behind regeneration and hypoxia-induced signaling, processes which are dependent on or underlying angiogenesis in physiology and pathology.

The vasculature in the zebrafish embryo starts to develop from hematopoietic progenitor cell differentiation and aggregation in bilateral strips in the posterior and anterior mesoderm from approximately 14–16 h after the egg has been fertilized (hours post fertilization, hpf). The two major axial blood vessels, the dorsal aorta (DA) and posterior cardinal vein (PCV) develop by classic and alternative vasculogenesis respectively by 20–24 h, which coincide with the morphogenesis and development of contractile functions of the heart. Thus at 24 hpf, the zebrafish: embryo has established a complete, albeit primitive, circulatory loop, a beating heart and circulation of red blood cells in the entire organism. This also largely marks the end of the vasculogenic period, the rest of the vascular expansion during development is largely achieved through angiogenesis. The first angiogenic wave of vascular expansion in the periphery takes place between 22 and 30 hpf, encompassing the sprouting and dorsal extension of arteries emanating from the DA and growing toward the dorsal aspect of the myotomes (embryonic muscle tissue). Their course takes a prototypical shevron-shaped path leading the formation of highly regular and conformed inter-somitic vessels (ISVs, see Fig. 17.1). Once the dorsal aspect of the myotomes have been reached, the ISVs branch laterally, anastomose with branches from neighboring ISVs and form the dorso-lateral anastomosing vessel (DLAV). The formation of this vessel complete the first wave of angiogenic expansion in the peripheral muscle tissues, and this simple vasculature does not change much throughout early development [2].

In the head and brain on the other hand, angiogenesis is a very dynamic process leading to progressive expansion of the vasculature starting with formation of the major arteries and veins already at 18–20 hpf, followed by development of the gill vasculature and from approximately 30–36 hpf the vasculature is rapidly expanding in the whole brain [2]. Also in peripheral organs including the liver and other digestive-tract organs, angiogenesis is apparent from approximately 48 h and onwards, with the characteristic growth of the sub-intestinal vessels (SIVs) being at a maximum at 72 hpf (see Fig. 17.1).

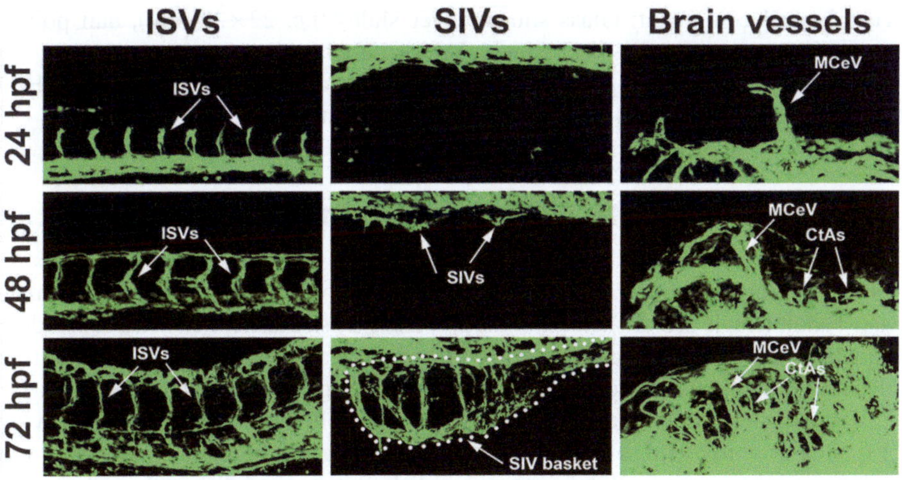

Fig. 17.1 Blood vessels in the muscle (inter-segmental vessels, ISVs), digestive tracts (sub-intestinal vessels, SIVs) and brain at 1, 2 or 3 days post fertilization (dpf). Images were acquired by confocal microscopy as described in this chapter from Tg(fli1a:EGFP)^{y1} transgenic zebrafish expressing GFP in endothelial cells and the vessels were pseudo-colored in green post-acquisition. Note that different vasculatures develop at different developmental times giving ample opportunities to study different aspects of tissue-specific or non-specific angiogenesis during development in the zebrafish embryo

Angiogenesis during zebrafish development can therefore be monitored at various time points in different vascular beds, and as such offers great flexibility to the researcher and the possibility to use zebrafish as an angiogenesis model in various ways depending on the particular research question under investigation. Most researchers have focused on the mechanisms behind development of the ISVs [3, 4], or brain vasculature [5], but also other vascular structures including retinal vasculature [6], intestinal vasculature [3] and lymphatic vasculature [7] has been studied in some detail.

In this chapter we will describe how to do genetic manipulation of the zebrafish embryo by injection of morpholinos, mounting and confocal imaging of the embryos as well as how to evaluate the extent of pathological angiogenesis in implanted tumors [8, 9].

17.2 Materials and Equipment

Fish breeding: Breeding aquaria, fish transfer nets, tea-sieves, petri dishes, squeeze bottles, disposable plastic Pasteur pipettes

Injection and manipulation tools: Sharp forceps (i.e. Dumont #5), metal probes, borosilicate capillaries (i.e. of 1 mm outer diameter, with filament, 10 cm long from World Precision Instruments, USA), needle puller (PC-10, Narishige, Japan), micromanipulator with capillary holder (Narishige, Japan), microinjection pump (MINJ-D, TriTech Research, USA), Pressurized nitrogen gas capsule, dissection microscope (i.e. SMZ1500, Bergman Labora, Sweden)

Histological equipment: Glass slides, cover slides (i.e. 22×22 mm), nail polish, mounting medium (i.e. VectaShield, Vector Laboratories)
E3 embryo medium: 290 mg NaCl, 13 mg KCl, 48 mg $CaCl_2$*$2H_20$, 82 mg $MgCl_2$*$6H_2O$ and 1 mg Methylene Blue per 1 L of dH_2O, adjust to pH 7.2 with NaOH

17.3 Injection of Morpholinos in Zebrafish Eggs

In the evening, at least 30 min. after the last feeding of the fish, transfer 2 adult male and 4 adult female zebrafish (>4 months old) using a fish transfer net, to a 2 L breeding aquarium filled half to two-thirds with system water and installed with a net-bottom insert. Put on the lid to prevent the fish from jumping out overnight.

Prepare an injection plate for the injections the following morning by dissolving 1 g of agarose in 50 mL of system or tap water, bring to a boil in a microwave oven and pour approximately 30 mL into a petri dish. Put the petri dish in the refrigerator until it has solidified and then add the remaining, still molten agarose solution on top. Place a mold consisting of 4–6 lines of approximately 1 mm in width and depth and 6 cm in length on a foundation (see Fig. 17.2) in the warm agarose and leave it to cool. Gently remove the mold and add E3 medium to the plate. Store at 28.5° in the incubator (for use tomorrow) or at 4° in the refrigerator (for later use).

The next morning, 15–30 min after lights-on in the facility, transfer the fish back to their home aquaria using a fish transfer net, remove the net-bottom insert and collect the eggs by pouring the water through a tea-sieve.

Rinse the eggs carefully to remove feces and other debris using E3 medium applied to the eggs from a squeeze bottle. Turn the sieve up-side down over a petri dish and apply more E3 medium over the eggs to in this way transfer the eggs to the petri dish.

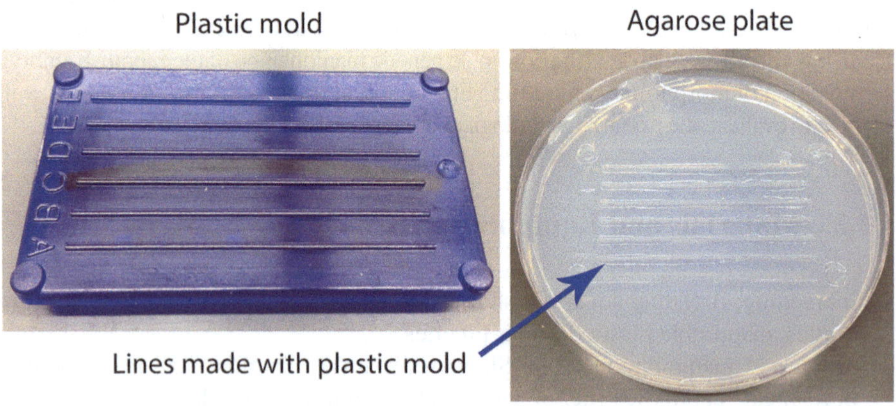

Fig. 17.2 Injection mold and – plate for morpholino injections. Notice the approximately 1 mm thick and 6 cm long lines in the blue mold to the *left*, which will give similar sized grooves in the agarose plate, which will facilitate non-harmful fixation of the embryos during injection

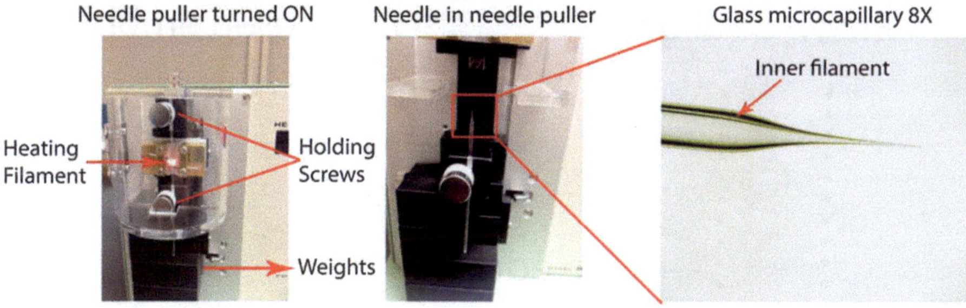

Fig. 17.3 Pulling needles in the PC-10 needle puller. This needle puller works by gravitational pulling where the pulling force is adjusted by putting on more or less weights. The capillary is held in place with screws and should be placed such that the *middle* of the capillary is inside the heating filament. After pulling the needle should have a long thin taper, the end of which is fused and should be broken off before injection

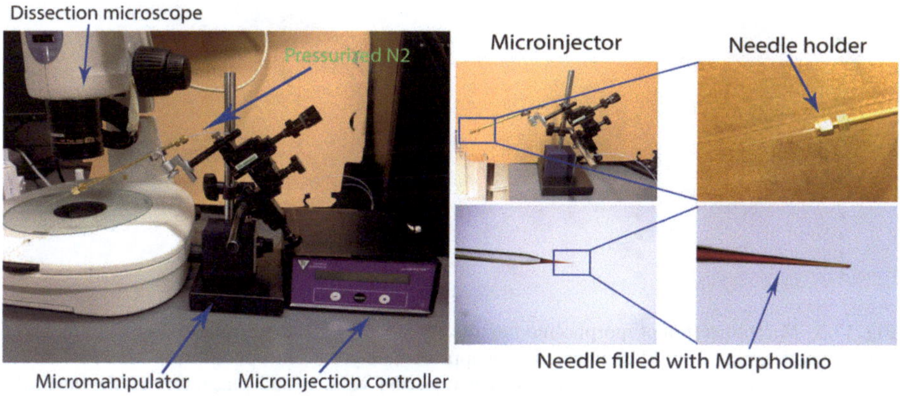

Fig. 17.4 The microinjection setup. All operations is done under a SMZ1500 dissection microscope equipped with fluorescence (Bergman Labora, Sweden). The needle is operated via a micromanipulator giving fine, 3-axis control by adjusting the three screws. Note the position of the micromanipulator such that the needle in pointed down-wards in an approximately 30° angle, in line with the fine and coarse advancing screws (on the *top* to the *right* on the manipulator). Air-pressure and thus injection volume is regulated by the MINJ-D pump/micro-injection controller. Make sure the needle is tightly inserted into the needle holder as shown in the figures to the *left* such that the pressurized air will go through the needle and not around it

Pull injection needles from the glass capillaries using an appropriate program in the needle puller – for example a one-stage pull using full weights and a heat-setting of 65 on the PC-10 puller (see Fig. 17.3).

Load the needles with 2 μL of morpholino solution, i.e. a 0.6 mM solution of standard control morpholino dissolved in 33 % injection buffer (9 μM spermine, 0.21 mM spermidine, 0.3 % phenol red in PBS) and 67 % Nuclease-free, non-DOPC-treated water.

Insert the loaded injection needle into the needle holder on the micro-manipulator. Turn on the pressure on the gas bottle and turn on the microinjection pump (see Fig. 17.4 for a complete view of the microinjection setup).

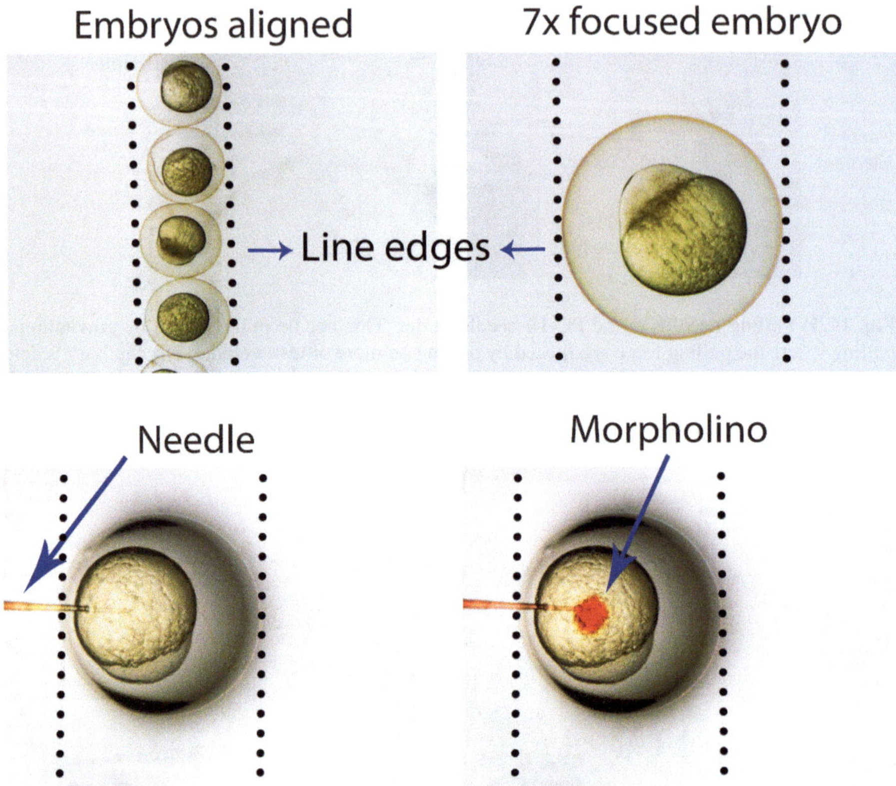

Fig. 17.5 Microinjection of morpholino into the zebrafish 1-cell stage embryo. Embryos mounted in the injection plate should be injected with approximately 1 nL morpholino solution (*red* due to phenol *red* in the injection medium). Correct injection volume and placement is shown in the *lower-right* figure

Using disposable Pasteur pipettes, transfer embryos to the injection plate. Under the microscope, gently push the embryos down into the groves taking care not to poke holes through the chorion (egg shell). A tip is to try and place the embryos tightly packed together in lines (see Fig. 17.5) as that will make the injections easier and faster.

Put a drop of halocarbon oil on a microscope ruler and place it centrally under the microscope. Zoom in and focus such that the grid-lines are clearly visible. Brake off the tip of the needle and use the micro-manipulator to place the tip of the needle in the oil, close to the grid lines. Inject once into the oil and measure the size of the drop. Adjust the injection pressure and time on the micro injection pump such that drops with a diameter of 0.12 mm are produced consistently. This corresponds to a volume of 1 nL.

Put the injection plate with the mounted embryos back under the microscope. Make sure the embryos are still at the 1-cell stage. Higher stage embryos might be less affected by the morpholino reducing the knock-down efficiency. Advance the tip of the needle to penetrate through the chorion and yolk sac and inject once.

You should see a small red drop with a clear border towards the yolk and which do not immediately disappear as a red cloud (see Fig. 17.5), as the latter indicate injection in the egg white rather than the yolk.

Pull the needle back, out of the embryo, and use the hand not engaged handling the micro manipulator to move up the injection plate slightly such that the next egg in the line is positioned right in front of the needle tip. Again advance the needle into the yolk of the egg and inject. Continue like this until all the eggs have been injected.

Carefully push out the eggs from the grooves in the injection plate and transfer the eggs with a disposable Pasteur pipette to a new petri dish containing E3 medium.

Place the eggs in the incubator for at least 1 h.

Put the embryos back under the microscope and transfer all fertilized embryos exhibiting more than one cell on top of the yolk (often 4–8 cells are clearly discernible after 1 h incubation following the morpholino injections) to new petri dishes with E3 water. There should be a maximum of 50 embryos per 10 cm. petri dish as higher densities will adversely affect the development. Put the plates back in the incubator at 28.5 °C.

The embryos can now be taken out at various time points and observed under the fluorescent or confocal microscope as described below in order to visualize their development. If transgenic embryos of the fli1a:EGFP or kdrl:EGFP strains are used, their blood vessels will be genetically labeled in green, and thus, the vascular development can be studied and quantified as shown in Fig. 17.1.

If the embryos are to be visualized at stages younger than 3 days, some or all of them may still be unhatched. In order to get a better view of the vasculature, hatching or dechorionation is advised. Dechorionation is best done manually by using sharp forceps and a probe. Pinch the chorion of the embryo with the sharp forceps taking care not to damage the embryo itself. Pinch the chorion at a second location close to the first one between a metal probe and the bottom of the petri dish. Gently tear open the chorion and release the embryo by nodding it out of the chorion, should it not have swum out itself.

If embryos older than 3 days are to be visualized, it is advised to first anaestitize them with 0.02 % MS-222 in E3 medium.

17.3.1 Injection of Substances or Cells in the Perivitteline Space

In order to stimulate or inhibit developmental angiogenesis, compounds, recombinant proteins, antibodies or cells producing such factors may be placed in a sub cutaneous-like region close to the developing vasculature known as the perivitelline space. This can be achieved following the protocol given here:

If cells are to be injected, it is advised to label them before hand with lipophilic membrane dyes such as DiI, following the manufacturer's instructions (i.e. incubation in PBS containing 2 μg/mL DiI for 30 min at 37°).

Prepare the solution to be injected, for example 100 million DiI labeled cells/mL of cell growth medium. Place it on ice until use.
Dechorionate 50 embryos of 48–50 h post fertilization and anaesthetize them in 0.02 % MS-222 in E3 water.
Load the needle, mount it in the needle holder and turn on the gas and the micro-injection pump. If cells are injected, a needle without an inner filament should be used as the filament may compromise the integrity of the cell membrane.
Calibrate the injection volume of the needle as described above. If cells are injected make sure that the holding pressure is relatively high such that there is a steady (slow) flow of cells out of the capillary. This will reduce the risk of cells clogging up the needle.
Transfer the embryos to an agarose-coated petri dish (prepared as in the first step of the morpholino injection plate, but without making the grooves, see Fig. 17.6). If the embryos react to being poked by a metal probe, add a few more drops of concentrated (i.e. 25×) anesthesia solution.
Remove most of the water such that the embryos are covered only by a thin film of water (see Fig. 17.6). This will increase their adherence to the agarose plate and ensure that they are not simply pushed away when attempting to insert the needle through the skin.
Advance the needle and gently insert it between the yolk and the fish body but only just under the skin (see Fig. 17.6). Do not push the needle in too deep as this will place the needle tip either in the yolk, in a vein or in the muscle.
Inject the cells by low pressure and high injection time to make sure they are staying at the injection site and not pressed out over the entire surface of the yolk ball. This also is important when injecting something other than cells (i.e. growth factors or antibodies). If done correctly a small ball of cells should have been created exactly at the injection site (see Figs. 17.6 and 17.7).
Transfer the embryos back in E3 medium and put them back in the incubator.

The angiogenic or anti-angiogenic effects of the injected material/cells can now be investigated over time. If for example the injected cells are producing angiogenic factors, this can be observed as a vascularization of the cell plug over time (see Fig. 17.7) by using confocal microscopy as described below.

17.3.2 Confocal Imaging of the Vasculature in Zebrafish Embryos

This protocol is based on a Zeiss LSM 700 confocal microscope with the ZEN software (2009 version). Many of the elements are however very similar to how other confocals work, but the details might differ, so please get advice from a technician or customer service representative on how to use your particular setup before starting the investigations in the zebrafish embryos.

First mount your sample by transferring the embryos to a viewing chamber (a normal microscope glass slide with spacers in each end made from 1 to 3 cover-slides

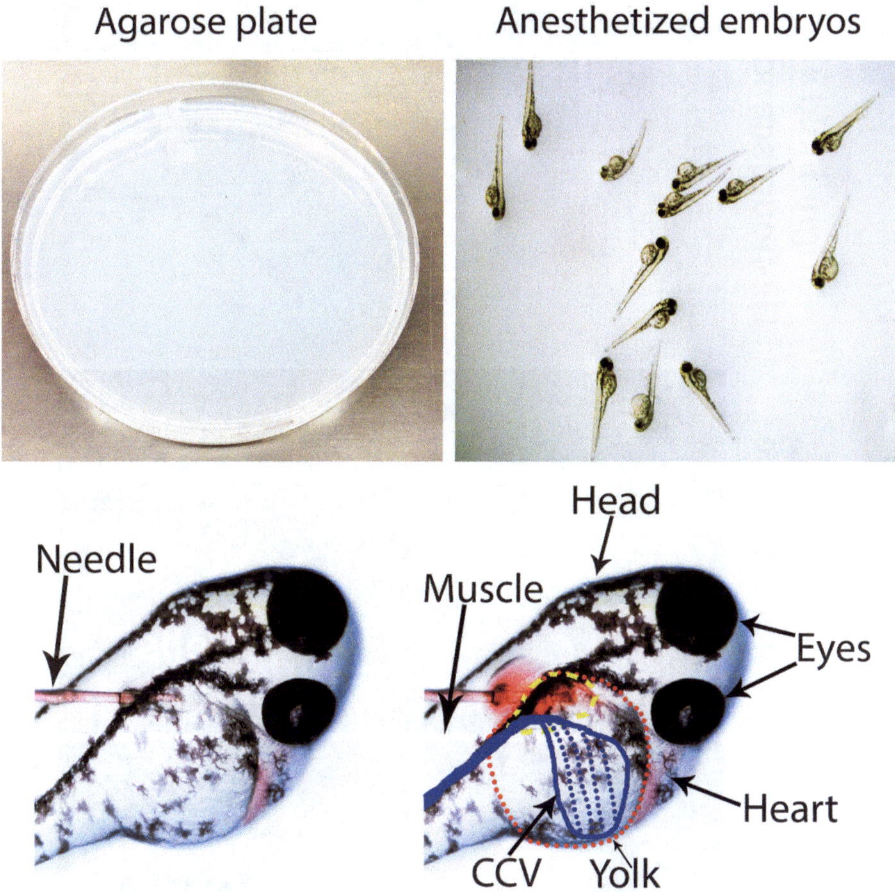

Fig. 17.6 Injection into the perivitelline space of 2 days old embryos. Embryos are placed on a plane agarose surface covered only by a thin film of anaestisia. The perivitelilne space covers the yolk, but is wider close to the muscle/body of the embryos, which is why this is a good region for injection. Make sure you do not insert the needle into the common cardinal vein (*shown in blue*) the yolk (*shown in red*) or the muscle, as this will disrupt the experimental conditions

attached to the microscope slide by with nail polish). Remove excess water and add half a drop of VectaShield. Orient the embryos under the microscope so they are lying in the desired position, i.e. lying on their side for lateral visualization or upright for dorsal visualization. Cover with a cover-slide and glue it to the spacers with additional nail polish. Leave at room temperature for fixation and infiltration of the mounting medium into the tissues of the embryo (this gives more clear visualization of the fluorescent blood vessels in non-fixed embryos). Store at –20° in the freezer unless they are visualized directly.

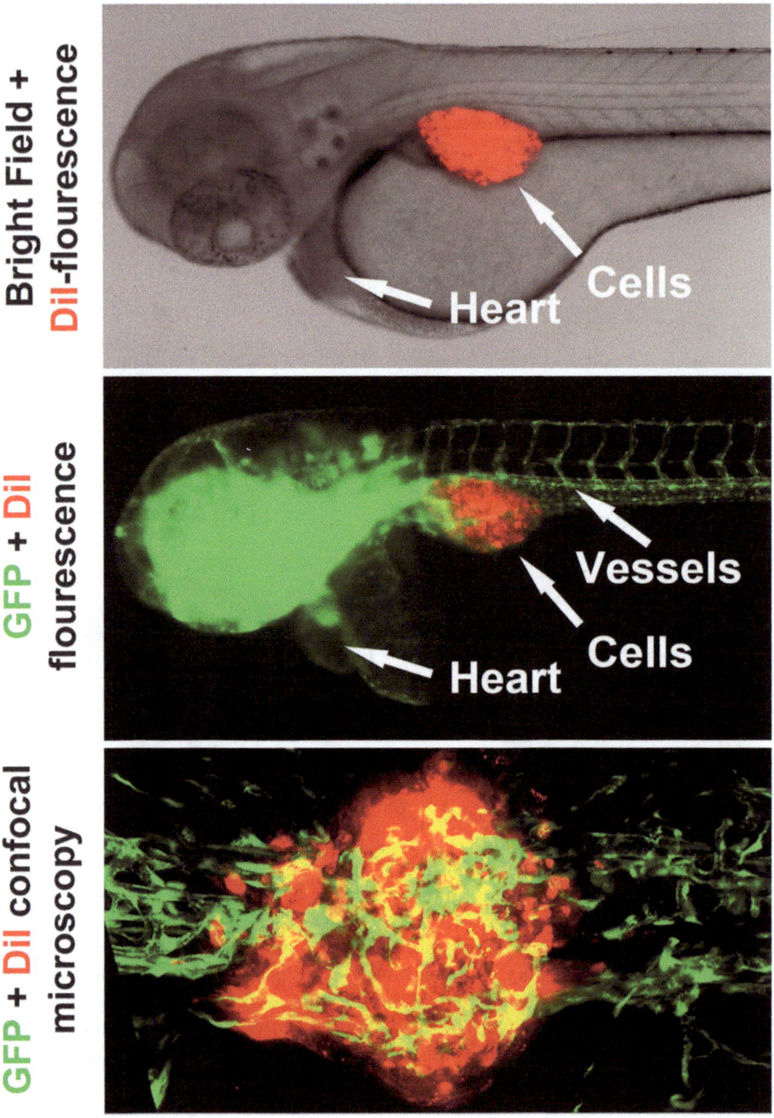

Fig. 17.7 Cells injected in the perivitelline space are imaged by fluorescent microscopy. Cells are shown in *red* and blood vessels in *green*. The correct implantation of cells should lead to a localized, coherent cell mass adjacent to the vasculature of the embryo and therefore at a suitable position for inducing angiogenic de-routing of these vessels in response to the cell-derived angiogenic factors. At the *bottom* is shown a representative image of confocal imaging of the induced angiogenic response 3 days after cell implantation

In the confocal setup, turn on all power sources associated with the microscope including the lasers, the controller box for the lasers, the microscope, the fluorescence module and the computer, in that order. Start the software (ZEN 2009) in the computer.

Once the software is turned, there will appear a new window in computer screen comprising many operational buttons. First of all click on the "OCCULOR", then click "ONLINE" and select an objective. It is good to start with the 10× objective and then change to higher magnification objectives if necessary. After selecting the objective, select the channel by selecting "FSET 10 WF" for GFP/flourescin or "FSET 20 WF" for RFP/dsRed/Rhodamine for example.

Use the binoculars to try to find a good region in the sample tissue. Once you have found a good area, click "OFFLINE" and select "ACQUISITION". By doing this you will be able to see an extension of the window appearing to right, which will allow you to open more program options. Here you can set parameters related to the image you want to take. If you have previously saved parameters it will be saved in the ACQUISITION program. You can bring those parameters into action by clicking "LOAD ACQUISITION CONFIGURATION" and click on already saved parameters. The first time you use the microscope, we advice you to have a skilled technician present, who can help you to optimize the parameters for your particular type of sample.

Now once you open your own parameters, click "LIVE" which will open a light path for the lasers to illuminate the sample, which you have already placed in focus. Tune in your parameters for example the exposure time, gain on the camera, depth of the tissue to select (Z. STACK) etc. the z stack should be defined, by adjusting the focus to the top of the sample and selecting "SET FIRST". Then adjust the focus to the bottom of the sample and select "SET LAST". Once you have selected the area and the z-stack in this way, decide on a number of averaging steps which will reduce background – for transgenic zebrafish tissues prepared as mentioned above, very little averaging is required – 1–4 will suffice. When you are ready to acquire your image, click "START EXPERIMENT". By doing this it will start grabbing different stack layers and can take up to several minutes depending upon the resolution, number of z-stack slices, colors and averaging steps selected. When the stacks are finished and image has be taken, it is possible to visualize the image in many formats e.g. 2D, 3D, combination of all the stacks, make a video of the turning of the 3D stack, etc. finally save the image in any format you want (usually .tif or .jpg is recommended).

17.4 Problems and Solutions

The most common problem in zebrafish facilities are keeping the fish healthy, strong and happy. This can only be achieved through rigorous control of the water quality, infection status, feeding regimes, group sizes and compositions. Please refer to existing guidelines or contact these authors for a suggestions related to solving such issues if you expect that any of these parameters could be optimized in your facility. Healthy fish should be good egg-layers, but asymptomatic health problems, poor feeding or stress brought about by insufficient water quality or social stress will cause the fish to stop laying eggs.

Young, virgin fish may not be as efficient egg-layers as fish with experience from several breeding rounds. We recommend that proxy-breedings is setup the first 2–5 times, in which the eggs are simply discarded and the fish put back, in order to stimulate the breeding behavior of the fish when it matters. Also, the first several rounds of breeding sometimes results in the fish laying their eggs in the evening rather than the morning. The fish usually figure this out by themselves after a few rounds of breeding, otherwise the males and females can be kept in the same aquarium but separated from interacting with each other by a wall. This wall can then be removed just at lights-on in the facility, enabling the interaction and breeding behavior of the fish.

Large embryo drop-outs during the experiment can be due to genetic polymorphisms in the fish, old age of the fish or that the experiment itself should be optimized. For example, embryos do not tolerate very large needles, as this will lead to loss of yolk through the hole generated, which will adversely affect development by itself. Also, weak embryos are susceptible to infections and killing by small water-living organisms such as colleps. This can be avoided by adding penicillin/streptomycin to the E3 medium, washing the embryos carefully before incubation or most effectively by frequent (at least once per day) sorting out dead or dying embryos, as such embryos pose the largest risk for contamination of the water.

For problems with perivitelline injections, please refer to previously published protocols, which go through all the issues related to this particular procedure [8, 10].

References

1. Jensen LD, Cao R, Cao Y (2009) In vivo angiogenesis and lymphangiogenesis models. Curr Mol Med 9:982–991
2. Isogai S, Horiguchi M, Weinstein BM (2001) The vascular anatomy of the developing zebrafish: an atlas of embryonic and early larval development. Dev Biol 230:278–301
3. Jensen LD, Cao Z, Nakamura M, Yang Y, Brautigam L, Andersson P, Zhang Y, Wahlberg E, Lanne T, Hosaka K, Cao Y (2012) Opposing effects of circadian clock genes bmal1 and period2 in regulation of VEGF-dependent angiogenesis in developing zebrafish. Cell Rep 2:231–241
4. Brautigam L, Jensen LD, Poschmann G, Nystrom S, Bannenberg S, Dreij K, Lepka K, Prozorovski T, Montano SJ, Aktas O, Uhlen P, Stuhler K, Cao Y, Holmgren A, Berndt C (2013) Glutaredoxin regulates vascular development by reversible glutathionylation of sirtuin 1. Proc Natl Acad Sci U S A 110:20057–20062
5. van Rooijen E, Voest EE, Logister I, Bussmann J, Korving J, van Eeden FJ, Giles RH, Schulte-Merker S (2010) von Hippel-Lindau tumor suppressor mutants faithfully model pathological hypoxia-driven angiogenesis and vascular retinopathies in zebrafish. Dis Model Mech 3:343–353

6. Alvarez Y, Astudillo O, Jensen L, Reynolds AL, Waghorne N, Brazil DP, Cao Y, O'Connor JJ, Kennedy BN (2009) Selective inhibition of retinal angiogenesis by targeting PI3 kinase. PLoS One 4:e7867
7. Hogan BM, Bos FL, Bussmann J, Witte M, Chi NC, Duckers HJ, Schulte-Merker S (2009) Ccbe1 is required for embryonic lymphangiogenesis and venous sprouting. Nat Genet 41:396–398
8. Rouhi P, Jensen LD, Cao Z, Hosaka K, Lanne T, Wahlberg E, Steffensen JF, Cao Y (2010) Hypoxia-induced metastasis model in embryonic zebrafish. Nat Protoc 5:1911–1918
9. Lee SL, Rouhi P, Dahl Jensen L, Zhang D, Ji H, Hauptmann G, Ingham P, Cao Y (2009) Hypoxia-induced pathological angiogenesis mediates tumor cell dissemination, invasion, and metastasis in a zebrafish tumor model. Proc Natl Acad Sci U S A 106:19485–19490
10. Nicoli S, Presta M (2007) The zebrafish/tumor xenograft angiogenesis assay. Nat Protoc 2:2918–2923

Chapter 18
Isolation and Expansion of Brain Microvascular Endothelial Cells

Stefania Elena Navone*, **Giovanni Marfia***, **and Giulio Alessandri**

18.1 Introduction

18.1.1 Brief History of Methods for Endothelial Cells Isolation

Until the end of 70 vascular pathophysiology were mostly studied using in vivo animal experimental models. However these methods did not permit the investigation of many phenotypic and functional properties of ECs because of lacking specific markers for endothelium (in particular monoclonal antibodies against endothelial antigens).

In addition EC cultures were poorly developed since the techniques for their isolation and purification were not enough efficient. The EC cultures were often contaminated by other cell types, in particular fibroblasts (or smooth muscle cells), which usually have higher growth capacity compare to ECs and therefore, they shortly became the predominant cell phenotype in culture.

Gimbrone [1] and Jaffe [2] bypassed this problem and obtained pure culture of ECs derived from large vessels like aorta or the human umbilical vein. They prepared EC cultures by detaching the endothelium of vessel wall with an enzymatic solution (containing generally collagenase-dispase) which was used to fill up the lumen of the vessel. The ECs were then successfully recovered by washing the ves-

*Author contributed equally with all other contributors.

S.E. Navone (✉) • G. Marfia
Laboratory of Experimental Neurosurgery and Cell Therapy, Neurosurgery Unit, Fondazione IRCCS Ca' Granda, Ospedale Maggiore Policlinico Milano, Milan, Italy
e-mail: stefania.navone@unimi.it

G. Alessandri
Cellular Neurobiology Laboratory, Cerebrovascular Diseases, IRCCS Foundation Neurological Institute "C. Besta", Milan, Italy

M. Slevin, G. McDowell (eds.), *Handbook of Vascular Biology Techniques*,
DOI 10.1007/978-94-017-9716-0_18

sel lumen and plated on culture dishes and grown as primary cultures that were serially passed. Gimbrone and Jaffe had been able to develop a method for ECs isolation from large vessels which is still used in many laboratories.

However, the difficult to separate ECs and stromal cells, particularly fibroblasts, from organs or tissues did not allow the isolation of ECs from small vessels or capillaries. As a consequence of this problem, therefore, the procedure did not permit to study microvascular ECs (MECs) which are, not only the most abundant ECs phenotype present in the body, but overall represent the preferential target for many hormones and growth factors which are involved in angiogenesis.

One of the first publication describing the successfully isolation of MECs was in 1979 in the laboratory of J. Folkman at Children Hospital in Boston [3]. The method led to isolate long-term culture of MECs from bovine adrenal gland (cortical part). Substantially the technique developed by J. Folkman and co-workers was based on the identification, under light microscopy, of MECs colonies rising in the gelatine coated plate after gland digestion. The procedure consisted of (1) preparation of gelatin coated dishes with 1 % (wt/vol) gelatin made up in magnesium and calcium free phosphate-buffered saline; (2) the removal from plate of all fibroblasts near the MECs colonies by scraping with Pasteur pipet tip; (3) the detachment of MECs colony with trypsine by using cloning rings and (4) the transfer of MECs into new culture wells for serial subcultures. In spite of this technique was long and difficult to apply, it was the first procedure described for isolation MECs and leaded the investigators, the possibility to examine many factors involved in tumour angiogenesis.

During the last years, the methods for isolation of MECs have improved enormously after the discovery of several endothelial cell specific markers.

One of the most commonly employed, when endothelial cells need to be identified, is the von Willebrand factor (vWF) antibody that is specifically secreted by Weibel Palade bodies of endothelial cells [4].

Other typical endothelial cell markers include: Ulex Europaeus type I lectin which selectively binds to L-fucose residues of the endothelial membrane, [5–7]; CD31, a membrane glycoprotein that is the platelet endothelial cells adhesion molecule PECAM 1 [8]; CD34, a surface glycophosphoprotein expressed on hematopoietic stem and progenitor cells, small-vessel endothelial cells, and embryonic fibroblasts [9].

At this regard the production of monoclonal antibodies (mAb) against ECs antigens could favour ECs antigen sorting purification in healthy and pathological tissues ([10, 11]).

18.2 Brain-Derived Endothelium Isolation and Culture

Focusing on central nervous system, brain microvascular endothelial cells (BMVECs) are the major element of the blood–brain barrier (BBB) and comprise the primary limitation for passaging of substances, both soluble and cellular, from the blood into the brain [12]. BMVECs are thus useful models for studying the biology of cerebral endothelium in the context of neuroinflammatory, neurodegenerative or infectious diseases and for large-scale screening of central nervous system (CNS) drug candidates.

In literature, multistep procedures including immunoselection, sieving/filtration, manual weeding, isopycnic centrifugation on primary culture or immortalized endothelial cells from brain microvasculature have been described with both advantages and limitations.

The primary BMVECs cultures, isolated from a living organism, in in vitro retain the closest similarity to phenotypic characteristics of brain endothelium [13], but their isolation from primary tissue is often difficult to perform. Primary BMVECs cultures grow in adherence conditions and are cultured in a flat un-coated plastic vessel, but sometimes a microcarrier, which can greatly increase the surface area, can be used. The attachment usually occurs about 24 h after initiation of the culture. A complete cell culture media, composed of a basal medium supplemented with appropriate growth factors and cytokines, is required. During establishment of primary BMVECs cultures, it may be useful to include an antibiotic in the growth medium to inhibit contamination introduced from the host tissue. However, long-term use of antibiotics is not advised, since some reagents may be toxic to cells over time. When primary culture have reached a desired per cent of cellular confluence (70–80 %) and are actively proliferating, they have to be subcultured. It is best to subculture primary cell cultures before reaching 100 % confluence, since post-confluent cells may undergo differentiation and exhibit slower proliferation after passaging.

Moreover, once adapted to in vitro culture conditions, primary cells undergo a limited, predetermined number of cell divisions before entering senescence. The number of times a primary cell culture can be passaged is minimal due to the Hayflick Limit, nutrient requirements and culture conditions, and the expertise by which they are manipulated and subcultured. Therefore, they are extremely time consuming and costly to generate, are easily contaminated with other cells and it is difficult to obtain large numbers of cells with a good proliferation and viability.

On the other hand, immortalized lines are easy to grow, have a low cost, allow the ECs characteristics to be maintained over many passages, permit the formation of functional barriers, are amenable to numerous molecular interventions, and facilitate the transfer of the model between different laboratories [14]. However, the introduction of immortalizing genes (i.e., telomerase and SV40 large T-antigen, hTERT) may affect a great variety of cellular functions as they may interact with numerous proteins, altering the physiological cell cycle [15]. Table 18.1 illustrates the main features of primary cell cultures and immortalized cell lines (Table 18.1).

This chapter describes in two paragraphs:

1. The methodology to isolate and establish long-term culture of human and mouse BMVECs without any genetic manipulation or immunoselection, but only main-

Table 18.1 Principal characteristics of primary cell cultures and immortalized cell lines

Features	Primary cell lines	Immortalized cell line
Cell proliferation	Limited to a small number of cell doublings	Infinite when handled properly
Genetic integrity	Retains in vivo tissue genetic makeup through cell doublings	Subject to genetic drift as cells divide
Consistency	Variability exists between donors and preparations	Minimal variability
Biological relevance	More closely mimics the physiology of cells in vivo	Relevance can drift over time as cells divide
Ease of use	Requires optimized culture conditions and careful handling	Well established conditions and robust protocols exist
Time & expense to use	More time and less abundance of cells	Less time and more abundance of cells

taining cultures of ECs in the new formulated Endothelial Proliferation Medium (EndoPM);

2. The protocol to establish BBB in vitro model to investigate ECs permeability.

18.2.1 Brain Microvascular Endothelial Cells Isolation and Expansion by EndoPM

EndoPM is a patented medium (MI2011A 000201, PCT and Italian patent) that allows and facilitates the isolation and expansion of long-term cultured BMVECs from human and murine brain biopsies in in vitro conditions. The newly formulated medium is based on our previous studies in the isolation, culture and characterization of ECs from different human organs [10, 16]. Such invention is suitable for the generation of an in vitro model of BBB in the field of cerebrovascular diseases' investigation. It guaranties the isolation and maintenance of BMVECs with high purity, in any laboratory and with minimal equipment and accessories. The properties of BMVECs in human and murine cultures are confirmed by immunofluorescence and immunophenotypical analyses (Fig. 18.1). BMVECs, so isolated, express endothelial markers (CD31, CD105, CD146, Tie-2 and vWF), have considerable tube-forming ability when cultured onto Matrigel, show intense endocytosis of Dil-Ac-LDL, eNOS and specific transporters (i.e., GLUT-1) and show P-gp expression. Moreover BMVECs are able to express markers of the junctional architecture such as Ve-cadherin, β-catenin and Claudin-5, demonstrating their in vitro capability to show typical features of BBB. In addition, BMVECs, in confluence culture condition, inhibit macromolecules transcytosis (as dextran) as demonstrated by vascular permeability assay [17].

Moreover BMVECs spontaneously organize in vascular-like structures and maintain the expression of endothelial markers in a xenograft model in the CD1

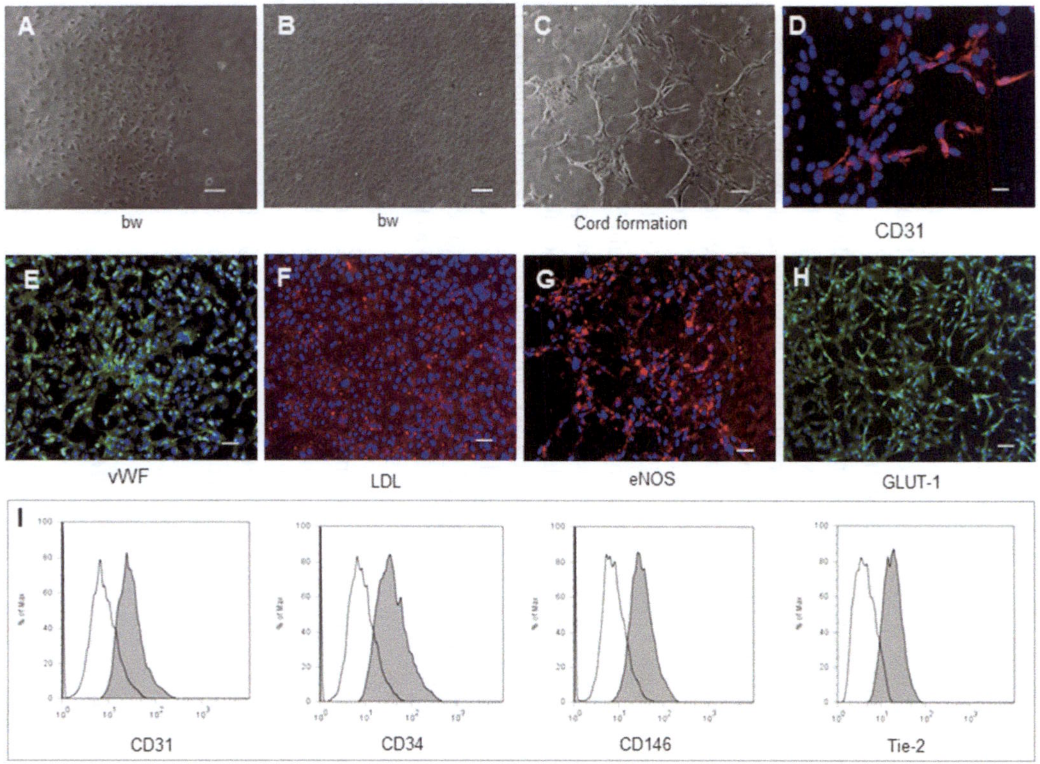

Fig. 18.1 Characterization and functional features of BMVECs cultured in EndoPM condition. (**a, b**): Phase-contrast micrographs of confluent monolayers of BMVECs, scale bars 200 µm. The cells present the typical 'cobblestone' appearance and (**c**) produce capillary tube-like structure formation 7 h after planting onto Matrigel, *scale bar* 100 µm. (**d**) BMVECs showed a clear cytoplasmic staining for CD31 (*Scale bar*, 100 µm), (**e**) an intense positive immunofluorescence for vWF (*Scale bar*, 50 µm) (**f**) accompanied by intense endocytosis of Dil-labeled acetylated low-density lipoprotein (Dil-Ac-LDL), (**g**) endothelial nitric oxide synthase (eNOS) and (**h**) glucose transporter 1 (Glut-1). Scale bars, 50 µm. (**i**) The flowcytometry analyses revealed positivity for CD31, CD34, CD146 and Tie-2. All nuclei were counterstained with DAPI (*blue*) (Figure published in [17])

mouse strain. The significant effect of EndoPM is confirmed studying the proliferation index and survival index. Moreover, BMVECs kept in co-culture condition with fibroblasts, enhance their number over the passages to the detriment of fibroblasts.

Our method can allow researchers to improve knowledge on phenotypical and functional features of brain microvasculature under physiological and pathological conditions [18].

Our protocol can be divided in four different steps:

1. Brain biopsies are finely minced using surgical scissors and scalpel and then incubated for 1 h at 37 °C in PBS containing 0.25 % Liberase (Roche) in order to obtain a heterogeneous cell suspension.
2. The digestion is stopped by resuspending the homogenate with 20 mL of sterile D-PBS at room temperature. The suspension is centrifugated at 276 g for 10 min at room temperature. The digestion medium is aspirated and discarded.
3. The cells are then plated onto flask previously coated with bovine collagen type I (a substrate that enhanced cellular adhesion) and cultured in EndoPM, containing basic Fibroblsat Growth Factor (bFGF), Epidermal Growth Factor (EGF), Brain Bovin Extract (BBE), Fetal Bovine Serum (FBS) at low concentration, hydrocortisone, heparin and all factors necessary to sustain isolation and growth of BMVECs. The cells are maintained at 37 °C in an atmosphere of 5 % CO_2.
4. After, more or less 24 h, plated cells are washed and the supernatant are transferred into another collagen-coated flask in EndoPM to favour adhesion of BMVECs. The medium are changed every 3 days. Once at confluence, cells are detached by trypsinization with 0.25 % Tryple (Gibco) and reseeded on collagen-coated culture dish at a split ratio of 1:3. This is assumed as the first in vitro passage of BMVECs.

This protocol allows to isolate human and mouse BMVECs from different mouse strains (male CD1, NOD/SCID and male wr/wr). BMVECs cultured for more than 60 days in vitro continuously maintained endothelial features without fibroblast contamination. Moreover, these BMVECs maintain high viability and express morphological and functional features after cryopreservation and thawing.

18.2.2 Endothelial Cells Permeability

The BBB formed by the cerebral microvascular endothelium, represents an essential component together with supporting structures for the health and function of the central nervous system (CNS). In several diseases, including stroke, encephalitis, and brain tumors, the permeability of the BBB is increased, and its loss of integrity may reflect the severity or progression of a disease [19]. The set up of a BBB in vitro model may significantly help scientists to understand the mechanisms involved in the cerebrovascular response to a number of physiological and pathological stimuli and, in turn, could provide new strategies to accelerate the development of novel CNS drug therapies and reduce the burden of major neurological disorders.

This is the rationale for trying to measure the BBB permeability, using either in vivo imaging, such as MRI [20], or in vitro models such as Fluorescein isothiocyanate (FITC)–dextrans [21].

In vitro model for measuring BBB permeability in situ, include the use of radiolabeled tracers in parenchymal vessels, measurements of transendothelial resistance (TEER) across the endothelial monolayer and low passive, non-specific paracellular permeability to small and large molecules binding fluorescent dyes such as Lucifer yellow (LY), hydrophobic compounds and FITC-labelled dextrans. Other features such as expression of receptors and transporters on the EC surface and intracellular transcytosis machinery must be maintained to allow transcellular transport pathways for ions, small molecules, peptides and proteins to be reconstituted in vitro.

Our procedure to set up an in vitro model for BBB permeability is based on three steps:

1. The plating of the cells: 2×10^4 BMVECs are plated on collagen coated insert of 6 wells Transwell with membrane filter (0.4-μm pore size) in EndoPM in the upper and in the lower chamber. Five days before assay the cells are grown in EndoPM without growth factors until they have reached confluence (day 3 = semiconfluence).
2. The confirming of the confluence: the confluence is determined and confirmed by hematoxylin-eosin staining of sentinel wells.
3. The evaluation of permeability by FITC-dextran efflux: FITC-dextran (4 μL, 25 mg/mL initial concentration) is filled to the upper insert. Every 30 min, 50 μL of medium is collected from the lower chamber. The aliquots are diluted to 1 mL with 1× PBS. 100 μL of each diluted sample are transferred into 96-well black plates and the fluorescent content at 492/520 nm absorption/emission wavelengths for FITC-dextran is measured.

In summary, our in vitro BBB model is characterized by expression of markers of the junctional architecture such as Ve-cadherin, β-catenin and Claudin-5, and low permeability. Our experiments have proven that the model is suitable for the study of different aspects of BBB function in basic research and for testing the interaction between the BBB and potential drug candidates (toxicity, permeability, interaction with efflux transporters) as well.

In conclusion, our protocol allows the isolation of human and murine BMVECs with extended lifetime, good viability and purity without alterations to their endothelial features.

Cell lines cultured with our method could be a useful model for studying the biology of cerebral endothelium in the context of tumor angiogenesis, neuro-inflammatory, neurodegenerative or infectious diseases and for large-scale screening of CNS drug candidates.

It should be noted that we have just begun application of BMVECs isolated with our method for the studies in cerebrovascular diseases area, and it is expected that further efforts to strengthen our knowledge of their functional properties will add significant value to the cells, which are known to be essential to the establishment of in vitro cerebrovascular diseases and BBB models.

Here we demonstrate improved efficiencies for BMVECs isolated specifically from human and mouse brain biopsies. In the future these conditions should be investigated for the isolation of ECs from other body districts, such as adipose tissue, skin or bone marrow.

References

1. Gimbrone MA Jr, Cotran RS, Folkman J (1974) Human vascular endothelial cells in culture. Growth and DNA synthesis. J Cell Biol 60(3):673–684
2. Jaffe EA, Hoyer LW, Nachman RL (1974) Synthesis of von Willebrand factor by cultured human endothelial cells. Proc Natl Acad Sci U S A 71(5):1906–1909
3. Folkman J, Haudenschild CC, Zetter BR (1979) Long-term culture of capillary endothelial cells. Proc Natl Acad Sci U S A 76(10):5217–5221
4. Peng G, Wen X, Shi Y, Jiang Y, Hu G, Zhou Y, Ran P (2013) Development of a new method for the isolation and culture of pulmonary arterial endothelial cells from rat pulmonary arteries. J Vasc Res 50(6):468–477
5. Holtfhofer H, Virtanen I, Kariniemi AL et al (1982) Ulex europaeus 1 lectin as a marker for vascular endothelium in human tissues. Lab Invest 47:60–66
6. Jackson CJ, Garbett PK, Nissen B, Schrieber L (1990) Binding of human endothelium to Ulex europaeus I-coated Dynabeads: application to the isolation of microvascular endothelium. J Cell Sci 96(2):257–262
7. Rafii S, Shapiro F, Rimarachin J, Nachman RL, Ferris B, Weksler B, Moore MA, Asch AS (1994) Isolation and characterization of human bone marrow microvascular endothelial cells: hematopoietic progenitor cell adhesion. Blood 84(1):10–19
8. Marelli-Berg FM, Peek E, Lidington EA, Stauss HJ, Lechler RI (2000) Isolation of endothelial cells from murine tissue. J Immunol Methods 244(1–2):205–215
9. Krause DS, Fackler MJ, Civin CI et al (1996) CD34: structure, biology, and clinical utility. Blood 87:1–13
10. Invernici G, Ponti D, Corsini E, Cristini S, Frigerio S, Colombo A, Parati E, Alessandri G (2005) Human microvascular endothelial cells from different fetal organs demonstrate organ-specific CAM expression. Exp Cell Res 308(2):273–282
11. Ricci-Vitiani L, Pallini R, Biffoni M, Todaro M, Invernici G, Cenci T, Maira G, Parati EA, Stassi G, Larocca LM, De Maria R (2010) Tumour vascularization via endothelial differentiation of glioblastoma stem-like cells. Nature 468(7325):824–828
12. Abbott NJ, Patabendige AAK, Dolman DEM, Yusof SR, Begley DJ (2010) Structure and function of the blood–brain barrier. Neurobiol Dis 37:13–25
13. Dorovini-Zis K, Prameya R, Bowman PD (1991) Culture and characterization of microvascular endothelial cells derived from human brain. Lab Invest 64:425–436
14. Li G et al (2010) Permeability of endothelial and astrocyte cocultures: in vitro blood–brain barrier models for drug delivery studies. Ann Biomed Eng 38:2499–2511
15. Sano Y et al (2010) Establishment of a new conditionally immortalized human brain microvascular endothelial cell line retaining an in vivo blood–brain barrier function. J Cell Physiol 225:519–528
16. Falchetti ML, Mongiardi MP, Fiorenzo P, Petrucci G, Pierconti F, D'Agnano I, D'Alessandris G, Alessandri G, Gelati M, Ricci-Vitiani L, Maira G, Larocca LM, Levi A, Pallini R (2008) Inhibition of telomerase in the endothelial cells disrupts tumor angiogenesis in glioblastoma xenografts. Int J Cancer 122(6):1236–1242

17. Navone SE, Marfia G, Nava S, Invernici G, Cristini S, Balbi S, Sangiorgi S, Ciusani E, Bosutti A, Alessandri G, Slevin M, Parati EA (2013) Human and mouse brain-derived endothelial cells require high levels of growth factors medium for their isolation, in vitro maintenance and survival. Vasc Cell 5(1):10–18
18. Navone SE, Marfia G, Invernici G, Cristini S, Nava S, Balbi S, Sangiorgi S, Ciusani E, Bosutti A, Alessandri G, Slevin M, Parati EA (2013) Isolation and expansion of human and mouse brain microvascular endothelial cells. Nat Protoc 8(9):1680–1693
19. Hawkins BT, Davis TP (2005) The blood–brain barrier/neurovascular unit in health and disease. Pharmacol Rev 57(2):173–185
20. Nagaraja TN, Ewing JR, Karki K, Jacobs PE, Divine GW, Fenstermacher JD, Patlak CS, Knight RA (2011) MRI and quantitative autoradiographic studies following bolus injections of unlabeled and (14)C-labeled gadolinium-diethylenetriamine-pentaacetic acid in a rat model of stroke yield similar distribution volumes and blood-to-brain influx rate constants. NMR Biomed 24(5):547–558
21. Watson PM, Paterson JC, Thom G, Ginman U, Lundquist S, Webster CI (2013) Modelling the endothelial blood-CNS barriers: a method for the production of robust in vitro models of the rat blood–brain barrier and blood-spinal cord barrier. BMC Neurosci 14:59–64

Part III
Imaging and Histological Analysis

Chapter 19
Adipose Angiogenesis

Carina Fischer, Sharon Lim, Jennifer Honek, and Yihai Cao

19.1 Introduction

Obesity and obesity-associated metabolic disorders such as cardiovascular disease, stroke, diabetes and even certain types of cancer, are on the rise and account for the majority of deaths worldwide [1, 2]. Conventionally, non-invasive strategies to combat obesity in humans include lifestyle changes and increased activity in order to achieve a negative energy balance. Those measures often only provide short term improvements. Furthermore, there are some pharmacological approaches such as the drugs Orlistat or Sibutramine, aiming at increasing metabolic rate, suppressing appetite or inhibiting nutrient absorption. However, these drugs often are associated with side effects that can result in low compliance rates [3–6]. If weight management cannot be achieved via these measures, gastric bypass surgery often proves effective in severely overweight patients [4, 7]. This invasive procedure however can lead to decreased quality of life as the patients often experience depression or other adverse effects. Consequently, there is an urgent need to develop novel potential therapeutic approaches for the treatment of obesity and metabolic disorders.

It has been hypothesized that targeting the adipose vasculature may be a promising approach to counteract obesity [2]. There are two distinct types of

---

Author contributed equally with all other contributors.

C. Fischer • S. Lim • J. Honek
Department of Microbiology, Tumor and Cell Biology, Karolinska Institute,
17177 Stockholm, Sweden

Y. Cao (✉)
Department of Medicine and Health Sciences, Linköping University, Linköping, Sweden

Department of Microbiology, Tumor and Cell Biology, Karolinska Institute,
Stockholm, Sweden
e-mail: yihai.cao@ki.se

M. Slevin, G. McDowell (eds.), *Handbook of Vascular Biology Techniques*,
DOI 10.1007/978-94-017-9716-0_19

adipose tissue, White Adipose Tissue (WAT) and Brown Adipose Tissue (BAT). The WAT stores excess energy in the form of triglycerides. When required, lipolysis can be induced, stored triglycerides are broken down and glycerol and free fatty acids are released into the blood stream [8]. Morphologically, adipocytes in the WAT consist of a large unilocular lipid droplet that is surrounded by a very thin layer of cytoplasm. Adipocytes in the BAT on the other hand are significantly smaller in size and they are densely packed with mitochondria expressing the BAT-specific Uncoupling Protein (UCP)-1 which can generate heat by uncoupling oxidative phosphorylation from ATP generation. This process termed Non-shivering Thermogenesis (NST) is exploited by hibernating animals and also keeps human babies warm [9–11]. Although it has long been assumed that BAT is only present in human infants, this heat-generating adipose depot has recently been identified in adults as well. It is now widely accepted that adult humans possess BAT and that it can be activated by cold-exposure [12–14]. However, a study has recently shown that human BAT resembles more closely to Beige Adipose Tissue in mammals than BAT, regarding morphology and molecular signature [15, 16].

The adipose tissue is a highly plastic tissue that undergoes regression and expansion even in adult individuals. Adipose tissue growth requires a switch of angiogenesis in order to supply the growing tissue with oxygen, nutrients, circulating stem cells and immune cells but also to aid in the removal of metabolic waste products which particularly in the highly metabolically active BAT. Angiogenesis is regulated by a myriad of pro- and anti-angiogenic factors such as Vascular Endothelial Growth Factor (VEGF)-A or Fibroblast Growth Factor (FGF) that are produced by adipocytes, inflammatory cells or stromal cells within the adipose depots [1, 2].

19.2 Methodology

This chapter describes an in vivo model to study adipose angiogenesis in mice. As opposed to in vitro angiogenesis assays, the described model recapitulates vascular alterations as well as the impact on whole body energy homeostasis. In order to detect microvessels in the adipose tissue, we describe immunodetection using whole-mount staining with freshly fixed tissues and paraffin-embedded tissue samples. When studying whole-mount tissue, the three-dimensional vessel structure can be visualized using confocal imaging. This technique allows for the combination of different markers and thereby provides information regarding the interplay of different cell types within the adipose tissue. Immunodetection of adipocytes and vascular cells in paraffin-embedded tissue samples provides quantitative information about angiogenesis and remodeling of the vasculature in adipose depots [17].

19.3 Materials (And Company Name)

19.3.1 Reagents

1×PBS (Hyclone, cat. no. SH30028.02)
4′,6-diamidino-2-phenylindole, dilactate (DAPI, dilactate) (Invitrogen, cat. no. D3571)
Alexa Fluor 488 donkey anti-rabbit IgG (H+L) antibody (Invitrogen, cat. no. A-21206)
Alexa Fluor 555 goat anti-rat IgG (H+L) antibody (Invitrogen, cat. no. A-21434)
Antigen unmasking solution, Citric Acid Based (Vector Laboratories, cat. no. H-3300)
Colorless nail polish (cosmetic shop)
Distilled water (dH_2O)
Dried fat-free milk (e.g., Semper, Sweden)
Dry ice
Ethanol 99.7 % (Solveco AB, cat. no. 200-578-6)
Hematoxylin Solution, Harris modified (Sigma-Aldrich, cat. no. HHS16)
Methanol (Sigma-Aldrich, cat. no. 32213)
Non-immune goat serum (Vector Laboratories, cat. no. S-1000)
Paraformaldehyde (PFA) (Sigma-Aldrich, cat. no. 441244) Pertex (Histolab Products Ab, cat. no. 00801)
Proteinase K (Invitrogen, cat. no. 25530–049)
Rat anti-mouse CD31 monoclonal antibody, MEC13.3 (BD Pharmingen, cat. no. 553370)
Rat anti-mouse CD31 monoclonal antibody (Dianova, DIA-310M)
Triton X-100 Polyoxyethylene [10] octylphenyl ether (Acros Organics, cat. no. 215680010) Vectashield mounting medium (Vector Laboratories, cat. no. H-1000)
Xylen (Histolab Products Ab, cat. no. 02080)

19.3.2 Equipment

Adobe Photoshop CS3 or later versions (Adobe)
BD falcon 50 ml polypropylene conical tubes (BD Biosciences, cat. no. 358206)
Confocal microscope (e.g., Nikon D-eclipse C1)
Confocal software (e.g., EZ-C1 3.9 Nikon digital eclipse)
Costar 6-well cell culture plates (Corning, cat. no. 3516)
Costar 96-well cell culture plates (Corning, cat. no. 3596)
Dry-line Oven (VWR, cat. no. DL53)
Fluorescence microscope (e.g. Nikon)

Fluorescence software (e.g. NIS Elements D 3.2)
Forceps (AgnTho's AB, cat. no. 08-060-120)
Microscope cover slips (VWR International, cat. no. 631–0135)
Microtome paraffin (Cellab, cat. no. Microm HM315)
Rocking board (VWR International, cat. no. 444–0341)
Scalpel blade (AgnTho's AB, cat. no. 02-040-010)
Scalpel blade holder (AgnTho's AB, cat. no. 02-030-030)
Spatula/microspoon (VWR International, cat. no. 231–1354)
Superfrost Plus microscope slides (Thermo Scientific, cat. no. 4951plus)
Timer (Fisher Scientific, cat. no. FB70232))
Vertical staining jar with glass lid (Electron Microscopy Sciences, cat. no. 70318–04)
Water bath up to 60 °C (e.g., Lauda Aqualine Al5)

19.4 Basic Protocol

19.4.1 Adipose Tissue Whole-Mount Staining Protocol

1. Sacrifice mice using carbon dioxide, remove blood from mice by cardiac puncture using a 1 ml syringe and 22 gauge needle.
2. Immediately dissect and isolate adipose tissues, e.g. epididymal white adipose tissue (eWAT), inguinal WAT (iWAT) or interscapular brown adipose tissue (iBAT)
3. Place the tissue in freshly prepared 4 % PFA solution and keep it at 4 °C up to 24 h (dependent on the size of the tissue)
4. Wash tissue twice 30 min with PBS
5. Slice the tissue into around 5 × 5 mm large pieces with a scalpel blade and transfer the pieces into a 12-well plate for further washing for 1 h in PBST

   (a) This step is necessary to remove remaining PFA

6. Incubate the tissue with proteinase K (20 μg/mL) in 10 mM Tris–HCl buffer (pH 7.4) for 5 min on room temperature (RT)
7. Incubate in 100 % methanol (30 min; RT; chemical fume hood) to dehydrate tissue
8. Wash three times 1 h with 1 × PBS on a rocking board
9. Block unspecific antibody binding sides by incubating the tissue in 3 % milk in PBST overnight (o/n)
10. Wash the tissue with PBS and incubate with primary antibodies o/n

    (a) For endothelial cell staining use CD31 antibody (1:200 dilution in PBS) and Perilipin antibody (1:400 in PBS) to label differentiated adipocytes

11. To wash away non-binding primary antibodies, the tissue is washed with PBST (1.5 h, 4 °C)

12. Incubate the tissue with a conjugated secondary antibody solution (3 % milk in PBS) for 2 h, RT
    (a) The dilution of secondary antibody depends on the strength of the primary antibody binding and the type of secondary antibody, but lies within a range of 1:100–1:500 dilution
13. Incubate the tissue with blocking buffer (0.15 % milk in PBST, 1 h, RT)
14. Wash the tissue for 2 h with 1×PBS
15. Transfer the tissue on a glass slide, mount it with Vectashield mounting medium and secure it with a cover slip.
    (a) Slides can be stored at −20 °C for several weeks
16. Take images using a confocal microscope with a 3D imaging software e.g. Nikon D-eclipse C1 with EZ-C1 3.90 software or any equivalent confocal microscope
    (a) Define a suitable magnification for your purpose
    (b) If the signals should be quantified, an universal layer depth and a defined amount number of layers should be used
17. Further quantitative analysis can be done using an *Adobe Photoshop* or *Image Quant* software

19.4.2 IHC Staining for Paraffin-Embedded Adipose Tissue

1. Isolate and fix the tissue as described above (step 1–4)
2. Use a microtome to section the tissue according to your needs (3–10 μm/section)
3. Incubate the slides for ≥2 h at 60 °C to melt the paraffin and let it cool down before proceeding with the staining
4. Dehydrate the slides using an xylene/ethanol gradient:

Substance	Incubation time	Repetitions
100 % Xylene	5 min	2×
99.7 % EtOH	5 min	2×
95 % EtOH	5 min	2×
70 % EtOH	5 min	2×
ddH_2O	5 min	2×

5. Boil the slides in antigen unmasking solution in a microwave (100 % power for 2:30 min, then 20 % power for 10 min)
6. Wash the tissue with PBS twice (5 min each)
7. Block the tissue with 4 % goat serum in PBS for 45 min at room temperature
8. Wash the tissue with PBS twice (5 min each)
9. Incubate for 15 min in avidin-blocking buffer

10. Wash tissue with PBS twice (5 min each)
11. Incubate for 15 min in biotin-blocking buffer
12. Wash tissue with PBS twice (5 min each)
13. Incubate in primary antibody at 4 °C o/n
    (a) For endothelial cell staining: biotinylated isolectin B4 (1:500 in PBS 4 % goat serum)
    (b) For adipocyte staining: guinea pig anti-Perilipin (1:400 in PBS 4 % goat serum)
14. Wash with PBS twice (5 min each)
15. Incubate the tissue with a conjugated secondary antibody solution (4 % goat serum in PBS) for 1 h at room temperature
    (a) The dilution of secondary antibody depends on the strength of the primary antibody binding and the type of secondary antibody, but lies within a range of 1:100–1:500 dilution
    (b) For detecting isolectin staining use streptavidin (SA)-Cy2 antibody
    (c) For perilipin staining any conjugated antibody against guinea pig can be used
16. Wash tissue twice with PBS (5 min each)
17. Mount with Vectashield mounting medium
    (a) This medium is available with and without DAPI for nuclear staining
18. Store samples at –20 °C until further analysis
19. Images can be taken with a fluorescence microscope in the appropriate magnification using NIS Elements D 3.2 software

19.5 Expected Results

The identification of microvessels in adipose tissue is essential to characterize the interplay between adipocytes and the vasculature. Whole-mount staining allows studying the alignment of vessels along adipocytes in 3D, which makes it possible to study their architecture, density and structure. To visualize endothelial cells in adipose tissue, the endothelial marker CD31 (red color) was used (Fig. 19.1). For quantitative studies, the area occupied by blood vessels in a defined field of adipose tissue can be assessed. This can be achieved using *Adobe Photoshop* software to quantify the number of pixels illustrating $CD31^+$ staining. With this technique, noticeable differences in the number of microvessels between WAT and BAT can be detected.

Another method is the immunohistochemical doublestaining of isolectin B4 (red color) with the adipose surface marker Perilipin (green color) on paraffin sections, which makes it possible to quantify parameters like number of vessels per adipocyte or the average cell diameter of single adipocytes (Fig. 19.2). This method gives more

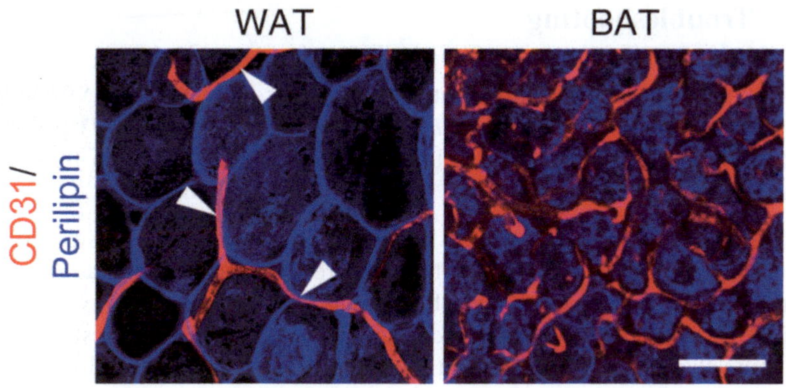

Fig. 19.1 Whole-mount staining of White and Brown Adipose tissue. WAT and BAT were stained with the endothelial marker CD31 (*red*) and Perilipin (*blue*). *White arrowheads* indicate CD31⁺ endothelial cells. Scale bar: 50 μm

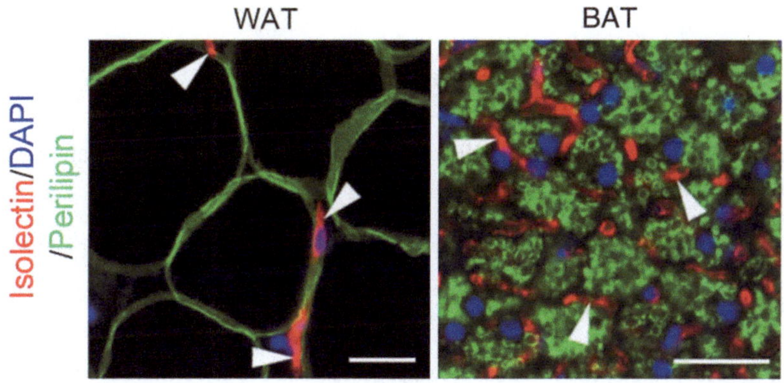

Fig. 19.2 Immunohistochemical staining of paraffin-embedded White and Brown Adipose tissue. Paraffin-embedded WAT and BAT was cut into 5 μm thick sections and co-stained with Isolectin B4 (*red*) and Perilipin (*green*). Nuclei are shown in *blue* color. *White arrowheads* indicate Isolectin B4⁺ endothelial cells. Scale bars: 25 μm

precise information regarding changes in the number of blood vessels after for instance treatment of mice with a certain drug. When only employing endothelial cell staining of whole-mount tissue, vessel density can be altered solely due to alterations in the adipocyte diameter and thereby in the intercapillary distance. Co-staining of adipocytes and endothelial cells on the other hand allows for quantitative studies of increased or decreased number of blood vessels per adipocyte. For the quantification of this type of immunostaining, adipocytes and vessels are counted. In this procedure, we have developed a scoring system where the vessels are attributed a score between 1 and 3 for short, intermediate and long vessels, respectively. The sum of the vessel is then divided by the number of counted adipocytes.

19.6 Troubleshooting

It can be difficult to localize different adipose depots (i.e. iWAT, eWAT and iBAT) in transgenic mice such as in *ob/ob*, *db/db* transgenic mouse strains. Therefore, one should be very careful when dissecting adipose tissue depots from these mouse strains.

Whole-Mount Staining Always use new blades or a sharp pair of scissors when cutting adipose tissues into thin sections to prevent damaging the adipose tissue structures. Non-specific fluorescent background can be prevented by using fresh 4 % PFA for tissue fixation and to increase the time for the blocking step with 3 % milk. Auto-fluorescence in adipose tissues can be prevented by adding sufficient Vectashield mounting medium to cover adipose tissues.

Paraffin Staining It is essential to replace new microtome blades when cutting paraffin sections to prevent damaged to the paraffin embedded adipose tissue sections. Tissue sections should never be dry out after antigen retrieval step.

19.7 Method Variations/Alternative Staining

The use of other secondary antibodies than the suggested should be titrated to ensure that the dilution is optimal.

References

1. Cao Y (2007) Angiogenesis modulates adipogenesis and obesity. J Clin Invest 117(9):2362–2368, Epub 2007/09/06
2. Cao Y (2010) Adipose tissue angiogenesis as a therapeutic target for obesity and metabolic diseases. Nat Rev Drug Discov 9(2):107–115, Epub 2010/02/02
3. Brower V (2002) Fighting fat. New drugs against obesity in the pipeline. EMBO Rep 3(7):601–603, Epub 2002/07/09
4. Cooke D, Bloom S (2006) The obesity pipeline: current strategies in the development of anti-obesity drugs. Nat Rev Drug Discov 5(11):919–931, Epub 2006/11/03
5. Finer N (2002) Sibutramine: its mode of action and efficacy. Int J Obes Relat Metab Disord 26(Suppl 4):S29–S33, Epub 2002/11/29
6. Foxcroft DR, Milne R (2000) Orlistat for the treatment of obesity: rapid review and cost-effectiveness model. Obes Rev 1(2):121–126, Epub 2002/07/18
7. Finer N (2002) Pharmacotherapy of obesity. Best Pract Res Clin Endocrinol Metab 16(4):717–742, Epub 2002/12/07
8. Cinti S (2005) The adipose organ. Prostaglandins Leukot Essent Fatty Acids 73(1):9–15, Epub 2005/06/07
9. Cannon B, Nedergaard J (2010) Metabolic consequences of the presence or absence of the thermogenic capacity of brown adipose tissue in mice (and probably in humans). Int J Obes (Lond) 34(Suppl 1):S7–S16, Epub 2010/10/12

10. Richard D, Picard F (2011) Brown fat biology and thermogenesis. Front Biosci (Landmark Ed) 16:1233–1260, Epub 2011/01/05
11. Sethi JK, Vidal-Puig AJ (2007) Thematic review series: adipocyte biology. Adipose tissue function and plasticity orchestrate nutritional adaptation. J Lipid Res 48(6):1253–1262, Epub 2007/03/22
12. Lee P, Zhao JT, Swarbrick MM, Gracie G, Bova R, Greenfield JR et al (2011) High prevalence of brown adipose tissue in adult humans. J Clin Endocrinol Metab 96(8):2450–2455, Epub 2011/05/27
13. van Marken Lichtenbelt WD, Vanhommerig JW, Smulders NM, Drossaerts JM, Kemerink GJ, Bouvy ND et al (2009) Cold-activated brown adipose tissue in healthy men. N Engl J Med 360(15):1500–1508, Epub 2009/04/10
14. Virtanen KA, Lidell ME, Orava J, Heglind M, Westergren R, Niemi T et al (2009) Functional brown adipose tissue in healthy adults. N Engl J Med 360(15):1518–1525, Epub 2009/04/10
15. Lidell ME, Betz MJ, Dahlqvist Leinhard O, Heglind M, Elander L, Slawik M et al (2013) Evidence for two types of brown adipose tissue in humans. Nat Med 19(5):631–634, Epub 2013/04/23
16. Wu J, Bostrom P, Sparks LM, Ye L, Choi JH, Giang AH et al (2012) Beige adipocytes are a distinct type of thermogenic fat cell in mouse and human. Cell 150(2):366–376, Epub 2012/07/17
17. Xue Y, Lim S, Brakenhielm E, Cao Y (2010) Adipose angiogenesis: quantitative methods to study microvessel growth, regression and remodeling in vivo. Nat Protoc 5(5):912–920, Epub 2010/05/01

Chapter 20
Assessing Tumor Angiogenesis in Histological Samples

E. Fakhrejahani and M. Toi

20.1 Introduction

Angiogenesis is a hallmark of cancer [1] and occurs in most human tumors. It has been shown that angiogenic tumors are more likely to develop metastasis and exhibit resistance to standard cancer therapies [2], making tumor angiogenesis a prognostic and sometimes predictive biomarker [3, 4]. Although new imaging technologies, such as dynamic contrast-enhanced magnetic resonance imaging (DCE-MRI) or positron emission tomography-computed tomography (PET-CT) scans, are clinically available to visualize tumor angiogenesis in vivo [5, 6], histological assessment of tumor angiogenesis remains a technique of interest, as it can provide information on the capillary level of newly developed microvessels in different parts of the tumor [7]. Via histological examination, the relationship between tumor microvessels and other clinicopathological tumor characteristics can be evaluated as well [8].

Although studies examining the grading of tumor angiogenesis date back to the 1980s [9], the introduction of more specific endothelial markers and the quantitative immunohistochemical study by Weinder et al. [10] in the early 1990s have made tumor angiogenesis a topic of active research by many investigators. However, due to a lack of standard methods for the identification and quantification of capillaries and inter- and intra-observer variation, the prognostic role of tumor angiogenesis has not been confirmed [11]. It is important to note that the angiogenic profiles of tumors do not always correspond to the histological grade in breast cancer and some other solid cancers [12, 13]. Additionally, the therapeutic effects of antiangiogenic agents in relation to angiogenesis activity are diverse [14]. These issues complicate understanding the role of tumor angiogenesis in tumor progression.

E. Fakhrejahani • M. Toi (✉)
Department of Breast Surgery, Graduate School of Medicine, Kyoto University,
54 Shogoin-Kawaharacho, Sakyo-ku, Kyoto 606-8507, Japan
e-mail: toi@kuhp.kyoto-u.ac.jp

M. Slevin, G. McDowell (eds.), *Handbook of Vascular Biology Techniques*,
DOI 10.1007/978-94-017-9716-0_20

Quantification of tumor angiogenesis by counting microvessels in immunostained tissue sections was ranked as category III (meaning: "all factors which are not sufficiently studied to demonstrate their prognostic value") based on the 1999 consensus of the College of American Pathologists [15–17], which has not yet been updated. However, with increasing evidence of the clinical usefulness of anti-angiogenic therapies in various cancers, the assessment of tumor angiogenesis to determine cases in which these new therapies are more likely to produce better results is transitioning from research laboratories to routine diagnostic pathological laboratories.

In this chapter, we briefly discuss the current optimal protocol based on the "Second international consensus on the methodology and criteria of evaluation of angiogenesis quantification in solid human tumors" [18]. We also briefly explain double staining for concurrent detection of mural and endothelial cells and evaluation of proliferating (Ki-67-positive) endothelial cells as markers of maturation [19]. In view of emerging digital pathology and new imaging technologies, we will also discuss the topic of computer-assisted image analysis for evaluating the morphology and characteristics of tumor microvessels.

20.2 Methodology

Immunohistochemistry (IHC) is a common technique for visualizing microvessels in tumor samples. For the details of basic IHC protocols, please refer to Current Protocols in Molecular Biology by Goldstein et al. [20].

In general, the IHC technique consists of antigen retrieval, selection of a specific antibody, a sensitive detection method, and negative and positive controls.

20.3 Materials

1. Silane-coated or charged microscope slides (e.g., Superfrost Plus®)
2. Slide staining tray
3. Coplin jars
4. Dry incubator or oven at 37 °C
5. Xylene or xylene substitute (e.g., HistoChoice®)
6. Ethanol (100 %, 90 %, and 70 %)
7. Normal goat serum
8. Phosphate-buffered saline (PBS) or Tris-buffered saline (TBS)
9. Tween 20
10. Methanol
11. 30 % hydrogen peroxidase
12. Tris-EDTA buffer, pH 9.0
13. Water bath at 95 °C, pressure cooker, or microwave oven

14. Antibodies to CD31 (JC70; Dako M0823) or CD34 (QBEND10; Dako M7165).[1]
15. Detection system: Avidin/biotin detection technology (VECTASTAIN Elite ABC system, Vector Laboratories) or Chain polymer detection systems (Dako EnVision™)
16. Chromogens: diaminobenzidine (Dako), 3-amino-9-ethylcarbazole (AEC, Dako), 5-bromo-4-chloro-3-indoxyl phosphate, and nitroblue tetrazolium chloride (BCIP/NBT, Dako)
17. Hematoxylin for counterstaining, if desired
18. Permanent (for DAB) or aqueous mounting medium (for Vector® Blue, Vector® Red or other alcohol-soluble chromogens)
19. Microscope coverslips
20. Chalkley graticule (25 dot)

20.4 Methods

It is important to note that similar to any other IHC method, optimization and standardization are necessary for obtaining high-quality staining with low nonspecific background. Specifically, when performing quantitative measurements by image analysis, minimum non-specific staining will yield more precise results. If IHC is being conducted in a non-histopathology research laboratory, we highly recommend that the basics of IHC protocols be reviewed before conducting the protocol. Otherwise, histopathology laboratories that have their own in-house IHC protocols can follow those protocols. However, attention must be paid to the antigen retrieval and antibody selection steps.

Basic Protocol

1. Cut 4-μm sections from formalin-fixed, paraffin-embedded (FFPE) tissue blocks and mount them onto silane-coated or charged slides. Using a pencil, label the slides with the specimen and primary antibody to be used.
2. Dry the slides at 37 °C in a dry incubator overnight.
3. Dewax the slides using two changes of xylene (or a xylene substrate) in a Coplin jar for 10 min each. Note: These jars can be stored and used for several cycles until they become cloudy. Xylene-containing jars with tightened lids should be kept in a fume cabinet between cycles. The sections should not dry out from this step onward.
4. Rehydrate the sections by passing them through graded ethanol solutions (100 %, 90 %, and 70 %) in Coplin jars for 2–5 min each. Then, rinse the

[1] CD31 and CD34 are the most commonly used endothelial markers due to their consistent and reliable results in paraffin-embedded tissues; however, depending on the objective of the study, other markers, such as vascular endothelial growth factor receptors (VEGFRs) or CD105, can be used.

sections in water for 5 min and PBS or TBS for 5–10 min. Note: Like xylene, ethanol can be reused before replacement with fresh solutions.

5. Pre-treatment with heat-induced epitope retrieval (HIER) is required and performed by placing sections in Tris EDTA buffer, pH 9.0.
   Note: The antibody specification sheet usually has a recommended method. Optimal results are obtained in our laboratory by using a water bath; pre-heat the water to 65 °C, perform epitope retrieval treatment at 95 °C for 20 min, and then cool to room temperature for 30 min in the same buffer. The same results can be obtained using the same buffer in a pressure cooker for 2 min. We have observed tissue damage resulting from microwave oven heating due to the sudden increase in temperature.
6. Rinse in PBS/TBS 3 times. Place the slide rack in a Coplin jar with PBS/TBS, drain, and refill with fresh solution 3 times for 5 min each.
7. Block endogenous peroxidase activity by incubating the sections in 0.3 % H_2O_2 in methanol for 30 min and 3 % H_2O_2 in methanol for 10 min. Note: This step is not required if an alkaline phosphatase detection system is used.
8. Rinse with PBS/TBS 3 times for 5 min each. Note: PBS with 0.1 % Tween-20 (PBS-T) can be used as a wash buffer from this step onward.
9. Block non-specific antibody binding sites by applying 200–300 μl of 5 % goat serum in PBS to each section after placing the sections in a staining tray. Make sure the tissue sections are thoroughly covered with blocking serum. Incubate for 30 min at room temperature.
10. Place the slides back in the rack and rinse with PBS/TBS 3 times for 5 min each.
11. Carefully dry the slides using KimWipes. Place the slides in a staining tray and carefully apply diluted primary antibody. Make sure the entire tissue section is covered with the antibody. Add IgG1 to the negative control slide. Note: 1:50–1:100 dilutions of primary antibodies can be applied at room temperature for 30–60 min. Lower antibody concentrations (1:200–1:300) can be used when the sections are kept at 4 °C overnight.
12. Rinse slides 3 times with PBS-T for 5 min each.
13. Utilize the appropriate detection system based on the primary antibody and sensitivity. Incubate the slides for 30 min at room temperature.
    Note: For CD31/CD34, chain polymer-conjugated technology, which avoids endogenous tissue biotin, is a fast one-step method with acceptable sensitivity. The Envision™ system can be used to detect any primary antibody of mouse or rabbit origin. However, due to the hydrophobic dextran backbone in this system, multiple washes with PBS-T are required. When the avidin/biotin system is used, endogenous biotin needs to be completely blocked before applying the appropriately diluted biotinylated secondary antibody. Attention should be paid to the species of the primary antibody.
14. Rinse the slides 3 times with PBS-T for 5 min each.
15. Apply the chromogen substrate for 5–10 min at room temperature.
16. Check the intensity of the internal control (i.e., medium-sized vessels) staining under a light microscope. Note: The color intensity needs to be optimized for image analysis.

17. Rinse the slides in tap water.
18. Place the slides in a jar of hematoxylin for approximately 1–2 min.
19. Wash the slides with normal tap water until the water is clear.
20. Dry the slides with four changes of 100 % ethanol, and clear them with three changes of xylene, each for a few seconds. Note: If an alcohol-soluble chromogen is used, skip this step. Use a separate set of ethanol and xylene jars for the dehydration step. Do not mix with the deparaffinization jars.
21. Apply a few drops of an appropriate mounting solution to each section, and place a coverslip on top. Note: If Aquamount medium is used, seal each coverslip with clear nail polish around the edges.

20.4.1 Identification of Tumor Blood Vessels

Examine the negative control slide to ensure the absence of nonspecific staining. Tumor microvessels have different sizes (diameters range from 10 to 200 μm) and morphologies and sometimes show a collapsed lumen (Fig. 20.1). Tumor-associated vessels can be concentrated within discrete areas due to the higher concentration of growth factors in these areas, which are called hot spots. The entire tumor should be scanned at low power (40× or 100×), and 3–5 hot spot areas should be selected. This step is very subjective. The tumor periphery usually exhibits higher microvessel density than the central areas. Avoid necrotic areas. The presence of red blood cells in the lumen is not a requirement. Many of the tumor-associated vessels have a collapsed lumen due to increased solid pressure from cancer cells and components of the microenvironment.

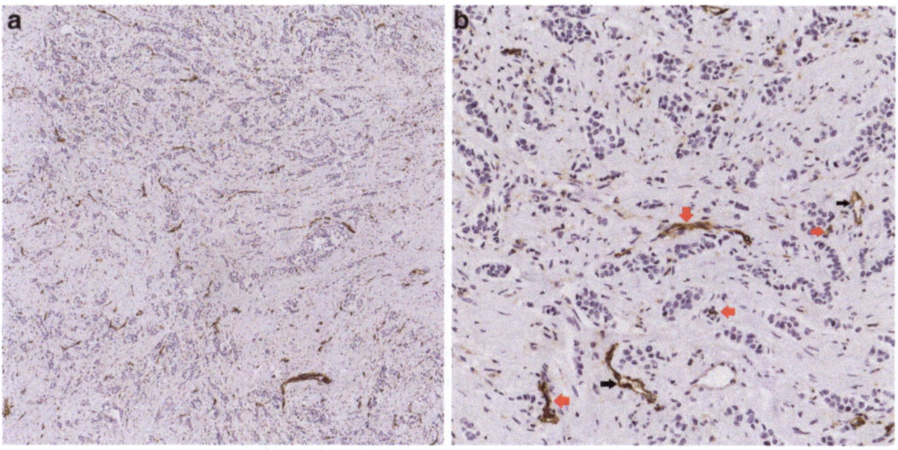

Fig. 20.1 Immunohistochemical staining of blood microvessels in breast cancer tissue. Intratumoral microvessels show strong CD31 staining (**a**) (5× objective). *Black arrows* show capillaries with open lumen, and *red arrows* show microvessels with collapsed lumen (20× objective) (**b**)

20.4.2 Measurement of Mean Microvessel Density (MVD)

Manual Counting Once hot spot areas are selected, use a higher magnification (200× or 400×) to count the stained structures, regardless of the size or presence of the lumen in the selected area, and record this number. Repeat this step for all hot spots. Calculate the mean of these counts. Check the microscope manufacturer's handbook to calculate the field of view in millimeters for each objective. The area of view can be calculated using the following formula: Area (mm^2) = π (Field of view/2) [2].

The MVD can be calculated using the following formula:

$$\text{MVD}\left(\text{mm}^{-2}\right) = \text{Mean vessel count} / \text{Area}\left(\text{mm}^{2}\right).$$

Chalkley Counting A Chalkley eyepiece graticule that fits the microscope is required. It is a 25-point eyepiece graticule and should be rotated such that the maximum number of dots overlaps stained microvessel structures. Record the number of overlapping dots (maximum score 25). Repeat this for all selected hot spot areas. The mean of these scores is the Chalkley count of that tumor section (Fig. 20.2).

20.4.3 Assessment of Microvessel Morphology and Patency

For this purpose, image analysis is very useful. The microscope should be equipped with a digital camera. After selecting hot spot areas, snapshots of these areas can be recorded. These pictures can then be analyzed by image analysis software. The perimeter, diameter, and area of highlighted tumor-associated vessels can be measured by the software tools (Fig. 20.3).

If a whole slide scanner is available, the accompanying software can usually provide morphological characteristics of selected structures after they are highlighted on the virtual slide (Fig. 20.4).

20.4.4 Measurement of Microvessel Proliferation

The ratio of tumor to endothelial cell proliferation (TCP/ECP) has been shown to be a marker of angiogenesis-independent tumor growth [21]. Angiogenesis inhibitors alone or in combination with other therapies are an accepted treatment in some cancers. To determine the tumors for which these therapies are most effective, endothelial cell proliferation has been shown to be a more reliable marker than MVD [22]. Double IHC staining on the same section with

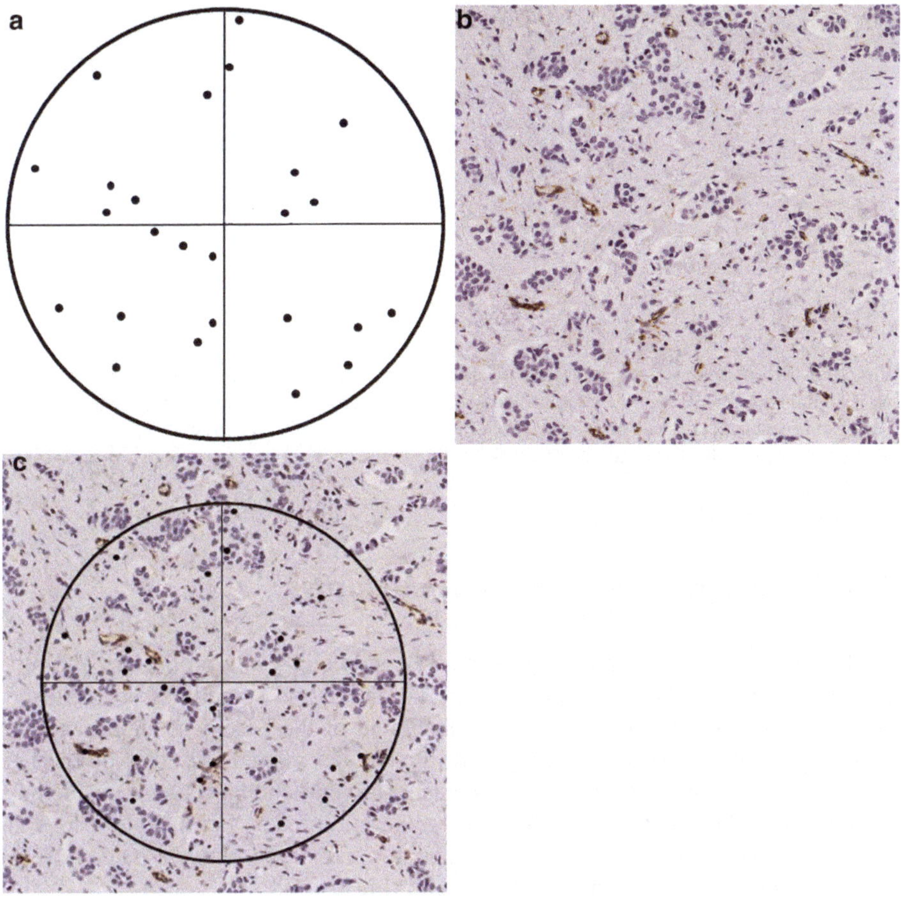

Fig. 20.2 The Chalkley method for estimating microvessel density. A Chalkley graticule (**a**). Intratumoral microvessels of breast cancer tissue stained for CD31 (**b**) are overlaid with a representation of the Chalkley grid (**c**). *Dots* that overlap microvessels are counted

antibodies to endothelial markers in conjunction with antibodies to proliferation markers (e.g., Ki67) can differentiate endothelial cells in proliferative versus quiescent states. The observer should be aware that tumor cells and inflammatory cells in the microenvironment contain abundant proliferating cells in addition to endothelial cells. Additional attention should be paid to detecting proliferating endothelial cells in microvessels with collapsed lumen. However, optimization and standardization of the double-staining IHC technique may not be simple in non-histology laboratories. This method can be easily substituted by staining for each marker on two separate serial sections. Virtual slides created by whole slide scanners allow simultaneous assessment of selected areas on both sections, therefore differentiating pericyte-positive versus -negative microvessels (Fig. 20.5).

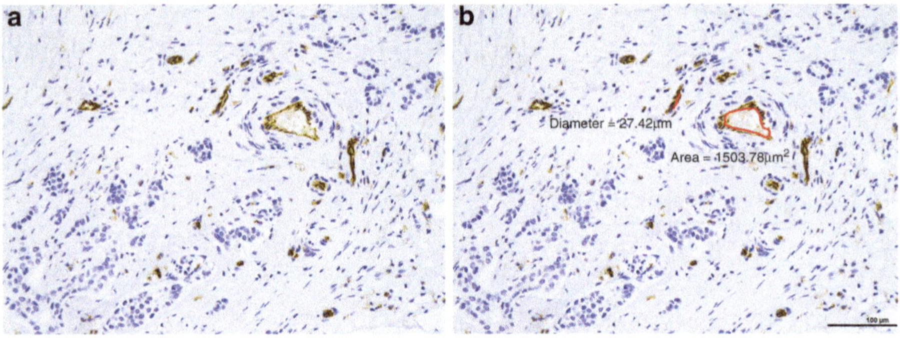

Fig. 20.3 Morphometric image analysis of microvessels. An image from a hot spot was captured by a digital camera attached to a light microscope (**a**). Microvessels stained for the expression of CD31 in breast cancer tissue can be analyzed by software programs to extract morphometric characteristics such as their area and diameter (**b**)

20.4.5 Assessment of Microvessel Maturity

In the early phases of tumor angiogenesis, endothelial cells proliferate and tube formation occurs. These microvessels are immature and sometimes not perfused. After recruitment of pericytes and establishment of the basement membrane, the endothelial cells become quiescent, and microvessel maturation takes place [19]. Evaluation of the presence of pericytes adjacent to endothelial cells is considered a useful tool in the calculation of the vessel maturation index. For this purpose, double staining of tumor sections for endothelial markers and pericyte markers (e.g., SMA) or serial section staining can be useful (Fig. 20.6).

20.4.6 Assessment of Hypoxic Markers

Despite the occurrence of active angiogenesis in most cancers and the increased number of microvessels, most of the newly formed vessels are not functionally normal, which makes hypoxia a common characteristic of tumors. Hypoxia can induce angiogenic pathways and, therefore, enhance angiogenesis [23]. Many hypoxia-related markers have been assessed in clinical settings and used as prognostic markers.

Carbonic anhydrase IX (CAIX), a hypoxia-induced enzyme that is overexpressed in tumor cells and has a pH regulatory function, has been shown to be a surrogate of hypoxia in some types of solid tumors [24] (Fig. 20.7).

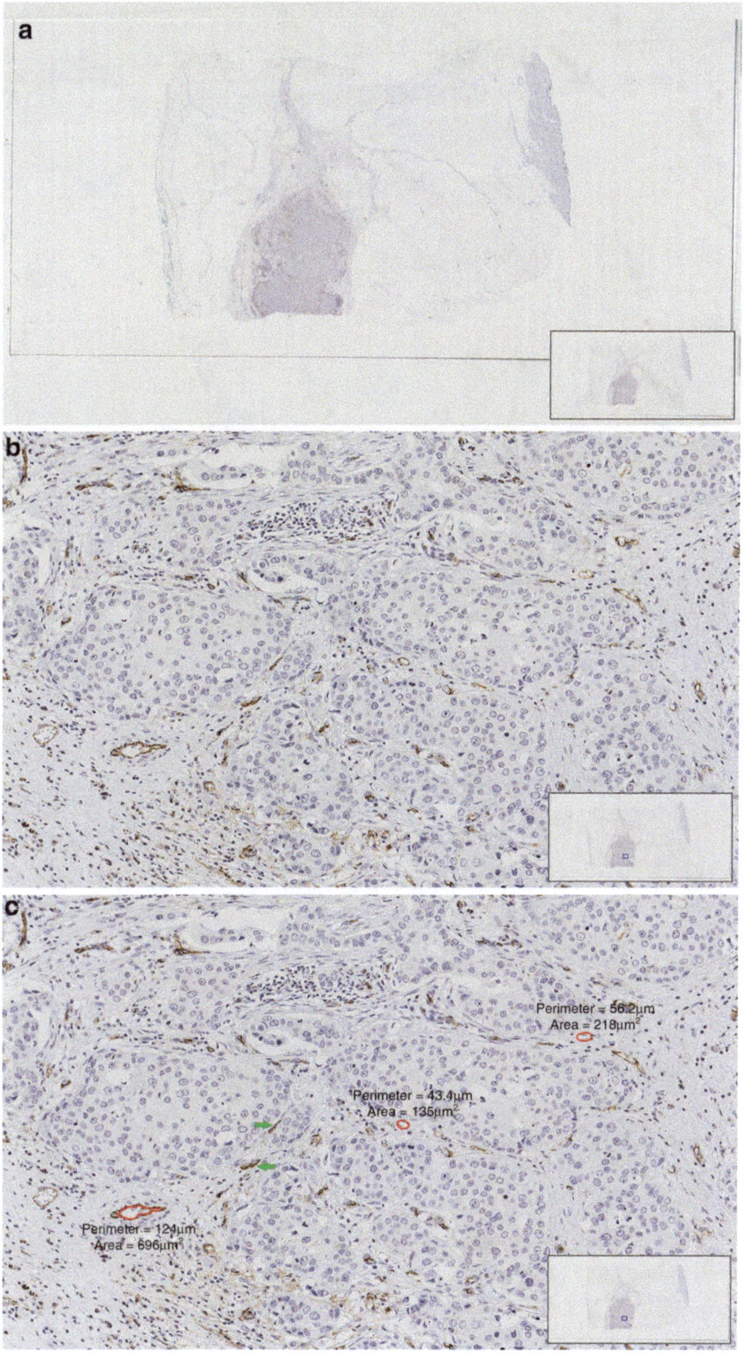

Fig. 20.4 Virtual slide of the entire tissue created by a slide scanner. The image of a breast cancer tissue section stained for CD31 was created using a Hamamatsu NanoZoomer slide scanner (**a**). Microvessels stained for CD31 in a hot spot (magnified digitally, equivalent to a 20× objective) (**b**). Morphometric analysis by Hamamatsu NDP Analyze software. *Green arrows* show microvessels with collapsed lumen (**c**)

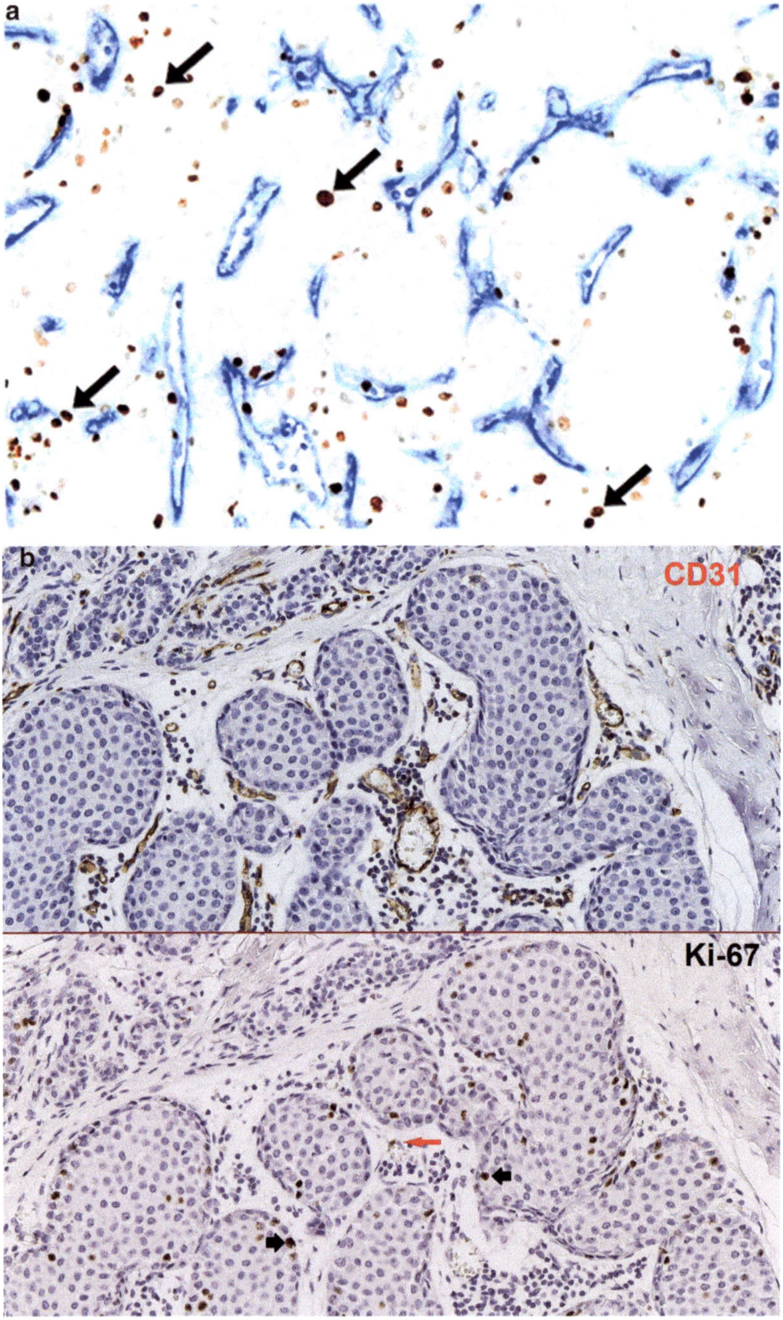

Fig. 20.5 Proliferation assessment of microvessels. Primary tumor tissue sections from patients with renal cell carcinoma were stained with a mixture of CD31/CD34 antibodies to visualize blood vessels (*blue*) and with anti-Ki-67 (*brown*, *arrows*) to monitor the proliferation status of both the endothelial and tumor cell compartments (**a**, adapted from Ref. [22]). CD31 and Ki-67 staining in two serial sections (4 μm) of breast cancer tissue (**b**). The *red arrow* indicates a proliferating endothelial cell. *Black arrows* indicate proliferating tumor cells

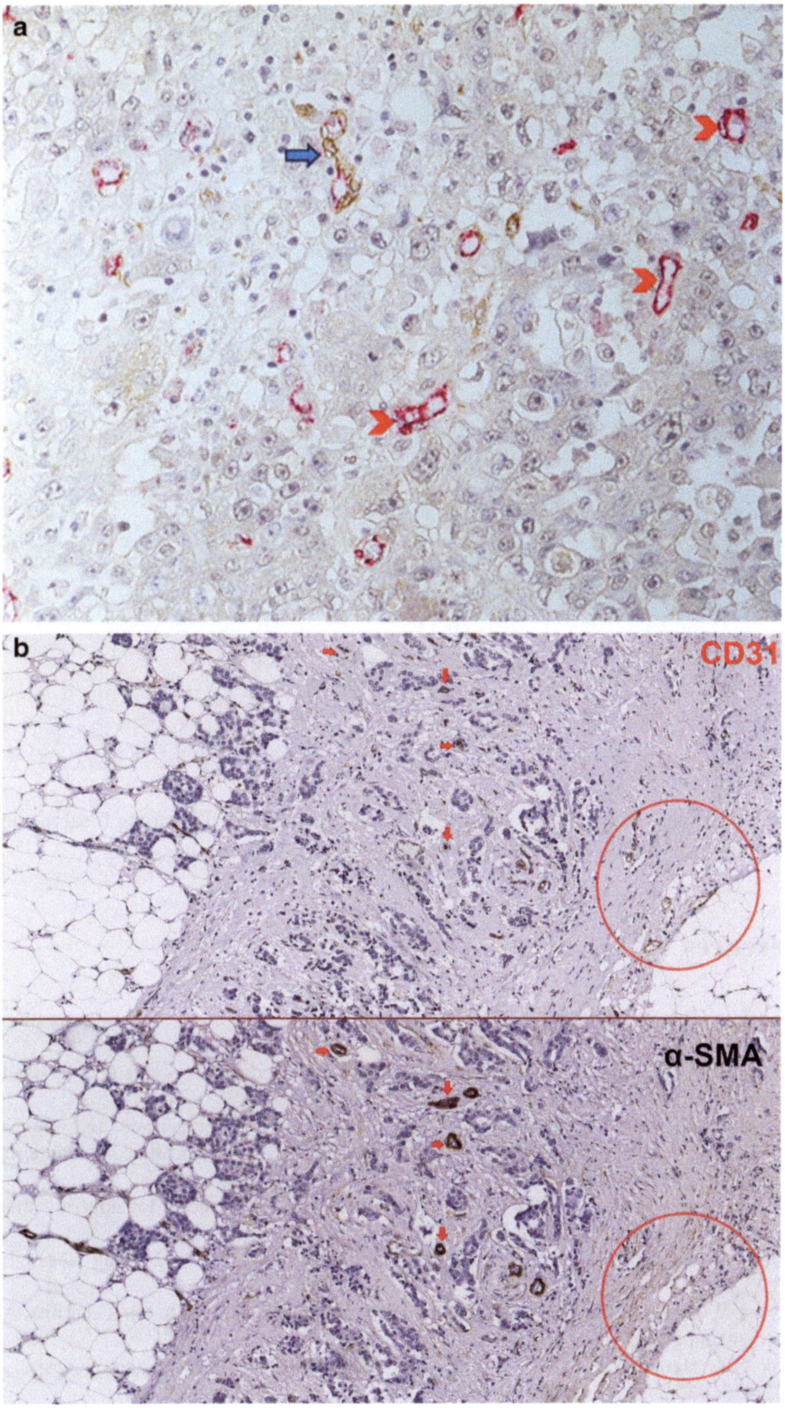

Fig. 20.6 Pericyte coverage of microvessels. Double-immunostaining of pericytes and endothelial cells in breast cancer tissue was performed; CD31 (*red*) and α-SMA (*brown*) staining are shown (**a**). The *blue arrow* indicates a vessel that is CD31+/α-SMA+; the *red arrowheads* indicate vessels that are only CD31+ (Adapted from Ref. [19]). CD31 and α-SMA staining in two serial sections (4 μm) of breast cancer tissue (**b**). *Red arrows* indicate microvessels with strong pericyte coverage. A *red circle* indicates an area where the microvessels are only CD31+

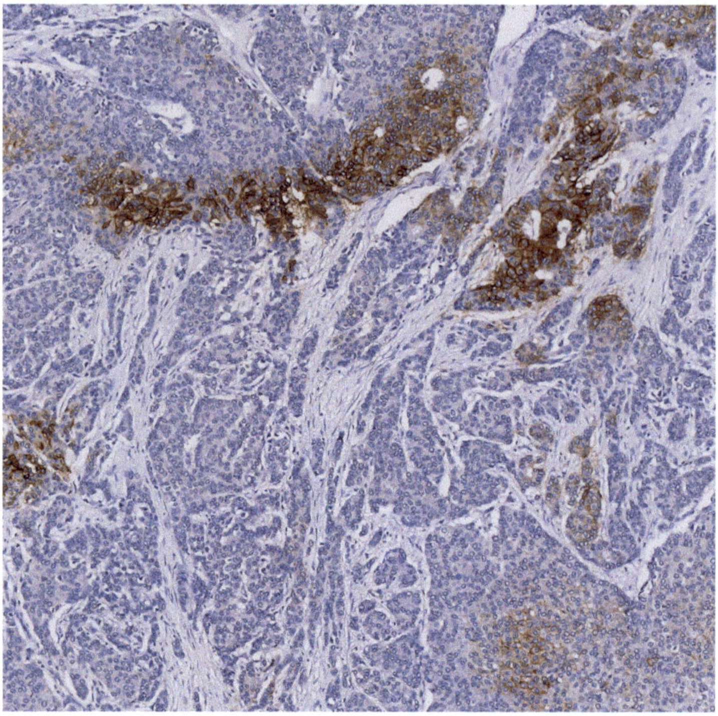

Fig. 20.7 Carbonic anhydrase IX staining of breast cancer tissue. The strong focal membranous and cytoplasmic staining of CAIX indicates hypoxia

References

1. Hanahan D, Weinberg RA (2011) Hallmarks of cancer: the next generation. Cell 144:646–674
2. Abdollahi A, Folkman J (2010) Evading tumor evasion: current concepts and perspectives of anti-angiogenic cancer therapy. Drug Resist Updat 13:16–28
3. Gasparini G, Toi M (1998) Prognostic significance of p53, angiogenesis, and other conventional features in operable breast cancer: subanalysis in node-positive and node-negative patients. Int J Oncol 12:1117–1125
4. Toi M, Bando H et al (2000) The predictive value of angiogenesis for adjuvant therapy in breast cancer. Breast Cancer 7:311–314
5. Barrett T, Brechbiel M et al (2007) MRI of tumor angiogenesis. J Magn Reson Imaging 26:235–249
6. Iagaru A, Gambhir SS (2013) Imaging tumor angiogenesis: the road to clinical utility. AJR Am J Roentgenol 201:W183–191
7. Fox SB, Harris AL (2004) Histological quantitation of tumour angiogenesis. APMIS 112:413–430
8. Bossi P, Viale G et al (1995) Angiogenesis in colorectal tumors: microvessel quantitation in adenomas and carcinomas with clinicopathological correlations. Cancer Res 55:5049–5053
9. Mlynek ML, van Beunigen D et al (1985) Measurement of the grade of vascularisation in histological tumour tissue sections. Br J Cancer 52:945–948

10. Weidner N, Semple JP et al (1991) Tumor angiogenesis and metastasis–correlation in invasive breast carcinoma. N Engl J Med 324:1–8
11. Mayers MM, Seshadri R et al (1998) Tumor microvascularity has no independent prognostic significance for breast cancer. Pathology 30:105–110
12. Horak ER, Leek R et al (1992) Angiogenesis, assessed by platelet/endothelial cell adhesion molecule antibodies, as indicator of node metastases and survival in breast cancer. Lancet 340:1120–1124
13. Chandrashud LM, Pendleton N et al (1997) Relationship between vascularity, age and survival in non-small cell lung cancer. Br J Cancer 76:1367–1375
14. Amit L, Ben-Aharon I et al (2013) The impact of Bevacizumab (Avastin) on survival in metastatic solid tumors–a meta-analysis and systematic review. PLoS One 8:e51780
15. Fitzgibbons PL, Page DL et al (2000) Prognostic factors in breast cancer. College of American Pathologists Consensus Statement 1999. Arch Pathol Lab Med 124:966–978
16. Compton CC, Fielding LP et al (2000) Prognostic factors in colorectal cancer. College of American Pathologists Consensus Statement 1999. Arch Pathol Lab Med 124:979–994
17. Bostwick DG, Grignon DJ et al (2000) Prognostic factors in prostate cancer. College of American Pathologists Consensus Statement 1999. Arch Pathol Lab Med 124:995–1000
18. Vermeulen PB, Gasparini G et al (2002) Second international consensus on the methodology and criteria of evaluation of angiogenesis quantification in solid human tumours. Eur J Cancer 38:1564–1579
19. Fakhrejahani E, Toi M (2012) Tumor angiogenesis: pericytes and maturation are not to be ignored. J Oncol 2012:261750
20. Goldstein M, Watkins S (2008) Immunohistochemistry. Curr Protoc Mol Biol 81:14.6.1–14.6.23
21. Stessels F, Van den Eynden G et al (2004) Breast adenocarcinoma liver metastases, in contrast to colorectal cancer liver metastases, display a non-angiogenic growth pattern that preserves the stroma and lacks hypoxia. Br J Cancer 90:1429–1436
22. Griffioen AW, Mans LA et al (2012) Rapid angiogenesis onset after discontinuation of sunitinib treatment of renal cell carcinoma patients. Clin Cancer Res 18:3961–3971
23. Jain RK (2013) Normalizing tumor microenvironment to treat cancer: bench to bedside to biomarkers. J Clin Oncol 31:2205–2218
24. Adams A, van Brussel AS et al (2013) The potential of hypoxia markers as target for breast molecular imaging – a systematic review and meta-analysis of human marker expression. BMC Cancer 13:538

Chapter 21
Whole-Mount Immunostaining Methods to Study the Blood and Lymphatic Vasculature in the Embryonic Mouse Skin and Adult Mouse Cornea

Anees Fatima, Kathryn Marie-Schultz, Seungwoon Seo, Ford Culver, Austin Culver, and Tsutomu Kume

21.1 Introduction

The blood and lymphatic vascular systems are vital units in mammals that carry blood and interstitial fluid. During development of an embryo, formation of these two vascular systems requires co-ordination of highly complex network of events. Development of blood vessels begins de-novo by formation of a primitive vascular network or primary plexus composed of mesoderm-derived endothelial progenitor cells by a process termed 'vasculogenesis'[1, 2]. Subsequently the primary plexus leads to formation of the mature vasculature by 'angiogenesis.' Shortly after the development of blood vessels, some of the blood endothelial cells differentiate to lymphatic endothelial cells by expression of *PROX-1*. These *PROX-1* positive endothelial cells then form the primitive lymph sacs leading to the formation of lymphatic vasculature [3]. Although both the vascular systems arise as primitive capillary plexus they are further reorganized into a highly complex network of vasculature in various organs through the process of pruning, reshaping, and vessel

---

AF & TK contributed towards development of the embryonic DS staining protocol and compilation/writing of the chapter.
KS, SS and TK contributed to development of the corneal flat mount staining protocol. KS contributed in writing the corneal flat mount part of the chapter.
FC and AC contributed to developing and writing the Matlab code for vessel width measurement.
All the above work was done under the supervision of TK.

A. Fatima • K. Marie-Schultz • S. Seo • F. Culver • A. Culver
Feinberg Cardiovascular Research Institute, Northwestern University Medical School, 300 E. Superior, Tarry 14-731, Chicago, IL 60611, USA
e-mail: a-fatima@northwestern.edu; fculver@nmh.org; austin.culver@gmail.com

T. Kume (✉)
Feinberg Cardiovascular Research Institute, Feinberg School of Medicine, Northwestern University, Chicago, IL 60611, USA
e-mail: t-kume@northwestern.edu

M. Slevin, G. McDowell (eds.), *Handbook of Vascular Biology Techniques*,
DOI 10.1007/978-94-017-9716-0_21

fusion [4]. Under pathological conditions such as tumor formation and injuries due to trauma, the existing blood/lymphatic vessels give rise to new vessels, in a similar fashion of complex vessel branching and morphogenesis as mentioned above [5, 6]. To study in-vivo angiogenesis and lymphangiogenesis in pathological and also normal physiological conditions, immunohistochemical techniques are employed wherein the blood/lymphatic vessels are detected with primary antibodies that interact with specific markers such as PECAM-1 (blood vessels) and Lyve-1 (lymphatic vessels), respectively, and second antibodies conjugated with a flouorophore/enzyme/biotin. Traditionally the tissue/organ wherein the vessel formation is intended to study is sectioned for immunohistochemical techniques. This method allows getting information on presence, increase/decrease of blood and lymphatic vessels in a particular tissue; however, it does not provide any information of the complex branching pattern of vessels. Additionally any quantification attempts from immunostained sections only give results based on the small area, so that one has to be very cautious to extrapolate the results to entire organ/tissue. Accordingly, it is necessary to stain whole tissue (whole mount) to better understand the processes of angiogenesis and lymphangiogenesis [7]. In this chapter we present detailed staining techniques using whole mount DS from mouse embryos and adult corneal tissues [8]. In addition we describe a new Matlab program specifically coded for purpose of vessel-width measurements following immunostaining. This method is more user-friendly and time-efficient compared to more traditional measuring tools such as Image-J (Fatima et al unpublished).

21.2 Part A: Whole Mount DS Immunostaining from Mouse Embryos

21.2.1 Materials

(a) Dulbecco's Phosphate-Buffered Saline (DPBS; Cellgro, Cat No: 21-030-CM)
(b) 4 % paraformaldehyde in PBS (PFA; Affymetrix, Cat No: 19943 1 LT)
(c) 100 % Methanol (EMD Millipore, Cat No: MX0475)
(d) 12-well tissue culture plate (Becton Dickinson, Cat No: 353043)
(e) Petri dish (Becton Dickinson, Cat No: 351029)
(f) Normal donkey serum (NDS; Sigma, Cat No: D9663)
(g) Triton X-100 (Sigma, Cat No: T8787)
(h) Rabbit anti-Lyve-1 (Abcam, Cat No: ab14917)
(i) AlexaFluor 488 donkey anti-rabbit IgG (Invitrogen, Cat No: A21206)
(j) R-Phycoerythrin (R-PE)-conjugated rat anti-mouse CD31 (BD Pharmingen, Cat No: 553373)
(k) Superfrost Plus microscope slides (VWR, Cat No:48311–703)
(l) Coverslips (VWR, Cat No:48393081)
(m) Vectashield mounting medium (Sigma, Cat No: F4680)

21.2.2 General Equipment

(a) Dissection microscope (Olympus)
(b) Fine forceps
(c) Small scissors
(d) Micropipettes and tips (10, 20, 100, 200, 1,000 μl)
(e) Pasteur-pipettes
(f) Refrigerator 4 °C and freezer –20 °C
(g) Tabletop centrifuge
(h) Nutator/rotating bench top shaker
(i) Confocal microscope (Zeiss LSM 510 Meta)

21.2.3 Methods

21.2.3.1 Preparation of Solutions/Buffers

(a) **0.2 % PBST:** To prepare 1,000 mL of 0.2 % PBST add 2 mL triton X-100 to 998 mL PBS. Stir until triton X-100 is completely dissolved using a magnetic stirrer.
(b) **Graded MeOH-PBST:** To prepare 100 mL of graded MeOH-PBST, add 75 mL MeOH to 25 mL PBST (to make 75 % MeOH-PBST), add 50 mL MeOH to 50 mL PBST (to make 50 % MeOH-PBST), add 25 mL MeOH to 75 mL PBST (to make 25 % MeOH-PBST).
(c) **Blocking buffer:** To prepare 10 mL of blocking buffer, add 1 mL normal donkey serum to 9 mL PBS (10 %).

21.2.3.2 Harvest of Mouse Embryonic DS

(a) Euthanize plugged pregnant dams (E14–E17) according to methods approved by institutional animal facility. All procedures were approved by Northwestern University's Institutional Animal Care and Use Committee (IACUC).
(b) Dissect intact uteri using scissors and place in cold PBS in a Petri dish.
(c) Carefully dissect out individual embryos in a six-well culture plate filled with cold PBS.
(d) Wash the embryos to remove residual blood.
(e) Fix in cold 4 % paraformaldehyde (PFA), overnight at 4 °C keeping the dish shaking on a nutator. Make sure to completely immerse embryos in PFA.
(f) Wash the embryos once in PBS for 5 min, transfer into cold 100 % methanol, and leave overnight at –20 °C.
(g) Carefully dissect the DS with help of fine forceps and scissors under the dissecting microscope (see note-1).

21.2.3.3 Immunostaining of Mouse Embryonic DS

(a) Following dissection rehydrate DS tissues in series of graded methanol (MeOH)/PBST (PBS+0.2 % Triton X-100). Use 75 %, 50 %, 25 % MeOH/PBST for 5 min each (see note-2). Finally wash the DS tissues in PBST, two times for 5 min each.
(b) Block the DS tissues in blocking buffer, for 2 h at room temperature on a nutator.
(c) Depending on requirement, incubate the DS tissues in appropriate primary antibodies for overnight at 4 °C (see note-3). For dilutions see note-4.
(d) Following overnight incubation of the tissues in primary antibodies, wash the tissues four times (10 min each) in PBST at room temperature on a nutator.
(e) Incubate the tissues in secondary antibody and/or a fluorescent conjugated second primary antibody for 1.5 h at room temperature on a nutator (see note-5). For dilutions see note-4.
(f) Wash tissues four times (10 min each) in PBST. Avoid exposure of tissue specimen to light during washing. For counter staining of nuclei see note-5.

21.2.3.4 Mounting the DS Tissues on Glass Slides

(a) Transfer tissues on glass slides. Under the dissecting microscope, remove residual artifacts using fine forceps. Keep the inner sides of the DS facing upwards.
(b) For mounting, blot dry excess PBST using a kimwipe. Add 2–4 drops of vectashield mounting media, carefully layover coverslip taking care not to create bubbles. The slides are then cured overnight before viewing under the microscope (see note-6).

21.2.3.5 Confocal Microscopy

(a) Following immunostaining, the tissues can be viewed under a fluorescent microscope. For documentation of data, the tissues can be subjected to Z-section imaging under a confocal microscope.
(b) We have used a Zeiss UV-LSM 510 Meta for acquiring data presented in this chapter (see Sample Fig. 21.1). To begin we set-up appropriate lasers and excitation for each fluorophore. Z-sectioning parameters were set-up depending on the thickness where fluorescence was observed. For further details refer to Zeiss manual.

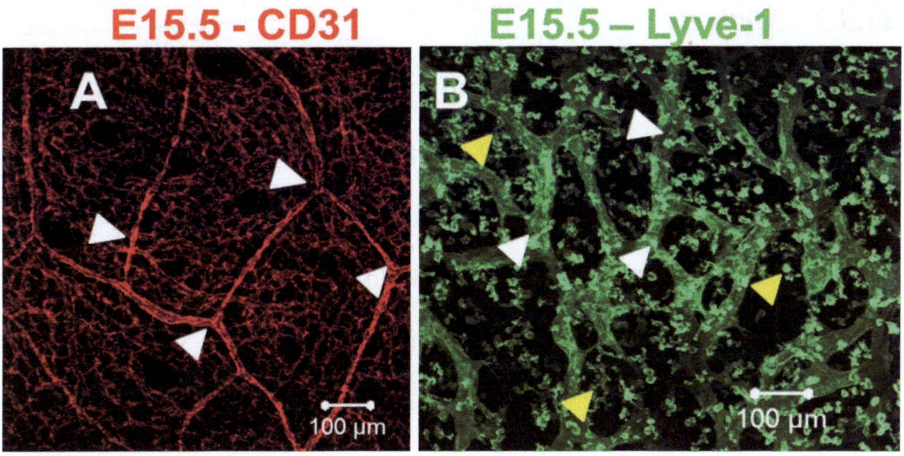

Fig. 21.1 Whole mount DS staining identifying blood/lymphatic vessels. (**a**), DS from E15.5 mouse embryos stained for blood vessel marker CD31. (**b**), DS from E15.5 mouse embryo stained for lymphatic vessel marker Lyve-1. Arrow heads (*white*) indicating the vessel branching pattern, arrow heads (*yellow*) indicate Lyve-1 positive macrophages in the skin (Scale bar = 100 µm)

21.3 Part B: Adult Mouse Corneal Flat Mount Immunostaining

21.3.1 Materials

(a) Dulbecco's Phosphate-Buffered Saline (DPBS; Cellgro, Cat No: 21-030-CM)
(b) 4 % paraformaldehyde in PBS (PFA; Affymetrix, Cat No: 19943 1 LT)
(c) Bovine serum albumin (BSA; Sigma, Cat No: A2153)
(d) 100 % Methanol (EMD Millipore, Cat No: MX0475)
(e) 12-well tissue culture plate (Becton Dickinson, Cat No: 353043)
(f) 48-well tissue culture plate (Becton Dickinson, Cat No: 353078)
(g) Petri dish (Becton Dickinson, Cat No: 351029)
(h) Nutator
(i) Normal donkey serum (NDS; Sigma, Cat No: D9663)
(j) Triton X-100 (Sigma, Cat No: T8787)
(k) Tween-20 (Sigma, Cat No: P9416)
(l) Rabbit anti-Lyve-1 (Abcam, Cat No: ab14917)
(m) AlexaFluor 488 donkey anti-rabbit IgG (Invitrogen, Cat No: A21206)
(n) R-Phycoerythrin (R-PE)-conjugated rat anti-mouse CD31 (BD Pharmingen, Cat No: 553373)
(o) Superfrost Plus microscope slides (VWR, Cat No: 483117032A)
(p) Coverslips (VWR, Cat No: 48393081)
(q) Fluoromount aqueous mounting medium (Sigma, Cat No: F4680)

21.3.2 Methods

21.3.2.1 Preparation of Solutions/Buffers

(a) **Dissection buffer:** To prepare 100 mL of dissection buffer, combine 100 mL of DPBS with 0.3 g of BSA (0.3 %). Store at 4 °C.
(b) **Permeabilization buffer:** To prepare 100 mL of permeabilization buffer, combine 100 mL of DPBS with 0.1 mL of Triton X-100 (0.1 %). Store at 4 °C.
(c) **Blocking buffer:** To prepare 10 mL of blocking buffer, combine 9.5 mL of DPBS with 0.5 mL of NDS (5 %).
(d) **Wash buffer:** To prepare 500 mL of wash buffer, combine 500 mL of DPBS with 0.5 mL of Tween-20 (0.1 %).

21.3.2.2 Dissection of Corneal Tissue

(a) Fill each well of a 12-well plate with 1.5 ml of DPBS.
(b) Enucleate (remove) the entire eye and place in one well of the plate. Swirl to rinse any blood away from the eye.
(c) Repeat for each eye.
(d) Remove DPBS from wells and replace with cold 4 % PFA.
(e) Place the plate on the nutator at 4 °C for 20 min.
(f) Wash the eyes with DPBS for 5 min, three times.
(g) Fill the Petri dish with dissection buffer.
(h) Transfer an eye to the Petri dish. Dissect the cornea by cutting the eye in half at the border of the limbus and conjunctiva. See note-8
(i) Cut the cornea four times from the outer to inner cornea to allow flat mounting onto the slide.
(j) Replace the DPBS in the 12-well plate with 100 % methanol. Put the plate at −20 °C for 20–30 min.

21.3.2.3 Immunostaining of the Corneal Tissue

(a) Incubate corneas in permeabilization buffer for 15 min at 4 °C, two times.
(b) Incubate in blocking buffer for at least 20 min at RT.
(c) Transfer corneas to the 48-well plate. Incubate overnight in primary antibody (150 μL) solution in each well of a 48-well plate at 4 °C. See note-4 for antibody dilution.
(d) Transfer the corneas back to a 12-well plate and wash with wash buffer for 5 min, four times.
(e) Incubate the corneal tissues in secondary antibody and/or a fluorescent conjugated second primary antibody for 1.5 h at room temperature on a nutator. For dilutions see note-4.

(f) Wash the corneas in a 12-well plate with wash buffer for 5 min, 4 times each.
(g) At this point DAPI can be used for counterstaining the nuclei. See note-6.

21.3.2.4 Mounting the Corneal Tissue

(h) Gently transfer each cornea to a microscope slide. Avoid bubbles.
(i) Use a kimwipe to wick away most of the liquid on the slide without touching the tissue.
(j) Apply two drops of fluoromount and carefully place a coverslip over the tissue.

21.3.2.5 Fluorescent Microscopy

1. We have used Zeiss Axiovision fluorescent microscope, for image documentation.
2. Lower magnification images (2.5×) are documented to quantify the entire blood/lymphatic vascular area (Sample Fig. 21.2).
3. Higher magnification images (10×) should be documented to study the branching pattern of the vasculature.
4. We used open source Image J software from NIH to quantify the vascular area. In this protocol, we have not provided details of quantifying vascular branching pattern for corneal flat mount.

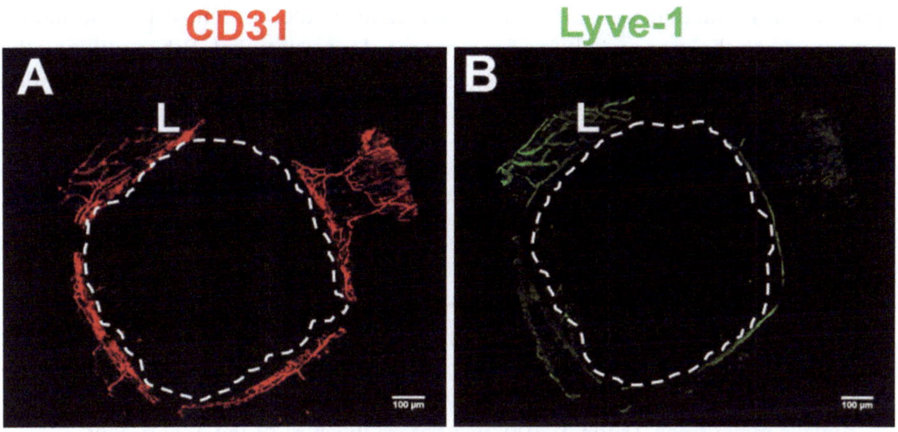

Fig. 21.2 Adult mouse corneal flatmount identifying blood/lymphatic vessels. (**a**), Adult mouse cornea stained for blood vessel marker CD31. (**b**), Adult mouse cornea stained for lymphatic vessel marker Lyve-1. Area within *dotted line*, avascular cornea, *L* Limbus (Scale bar = 100 μm)

21.4 Part C: Quantification of Lymphatic Vessel Width via Digital Imaging Software

Traditionally, vessel width (VW) has been measured manually via ImageJ or similar software, which requires the user to load each image file individually and a use a "drag and drop" approach to linear width measurement. ImageJ exports values in terms of pixels, which then need to be converted to SI units by the user, after obtaining an appropriate scaling factor from the microscope software that captured the initial image. With a large volume of images, the amount of work for a single user can become prohibitive. In our experience, having a single, preferably blinded user of the software allows for the most accurate measu rements of LVW, as it eliminates interoperator variability and systemic bias, so we created a more streamlined method of LVW measurement to cut down on the time required for analysis. Matlab is a highly versatile development package which includes a number of helpful scientific function libraries, which suited our needs. Our software was based on the same fundamental algorithm for distance determination as ImageJ, namely recording the vertical and horizontal displacement between the two selected points on each edge of the lymphatic vessel of interest and applying the Pythagorean theorem to determine the diagonal distance. However, we built in batch image processing, which allows the user to queue a number of images for analysis by designating a directory in which they are located. With the batch processing feature, the user sets a given number of vessels to measure in each image, and once they have completed their measurement of an image, the software pulls up the subsequent image, which significantly streamlines the workflow. The data is output to an Excel file, with averages and standard deviations calculated automatically, as well as SI unit conversion once a scaling factor is provided. We also opted for a simpler "click and click" approach to the actual measurement, which reduces the burden of precise mouse movements on the user and allows for more rapid measurement. When utilizing the Matlab software, we found no difference between previously and newly recorded VWs and users noted significantly easier use and decreased time expenditure.

21.5 Notes

1. After dissection of the embryonic DS, examine it under the dissecting microscope and carefully remove muscle tissue without disturbing the visible blood vessels. If excess muscle tissue is left unattended, staining and visualization of lymphatic vessels will be hampered.
2. Always make fresh Triton X-100, just before the start of the experiment.
3. Make sure the DS tissues are completely immersed in antibody solution. Use 1.5 ml tube or 24-well tissue culture dish for overnight incubations with 800 μL of antibody.

4. Antibody dilutions: Rabbit anti-Lyve-1 (10 μg/ml), Anti-CD31 (1:100), Donkey anti-rabbit Alexa 488 (1:500). Dilute primary antibodies in blocking buffer, the Alexa secondary antibodies should be diluted in PBST.
5. Fluorescent conjugated primary and secondary antibodies should be protected from light. All incubations should be carried out in dark.
6. The nuclei in DS/corneal tissues could be counterstained with DAPI (10 μg/mL in PBS). We have not used DAPI counterstaining in this protocol.
7. Store the slides with stained tissues at 4 °C, until ready for documentation.
8. Remove any other ocular tissues (lens, retina, and iris) and discard. Very carefully remove the iris from behind the cornea.

References

1. Kume T (2010) Specification of arterial, venous, and lymphatic endothelial cells during embryonic development. Histol Histopathol 25:637–646
2. Park C, Kim TM, Malik AB (2013) Transcriptional regulation of endothelial cell and vascular development. Circ Res 112:1380–1400
3. Marcelo KL, Goldie LC, Hirschi KK (2013) Regulation of endothelial cell differentiation and specification. Circ Res 112:1272–1287
4. Chauvet S, Burk K, Mann F (2013) Navigation rules for vessels and neurons: cooperative signaling between VEGF and neural guidance cues. Cell Mol Life Sci 70:1685–1703
5. Wang Y, Oliver G (2010) Current views on the function of the lymphatic vasculature in health and disease. Genes Dev 24:2115–2126
6. Eklund L, Bry M, Alitalo K (2013) Mouse models for studying angiogenesis and lymphangiogenesis in cancer. Mol Oncol 7:259–282
7. Mukouyama YS, James J, Nam J, Uchida Y (2012) Whole-mount confocal microscopy for vascular branching morphogenesis. Methods Mol Biol 843:69–78
8. Seo S, Singh HP, Lacal PM, Sasman A, Fatima A, Liu T, Schultz KM, Losordo DW, Lehmann OJ, Kume T (2012) Forkhead box transcription factor FoxC1 preserves corneal transparency by regulating vascular growth. Proc Natl Acad Sci U S A 109:2015–2020

Chapter 22
Computed Tomography Angiography: Fundamental Techniques and Data Interpretation

Cristina Corbella Sala, Laura Susana Goiburú González, and Josep Lluis Dolz Jordi

22.1 Introduction

Computed Tomography Angiography (CTA) has become a fundamental diagnostic tool for evaluating the intra- and extracranial vasculature, owing to advances in multi-slice computed tomography (MSCT). The ability to obtain isotropic volume images enables physicians to assess the entire arterial tree, from the aortic arch to the intracranial circulation – not only in the axial plane, but also in the coronal and sagittal planes. Three-dimensional (3D) reconstructions offer a wealth of information for planning surgical procedures. As CTA is non-invasive, fast and cheaper than conventional angiography, it has become increasingly important for clinical and surgical evaluation of cerebrovascular diseases, especially of ischaemic pathologies and of subarachnoid haemorrhage. It is well tolerated in patients with few contraindications and is advantageous over Magnetic Resonance Imaging (MRI) for patients with pacemakers or that are claustrophobic. Furthermore, unlike conventional angiography, CTA enables physicians to rule out non-vascular pathologies during the examination.

22.2 Multi-Slice Computed Tomography (MSCT)

In MSCT scanners, the data are acquired as the table moves and the gantry rotates in continuous fashion. Data for CTA studies are acquired in the axial plane. The quality of these studies depends on the following CT *data acquisition parameters*:

C.C. Sala (✉) • L.S. Goiburú González • J.L. Dolz Jordi
Servicio de Radiodiagnóstico, Hospital Mútua de Terrassa,
Plaza Dr Robert 5, 08221 Terrassa (Barcelona), Spain
e-mail: corbi555@hotmail.com

M. Slevin, G. McDowell (eds.), *Handbook of Vascular Biology Techniques*,
DOI 10.1007/978-94-017-9716-0_22

22.2.1 Acquisition Time or Speed

This depends on the number of CT detectors, and is influenced by the slice thickness and by the pitch.

22.2.2 Spatial Resolution

This depends on the detector geometry and on the image processing (the convolution kernel).

22.2.2.1. **Table speed**
22.2.2.2. **Collimation**
22.2.2.3. **Pitch:** table speed/collimation divided by gantry rotation speed.

The resolution can be increased by decreasing the collimation or the table speed.

22.2.3 Injection of the Contrast Medium

The injection protocol should be simple and standardised to ensure reproducibility and guarantee good results.

Basic protocol for a CTA study of the cranium and supra-aortic trunks:

- Use of a non-ionic, iodinated contrast medium (concentration: 300–350 mmol/mL)
- Injection speed: 4–5 mL/s
- Contrast medium bolus (100 mL at 5 mL/s) followed by a saline serum bolus (40 mL)
- Injection method: test bolus, bolus tracking or CARE (Combined Applications to Reduce Exposure) bolus.

Absolute contraindications for administration of an iodinated contrast medium:

- Kidney failure (always check the patient's creatinine levels and glomerular filtrate);
- Allergy to iodinated contrast media.

22.2.4 Reconstruction Interval

This is the distance between the reconstructed images. In order to minimise partial volume artefacts, the reconstruction interval should be smaller than the slice thickness. It provides greater anatomic detail and more information in the 2D/3D post-processing. The recommended value for the reconstruction interval is 50 % of the slice thickness.

22.2.5 Morphologic Post-processing Techniques

Table 22.1 The traditional CTA Imaging process

Multiplanar reconstruction (MPR)	
Volume reconstruction tools	Average Intensity Projection (AIP)
	Maximum intensity projection (MIP)
	Minimum Intensity Projection (MinIP)
	Volume rendering (VR)
	Surface rendering (SSR)
Volume post-processing tools	**Volume segmentation**
	1. Defining parameter threshold values
	2. Virtual surgical tools
	3. Painting tools
	4. Automatic selection tools
Other tools	Vessel analysis tools (based on the centreline approach)
	2D transfer functions
Bone-subtraction computed tomography angiography (BSCTA)	

The traditional CTA imaging processes comprise *MPR* Multiplanar Reconstruction, *MIP* Maximum Intensity Projection, *VR* Volume Rendering and *SR* Surface Rendering

22.2.5.1 Multiplanar Reconstruction (MPR)

Although MPR is not a 3D reconstruction per se, it does enable physicians to study the acquired image in multiple planes (axial, coronal and sagittal), and allows for reconstruction of oblique or curved planes with no loss of information. Curvilinear reconstruction is a variation of MPR that is used to study highly tortuous blood vessels. In clinical radiology practice it is not used for studying brain aneurysms, although it can serve as a starting point (Fig. 22.1).

22.2.5.2 Maximum Intensity Projection (MIP)

This technique creates a 3D representation of the maximum value in each voxel (the unit of volume of the generated image). The voxels of maximum intensity (typically due to contrast medium or calcium) appear in a greyscale. Thus, MIP is ideal for studying vascular structures, which are illustrated with high precision. The main drawback in MIP studies is the generation of artefacts by calcium (namely, from blood vessels located near bone structures or from calcified atheromatous plaques). The limitations of MIP for studying aneurysms are calcified atheromatous plaques and the carotid arteries (the segments that travel through the base of the cranium) (Fig. 22.2).

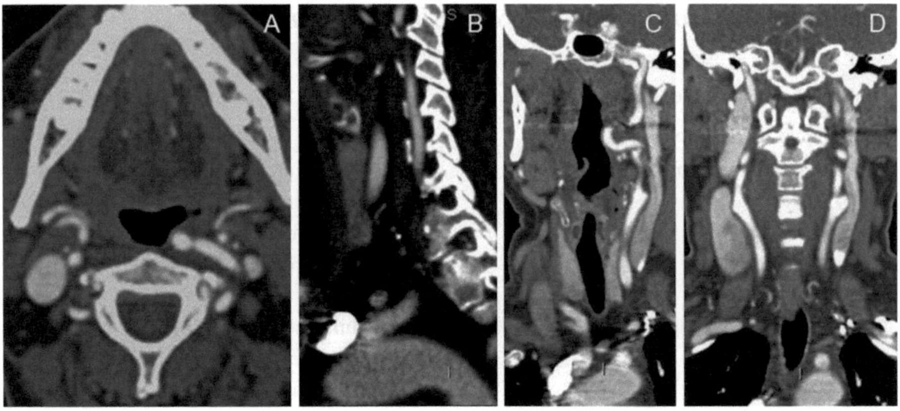

Fig. 22.1 MPR of the supra-aortic trunks: (**a**) axial; (**b**) sagittal; (**c**) oblique-coronal; and (**d**) coronal

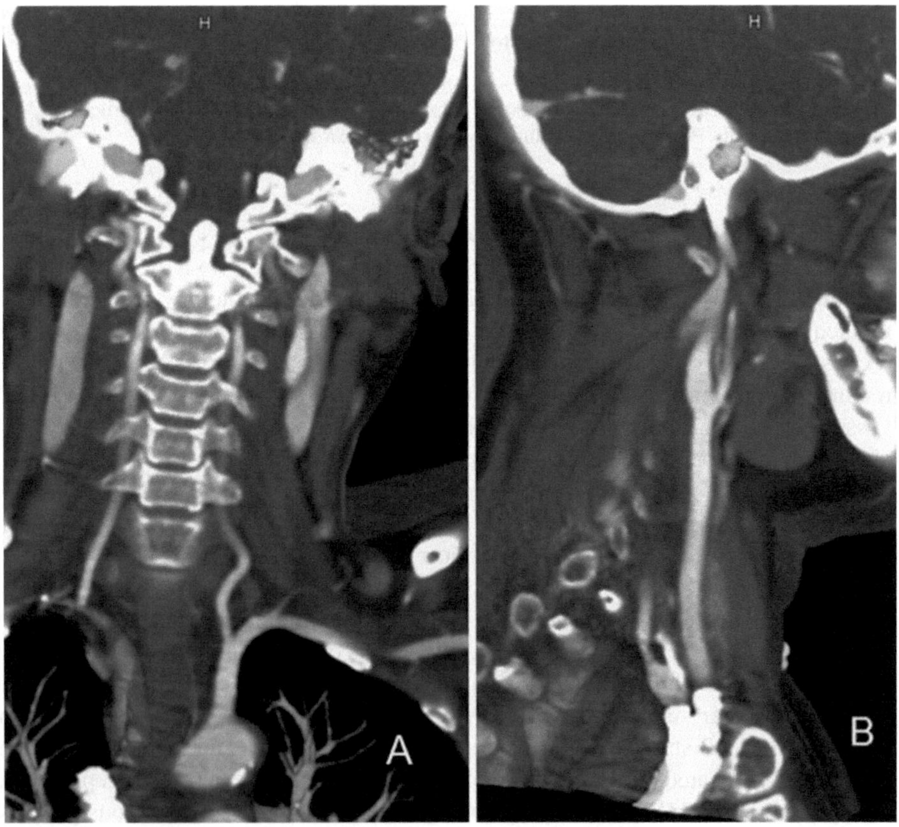

Fig. 22.2 MIP of the supra-aortic trunks: (**a**) coronal; and (**b**) sagittal

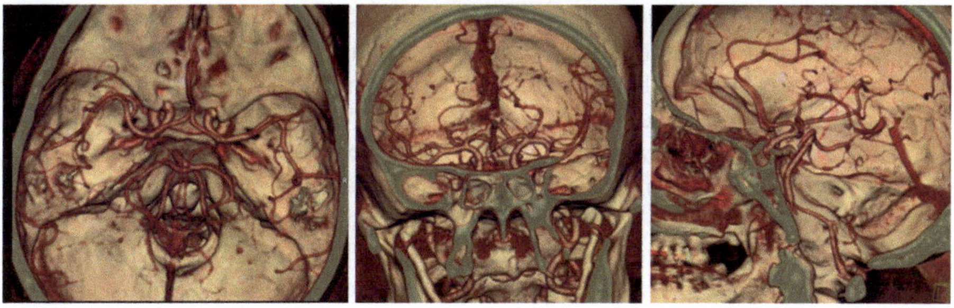

Fig. 22.3 Intracranial VR of the circle of Willis: axial, coronal and sagittal planes

22.2.5.3 Volume Rendering (VR)

Volume Rendering refers to creation of a 3D reconstruction using all of the information in a voxel. It functions according to the proportions of different tissue types in a given voxel, assigning a different colour and transparency level to each type. This approach provides more and better data than all other techniques and therefore, is superior to MIP and SSR. In fact, VR is the most widely used technique in clinical practice, especially for studying aneurysms. The main drawback of VR is that it demands powerful computers (Fig. 22.3).

22.2.5.4 Surface Rendering (SR)

In SR, signal values are assigned to the voxels according to a reference threshold value, such that each voxel is classified as either above or below the threshold (binary system). Based on this classification, the surface is then reconstructed. Surface rendering is fast and requires little computing power, as only a small amount of image data is used; however, as it generates an "average" representation based on the signal of the voxels, it can cause major artefacts for heterogeneous tissue such as bone. In general practice it is only used for simple bone reconstructions; it is not used to study aneurysms, as it implies a massive loss of information.

22.2.5.5 Segmentation

Segmentation is the process used to choose the data to be included in a 3D image. It enables the user to selectively exclude specific portions of the image. This process requires the identification and delineation of different tissue types, which can be done manually or automatically.

There are four principal methods for extracting a 3D object or volume of interest from a given 3D model:

22.2.5.5.1 *Defining parameter threshold values:* The minimum opacity to be displayed can be set by defining a threshold value for opacity. Once the threshold has been established, all of the data below it will disappear from the image.

22.2.5.5.2. *Virtual surgical tools:* These tools enable the user to make "cuts" in, define a region of interest in, or eliminate a region from the 3D volume, or to divide an object into separate components.

22.2.5.5.3. *Painting tools:* These enable the user to highlight in 3D and in colour a region of interest in the 3D volume, in order to subsequently isolate it (i.e., to erase the area surrounding it).

22.2.5.5.4 *Automatic selection tools:* These enable the user to select an object in the 3D volume and automatically add it to, or subtract it from, the primary view.

22.2.5.6 Bone Subtraction

Since bone and calcifications are problematic for studying the vasculature of certain cranial regions, some CT scanners are equipped with bone subtraction methods that enable calcium-rich structures to be removed from the image. The use of bone subtraction methods requires that a volume be acquired before and after administration of the contrast medium (Fig. 22.4).

22.3 Advantages and Limitations of CTA

22.3.1 Advantages

Enables the study of blood vessel walls and adjacent structures;

Enables the assessment of atheromatous plaques and provides visual evidence of calcifications.

22.3.2 Limitations

Contraindicated for patients that are allergic to the contrast medium or that are suffering from kidney failure;

Of limited use for studying the severity of stenoses, because of the presence of calcifications;

Of limited use for studying the circle of Willis, because of the overlap of bone and blood vessel at the base of the cranium.

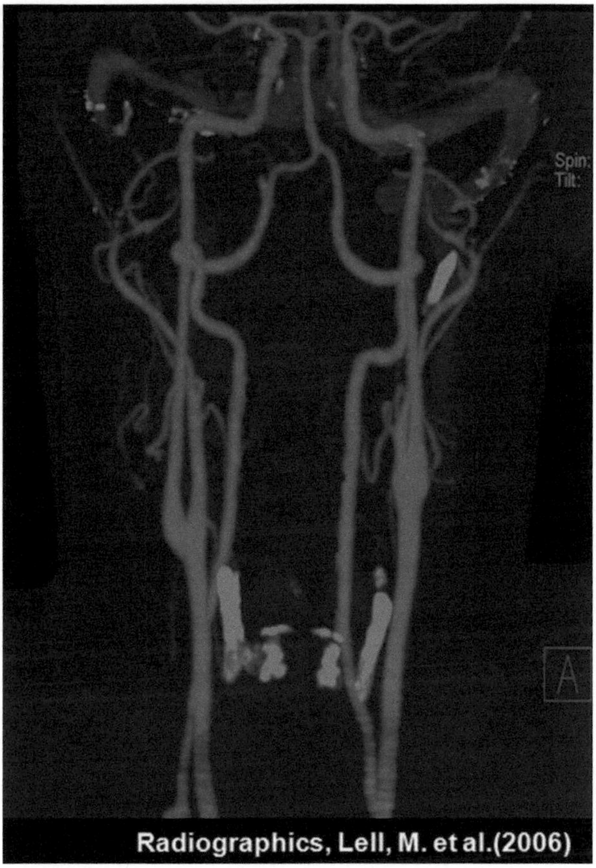

Fig. 22.4 Coronal reconstruction of the supra-aortic trunks using bone subtraction [2]

22.4 Clinical Applications of CTA

22.4.1 Indications for Intracranial CTA

22.4.1.1. **Study of the circle of Willis and its anatomic variants, for intra- and extracranial pre-surgical planning** (Fig. 22.5)

22.4.1.2. **Identification of aneurysms and vascular malformations in cases of subarachnoid haemorrhage**

Computed Tomography Angiography is the technique of choice for the diagnosis and monitoring of aneurysms (above all, in the Emergency Department). However, if a negative CTA is obtained in a patient with subarachnoid haemorrhage in whom an aneurysm is suspected, then Digital Subtraction Angiography (DSA) should be performed, as it provides greater sensitivity and specificity (Fig. 22.6, 22.7, 22.8, 22.9 and 22.10).

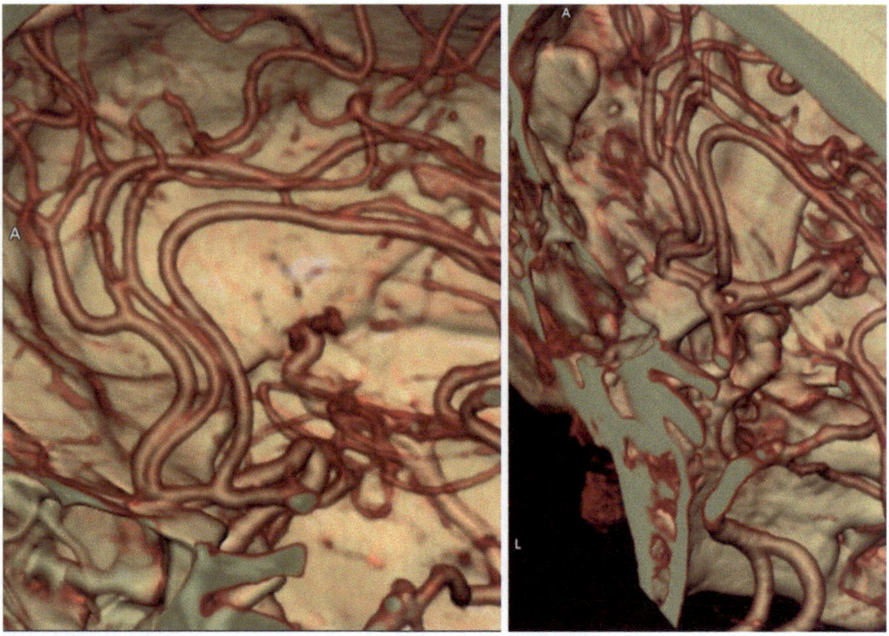

Fig. 22.5 VR of the circle of Willis. Anatomic variant: triple A2 segment with fenestration

22.4.1.3. **Initial evaluation of the intracranial circulation in case of acute ischaemic stroke**

Multimodal CT comprises three exams:

(a) A standard cranial CT, without contrast medium, performed to rule out haemorrhage or other causes of ischemic stroke-like symptoms;
(b) A CTA to determine if any major blood vessels are occluded (based on the location and extent of a thrombus or evidence of collateral circulation);
(c) A perfusion-CT to reveal the area of irreversible brain parenchyma and the penumbra.

22.4.1.4. Study of the vascularisation of brain tumours (arterial and venous phases) (Fig. 22.11).

22.4.1.5. Carotid-cavernous fistulae: These can be studied by CTA, MRA or conventional angiography, depending on the availability of the equipment.

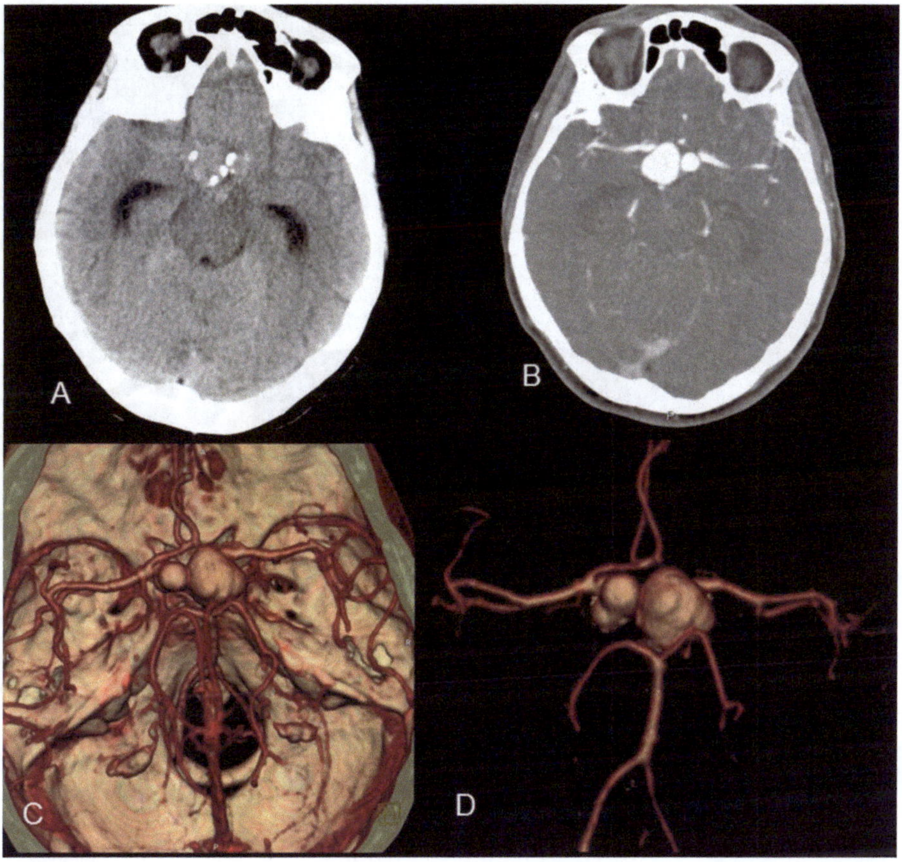

Fig. 22.6 (**a**) CT without contrast medium: partially calcified aneurysm suspected on the *right side* of the circle of Willis. (**b**) CT with contrast medium: bilateral nodular uptake. (**c**) Intracranial VR: confirmation of double aneurysm in the intracranial internal carotid arteries. (**d**) VR with segmentation and bone-subtraction (removal of the base of the cranium)

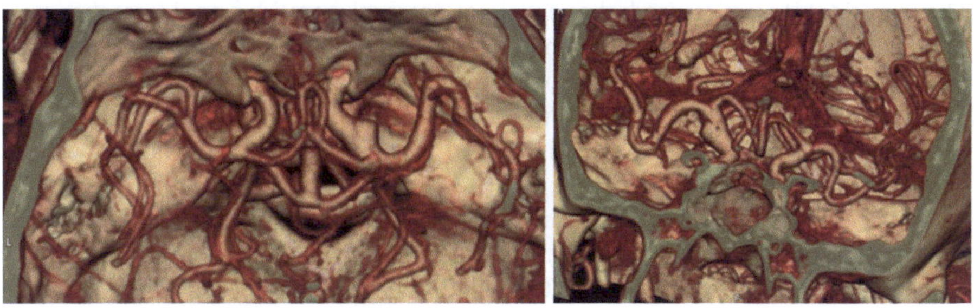

Fig. 22.7 Intracranial VR: carotid bifurcation saccular aneurysm with the left MCA and ACA

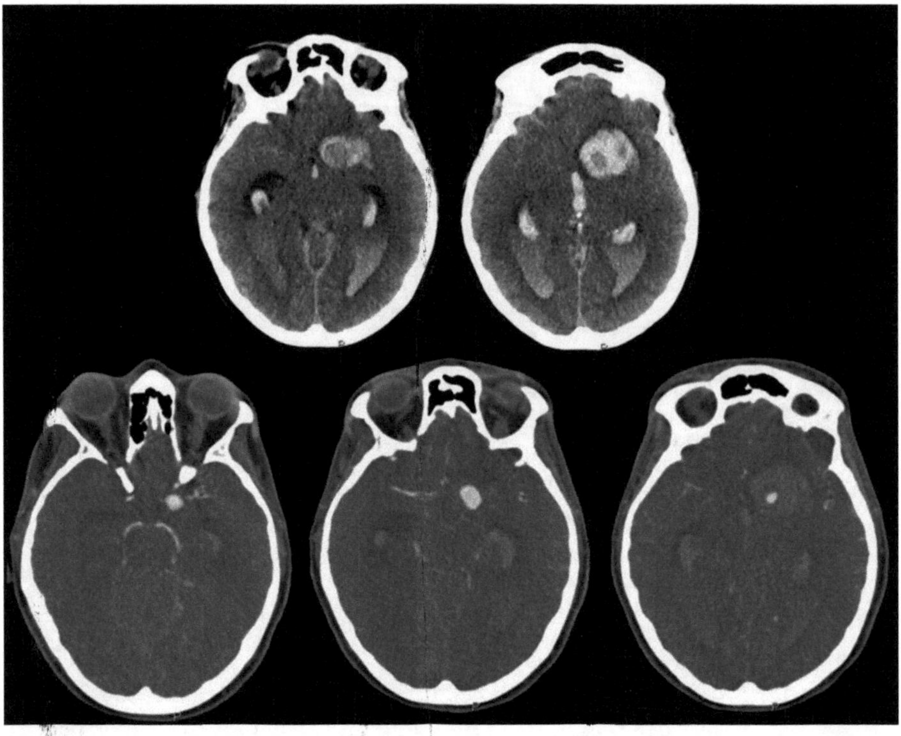

Fig. 22.8 (**a** and **b**) CT without contrast medium: large haematoma with hypodense central nodular area. (**c**, **d** and **e**) CT with contrast medium: nodular uptake inside the haematoma. (**f** and **g**) Intracranial VR: carotid bifurcation aneurysm

22.5 Indications for Extracranial CT

22.5.1 Study of Carotid Stenoses

The carotid bifurcation and the proximal internal carotid artery are the most frequently affected regions. Computed Tomography Angiography has enabled major advances in the study of blood vessels: through acquisition of volumetric data, and subsequent post-processing (at a workstation) of these data using MPR and MIP methods, it enables visualisation of the vessel lumen and walls as well as of the surrounding tissue.

Physicians often wish to ascertain the grade of stenosis and characterise the plaque, for which the preferred techniques at many centres remain Doppler ultrasound and MRI. Using either CTA or MRA, one can measure the area of a blood vessel in the axial plane and determine the grade of stenosis in the vessel. Using CT, one can also identify calcifications. Any one of these techniques can be used, depending on the experience of the staff and on the equipment available.

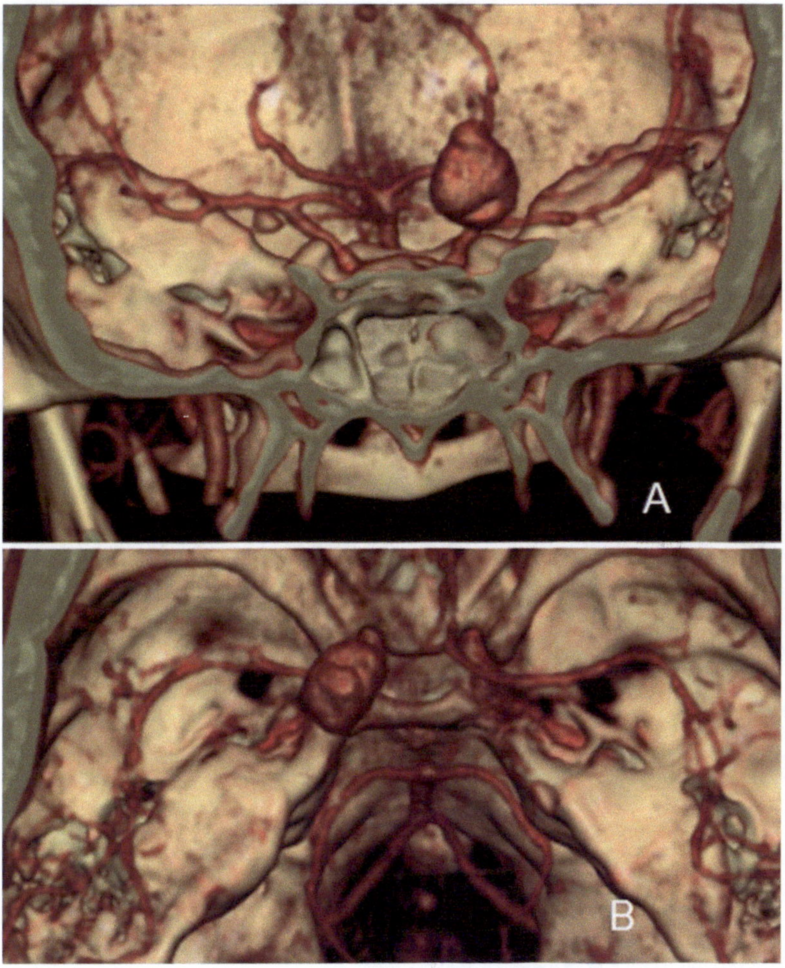

Fig. 22.9 (**a** and **b**) CT without contrast medium: large haematoma with hypodense central nodular area. (**c**, **d** and **e**) CT with contrast medium: nodular uptake inside the haematoma. (**f** and **g**) Intracranial VR: carotid bifurcation aneurysm

22.5.2 Post-operative Monitoring of Supra-Aortic Trunk Surgeries

The most widely used techniques are Doppler ultrasound and CT. Magnetic Resonance Imaging is less reliable, since magnetically active items remaining from surgical procedures (e.g., clips, etc.) can generate artefacts.

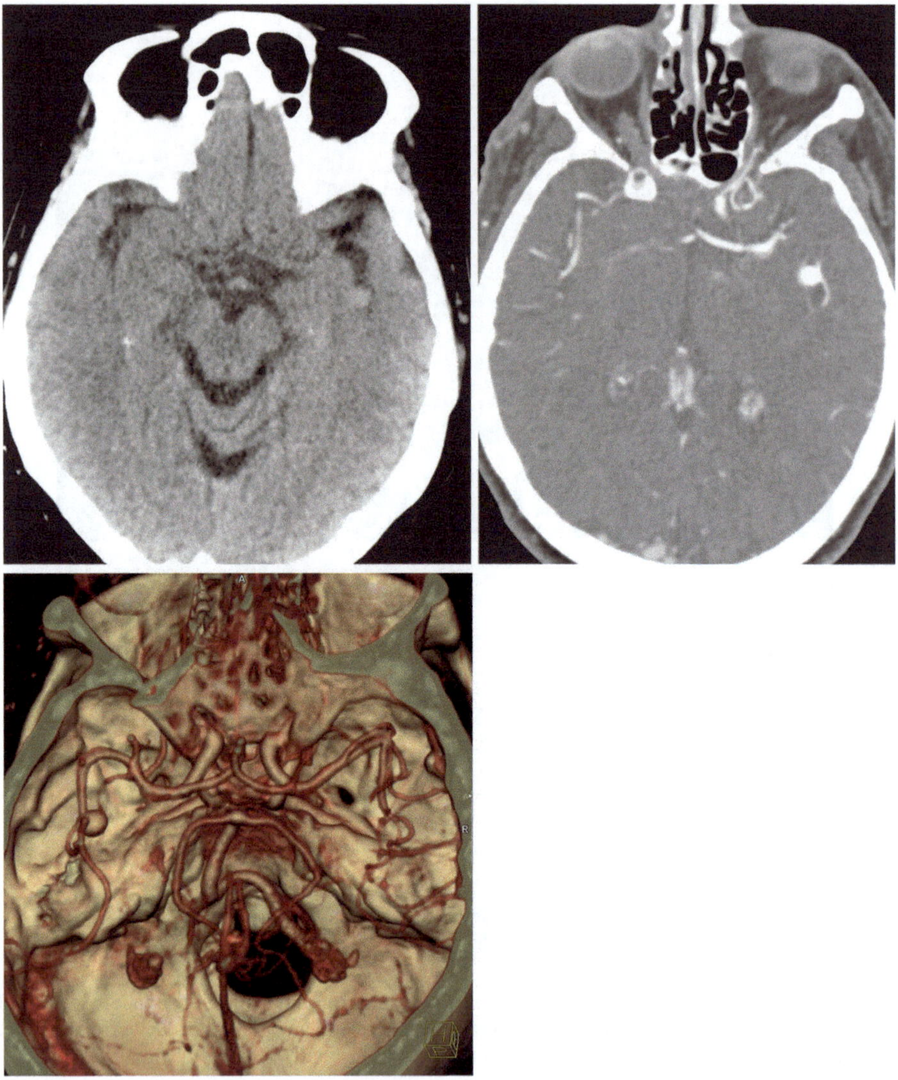

Fig. 22.10 CTA and basal CT. Intracranial VR: aneurysm in the A2 segment of the left MCA

22.5.3 Examination of Craniocervical Trauma

The preferred technique is CTA, owing to its speed, efficacy and ability to reveal occlusions, extravasation of the contrast medium and associated bone lesions.

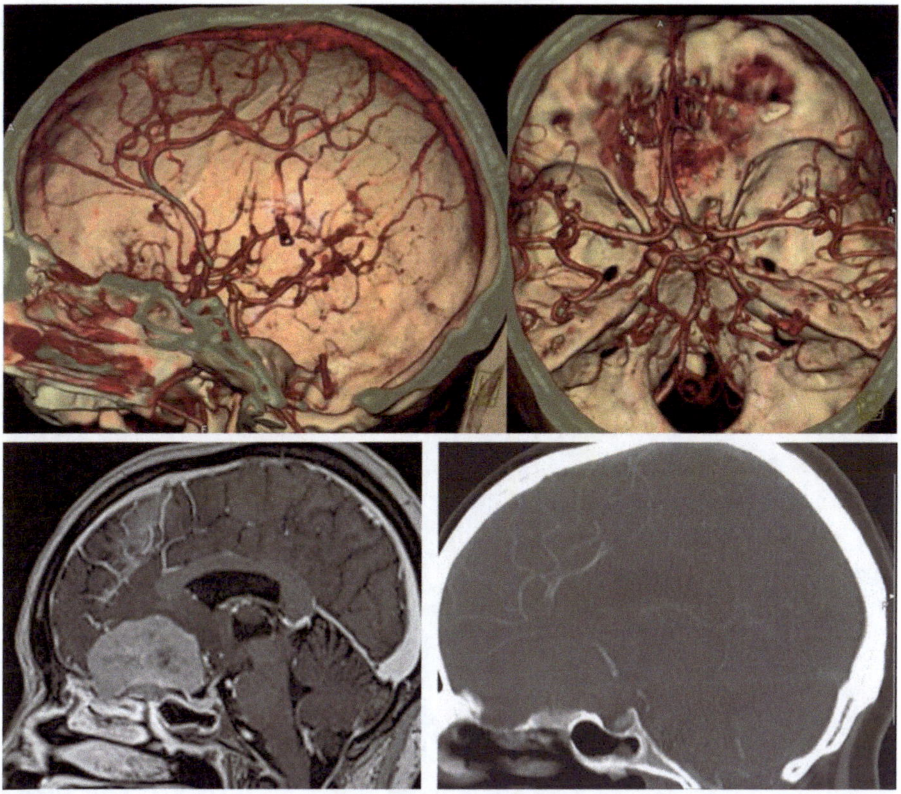

Fig. 22.11 Olfactory groove meningioma. VR: Note the displacement of the A2 segment of the ACA. T1-weighted sagittal MRI with paramagnetic contrast medium: intense and homogeneous contrast. CT-MIP: upwards displacement of the A2 segment of the ACA and associated hyperostosis of the cortex

22.5.4 Study of Cervical Spinal Cord Vascular Malformations Such as Paragangliomas or Carotid Bodies

Computed Tomography Angiography can also be used to diagnose the following conditions, although it is not the method of choice:

- Carotid or vertebral artery dissection: MRA is the preferred technique (Fig. 22.12).
- Fibromuscular dysplasia: can be diagnosed by CTA or MRA, by identification of alternating dilatations and stenoses.
- Subclavian steal: the preferred techniques for initial examination are Doppler ultrasound and MRI.

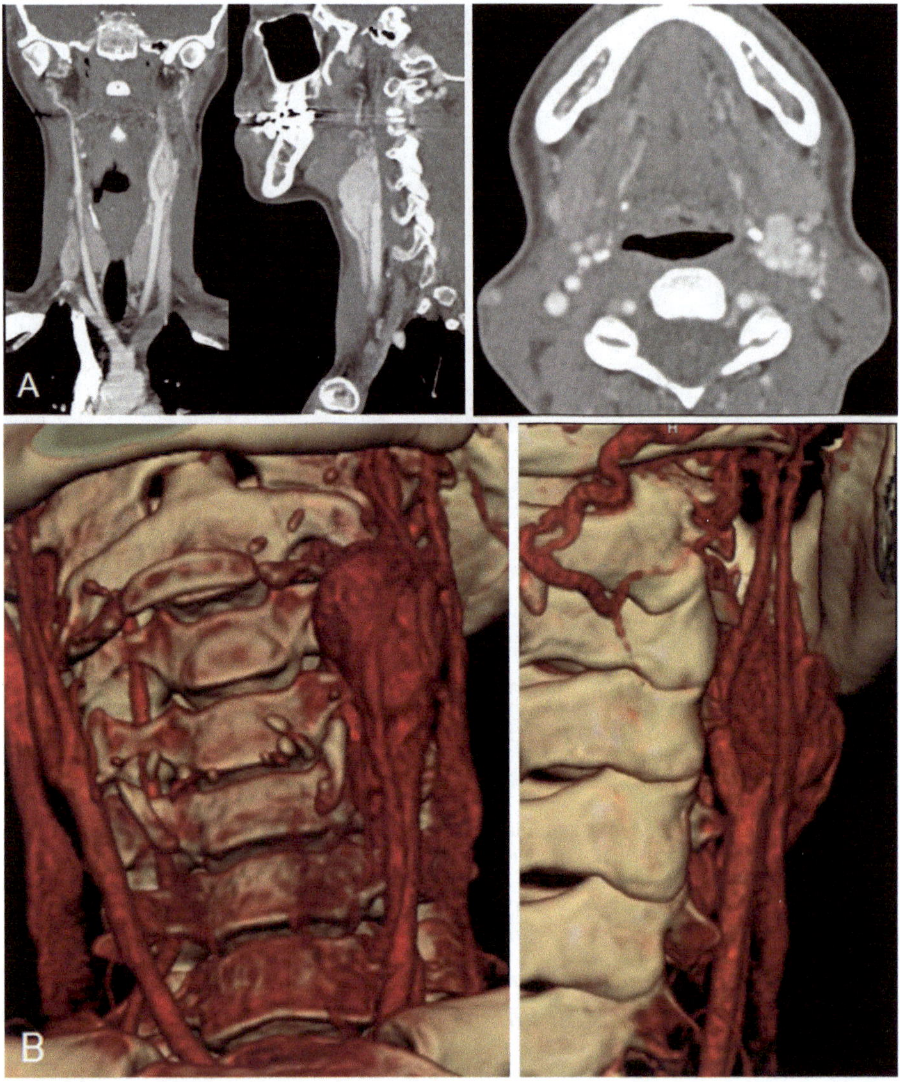

Fig. 22.12 (**a**, **b** and **c**) MIP of the supra-aortic trunks. Carotid body: coronal, sagittal and axial planes. (**d** and **e**) VR of the supra-aortic trunks

22.6 Venous CTA

Venous CTA is recommended for suspected cases of cerebral venous thrombosis, especially when MRI has given ambiguous results, is unavailable or is contraindicated. In patients that show signs of cerebral venous thrombosis on CT without contrast medium, venous CTA should be realised immediately in order to confirm the diagnosis and begin adequate treatment.

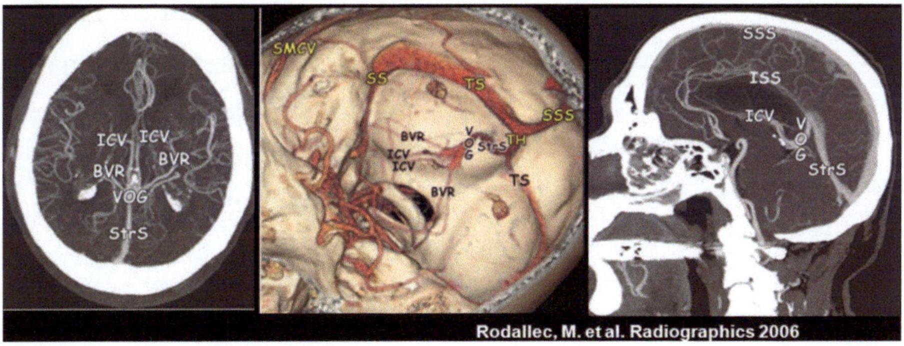

Fig. 22.13 MPR and MIP of the venous sinuses [4]

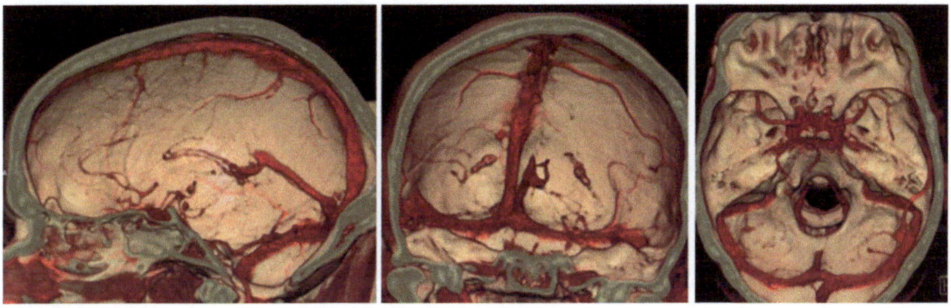

Fig. 22.14 VR of the venous sinuses in three planes

A CT scanner with 16 detectors provides data with a minimum thickness of 0.6 mm and a pitch of 0.9 mm. The contrast agent (non-ionic iodinated) is administered at a concentration of 300 mg/mL, a speed of 3 mL/s and with an injection delay of 45 s pre-acquisition. An injection delay shorter than 30 s implies a greater risk of failed diagnosis due to insufficient contrast in venous structures and to flow-related artefacts. The study area extends from the apex of the cranial vault to the first spinal vertebra (C1), such that it includes the internal jugular veins.

Given that all 3D methods entail some loss of information, none of them is an adequate substitute for exhaustive analysis of the source images. The most widely used post-processing techniques are 2D MPR (for visualisation of the dural venous sinuses and the cerebral veins), 2D/3D MIP and VR (Figs. 22.13 and 22.14).

References

1. del Cura JL et al (2010) Radiología esencial; Diagnóstico no invasivo de la patología de los troncos supraaórticos. Médica Panamericana, Madrid
2. Lell M et al (2006) New techniques in CT. Radiographics 26:45–63
3. Osborn AG (2013) Osborn's brain: imaging, pathology, and anatomy. Amirsys, Utah
4. Rodallec M et al (2006) Cerebral venous thrombosis and multidetector CT angiography: tips and tricks. Radiographics 26:5–19

Chapter 23
Magnetic Resonance Angiography: Fundamental Techniques and Data Interpretation

Josep Lluis Dolz Jordi, Laura Susana Goiburú González, and Cristina Corbella Sala

23.1 Introduction

Magnetic Resonance Angiography (MRA) techniques have become well established thanks to their increased use and broader application, made possible by several factors such as the use of high-field technology and stronger magnetic gradients. These advances, combined with faster processors and specific coils, have been employed to optimise the spatial resolution and resonance signal in MRA. Another important development has been the simplification of post-processing done at stand-alone workstations, which has enabled shorter examination times. Thus, MRA techniques are now a viable option for imaging of vascular structures throughout the body, as well as for characterisation of plaques and analysis of related complications.

23.2 MRA Techniques

Magnetic Resonance (MR) is a highly sensitive technique based on the movement of protons (hydrogen nuclei). This motion is harnessed to generate images in which voxels containing moving protons are distinguished from those containing stationary protons. In MRA, information is acquired by exploiting the differences between moving protons and stationary ones: either in terms of their respective absorption of radiofrequency (RF) pulses, or in terms of their respective dephasing when moving under weak magnetic gradients. The former is analysed using Time of Flight (TOF) sequences and the latter, using Phase Contrast (PC) sequences. (Table 23.1).

J.L. Dolz Jordi (✉) • L.S. Goiburú González • C.C. Sala
Servicio de Radiodiagnóstico, Hospital Mútua de Terrassa,
Plaza Dr Robert, 5, 08221 Terrassa (Barcelona), Spain
e-mail: jlldolzjordi@gmail.com

M. Slevin, G. McDowell (eds.), *Handbook of Vascular Biology Techniques*,
DOI 10.1007/978-94-017-9716-0_23

Table 23.1 MRA techniques

2D PC
PC 3D
Cine PC
2D TOF
TOF 3D
MOTSA
CE-MRA
Black-blood techniques

23.2.1. Time of Flight sequences are based on the detection of unsaturated protons as they enter into a plane or volume in which the protons are saturated. When the plane or volume is subjected to repeated RF pulses using a short Repetition Time (TR), the saturated protons will exhibit a gradually decreasing longitudinal relaxation time and therefore, will give a much weaker resonance signal than will the unsaturated protons that have just entered.

23.2.2. Acquisition of PC images is based on the use of magnetic gradients to differentiate between stationary spins and moving ones. The dephasing between moving spins and stationary spins is proportional to the speed at which the former move. Thus, information on blood flow can be obtained by adjusting the acquisition speed to the blood flow speed in the desired area.

23.2.3. There are two options for studying a given anatomical region by MRA: either choosing a single volume of appropriate size (3D TOF or PC), or dividing the region into slices of an appropriate thickness for consecutive imaging of each one (2D TOF or POC) and subsequent reconstruction of the volume of interest.

23.2.4. Each one of the aforementioned approaches has its respective advantages and disadvantages. It is important to understand each approach well in order to choose the best one for each examination (Tables 23.2 and 23.3). In 2D TOF MRA sequences, the flow signal reaches its maximum intensity when the acquisition plane is perpendicular to the direction of the blood vessel, but it loses intensity in arterial branches and in arterial courses that are parallel to the plane. The signal also diminishes due to a phenomenon known as *intravoxel dephasing*, which is caused by turbulent flow related to ectasia or distal to stenosis. However, 2D TOF MRA can prove useful for studying slow venous flow in the intracranial venous sinuses or in the veins of the neck when contrast-enhanced MRA (CE-MRA) cannot be used. Three-dimensional TOF MRA is not as strongly influenced by vessel branching or directional changes and therefore, offers full potential for assessing the intracranial arteries. However, use of this sequence in patients with slow blood flow (such as those with low cardiac output or with other pathological conditions that affect flow) leads to a major increase in the saturation of moving spins, causing a gradual loss in visualisation of the fine distal blood vessels. To partially overcome this saturation effect, one can use a variant of 3D TOF known as Multiple Overlapping Thin-Slab Acquisition (MOTSA). In this technique, the volume of interest is divided

Table 23.2 Advantages and disadvantages of MRA techniques

2D PC:
Fast
Adjustable speed range for arterial or venous examination
Imaging of vascular topography for planning MRA studies
Rapid, motion-based evaluation of the circle of Willis
Detects flow in the large intracranial venous sinuses
Low spatial resolution
Random encoding of the speed range
3D PC:
Adjustable speed ranges
Good background suppression
Very slow
Very sensitive to signal loss caused by turbulent flow distal to severe stenosis
Very rarely used, except for studying the renal arteries
2D TOF:
Minimal saturation
Useful for slow flow
Used for MR Phlebography (MRP)
Short study time
Low spatial resolution
Difficult to distinguish material with short T1 times (hyperintense in T1-weighted sequences)
Very sensitive to signal loss in the direction parallel to the acquisition plane
3D TOF:
High spatial resolution
Intermediate study time
Not directionally dependent
Enables examination of stationary tissue
Flow saturation due to signal loss from slow flow
Difficult to distinguish material with short T1 times (hyperintense in T1-weighted sequences)
CE-MRA
High spatial resolution
3D acquisition
Short study time
Does not depend on the flow direction
Does not suffer from the flow artefacts of traditional MRA sequences
Does not allow examination of stationary tissue

into a series of smaller overlapping volumes that are independently imaged and subsequently reconstructed.

23.2.5. Two-dimensional PC sequences offer rapid acquisition, making them ideal for planning a more detailed MRA study. For example, they enable ready location of the carotid bifurcations and of the intracranial vessels for the initial assessment of any altered flow.

Phase-Contrast sequences can be synchronised with the cardiac cycle (this is known as *cine PC*) in order to provide information on vessel elastic-

Table 23.3 Applications of MRA sequences

2D PC:
Imaging of vascular topography for planning MRA studies
Rapid, motion-based evaluation of the circle of Willis
Detection of flow in the large intracranial venous sinuses
3D PC:
Rarely ever used, although it can be utile for studying the renal arteries
Cine PC:
Imaging of vascular topography for planning MRA studies
Superior-inferior flow direction encoding for studying subclavian steal syndrome
Anterior/Posterior and Right/Left flow-direction encoding to detect reversed flow in the branches of the circle of Willis or in the ophthalmic artery, due to carotid artery occlusion
2D TOF:
Useful for slow flow
Used for MR Phlebography (MRP)
3D TOF and variants (MOTSA):
Exploration of stationary tissue
Amenable to the study of carotid artery bifurcations
Enables study of:
Vessel walls, plaques, dissections and intramural haematomas
Circle of Willis and the basilar artery
Neurovascular contact between the trigeminal nerve and the basilar artery, or between the facial nerve and the anterior inferior cerebellar artery
Cerebral aneurysm, flow and mural thrombus
Cerebral AVM
CE-MRA:
Examination of patients with extracranial occlusive vascular disease
Couple to MOTSA or T1 sequence with fat suppression for carotid or vertebral artery dissection
Enables study of:
Congenital defects of the aortic arch
Obstructive, inflammatory or arteriosclerotic pathology of the aortic arch and its branches
Soft-tissue space-occupying lesions of vascular origin
The relationship between a soft tumour and adjacent blood vessels
Enables exploration of the intracranial venous sinuses, venous thrombus and surrounding tumour

ity and flow direction, which can be used to determine the normal flow direction or identify any flow reversal or collateral routes. Arterial or venous vessels can be studied by either adjusting the speed in a PC sequence or by using saturation pulses in a TOF sequence, each of which will suppress the flow from a given direction. Although the advent of contrast-enhanced MRA (CE-MRA) relegated some conventional (or “flow-based”) MRA techniques to a minor role, the authors of this chapter believe that these techniques remain valuable for studying certain anatomical regions, especially the cerebral arteries.

23.2.6. Contrast-enhanced MRA

A diverse array of vascular structures can be studied by 3D (volume) imaging, by using very fast spin-echo gradient sequences, gadolinium (Gd)-based paramagnetic contrast agents to intensify the blood signal, and an infusion pump. These contrast agents enable study of blood vessels based on shortening of the T1 and T2 relaxation times of the blood, rather than according to blood flow. Gadolinium causes the protons surrounding each contrast agent molecule to relax more quickly (shortening of T1) while simultaneously influencing the local magnetic susceptibility (shortening of T2). In T1-weighted spin-echo gradient sequences, these agents give positive contrast (stronger blood signal), whereas in T2*-weighted sequences, they give negative contrast. Thus, for MRA, a T1-weighted, fast spin-echo gradient sequence is used with 3D (volume) acquisition. For the best results, the T1 values for blood should be lower than for other tissue (including fat).

Tables 23.4, 23.5 and 23.6 list the technical specifications for the sequences used at the two centres in which the authors work.

Table 23.4 Technical characteristics of MRA sequences

Scanner model and manufacturer	Symphony 1.5T, Siemens (Erlangen, Germany)	
Sequence	Multislab TOF 3D	MOTSA
TR (ms)	39	22.0
TE (ms)	7.12	Minimum
Angle	25°	20°
Field of view (FOV) (X)	200	200
Slice thickness (mm)	1.5	2.4
Acquisition plane	Axial	Axial
Matrix	192×512	256×192
Coils	Head array and neck array	Neurovascular
Saturation band (placement)	Craniocaudal (above)	Craniocaudal (above)
Acquisition time (min)	5.26	7.14

Table 23.5 Technical characteristics of MRA sequences

Symphony 1.5T/Siemens (Erlangen, Germany)	
Sequence	CARE bolus
TR (ms)	3.24
TE (ms)	1.08
Angle	20°
Field of view (FOV) (X)	350
Slice thickness (mm)	60
Acquisition plane	Coronal
Matrix	144×192
Coils	Head array, neck array
Scan rate	1 scan per second (60 scans per minute)

Table 23.6 Technical characteristics of CE-MRA sequences

Scanner model and manufacturer	Symphony 1.5T, Siemens (Erlangen, Germany)	SIGNA 1.5T, GE (Milwaukee, USA)
Field strength	1.5T	1.5T
Sequence	Fl 3D-CE	3D-fast SPGR (TOF)
Bolus tracking technique	CARE bolus	Real-time, smart-prep, test bolus
Bolus tracking region-of-interest	Aortic arch (contrast agent reaches supra-aortic vessels)	Aortic arch and supra-aortic vessels
Delay	Up to the supra-aortic vessels	Obtained from the test bolus
TR (X)	3.77	Minimum
TE (X)	1.54	Minimum
Angle	30°	30°
Matrix	168×384	384×256
Field of View (FOV) (mm)	300	340
Slice thickness (mm)	0.8	2
Acquisition plane	Subclavian-axillary vessels and supra-aortic vessels: coronal	Subclavian-axillary vessels and supra-aortic vessels: coronal
Coils	Head array & neck array	Supra-aortic vessels: neurovascular coil
Contrast medium	Gd: Gadovist® 1 M	Gd: MultiHance®
Concentration (M)	1.0	0.5
Delivery system	Infusion pump	Infusion pump
Injection speed (mL/s)	2	1.5
Acquisition time (s)	40	45
Stations	1: Supra-aortic vessels	1: Supra-aortic vessels
Mask	Supra-aortic vessels	Supra-aortic vessels
Repeated acquisitions	–	Vascular malformations
		Vascular tumours
		Carotid-cavernous fistula
Apnoea (breath-hold) required?	No	No
Elliptic-centric techniques	Yes	Yes

23.2.7. Contrast-enhanced MRA techniques are especially utile for patients that are contraindicated to iodinated contrast agents due to kidney failure or to allergy: the Gd-based contrast agents used in CE-MRA exhibit lower nephrotoxicity than do the iodinated contrast agents used in conventional angiography, DSA or CT. The reader is reminded that paramagnetic contrast agents very rarely cause allergic reactions, and these are less severe than those caused by iodinated contrast agents. In order to avoid the risk of the patient developing nephrogenic systemic fibrosis related to chronic kidney failure, the healthcare professional must take into consideration the patient's glomerular filtration rate (GFR). Patients with stage 4 or 5 chronic kidney disease (CKD) and a GFR lower than 30 mL/min are at high risk, whereas patients with stage 3 CKD and a GFR of 30–59 mL/min are at lower risk. The recommended type of contrast medium to be used is stable gadolinium macrocycles (gadobutrol, gadoterate meglumine, gadoteridol, etc.). Regardless, the healthcare professional must never deny any well-indicated patient a CE-MRA. Another advantage of MRA compared to certain other imaging methods is that it does not expose the patient to any ionising radiation.

23.2.8. The only limitations and contraindications of these techniques are those inherent to the use of magnetic fields: they are not adequate for patients with pacemakers, neural stimulators, inner ear prostheses or older models of cardiac prostheses, or with vascular clips made of materials that incompatible with MR. Furthermore, before proceeding to examine any patient that regularly works with metal, the healthcare professional must first confirm that the patient's eyes do contain any residual metal fragments. One common problem among patients is claustrophobia, which can be avoided by sedating the patient.

23.3 Clinical Applications and Indications

Whilst the previous section described the nomenclature of MRA, this section outlines its diagnostic utility and indications. For examination of patients with a suspected cerebrovascular pathology that have been screened by appropriate Doppler ultrasound techniques, MRA has replaced conventional angiography.

23.3.1 Supra-aortic Vessels

Among the most frequent applications of MRA is the examination of carotid bifurcations. According to numerous studies, MRA offers comparable results to those obtained with Doppler ultrasound or even DSA. To study the supra-aortic vessels by CE-MRA, elliptic-centric K-space filling is employed. In this approach, those central lines of the K-space that are responsible for the contrast are filled, thereby reducing signal interference from jugular vein overlap. The recommended screening method is ultrasound; if a haemodynamically significant stenosis (>50 %) is identified, then the patient should be examined by Computed Tomography Angiography (CTA) or MRA. The authors of this chapter use DSA primarily for those cases in which the results from Doppler ultrasound are inconsistent with those of MRA, and in which re-examination of the results previously obtained with each technique does not enable any conclusive diagnoses, although to date such cases are only anecdotal (Fig. 23.1). Another important application of DSA is for definitively ruling out any pseudo-occlusion of the carotid artery caused by a critical stenosis that has blocked the flow of blood to the cervical portion of the internal carotid artery (ICA). However, for certain patients some healthcare teams prefer surgical exploration of the carotid sinus, in order to avoid the vascular intervention and contrast media required for DSA.

The reader is reminded that some types of MRA sequences can lead to overestimation of the grade of stenosis (from low to moderate), due to dephasing of the spins inside the vessel caused by turbulent flow; however, this rarely causes any alteration in treatment plans. Regardless, use of CE-MRA has minimised this problem. There are three methods for grading stenosis: the North American Symptomatic Carotid Endarterectomy Trial (NASCET), European Carotid Surgery Trial (ECST) and Common Carotid (CC) methods. The most common method is the NASCET method (see Step 23.4.2.8), but the CC method can also be useful (Fig. 23.2).

If a carotid or vertebral artery dissection is suspected, than a combination of CE-MRA and MOTSA or 3D TOF MRA sequences can prove highly valuable: it can reveal not only the compression of the arterial lumen, but also vessel wall haematoma; double lumen caused by detachment of the intima; vessel occlusion; or recanalisation of previously occluded vessels (Figs. 23.3 and 23.4). Another option is to image the suspected arterial region by acquiring T1-weighted slices with fat suppression.

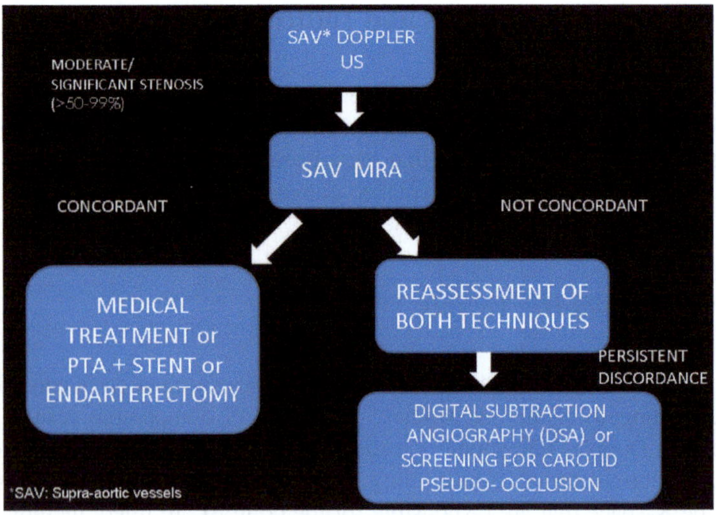

Fig. 23.1 Clinical algorithm for carotid artery stenosis

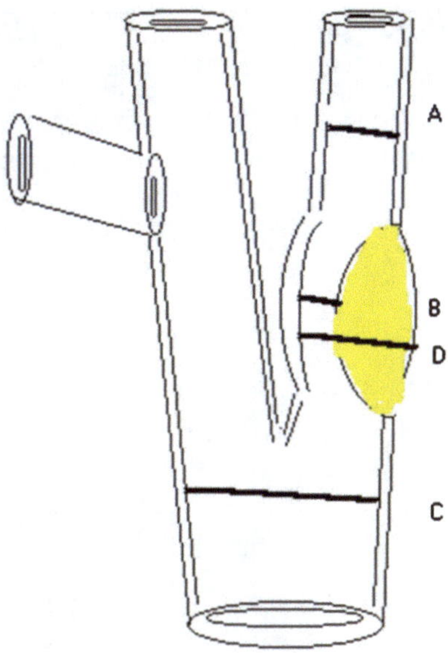

Fig. 23.2 Common methods for grading carotid artery stenosis. *A*: Normal diameter of the internal carotid artery distal to the stenosis. *B*: Diameter of the minimum residual lumen. *C*: Visually estimated diameter of the internal carotid artery. *D*: Diameter of the common carotid artery. Formulas for grading carotid artery stenosis: 1. NASCET: (A–B)/A × 100 % stenosis, 2. ECST (D–B)/D × 100 % stenosis, 3. Common Carotid: (C–B)/C × 100 % stenosis

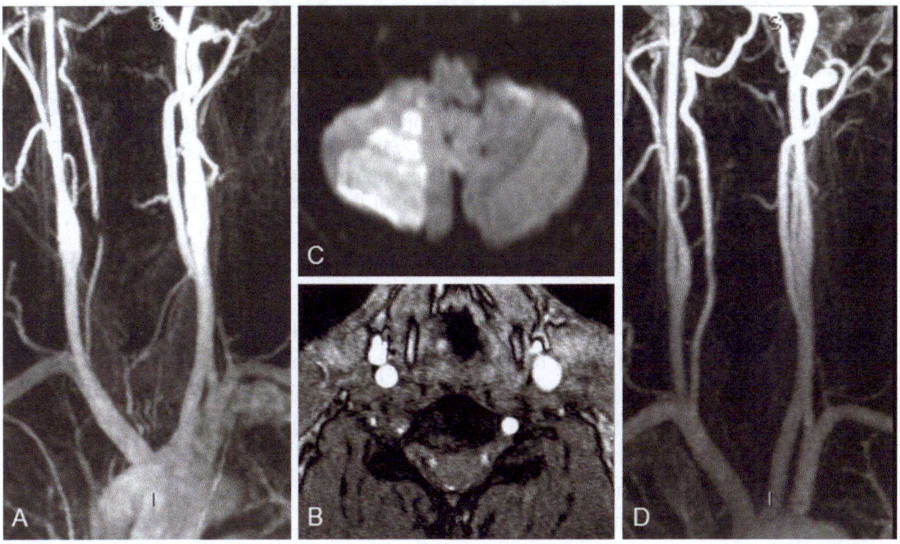

Fig. 23.3 (**a**) Coronal MIP reconstruction of CE-MRA of the supra-aortic vessels, revealing stenosis in the V2 segment of the right vertebral artery. (**b**) Partition of the MOTSA sequence: the intramural haematoma appears hyperintense, indicating a vertebral artery dissection, which has led to the reduction in the arterial lumen. (**c**) DWI B1000: restricted diffusion in the basal aspect of the right cerebellar hemisphere secondary to an acute infarct. (**d**) Coronal MIP reconstruction of CE-MRA of the supra-aortic vessels performed 3 months later, revealing near-complete resolution of the previously observed stenosis in the V2 segment of the right vertebral artery

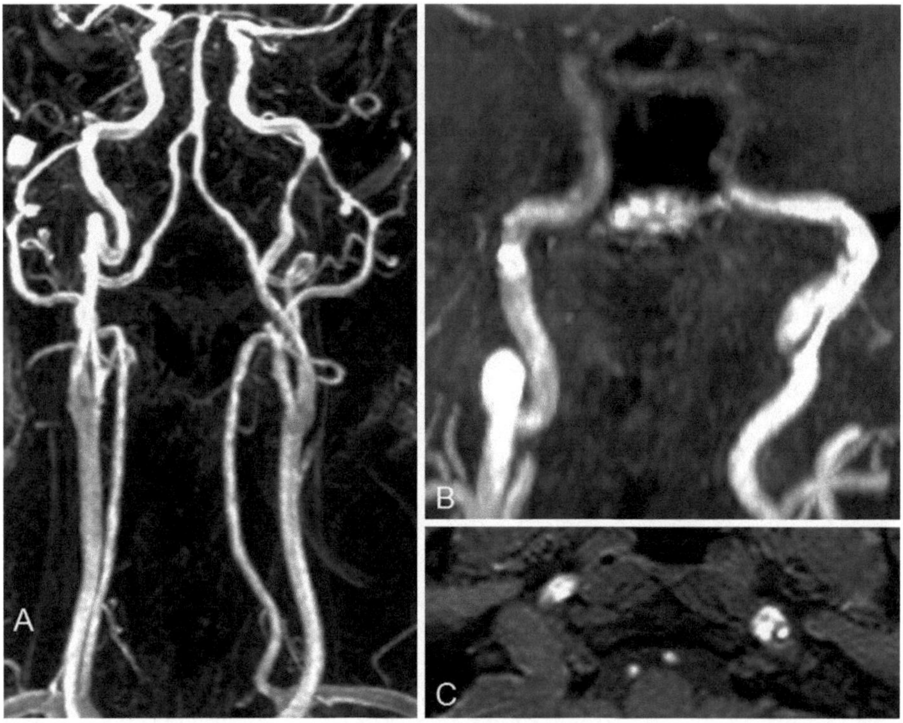

Fig. 23.4 (**a**) Coronal MIP reconstruction of CE-MRA of the supra-aortic trunk, revealing reduction in the diameter of the cervical portion of the left internal carotid artery. (**b**) Oblique-coronal MIP reconstruction of 3D TOF MRA of the intracranial and upper cervical segments of the carotid arteries, revealing an intramural haematoma and showing a double lumen. (**c**) Partition of a 3D TOF MRA sequence: the intramural haematoma appears hyperintense, indicating a carotid artery dissection, which has led to the reduction in the lumen of the internal carotid artery

Likewise, for carotid artery obstructions or critical stenoses, use of cine PC sequences enables evaluation of the flow direction in the arterial segments comprising the circle of Willis or in the ophthalmic artery, for use in determining their role in compensation in the distal vascular bed. Cine PC sequences are also valuable in cases of Subclavian Steal Syndrome, as they can reveal reversed flow in the vertebral artery ipsilateral to the subclavian stenosis.

Severe obstruction or stenosis of the prevertebral subclavian artery can be hidden behind other blood vessels. Therefore, detection of any occult stenoses requires MIP reconstruction of the MRA studies in different planes (Fig. 23.5).

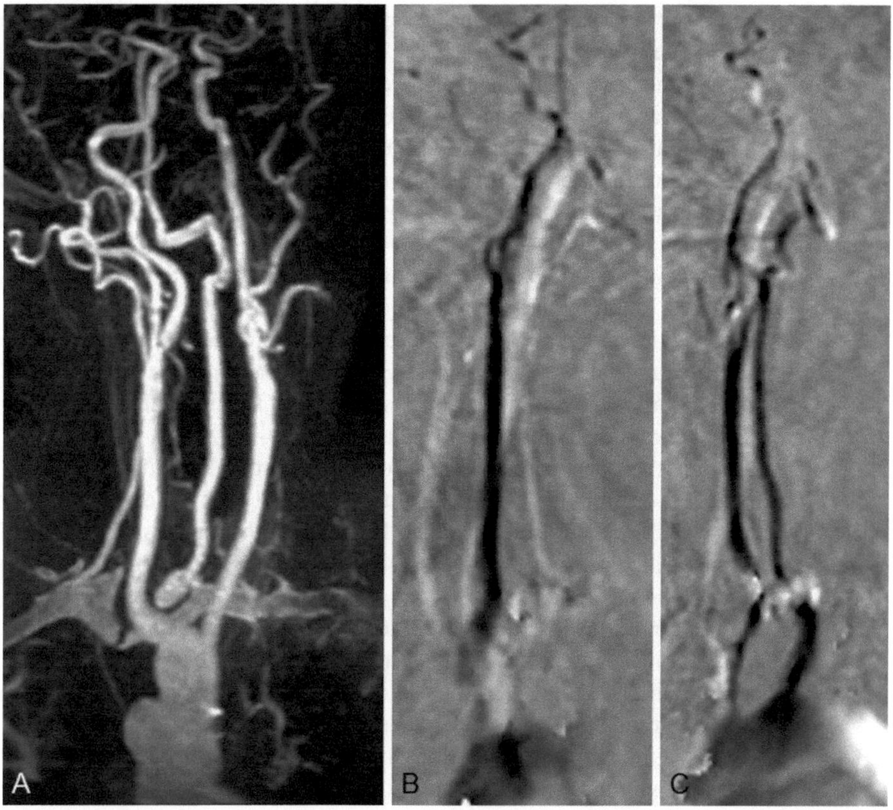

Fig. 23.5 (**a**) Oblique-coronal MIP reconstruction of 3D SPGR CE-MRA, evidencing obstruction of the prevertebral segment of the left subclavian artery. (**b**) Sagittal cine PC with superior-to-inferior flow encoding on the right axis: caudocranial venous flow appears hypointense (*dark*) and craniocaudal venous flow appears hyperintense (*light*). Cranial flow is observed in the right carotid axis and from the right verterbral artery, and craniocaudal venous flow is observed in the internal jugular vein. (**c**) Sagittal Cine PC with superior-to-inferior flow encoding on the left axis: cranial flow is observed in the left carotid axis; craniocaudal venous flow, in the internal jugular vein; and craniocaudal venous flow from the left vertebral artery indicates reverse flow in the vertebral artery

23.4 Basic Protocol for MRA of the Cranium and the Supra-aortic Vessels

23.4.1 Image Acquisition

23.4.1.1. Before entering the MR examination room, confirm that the patient does not have any contraindications to MR.

23.4.1.2. Describe the procedure to the patient. Remember that the patient's collaboration is fundamental: by ensuring that the patient does not move during the acquisition, the healthcare professional can minimise the generation of artefacts in the image (these artefacts seriously degrade image quality).

23.4.1.3. Choose the appropriate head array coil and a neck array coil.

23.4.1.4. The contrast medium is administered via intravenous (IV) injection into the right arm, in order to prevent the brachiocephalic venous trunk causing any artefacts in the images of the aortic arch and the origin of the supra-aortic vessels.

23.4.1.5. Multiplanar localiser images

23.4.1.6. Sagittal cine PC or 2D PC MRA images of each carotid axis are acquired for use in planning the MRA examination.

23.4.1.7. To study the circle of Willis, choose the following axial 3D TOF volume (with saturation of craniocaudal venous flow) on the sagittal 2D PC localiser image: from the cavernous portion of the internal carotid arteries to beyond the anterior cerebral arteries.

23.4.1.8. To study the cervicocranial transition of the internal carotid artery (segments C2–C5), the V4 segment of the vertebral arteries, and the basilar artery, choose the following axial 3D TOF volume on the sagittal 2D PC localiser image: from below the V4 segment of the vertebral arteries to beyond the basilar artery.

23.4.1.9. Another option is to acquire a single MOTSA volume in the axial plane, from below segment V4 of the vertebral arteries to beyond the anterior cerebral arteries; however, in case of slow flow, there will be excessive phasing at the boundaries between slabs.

23.4.1.10. To study the carotid bifurcations in detail, choose the following axial MOTSA volume (with saturation of craniocaudal venous flow) on the sagittal 2D PC localiser image: from below the carotid bifurcations to the cervical portion of the internal carotid arteries.

23.4.1.11. Choose the CE-MRA volume (in the coronal plane) on the sagittal 2D PC localiser image, making sure to include the following area: from the aortic arch to above the C5 segment of the internal carotid arteries, in front of the C4 segments of the internal carotid artery and behind the vertebral arteries.

23.4.1.12. A 3D SPGR (CE-MRA) volume without contrast medium is acquired for use as a mask in subsequent subtraction studies. For the highest quality images, use the thinnest possible slice thickness.

23.4.1.13. The contrast medium bolus is injected by infusion pump (2 mL/s), followed by a saline flush (30 cc) to ensure that the contrast medium is pushed through the venous circulation. The standard gadolinium dose is 0.1 mmol/kg.

23.4.1.14. The authors of this chapter use the CARE (Combined Applications to Reduce Exposure) bolus technique to optimise arrival of the contrast medium bolus to the study area and synchronise it with the acquisition sequence. Another option is to run a test bolus in order in order to calculate the delay between the time of injection and the time of arrival. This is done by first injecting 1 mL of gadolinium, followed by a 20 mL saline flush, while simultaneously performing a rapid acquisition (one low-resolution image per second) up to the area of interest. The delay is calculated using the following formula:

$$D = CT - Kt + (AT/2)$$

Where D is the delay; CT is the circulation time; Kt is the time at the centre of K-space (specific value for each sequence); and AT is the acquisition time.

23.4.1.15. Acquisition of a 3D SPGR (CE-MRA) volume with contrast equal to the mask volume

23.4.1.16. Quality control of the acquired images: working directly from the acquisition console

Listed below are several basic parameters for different MRA sequences, the values of which must be adjusted according to the scanner used, as they depend on the magnetic field and gradient strength, the software used and the manufacturer (see Tables 23.4, 23.5 and 23.6).

23.4.2 Reconstruction and Manipulation of the Images

23.4.2.1. Once the images have been acquired, they must be processed.

23.4.2.2. The 3D volume must be manipulated at an independent workstation. This entails subtracting the masks from the images with contrast in order to obtain enhanced images. Subtractions are especially utile for studying highly peripheral vessels. Subtraction can also be performed directly at the acquisition console; the resulting images are then sent off for processing.

23.4.2.3. Reconstructions are performed using the technique of Maximum Intensity Projection (MIP), which provides images similar to those obtained with conventional angiography and enables visualisation of the vascular segments in different planes and orientations. If signal interference from vascular overlap complicates the arterial examination, then individual MIP reconstructions should be generated for each arterial segment.

23.4.2.4. In order to adequately assess the anatomical relationships among and between the different vascular structures, to characterise any plaque present or to identify a carotid aneurysm, an ulcerated plaque (see Fig. 23.6), intramural haematoma, intimal flap or mural thrombus, every partition of the acquired volume must be visualised. Plaque should not be confused with the typical artefacts caused by complex flow in the carotid sinus. These artefacts extend from the posterior wall of the artery to the central portion of the vessel lumina (Fig. 23.7).

23.4.2.5. Another post-processing option is to perform Multiplanar Volume Reconstruction (MPVR), which is especially useful for anatomical

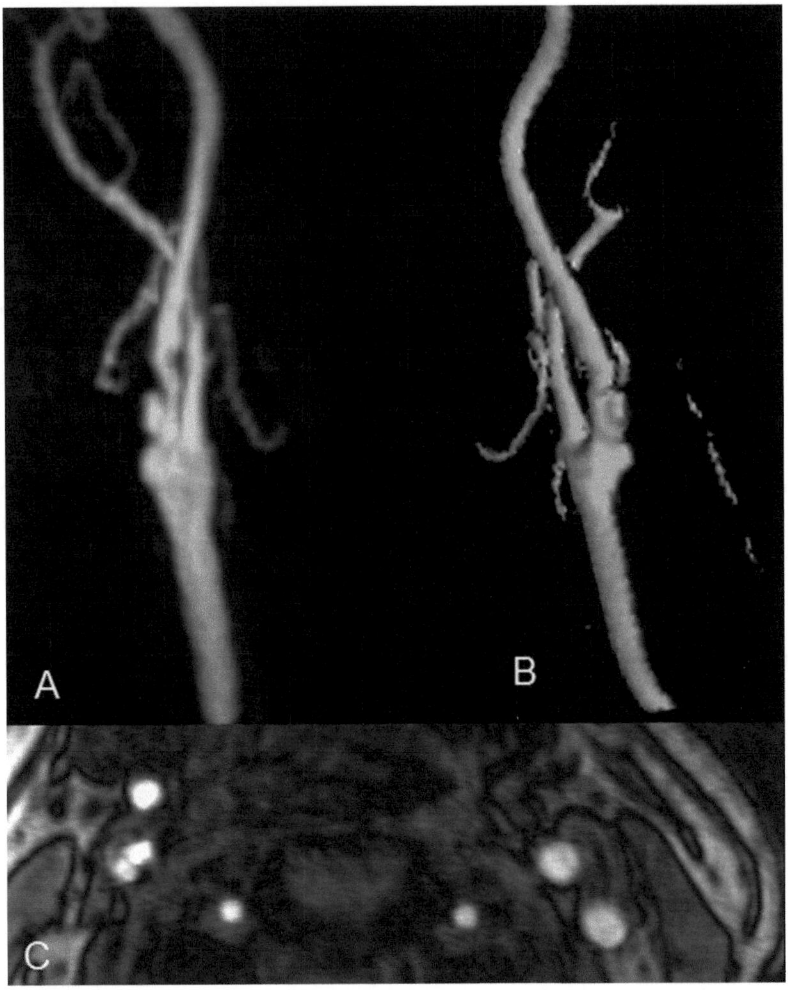

Fig. 23.6 (**a**) Oblique-coronal MIP reconstruction of CE-MRA, revealing an ulcerated plaque in the right internal carotid artery bulb. (**b**) Surface reconstruction, posterior view. This view provides better visualisation of the ulcer, thanks to the volume effect. (**c**) Partition of the MOTSA sequence, showing the double image produced by the arterial lumen and the ulcer crater in the posterior wall of the internal carotid artery

studies of complex vascular segments. This type of reconstruction also provides the measurements required for grading stenosis. To obtain the most accurate measurements possible with the software, the plane of study should be made perpendicular to the artery.

23.4.2.6. Surface reconstructions are another option for vascular imaging. Whilst they can exaggerate the grade of stenosis, they are invaluable for detecting ulcerated plaques or small aneurysms and for discovering concomitant occult stenosis in an arterial elongation.

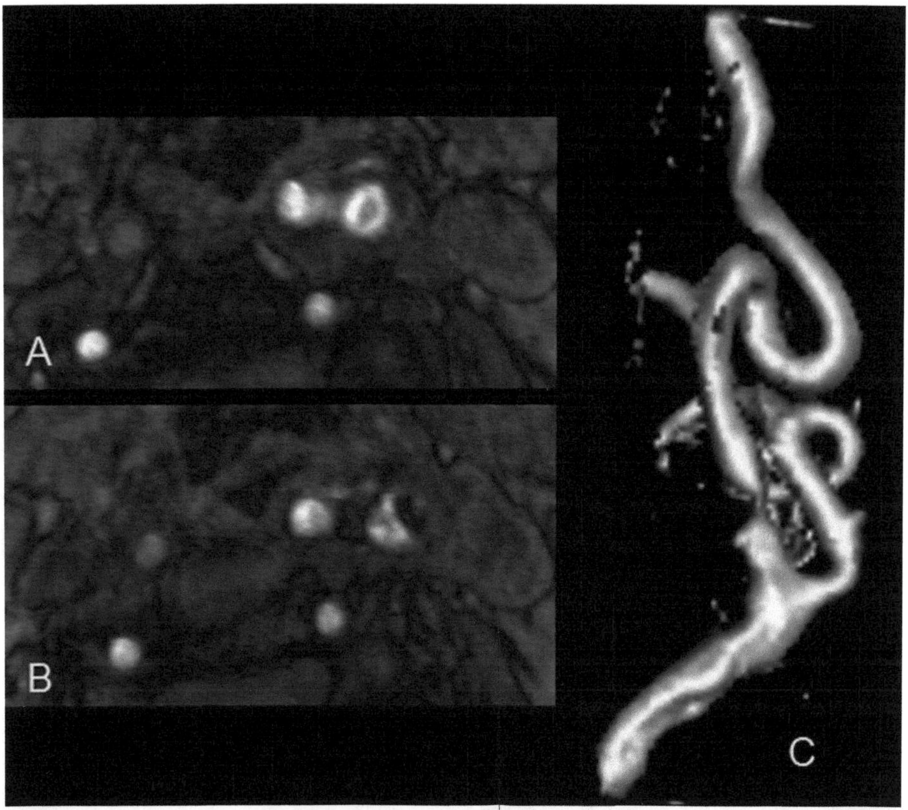

Fig. 23.7 (**a**) Partition of the MOTSA sequence, showing evidence of complex flow in the arterial lumen of the left carotid sinus. (**b**) Partition of the MOTSA sequence, showing a complex plaque formed by calcification (eccentric marginal hypointensity), a haemorrhagic plaque (central hyperintensity) and the residual lumen of the artery. (**c**) Surface reconstruction, lateral view: severe carotid artery stenosis

23.4.2.7. In order to avoid the errors inherent to collapse images, every partition must be closely examined. The partitions also provide information for characterising atherosclerotic plaques, as they can reveal signs of intraplaque haemorrhage, whether acute (hyperintensity) or chronic (hypointensity). One important factor to consider is that although MR is not sensitive to calcium (and therefore, cannot directly detect arterial calcifications), calcifications do appear hypointense in a characteristic signal on MR images (see Fig. 23.7).

23.4.2.8. The partitions also provide the arterial lumen diameter values for grading stenosis (based on percentage reduction in vessel diameter). Grading this way avoids the problem of overestimation of stenosis severity that occurs when grading based on MIP reconstructions (see Fig. 23.11). The most widely used grading method is the NASCET method, which is

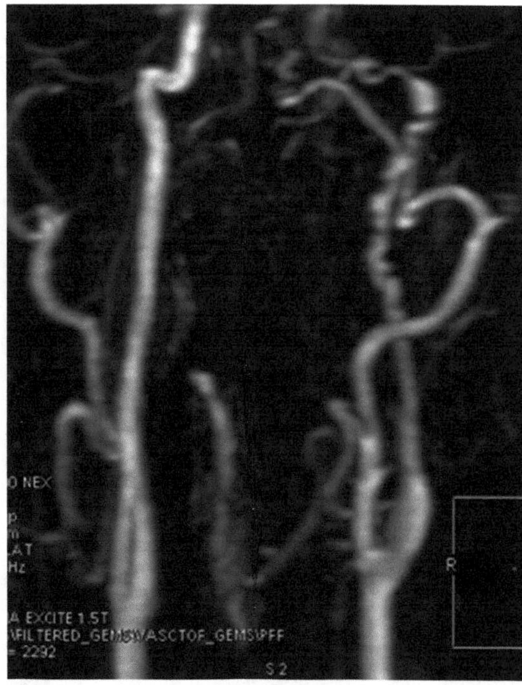

Fig. 23.10 Coronal MIP reconstruction of CE-MRA: the classic "string of beads" appearance indicates a fibromuscular dysplasia affecting the cervical portion of the left internal carotid artery

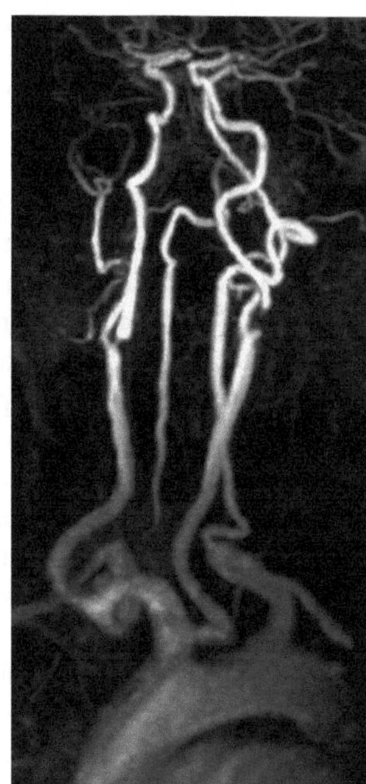

Fig. 23.11 Coronal MIP reconstruction of CE-MRA, showing bilateral stenosis of the internal carotid arteries. The grade of the left stenosis is overestimated (from severe stenosis to segmental obstruction) due to turbulent flow; however, this will not alter the course of treatment

23.5 Black-Blood Techniques

The vessel lumen appears hypointense, whereas the vessel wall appears hyperintense. These techniques are excellent for studying vessel walls. Used to study arterial dissections, to reveal intramural haematomas, ulcers or aortitis, and to analyse atheromatous plaques.

References

1. Tardáguila Montero F, Ferreiros Domínguez J (2003) Imagen cardiovascular avanzada: RM y TC. Monografía SERAM. Panamericana, Madrid
2. Osborn A (1999) Diagnostic cerebral angiography. Lippincott, Williams and Wilkins, Philadelphia
3. Prince MP, Grist TM, Debatin JF (2003) 3D contrast MR angiography. Springer, Berlin
4. Watanabe Y, Dohke M, Okumura A, Amoh Y, Ishimori T, Oda K, Hayashi T, Hiyama A, Dodo Y (2000) Dynamic subtraction contrast-enhanced MR angiography: technique, clinical applications, and pitfall. Radiographics 20:135–152
5. Randoux B, Marro B, Koskas F et al (2001) Carotid artery stenosis: prospective comparison of CT, three-dimensional gadolinium-enhanced MR, and conventional angiography. Radiology 220(1):179–185
6. Heiss SG, Shifrin RY, Sommer FG (2000) Contrast-enhanced three-dimensional fast spoiled gradient-echo renal MR imaging: evaluation of vascular and nonvascular disease. Radiographics 20:1341–1352
7. Osborn A (2013) Arterial anatomy and strokes. Osborn's: brain imaging, pathology, and anatomy, vol 8. Amirsys, Altona/Manitoba, pp 170–214
8. Remonda L, Senn P, Barth A (2002) Three-dimensional MR angiography of the carotid artery: comparison with conventional digital subtraction angiography. AJNR Am J Neuroradiol 218:138–143
9. Huston J III, Fain SB, Wald JT (2001) Carotid artery: elliptic centric contrast enhanced three-dimensional MR angiography compared with conventional angiography. Radiology 226:798–811
10. Saba L, Anzideib M, Sanfilippo R, Montisci R, Lucatelli P, Catalano C, Passariello R, Mallarini G (2012) Imaging of the carotid artery. Atherosclerosis 220(2):294–309
11. Qiao Y, Etesami M, Malhotra S, Astor BC, Virmani R, Kolodgie FD, Trout HH III, Wasserman BA (2011) Identification of intraplaque hemorrhage on MR angiography images: a comparison of contrast-enhanced mask and TOF techniques. AJNR Am J Neuroradiol 32(3):454–459
12. Rodallec MH, Marteau V, Gerber S, Desmottes L, Zins M (2008) Craniocervical arterial dissection: spectrum of imaging findings, and differential diagnosis. Radiographics 28(6):1711–1728
13. Byrenes KR, Ross CB (2012) The current role of carotid duplex ultrasonography in the management of carotis atherosclerosis: foundations and advances. Int J Vasc Med 2012:187872
14. Fortuño JR, Perendreu J, Branera J (2008) Diagnóstico no invasivo de la patología vascular cervical. En Radiología vascular no invasiva y Radiología del cuerpo entero. Editorial Medica Panamericana 5:37–44
15. Oleaga L (2008) Estudio vascular con resonancia magnética. En Radiología vascular no invasiva y Radiología del cuerpo entero. Editorial Medica Panamericana 2:11–19

oxime (HMPAO), also known as *exametazime*, (Ceretec®; General Electric Healthcare) and *^{99m}Tc-ethyl cysteinate dimer* ([ECD]; Neurolite®). These compounds can be delivered proportionally to regional CBF. Their lipophilic character, in addition to enabling them to cross the BBB, enables them to penetrate into the neuronal cytoplasm, where they are converted into hydrophilic species that cannot be transported out to the extracellular space. Owing to these pharmacokinetic characteristics, ^{99m}Tc-HMPAO and ^{99m}Tc-ECD reflect blood flow and cellular viability in the brain.

Single-Photon Emission Tomography data are collected via acquisition of multiple 2-dimensional (2D) images of the organ being studied at various angles, as the gamma camera rotates about it. These images (also known as *projections*) are acquired and stored in digital format, and then processed together to yield 2D tomographic slices along the three anatomic planes (axial, sagittal, and coronal).

24.2 Radiopharmaceuticals in Brain SPECT

24.2.1 ^{99m}Tc-HMPAO

Retention of this drug in the body is poorly understood, although it is thought to involve conversion of the drug into a hydrophilic analogue via interaction with intracellular glutathione. Directly following injection, approximately 70–80 % of the drug localises to the brain (approximately 6 % of the total dose within 2 min post-injection). The physiologic distribution of ^{99m}Tc-HMPAO is proportional to the CBF value estimated using other techniques. However, the distribution tends to be underestimated in cases of high CBF, due to incomplete uptake by the brain and to reverse diffusion in the vascular compartment that occur during the first few minutes post-administration.

^{99m}Tc-hexamethylpropyleneamine oxime does not undergo any significant level of redistribution, and its distribution remains proportional to the CBF during for 4 h post-injection. This enables the study of brain perfusion through drug treatment or other interventions: the radioactive tracer is injected at the point of maximum intervention and, once the patient has stabilised, the SPECT is then collected.

The principal drawback of ^{99m}Tc-HMPAO is its radiochemical instability: it must be administered within 30 min of being prepared.

24.2.2 ^{99m}Tc-ECD

This compound has a similar pharmacokinetic profile and distribution to that of ^{99m}Tc-HMPAO; however, once prepared, it is far more stable in vitro, making it much easier to handle.

It rapidly crosses the BBB: 7 % of the total dose localises to the brain within 2 min post-injection. It is distributed in proportion to the CBF, except for the most perfused regions, where the distribution follows a non-linear proportion. As such, the distribution values for hypoperfused regions tend to be overestimated (this is also true for ^{99m}Tc-HMPAO).

Retention of ^{99m}Tc-ECD in the brain depends chiefly on tissue-specific metabolism, which apparently occurs through its esterification. Its retained form is less stable than that of ^{99m}Tc-HMPAO; in fact, it undergoes a minor level of redistribution. Therefore, the recommended time limit for a SPECT study is 2 h. It is rapidly cleared by the renal pathway, which implies less irradiation for the patient. Furthermore, it has a higher grey-matter-to-white-matter uptake ratio, which translates to better images.

There is some controversy over the choice of which radioactive tracer to use for the study of cerebrovascular pathology. ^{99m}Tc-HMPAO can exhibit luxury perfusion in subacute stroke, whereas ECD leads to hypoperfused areas in ischaemic lesions. Luxury perfusion is problematic because in acute stroke, it can lead to overly positive prognoses, whereas in the subacute phase, it can mask the true size of regions damaged by ischaemic lesions.

24.3 Brain SPECT with a Stimulation Test

This method comprises the study of changes in the CBF that are induced by specific stimuli, whether sensory, motor, cognitive (neuroactivation), drug-induced, or generated through invasive procedures (occlusion of the carotid artery).

The most commonly used type of stimulus for studying cerebrovascular pathology is drug-induced stimuli. For instance, acetazolamide is used to study the brain's vascular reserves. In this procedure, two SPECT studies are performed: the first, under basal conditions; and the second, after the drug has been administered but 20 min before the radioactive tracer has been injected (dose: 15 mg /Kg) Acetazolamide causes healthy brain arteries to dilate intensely and causes the CBF to increase (relative to that under basal conditions), which is a direct sign of cerebrovascular reserves. The patient's blood pressure should be monitored.

24.4 Procedure for Brain-Perfusion SPECT

24.4.1 Preparing the Patient

The following precautions should be taken:

- Avoid the use of CNS-active drugs.
- The radiopharmaceutical should be injected while the patient is at rest (free of any minimal cognitive or sensory tasks).
- The radiopharmaceutical should be injected through a previously established IV catheter, so that the patient does not feel any pain at the moment that it is administered.

24.4.2 Radiopharmaceuticals, Dosing and Administration

- Radiopharmaceuticals: ^{99m}Tc-HMPAO or ^{99m}Tc-ECD
- Dosing: 20–25 mCi (740–925 MBq)
- Administration: IV.

24.4.3 Instrumentation

- Collimator: fan beam or other high-resolution type.
- Energy window (width/photopeak): 10 %/140 KeV

24.4.4 Acquisition Protocol

- Patient position: supine. The patient's head rests on a special support and is immobilised. To ensure that the cerebellum is included in the gamma camera's field of vision, the patient's head should be tilted forward slightly. The detector and the gamma camera should be as close as possible.
- Acquisition mode: SPECT
- Acquisition time: 5–10 min post-injection. The images can be acquired up to 2 h (when using ECD) or 4 h (when using HMPAO) post-injection.
- Acquisition parameters:
  - Matrix: 128 × 128 (pixel size: 3 mm)
  - Orbit: circular (360°)
  - Number and duration of images: 60 images (20 s each)
  - Scan rate: 1 image per 3° (120 images)

24.4.5 Data Processing

- Reconstruction: the images are reconstructed into cortical, sagittal and axial slices.
- Reconstruction method: iterative or filtered (choice of filter).

24.4.6 Attenuation Correction Algorithms

These algorithms correct for the loss of photons that results from attenuation due to the distance between the detector and the camera, and to the density of the tissue crossed. They include Sorenson's method and Chang's method.

24.4.7 SPECT/CT

These hybrid systems combine a gamma camera with Computed Tomography. They use the transmitted image to correct the SPECT attenuation.

24.4.8 Image Presentation

The images are represented in a false colour-scale, whereby each colour corresponds to a value for perfusion intensity. The reason for using colour is that grey scales tend to lead to underestimation of alterations in distribution. The use of continuous colour gradients is recommended, since harsh colour changes can lead to overestimation of physiologic variations in perfusion.

24.4.9 ^{99m}Tc: Emission Parameters

- Half-life: 6.01 h
- Radiation type: gamma
- Average decay: 89.07 %
- Average energy: 140 Kev

Dosimetry for HMPAO	
Organ	Concentration (rad/mCi)
Lachrymal glands	0.26
Gallbladder	0.19
Kidneys	0.13
Thyroid	0.10
Large intestine	0.08
Liver	0.05
Small intestine	0.04
Bladder	0.04
Brain	0.5
Ovaries	0.5
Total body	0.3
Testicles	0.1

Dosimetry for ECD	
Organ	Concentration (rad/mCi)
Brain	0.02
Gallbladder	0.09
Lower large intestine wall	0.05
Small intestine	0.04
Upper large intestine wall	0.06
Kidneys	0.03
Liver	0.02
Lungs	0.01
Ovaries	0.02
Bone marrow	0.01
Testicles	0.01
Bladder	0.11
Total body	0.01

24.5 Data Interpretation

Among the most important aspects of data interpretation in SPECT is the ability to identify normal variations in perfusion. Under normal conditions the grey matter, cortex and subcortex are the most perfused regions, the subcortical white matter is less perfused, and the anatomic regions that contain cerebrospinal fluid (i.e. the ventricles, fissures and sulci) do not show any uptake.

Perfusion patterns can vary by patient. Therefore, the healthcare professional must take into consideration the various factors that can alter perfusion in the brain, including.

24.5.1 Anatomic Variations

The size, shape and symmetry of the brain vary among patients. Thus, when interpreting SPECT data, the healthcare professional should have the appropriate structural diagnostic studies on hand.

24.5.2 Age

This is an important factor, as the infant brain can differ dramatically from the adult brain. Thus, at birth brain perfusion is high in the cerebellum, the basal ganglia, the sensorimotory cortex and the calcarine fissure. It gradually increases in other cortices with age, and by 1–2 years of age, the perfusion pattern overlaps with that of an adult brain. Around the age of 36, brain perfusion begins to slowly diminish. The greatest changes occur in the frontal lobe and the left hemisphere.

24.5.3 Sex

There are only minimal differences in brain perfusion between the sexes, as women show greater neuronal activity in the right parietal cortex, whereas men show greater activity in the cerebellum, anterior temporal cortex and left orbitofrontal cortex. Nevertheless, observation of these differences should not pose any major problems for the experienced specialist.

A systematic approach to interpretation of results is recommended.

1. Choose slice type adequate for identification of perfusion patterns: for example, transaxial slices or oblique fronto-occipital slices (crossing the lower limit of both lobes). Slices from the other planes (coronal and sagittal) can then be used to confirm, locate and assess the findings. To evaluate the temporal lobe, the planes parallel to its longest axis must be used.
2. Identify all possible anatomic structures, using findings from structural studies or, if these are unavailable, an atlas.
3. Confirm that the perfusion patterns of all cortical structures adhere to the following order (from greater to lesser):
   - For HMPAO: calcarine fissure → cerebellum → frontal cortex → temporal cortex → parietal cortex
   - For ECD: cerebellum → occipital cortex → frontal cortex → temporal cortex → parietal cortex
4. Analyse all subcortical structures in order to rule out any asymmetries or alterations in perfusion.
5. Compare the SPECT images with Magnetic Resonance (MRI) and/or CT images in order to rule out hydrocephalus or other lesions that would explain certain alterations in perfusion (Figs. 24.1 and 24.2).

24.6 Brain Perfusion SPECT in Acute Stroke (Within 3 Hours of Onset)

Within the first 3 h of symptoms onset, SPECT is clearly superior to CT without contrast medium for the detection of ischaemic stroke. Moreover, SPECT enables one to delineate the vascular territory involved in the stroke and to corroborate the CT findings, which is especially important for subtle or unclear alterations. It also provides information on the brain parenchyma at risk, which comprises the necrotic

Recombinant Tissue Plasminogen Activator (rt-PA) Study. During the first few hours following a stroke, various medical imaging techniques are used to rule out intracranial haemorrhage or other non-ischaemic phenomena, identify any occluded blood vessels, and help the healthcare professional to determine which patients are the best candidates for thrombolytic therapy.

In this context, functional neuroimaging is especially useful, as it can identify the ischaemic penumbra as a therapeutic target and differentiate this area from necrotic tissue. Imaging techniques should help in ensuring the safety and efficacy of thrombolytic therapy, reducing any haemorrhagic complications of the therapy, and guaranteeing that all eligible patients receive treatment. Knowledge on the functional state of the brain parenchyma is essential. However, the therapeutic window limits the number of studies that can be performed.

Computed Tomography without contrast medium is the most widely used and accessible diagnostic technique. However, whilst it is highly sensitive for detection of haemorrhagic processes, its sensitivity in acute ischaemic stroke is low. Characteristic early signs of ischaemic stroke on CT have been described in the literature, including: loss of grey/white matter differentiation; effacement of the internal capsule and the cortical sulci; hypodensity of the insula; focal hypoattenuation; mass effect; and hyperdensity of the middle cerebral artery or the internal carotid artery. These alterations are often very subtle and therefore, difficult to identify. Consequently, up to 30 % of CT scans in the acute phase of stroke are misinterpreted. Characteristic early signs of ischaemic stroke have been described in 31–53 % of patients within the first 3 h of stroke. The clinical trials ECASS I and ECASS II were pioneers in recognising the value of early signs of ischaemic stroke for predicting the benefits of thrombolytic therapy. In both trials, thrombolytic therapy was only administered to patients whose stroke involved ***more than one-third*** of the middle cerebral artery territory. However, even experience radiologists and neurologists can have difficulties identifying these changes. The Alberta Stroke Programme Early CT Score (ASPECTS) facilitates the evaluation of early signs of ischaemic stroke in the middle cerebral artery territory and therefore, can assist in identifying those patients that are eligible for thrombolytic therapy.

24.8 Use of SPECT in Selecting Patients for Thrombolytic Therapy

Thrombolytic therapy is used to recanalise the occluded artery to obtain nutritional reperfusion, such that metabolism can be re-established in the ischaemic tissue before the brain lesion becomes irreversible. The aim is to improve the patient's long-term clinical progress, prevent infarction of the affected brain tissue and reduce the area of this tissue. In 2002 the EMEA conditionally approved the use of rt-PA for acute ischaemic stroke, but only when used within 3 h of symptoms onset and in dedicated stroke units or in centres that are highly qualified for stroke treatment. The two principal conditions for approving rt-PA treatment were that all treated patients would have to be monitored (the data were collected in the Safe Implementation of Thrombolysis in Stroke Monitoring Study [SITS-MOST]) and that the participating professionals would have to contribute to a clinical trial (ECASS III) on the efficacy of rt-PA for patients that exhibit symptoms up to 4.5 h after stroke onset. The SITS-MOST study confirmed that rt-PA treatment was safe and effective when used within 3 h of stroke onset, even in centres that lacked previous experience in thrombolytic therapy. *In ECASS III, thrombolytic treatment with alteplase 3–4.5 h after acute ischemic stroke was associated with modest but significant improvement in clinical outcome.*

Use of SPECT/CT can facilitate the identification of those patients that are eligible for thrombolytic therapy. During ECASS I and II, SPECT was used together with clinical data and CT without contrast medium. Clinical responses to thrombolytic treatment enabled establishment of perfusion patterns. Based on these patterns, one can establish whether thrombolytic therapy is indicated, predict whether the treatment will be effective, and determine the risk of haemorrhage, as follows:

1. If the hypoperfused areas on SPECT are of moderate intensity, and CT does not show any early signs of ischaemic stroke, then thrombolytic therapy is indicated.
2. If the area of the early ischaemic oedema observed on CT correlates exactly with the perfusion defect on SPECT, then thrombolytic therapy will not be effective.
3. Lastly, if the alterations in perfusion are severe and involve nearly all of the middle cerebral artery territory, then there will be a risk of haemorrhagic transformation, even if no early signs of stroke are observed on CT.

In the ECASS and NINDS studies, the clinical and CT data on patients that experienced a haemorrhagic transformation after having received thrombolytic therapy suggest that CT should be complemented with a perfusion study (e.g. by SPECT). One option for such an approach is to administer ^{99m}Tc-HMPAO immediately before the thrombolytic agent. Since radiopharmaceuticals provide a snapshot of CBF at the moment they are administered, SPECT can be performed once the thrombolytic treatment has been completed and the patient is clinically ready. This approach enables the healthcare professional to know the pre-treatment status of the CBF. A second test with ^{99m}Tc-HMPAO, performed 24 h later, will report on the status of brain perfusion after the thrombolytic treatment, enabling the healthcare professional to determine whether or not reperfusion has occurred.

During the subacute phase of stroke, in patients that have not received thrombolytic therapy, SPECT will sometimes reveal an increase in perfusion in the region involved in the stroke. This increase in regional flow represents a state of non-nutritional reperfusion, which translates into a discrepancy between the CBF and the cerebral metabolism. However, this alteration disappears after several days. The result is an area of compromised perfusion that correlates to the infarcted area that is observed on CT during the chronic phase. This reperfusion does not improve the patient's clinical progress; in fact, on some occasions, it is associated with a haemorrhagic transformation in the infarcted area. Increased flow can also be observed in acute stroke, as a consequence of spontaneous or thrombolysis-induced reperfusion. However, the significance of this finding remains controversial. Whilst some researchers have reported that these patients do not show any signs of major metabolic alterations according to PET (which gave a favourable prognosis), others have affirmed that an increase in CBF revealed by SPECT with ^{99m}Tc-HMPAO within the first 6 h of stroke onset ultimately evolves into an infarct. The increase in perfusion that results from early reperfusion can be associated with reversible or irreversible damage to brain tissue. Estimating the viability of tissue based on only one exploration is difficult; thus, a second study should be performed as control several days later.

24.9 Brain SPECT in Assessment of Diaschisis

In 1914 von Monakow defined the concept of diaschisis: a loss in cerebral blood flow and metabolism resulting from deafferentation and transneuronal metabolic reduction in a region that is not directly involved in an infarct. The region of diaschisis can be proximal to the infarct (*cortico-cortical diaschisis*) or distal to it (*crossed cerebellar diaschisis*; *crossed supratentorial diaschisis*; or *transcallosal/transhemispheric diaschisis*).

Infarcts typically appear larger when measured by brain perfusion SPECT than when measured by CT or MRI. This difference is down to the pronounced reduction in CBF caused by the combination of cortico-cortical diaschisis, the penumbra and the ischaemic core. The boundary between the necrotic core and the penumbra can be delineated by stimulation studies using the potent vasodilator acetazolamide. In this approach, an increase in CBF is observed in the area surrounding the lesion (vascular reserves), indicating favourable clinical evolution of the ischaemic stroke.

Crossed cerebellar diaschisis refers to a loss in blood flow and in metabolism in the cerebellar hemisphere contralateral to a focal supratentorial lesion. It occurs when a cortical infarct causes an interruption in the cortico-ponto-cerebellar pathway, which provokes deafferentation and transneural metabolic reduction of the contralateral cerebellar hemisphere. These functional consequences do not exhibit any characteristic morphological signs on conventional imaging (CT or MRI).

Crossed cerebellar diaschisis is closely associated to infarcts that affect the frontoparietal region (especially the precentral, postcentral and supramarginal gyri), and is infrequently associated with infarcts in the occipital lobe. However, there is no correlation between this condition and the size of the infarct. Also, it can be detected early on during the acute phase of stroke.

Patients with crossed cerebellar diaschisis do not exhibit any clinical cerebellar symptoms. They often present with hemiparesis; however, this type of diaschisis is not always detectable in these patients and is often observed in patients without hemiparesis. The absence of crossed cerebellar diaschisis correlates with better clinical progress.

Crossed cerebellar diaschisis is a metabolic phenomenon that tends to persist, even in patients that have recuperated clinically. Diffusion-Weighted Imaging (DTI) studies on patients with chronic infarcts and CCD have shown that these patients undergo a reduction in fractional anisotropy, which translates into microstructural morphologic changes in neuronal pathways in the middle cerebellar peduncle.

A less frequent condition is *crossed supratentorial diaschisis* secondary to an infratentorial infarct, which implies contralateral fronto-parietal hypoperfusion and which clinically manifests as a syndrome of compromised cognitive ability (Figs. 24.3 and 24.4).

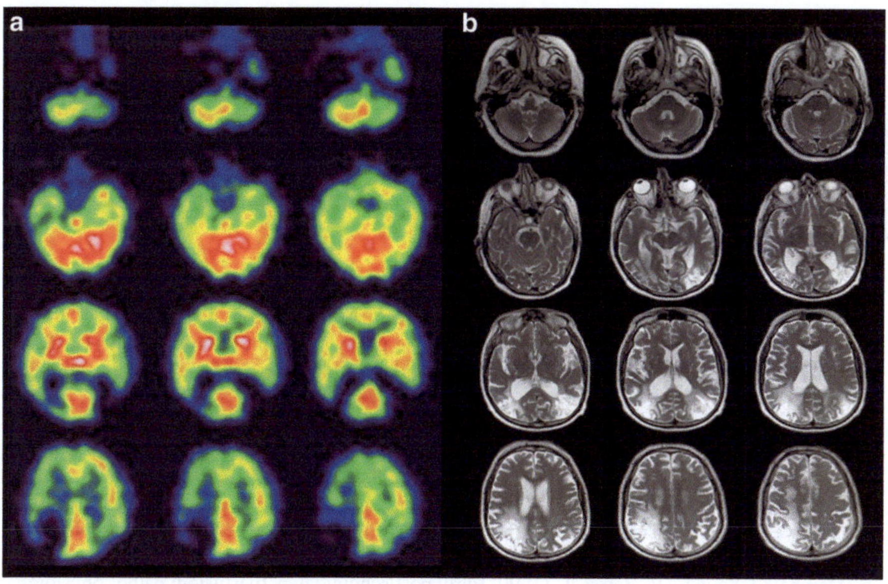

Fig. 24.3 (**a**) Brain perfusion SPECT (^{99m}Tc-ECD), axial slices. Intense bilateral temporo-parietal hypoperfusion, predominant on the right side. The left cerebellar hemisphere is hypoperfused (crossed cerebellar diaschisis). (**b**) T2-weighted MRI, cortical slices. Two chronic infarcts that involve part of the surface of each middle cerebral artery. The cerebellar hemispheres do not show any major alterations

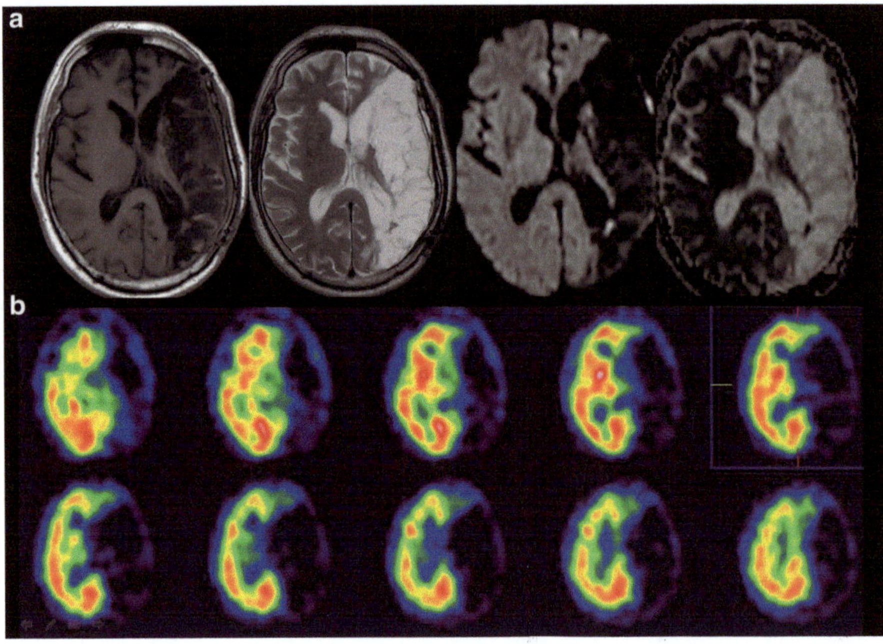

Fig. 24.4 Malignant left middle cerebral artery infarction. (**a**) MRI. Extensive area of encephalomalacia relative to a chronic stroke in the left hemisphere that is compromising the surface and internal territory irrigated by the left middle cerebral artery, associated to Wallerian degeneration of the cortico-spinal tract. Left fronto-parietal craniotomy with subdural accumulation of and hemosiderin deposits in subarachnoid spaces. (**b**) Brain perfusion SPECT: absence of extensive cortico-subcortical perfusion in the left hemisphere

References

1. Catafau AM (2001) Brain SPECT in clinical practice. Part I: perfusion. J Nucl Med 42:259–271
2. Andersen AR (1989) Tc99m- D, L-hexamethylene-propyleneamine oxime (Tc99m- HMPAO): basic kinetic studies of a tracer of cerebral blood flow. Cerebrovasc Brain Metab Rev 1:288–318
3. Holman BL, Hellman RS, Goldsmith SJ, Mena IG, Leveille J, Gherardi PG et al (1989) Biodistribution, dosimetry and clinical evaluation of technetium-^{99m} ethyl cysteinae dimer in normal patients and in patients with chronic cerebral infarction. J Nucl Med 30:1018–1024
4. Mountz JM, Liu HG, Deutsch G (2003) Neuroimaging in cerebrovascular disorders: measurement of cerebral physiology after stroke and assessment of stroke recovery. Semin Nucl Med 23(1):56–76
5. Marchal G, Young AR, Baron JC (1999) Early postischemic hyperperfusion: patho- physiologic insights from positron emission tomography. J Cereb Blood Flow Metab 19:467–482
6. Latchaw RE (2004) Cerebral perfusion imaging in acute stroke. J Vasc Interv Radiol 15:S29–S46
7. Mountz JM, Deutsch G, Khan SH (1993) Regional cerebral blood flow changes in stroke imaged by Tc-^{99m} HMPAO SPECT with corresponding anatomic image comparison. Clin Nucl Med 18:1067–1082

8. Alexandrov AV, Black SE, Ehrlich LE, Bladin CF, Smurawska LT, Pirisi A et al (1996) Simple visual analysis of brain perfusion on HMPAO SPECT predicts early outcome in acute stroke. Stroke 27:1537–1542
9. Sá de Camargo EC, Koroshetz WJ (2005) Neuroimaging of ischemia and infarction. NeuroRx 2:265–276
10. Sarharati D, Reisdorff EJ (2002) Emergent CT evaluation of stroke. Emerg Med Clin N Am 20:553–581
11. Ozdemir O, Leung A, Bussiére M, Hachinski V, Pelz D (2008) Hyperdense internal carotid artery sign: a CT sign of acute ischemia. Stroke 39:2011–2016
12. Pexman JHW, Barber Ph A, Hill MD, Sevick RJ, Demchuk AM, Hudon ME et al (2001) Use of the Alberta stroke program early CT score (ASPECTS) for assessing CT scans in patients with acute stroke. AJNR Am J Neuroradiol 22:1534–1542
13. Barthel H, Hesse S, Danenberg C, Rössler A, Schneider D, Knapp WH et al (2001) Prospective value of perfusion and X-ray attenuation imaging with single-photon emission and transmission computed tomography in acute cerebral ischemia. Stroke 32:1588–1597
14. Wahlgren N, Ahmed N, Dávalos A, Ford GA, Grond M, Hacke W et al (2007) Thrombolysis with alteplase for acute ischaemic stroke in the safe implementation of thrombolysis in stroke-monitoring study (SITS-MOST): an observational study. Lancet 369:275–282
15. Alexandrov AV, Black SE, Bladin CF, Smurawska LT, Caldwell CB (1996) Simple visual analysis of brain perfusion on HMPAO SPECT predicts early outcome in acute stroke. Stroke 27:1537–1542
16. Amemura A, Suzuka T, Yamaka K (2000) Quantitative measurement of cerebral blood flow by ^{99m}Tc-HMPAO SPECT in acute ischaemic stroke: usefulness in determining therapeutic options. J Neurol Neurosurg Psychiatry 69:472–478
17. Alexandrov AV, Masdeu JC, Devous MC Sr, Black SE, Grotta JC (1997) Brain single-photon emission CT with HMPAO and safety of thrombolytic therapy in acute ischemic stroke. Stroke 28:1830–1834
18. Bonaffini N, Altieri M, Rocco A, Di Piero V (2002) Functional neuroimaging in acute stroke. Clin Exp Hypertens 24:647–657
19. Wintermark M, Bogousslavsky J (2003) Imaging of acute ischemic brain injury: the return of computed tomography. Curr Opin Neurol 16:59–63
20. Scharamm P, Schellinger PD, Klotz E, Kallenberg K, Fiebach JB, Külkens S et al (2004) Comparison of perfusion computed tomography and computed tomography angiography source images with perfusion-weighted imaging and diffusion-weighted imaging in patients with acute stroke of less than 6 hours duration. Stroke 35:1652–1658
21. Bisdas S, Donnerstag F, Ahl B, Bohrer I, Weissenborn K, Becker H (2004) Comparison of perfusion computed tomography with diffusion-weighted magnetic resonance imaging in hyperacute ischemic stroke. J Comput Assist Tomogr 28:747–755
22. Januel AC, Tailleur T, Loubes-Lacroix F, Catalaa I, Irsutti-Fjortoft M, Molinier S et al (2005) Imaging of cerebral ischemia within first hours: computed tomography (CT). J Radiol 86:1091–1101

Chapter 25
Enhancing Endothelialisation of Artificial/Engineered Blood Vessels Using Structural Cues

Kirstie Andrews and Amir Keshmiri

25.1 Introduction: The Need for Artificial Blood Vessel Analogues

We will describe the importance of endothelialisation of artificial blood vessels, discuss the techniques that can be utilised to enhance this cell layer, with a focus on using the underlying structural cues from the engineered materials, and detail methodology used to examine the effects of these properties upon the endothelial cell behaviour.

The processes of angiogenesis and vasculogenesis are important for the formation and regeneration of tissue and organ structures. However, they are not always naturally possible due to these functions being impaired by damage or disease. The fabrication of engineered blood vessels to artificially create angio- or vasculogenesis can therefore be employed [1, 2].

Problems requiring the use of an engineered/artificial blood vessel (graft) include occluded vessels due to stenosis, damaged vessels resulting from trauma or aneurysm, and the formation of a new tissue structure through regenerative therapies. Both the replacement and the repair of blood vessels are viable treatment options.

Currently, the "gold standard" option is to use naturally occurring vessels such as the saphenous vein; however, this brings inherent problems including additional

K. Andrews (✉)
Division of Mechanical Engineering, School of Engineering, Manchester Metropolitan University, Manchester M1 5GD, UK
e-mail: k.andrews@mmu.ac.uk

A. Keshmiri
School of Engineering, Manchester Metropolitan University, Manchester M1 5GD, UK

School of Mechanical, Aerospace and Civil Engineering, The University of Manchester, Manchester M13 9PL, UK
e-mail: a.keshmiri@mmu.ac.uk

M. Slevin, G. McDowell (eds.), *Handbook of Vascular Biology Techniques*,
DOI 10.1007/978-94-017-9716-0_25

surgery for the patient, and the frequent unsuitability or limited availability of their veins due to systemic disease. There is also a lack of viable treatment options when the blood vessel is less than 6 mm in diameter [2–4, 6]. Hence, artificial blood vessels, following either biomaterial or tissue engineered approaches, are utilised.

The artificial vessel, regardless of its construction method/material, should closely match the structure and properties of the natural tissue it is replacing. This is to maximise the functionality and viability of the substitute tissue and the contained blood flow. A significant contributing factor towards achieving this is the vascular biology technique of endothelialisation.

25.2 Endothelialisation of (Artificial) Blood Vessels

Endothelialisation is the formation of a layer of endothelial cells on the luminal surface of natural blood vessels; with regard to engineered grafts, these cells form a confluent barrier between the artificial material and the circulating blood to prevent the initiation of the clotting cascade, immune response, thrombus/plaque formation and stenosis. The ultimate aim is for cells to align preferentially in the direction of blood flow, with non-activated, non-inflamed, stable phenotype, thus allowing the blood vessels to function naturally demonstrating thromboresistivity and high patency [2–4, 7, 8].

25.3 Current Methodologies in Optimising Artificial Blood Vessels: Use of Endothelialisation

In order to achieve this, endothelialisation has to be addressed as a vascular biology strategy- although promising animal trial data has been produced, there is a lack of inherent mechanism in humans for re-endothelialisation to occur over sustained surface areas. There is also evidence that human endothelial cells show low adhesion, are slow to populate the disturbed vessel surfaces, and are easily removed and damaged by blood shear stresses [2, 4, 7, 9].

This endothelialisation of the artificial vessels can be approached using three potential strategies (or a combination of): preseeding; precoating; in situ endothelialisation. Each has advantages and disadvantages to their use:

Preseeding is an in vitro approach, which uses the structure and/or chemistry of the graft material to produce a layer of adhered, oriented and spread endothelial cells on the luminal surface. Significant bodies of research have demonstrated a direct link between the underlying material and the induced resulting cellular behaviour [1, 2, 4, 10]. The cells are seeded onto the materials, in optimised densities, in sterile laboratory conditions and cultured. This seeded material is then implanted as a whole construct. Some researchers are trying to move away from/modify this approach due to its associated cost, time and cell sourcing issues [6, 11].

Precoating is the use of a covering layer upon the luminal surface of the material comprising the blood vessel wall, usually a biological material such as collagen.

This enhances the cellular adhesion, growth and proliferation upon the artificial structure; however, the underlying structure is still ideally detectable through this layer to aid in directing the cell growth. This coating itself introduces material effects, such as hydrolysis, and is potentially unstable in its attachment to the material; it is also difficult to achieve with small diameter grafts. This strategy can be used in combination with preseeding or implanted directly [2, 4, 12, 13].

The in situ method aims to coat the artificial materials with a natural endothelial layer after implantation; either using uncoated material or in combination with the precoating technique. However, this approach is subject to the previously mentioned issues for of slow cell adhesion, coverage of only small distances and shear stress effects from the flowing blood. Hence, enhancement of the inherent regrowth mechanisms of the endothelial cells is required. There is also a risk of, in the initial implantation stages, that the uncovered material will induce clotting and/or immune responses [2, 4, 14, 15].

It can be seen that all of these methods, from some aspect, require the contribution of the underlying material structure and chemistry, in order to produce the required cellular mechanisms and growth/adhesion. There is a large bulk of research looking at the effect of the material chemistry, and which, in combination with surface coatings, has shown the benefit of carefully selecting the material (and its associated surface charges and hydrophobicity) to produce the optimum cell adhesion and growth patterns [6, 10, 16–19]. However, there are also studies that indicate the critical effect of the structural cues upon the same cell behaviour [8, 20–27]. Structural cues can be used regardless of material choice, and are a significant research area within cell biology; the effect of a cell's environment upon its behaviour has been widely reported, with the nano- and micro-environments of the natural extracellular matrices and basement membranes providing contact guidance for the cell responses [28, 29]. These properties are also an important factor if considering the performance of currently used clinical vascular grafts, with protein coatings such as collagen/albumin, used to reduce blood loss through the pores; although the structures are still detected and are an influencing factor upon the contacting endothelial cells [5–8, 12, 19, 23, 30]. The ultimate aim is to replicate the signalling and guidance provided in the natural environment through the manipulation of the underlying artificial material [9, 31–34].

Hence, here we describe basic protocols for developing improved endothelialisation of artificial/engineered blood vessels through the use of structural cues, and how to analyse the effects of these modifications. The chapter will end with a focussed example concerning the topical, and increasingly used, electrospinning technique [8, 20, 35–41].

25.4 Methodologies Using Structural Cues

The surface properties most identified in the literature as having significant effects upon the contacting cellular behaviour are: pore size, "fibre" diameter, surface roughness, porosity (including interconnectivity), fibre orientation and material stiffness (Young's Modulus). The significance of the property varies with the fabrication method used. In general terms, studies have shown that endothelial cell proliferation is affected by the amount of material (including void space and fibre diameter; increased material raises the proliferation); material fibre orientation affects cellular alignment (increased orientation increases alignment) but not proliferation; cell retention under shear stress (blood flow) is also influenced by fibre orientation although there is an additional link to the amount of material present, with more complicated dual-effects in place (increased fibre orientation increases cell retention although this varies with combined increased surface area of material); lower surface roughness increased endothelial cell adhesion; increased topography increases the cytoskeletal involvement of the cells and so the ability to adapt to shear stress effects (again there are complex inter-linked effects from the different properties- the aim is to produce high amounts of both focal contacts and F-actin to enable adhesion but also movement in combination with low height profiles) [3, 7, 8, 20]. These overall trends are affected by the class of structure produced by the selected fabrication method, e.g. fibrous electrospun scaffolds, or porous extruded foams [19, 21, 23, 24, 35, 39].

In order to determine the specific material effects upon induced/controlled cell responses, a wide range of cell behaviours can be investigated: changes in cell morphology, adhesion, proliferation, viability, cell area and expressed signalling/inflammatory markers. These can be analysed within a simple panel of laboratory tests performed upon the cell-seeded material samples, as outlined below. The experiments performed are applicable for static and dynamic (peri- and/or post-culture) cell culture samples, as an initial stage of testing before animal models and tissue samples are used, and can be used as whole panel or as individual tests (depending on the focus of the experiments/research).

25.5 Basic Protocol

All materials used and their source companies are listed throughout the detailed protocol sections. For all sections, a minimum of four repeats is recommended to increase statistical accuracy and the assessment of relationships between the altered structural parameters and the induced cellular behaviour.

Step 1: Fabricate a range of material samples with systematically altered and controlled structures

Fabricate the selected polymer into an artificial blood vessel construct using the production method of choice (e.g. extrusion, electrospinning); systematically alter the fabrication parameters to produce a range of structures with variation across the property under focus (e.g. pore size) (for guidance regarding potentially significant structural properties to initially investigate please see Sect. 25.4 of this chapter).

Step 2: Analyse the pre-culture material properties

Cut 1 cm^2 sized pieces of the material- these are then analysed to determine the pre-culture analysis of the material structural properties (see later sections for more detail).

Step 3: Perform cell culture tests on the material samples

Cut appropriate sized samples for cell culture experiments; these will typically consist of 1 cm^2 flat pieces for static or flat dynamic tests, or 5–10 cm lengths of tubes for dynamic (e.g. bioreactor) tests.

Sterilise the samples using a non-reactive method (polymer-dependent), i.e. your samples should not shrink/swell during the process.

Secure the sample in place in a 24-well plate (these experiments should be run simultaneously across the range of selected structural property, and further repeats performed at future dates); cell suspension rings or sterile glue can be used.

Seed the samples with endothelial cells; a seeding density of 5×10^4 cells per sample allows a range of culture periods to be analysed without over-proliferation of the cells. Seeding with the cells suspended in a very small volume of media (approximately 100–200 μl), leaving to settle in the incubator for 20–30 min then adding the required 1–2 ml of media ensures the adhesion to the material rather than tissue culture polystyrene well surface. (Medium consists of 40 % DMEM (Dulbecco's Modified Eagles Medium), 40 % 199 (containing modified Earle's salts) (Gibco™, Invitrogen Corporation, Paisley, UK), 20 % bovine fetal calf serum (FCS) (Cambrex, Nottingham, UK), 1 % non-essential amino acids (NEAA), 1 % sodium pyruvate, 1 % streptomycin and penicillin.)

Samples are cultured for periods of 3, 7, 14 and 28 days, with media changes every 3 days.

Step 4: Remove the cell-seeded material samples from culture conditions and perform post-culture analysis to acquire data

After the designated time period, samples are removed from the culture conditions, the media removed and a gentle rinse with phosphate buffered saline (PBS) (Dulbecco's, d.PBS) performed.

Samples are prepared according to their method of testing- see separate sections in Sect. 25.6 for a more detailed, focussed example of their use. Cell-material analysis would typically consist of assays for: morphology; viability; cell number; cell proliferation; cell coverage/area; cell spreading; cell orientation; cell signalling and marker expression (including examination of activation, inflammation and endothelial phenotype) (all or a selection of these studies depending on the focus of the investigation).

Statistical analysis is then performed, with averages and standard deviations calculated, and one-way ANOVA tests of variance executed to determine significant differences between sample groups and observed behaviour. Tukey post-hoc analysis with a 95 % confidence interval is used to rank these significances; Pearson correlation coefficient is utilised to determine correlative effects from the underlying structural properties upon the induced material-cell interactions.

25.6 Detailed Protocol with a Focus on the Investigation of the Effect of Structural Cues Provided from Electrospun Scaffolds on Endothelial Cells (Human Umbilical Vein Endothelial Cells (HUVECs)) (Some Details Adapted from [8, 20, 41])

25.6.1 Production of a Range of Fibrous Scaffolds Using Electrospinning

The polymer of choice was dissolved in an appropriate solvent (e.g. 12.5 w/v % Tecoflex® SG-80A polyurethane (Thermedics, Woburn, USA) with 1:1.68 dimethylacetamide (DMAC): 2-butanone (methyl ethyl ketone) (MEK) (Aldrich, Gillingham, UK)).

The spinning (process) parameters were systematically altered between spinning runs to produce a predictable range of scaffolds with varying structural components [41]. Parameters included: flow rate (flow of solution from nozzles); spray height (the relative vertical distance of the nozzles from the mandrel); spray distance (the horizontal distance of the nozzles from the mandrel); traverse speed (the constant linear speed of the traverse); mandrel speed (the constant rotational speed of the mandrel); applied voltage and temperature/relative humidity. 1 cm^2 samples were cut for use in further analysis and tests.

25.6.2 Characterisation of Pre-culture Scaffold Structural Properties

The pre-culture scaffolds were characterized for inter-fibre separation (ifs), fibre diameter (f.dia), void fraction (VF), surface roughness (SR) and fibre orientation (f.orn).

Ifs, f.dia and f.orn measurements were taken using a field emission scanning electron microscope (SEM), with a working distance of 8 mm, an acceleration gun voltage of 5 kV and the secondary electron detector. Samples were first sputter-coated with chromium (approximately 50 nm coating) for 2 min at 125 mA in an argon atmosphere. Measurements were taken of the structures at approximately ×1,000 magnification, using SEM digital measuring tools, from the top layer of fibres as indicated by their overlapping on the images.

VF was also analysed using the SEM images, combined with the use of the imaging software, using macro programs to analyse the stored microscope images (software can include the proprietary microscope package such as KS400 or AxioVision (Zeiss, Welwyn Garden City, UK) or ImageJ (NIH, USA)).

Atomic force microscopy (AFM) was used to quantify the SR of the scaffolds. The probe was fitted with a silicone-nitride (Si2N4) tip, operated in contact mode.

Measurements were taken on an individual fibre basis, with scan sizes set to 15 μm, line scanned at a rate of 1 Hz, and a z range of 4.69 μm.

Ifs measurements were repeated 20 times, f.dia 14 times, SR and VF 5 times and f.orn 20 times per sample.

25.6.3 Cell Culture of Electrospun Material Samples

As described in Steps 2 and 3 in Sect. 25.5 of this chapter.

25.6.4 Analysis of Post-Culture Cell-Material Samples with Induced Cell-Material Interactions

25.6.4.1 General Cell Morphology, Area and Spreading Imaging and Analysis

SEM Analysis Cell-samples were immersed in 2.5 % glutaraldehyde (VWR, Poole, UK) solution for 1 h, washed in PBS and distilled water, and dehydrated in alcohol (70, 90 and 100 % ethanol for 15 min respectively, each stage performed twice). Samples were then dried, preferably critical point dried to enhance the surface features of the cells (although careful air-drying can be used), followed by sputter-coating with chromium (details previously described). The same SEM parameters as detailed in Sect. 25.6.2 of this chapter were used.

Confocal Staining and Analysis Cell-seeded samples were fixed (after rinsing with d.PBS) using 4 % formaldehyde, 2 % sucrose fixative solution at 37 °C, 5 % CO_2, humidified, for 10 min, then rinsed with PBS. Samples were permeabilized using 0.5 % Triton X100 solution at 4 °C for 5 min, then rinsed again with PBS. Primary antibodies were added and incubated for 1 h at 37 °C: 0.22 mg/ml monoclonal mouse anti-human vinculin (Serotec, Oxford, UK). This was followed by a trio of wash steps performed in the dark at room temperature for 3 min each, using 0.1 % Tween 20® solution (ICN Biomedicals Inc., Aurora, Ohio, USA). Secondary antibodies were added for 1 h at 37 °C: 0.1 mg/ml rhodamine (rhodamine-conjugated goat IgG fraction to mouse IgG whole molecule) (Sigma, Gillingham, UK). The Tween 20® wash step was then repeated. Oregon-green phalloidin (Molecular Probes, Leiden, The Netherlands) was added at a concentration of 5 μg/ml for 30 min at 4 °C. Samples were washed with PBS before mounting with fluorescence stabilising mountant containing DAPI stain (Vectashield® with DAPI H-1200) (Vector Laboratories Inc., Burlingame, CA, USA). Samples were kept in the dark, at 4 °C, until analysis.

Images were obtained from a confocal laser scanning microscope (LSM 510) (Zeiss, Welwyn Garden City, UK). The DAPI nuclear stain (both cell types) was visualised by UV excitation at $\lambda = 364$ nm, a HeNe, $\lambda = 568$ nm laser was used to visualise vinculin for focal contacts (L929 cells), $\lambda=543$ nm for HUVECs, and oregon green phalloidin for F-actin was excited with an Argon laser $\lambda=488$ nm. All samples were examined at x100 magnification. Z-stack (imaging through the vertical plane) measurements, through the use of the confocal computer software, were also taken of the cells to determine their height profiles on the three different materials.

Analysis of all microscope images was performed using an imaging software package (as Sect. 25.6.2). Programmed macros assessed and calculated cell area, coverage (defined as the percentage of scaffold surface covered with cells), spreading (the index of the degree of spreading of the cells, calculated by dividing the cell coverage by the cell number), and orientation (the angle of the cell measured along the long axis; degrees).

25.6.4.2 Cell Numbers and Proliferation Analysis

Samples were washed with d.PBS and stored (dry) at −80 °C. Upon analysis, samples were thawed at room temperature, and the CyQuant assay performed to determine cell number according to manufacturer's instructions (Molecular Probes, Eugene, Oregon, USA); fluorescence was read using a plate reader. Standard curves (fluorescence versus cell number) were produced from cell pellet data, and used to convert fluorescence into cell number for the material-cell samples.

25.6.4.3 Cell Viability Analysis

Samples were washed with PBS and kept in a wet state under a layer of PBS whilst preparing reagents. Live:dead assay (Life Technologies, UK) is used to produce a ratio of percentage live and dead cells through two colour fluorescent staining and image analysis, or fluorescent plate reader data acquisition. Full details should be followed according to the manufacturer's protocol.

25.6.4.4 Analysis of Cell Signalling and Marker Expression

Cell-seeded material scaffolds were rinsed in d.PBS, fixed in 4 % formaldehyde 2 % sucrose solution (VWR, Poole, UK) for 10 min at 37 °C, 5 % CO_2, humidified, then rinsed again with d.PBS. Samples were then stained with sterile filtered 0.4 % methylene blue for 12 min (VWR, Poole, UK).

Each sample was cut into smaller areas and incubated with rabbit serum for 30 min at room temperature, then with a 1:200 dilution of mouse anti-human primary antibodies in PBS containing 1 % bovine serum albumin (BSA) at room temperature for 1 h. The primary antibodies used were: collagen I, elastin, fibronectin, CD54, CD106, CD51/61, CD49c; CD31, CD62E/P, vWF (see Table 25.1 for more details on this immunostaining panel). An IgG1 isotype control and PBS negative control were used throughout all staining procedures. Two PBS buffer rinses followed. Secondary antibody solution of rabbit anti-mouse immunoglobulin (biotinylated)

Table 25.1 Function of exploratory immunostaining panel for endothelial cells cultured on fabricated material samples

Antibody used	Information provided from positive staining
Collagen I	Fibrous extracellular matrix protein
Elastin	Elastic extracellular matrix protein
Fibronectin	Component of extracellular matrix, that binds other proteins including collagen
CD54	Intercellular Adhesion Molecule 1 (ICAM-1); cell-cell interaction and adhesion molecule, expressed by stimulated cells
CD106	Vascular cell adhesion molecule 1 (VCAM-1); cell adhesion molecule
CD51/61	Cell-cell adhesion molecule
CD49c	Cell-cell and cell-matrix adhesion molecule
CD31	Platelet endothelial cell adhesion molecule (PECAM-1); an endothelial cell marker, particularly for intercellular junctions
CD62E/P	E-Selectin/P-Selectin; expressed on the surface of stimulated endothelial cells (at different rates); affected by applied shear stresses
vWF	von Willebrand factor; an endothelial cell phenotypic marker

(E0464) (Dako A/S, Denmark) (25 μl in 5 ml PBS) was added for 30 min at room temperature; again followed by PBS rinses. Samples were then incubated with Vectastain ABC-AP kit (AK-500) (Vector Laboratories Inc., Burlingame, CA, USA) for 30 min. This was followed by a PBS wash. Incubation in Alkaline Phosphatase substrate (Kit 1, SK-5100) (Vector Laboratories Inc., Burlingame, CA, USA) followed (Trizma® Base (Tris[hydroxymethyl]aminomethane) (Tris) (Sigma, Gillingham, UK) solution of 1.2 g Tris in 100 ml distilled water (pH range 8.2–8.5) combined with kit reagents). Samples were incubated with this solution for 30 min in the dark, followed by a final wash step in distilled water prior to mounting in a fluorescence stabilising mountant containing DAPI nuclear stain (Vectashield® with DAPI) (Vector Laboratories Inc., Burlingame, CA, USA). Samples were kept in the dark, at 4 °C, until analysis.

Positive/negative expression of extracellular matrix and adhesion molecules was determined by reflective light microscopy and laser scanning confocal microscopy (methodology already detailed) (methylene blue and DAPI nuclear stains confirmed the correct location of positive staining and provided the ability to co-image cell area and quantity/location of positive marker expression); quantified results were obtained through the image analysis of the images (described in Sects. 25.6.2 and 25.6.4.1).

25.6.5 *Further Analysis to Determine Correlative Relationships and Develop the Data Obtained*

Statistical analysis was then performed as detailed in Sect. 25.5 (step 4) to determine significant differences in the produced cell behaviour, and then specific correlative relationships, i.e. an induced effect upon the cell behaviour directly from the changes made to the underlying material structures.

Optimised structural properties to enhance endothelialisation can then be selected and the "designed", fabricated grafts further investigated for specific aspects of functionality.

25.7 Sample Results

The examination of two commonly used clinical vascular graft materials (PET and ePTFE) showed significant changes to the cellular behaviour resulting from the underlying substrate structures. This altered the cellular ability to adapt to the shear stresses and so the retention of the endothelial layer, having a direct effect on the performance and functionality of the artificial vessel [8] (Figs. 25.1 and 25.2).

As a result of these findings, electrospun materials (PU) were investigated in more detail, with a range of structures included in the tests. Fundamental mechanisms relating the scaffold features to the induced contacting cell behaviour were established (based on data and statistical analysis): initial cell adhesion relied on the physical amount of material present; increased cell proliferation was affected by scaffold material and orientation of physical features; features influencing cell spreading and coverage included void size, surface roughness and fibre orientation; cell-cell interactions were influenced to different degrees by all scaffold features [20].

25.8 Developing the Application of These Methodologies

The results of using these methodologies can be extended and applied to further research. For example, in order to determine the optimum structure for a specific vascular application/location, these tests can be used to find the best properties to fabricate and those parameters then used and researched in greater detail regarding the functionality, biocompatibility and immunological responses provoked (for example using ELISAs, Western Blotting, or Real-Time PCR). The ideal aim is to use these tests as a way of determining the optimum fabrication parameters to produce a material that has (structural) properties that enhance endothelialisation. Hence, the artificial blood vessel can be produced in such a way to mimic the natural tissue it is replacing, i.e. replicates the natural basement membrane that produces and sustains endothelial layers.

Another important area directly connected to this development of artificial blood vessels is that of haemodynamics. This field is worthy of further attention, with significant potential and need for advancement alongside the use of material structural cues to enhance endothelialisation. In order to achieve this, the haemodynamic parameters applied to the cells from the contacting blood flow should be known as fully as possible [5, 7, 8, 16, 17, 19, 42].

As mentioned throughout this chapter, endothelialization in engineered/artificial grafts is a crucial process to maximise the functionality and viability of the vessel

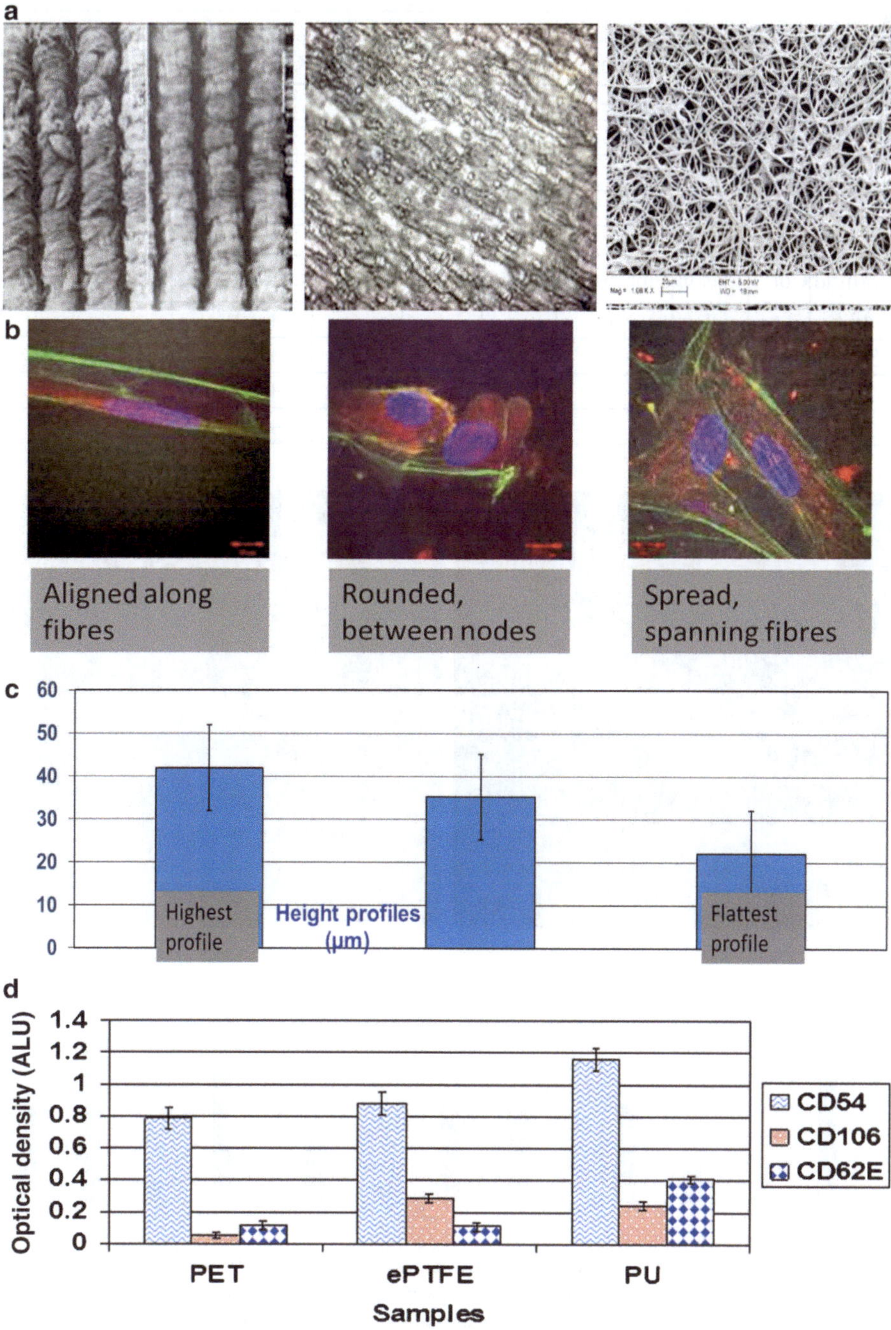

Fig. 25.1 Investigation of two commonly used vascular graft materials and an electrospun scaffold. (**a**) Images (light microscope and SEM) showing surfaces of PET, ePTFE and electospun PU (*left* to *right*); (**b**) confocal microscope images showing cell morphology and cytoskeletal involvement across the three materials; (**c**) graph showing the cell height profiles; (**d**) graph showing the comparative levels of three cell markers (Adapted from [8])

a

Scaffold	Col. I	Elastin	Fibronectin	CD54	CD106	CD51/61	CD49c	CD31	CD62E/P	vWF
48A	-	-	+	+	+	+	-	+	-	+
507A	+	-	+	+	-	+	+	+	+	+
22D	+	+	+	+	-	+	+	-	+	+
29D	-	+	+	+	+	-	+	+	-	+

Scaffold	Col. I	Elastin	Fibronectin	CD54	CD106	CD51/61	CD49c	CD31	CD62E/P	vWF
48A	+	+	+	-	+	+	+	+	-	+
507A	+	+	+	+	+	+	+	+	+	+
22D	+	+	+	+	+	+	+	+	+	+
29D	+	+	+	-	+	+	+	+	+	+

b

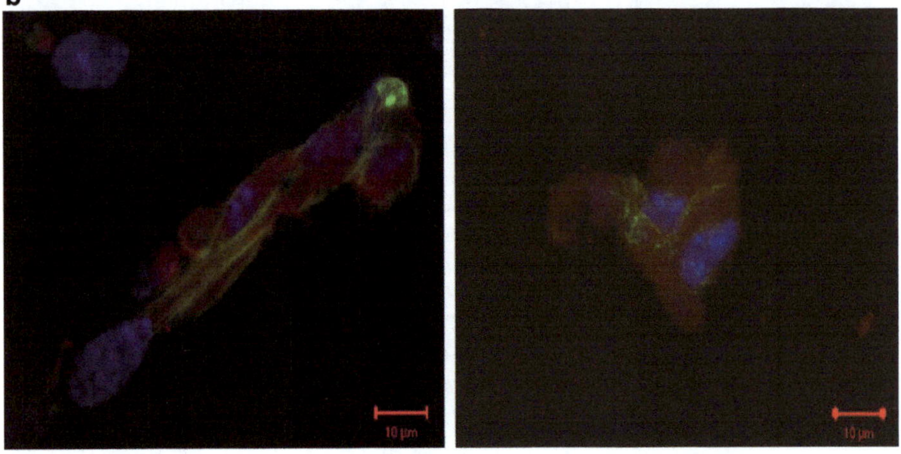

c

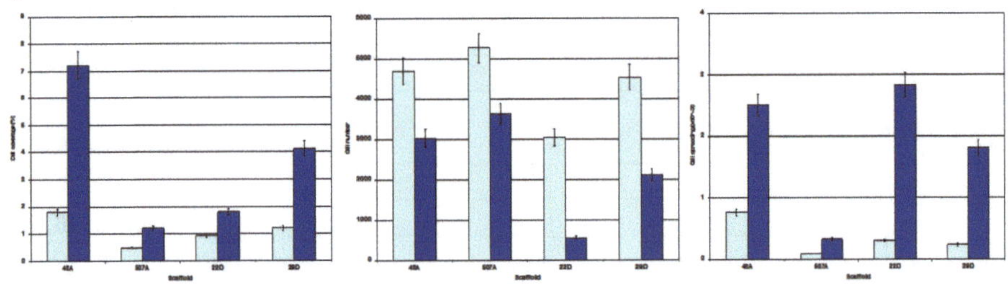

Fig. 25.2 Further investigation of the electrospun materials, using a range of scaffolds with varying structures. (**a**) Charts to show positive/negative expression of cell markers at 7 and 28 days (*top* to *bottom*); (**b**) confocal microscope images showing cell morphology, cell orientation and cytoskeletal involvement across two of the electrospun materials (*left*- 22D; *right* 48A); (**c**) graphs showing the cellular behavior (*left*- cell coverage; *centre*- cell number; *right*- cell spreading) (Adapted from [20])

and the contained blood flow and has significant effects on the success of implantation. Endothelial cells lining the artery wall have the ability to act as the fluid dynamic wall-shear biosensor and can minimize the development of plaque.

Currently, one of the most important haemodynamic parameters in cardiovascular problems is the 'shear stress'. In fluid mechanics, shear stress is defined as the friction between two layers of the fluid moving at different velocities. In blood vessels, the shear stress also arises at the interplay between blood and endothelial layer where it induces a shearing deformation of the endothelial cells. Nowadays, thanks to the increasing power-to-cost ratio of computers and the advent of methods for subject-specific modelling of cardiovascular mechanics, it is possible to calculate shear stress-related forces using Computational Fluid Dynamics (CFD).

In general, CFD is a technique to analyse fluid flow, heat transfer and associated phenomena, using computer-based simulation and has recently shown great potential for the calculation of various haemodynamic parameters in patient-specific models including Time-Averaged Wall Shear Stress (TAWSS), Oscillatory Shear Index (OSI) and Relative Residence Time (RRT), amongst others. CFD also provides an alternative to invasive or non-invasive flow measurements of blood flow by and/or ex vivo experimental flow measurement techniques, which can be very expensive and time consuming. Further details on the use of CFD in cardiovascular problems and haemodynamic parameters can be found in Chap. 27.[1]

In artificial grafts, the shear stress at the endothelial wall (also known as the Wall Shear Stress – WSS), has a fundamental role in the endothelialization process. For example, it is known that high shear stress could limit the endothelialization process, therefore, the calculation of shear stress within the whole artificial blood vessel is considered to play an important role in its design and optimization [42].

Apart from the importance of WSS in the endothelialization process, WSS can also help to better explain some of the common graft failures. For instance, in their computational simulations, Kouhi et al. [43] reported a dramatic drop in the magnitude of WSS in critical areas of graft anastomosis such as toe, heel, and suture lines which could initiate the promotion of intimal hyperplasia and cause early graft failure after coronary artery bypass graft.

Another graft failure is due to atherosclerosis, in and around the downstream anastomosis. Atherosclerotic lesions occur predominantly at sites of low shear, whereas regions of the vasculature exposed to a physiologic shear are protected [44]. A developing plaque can modify the local shear stress around the lesion. Lumen narrowing would result in an increase in the blood velocity and shear stress over the plaque, low shear stress in the upstream region, and disturbed flow in the form of directionally OSI in the downstream zone of the plaque [45, 46].

[1] *Note for the editor*: this is a reference to another Chapter submitted as part of this book entitled "Vascular Flow Modelling using Computational Fluid Dynamics", authored by A. Keshmiri and K. Andrews.

25.9 Conclusions

There is a clinical need to enhance endothelialisation of artificial blood vessels used to repair or replace damaged/diseased blood vessels. The increased understanding of successful techniques that can be used to achieve this, regardless of the material, fabrication method or design used, would be a significant achievement in the fields of vascular biology, biomaterials and tissue engineering. This chapter has detailed a variety of methodologies that can be used as individual or a panel of tests to determine the effect of altering the underlying material structures upon the endothelial cells. These results provide the means to determine correlative relationships between the structural properties and the cell behaviour; these can then be incorporated into the fabrication design and progressed to develop improved strategies for artificial blood vessels with enhanced functionality, creating engineered tissue analogues.

References

1. Tian L, George SC (2011) Biomaterials to prevascularize engineered tissues. J Cardiovasc Transl Res 4:685–698
2. Vara DS, Salacinski KJ, Kannan RY et al (2005) Cardiovascular tissue engineering: state of the art. Pathol Biol 53:599–612
3. Andrews KD, Hunt JA (2009) Developing smaller-diameter biocompatible vascular grafts. In: Di Silvio L (ed) Cellular responses to biomaterials (section II: cell responses & regenerative medicine). Woodhead, Cambridge
4. Menu P, Stoltz JF, Kerdjoudf H (2013) Progress in vascular graft substitute. Clin Hemorheol Microcirc 53:117–129
5. Spadaccio C, Rainer A, Barbato R et al (2013) The fate of large-diameter Dacron® vascular grafts in surgical practice: are we really satisfied? Int J Cardiol 168:5028–5029
6. Sarkar S, Schmitz-Rixen T, Hamilton G et al (2007) Achieving the ideal properties for vascular bypass grafts using a tissue engineered approach: a review. Med Biol Eng Comput 45:327–336
7. Feugier P, Black RA, Hunt JA et al (2005) Attachment, morphology and adherence of human endothelial cells to vascular prosthesis materials under the action of shear stress. Biomaterials 26:1457–1466
8. Andrews KD, Feugier P, Black RA et al (2008) Vascular prostheses: performance related to cell-shear responses. J Surg Res 149:39–46
9. Cittadella G, de Mel A, Dee R et al (2013) Arterial tissue regeneration for pediatric applications: inspiration from up-to-date tissue-engineered vascular bypass grafts. Artif Organs 37:423–434
10. Kurobe H, Maxfield MW, Breuer CK et al (2012) Concise review: tissue-engineered vascular grafts for cardiac surgery: past, present, and future. Stem Cells Transl Med 1:566–571
11. Fernandez CE, Achneck HE, Reichert WM et al (2014) Biological and engineering design considerations for vascular tissue engineered blood vessels (TEBVs). Curr Opin Chem Eng 3:83–90
12. Fernandez P, Deguette A, Pothuaud L et al (2005) Quality control assessment of ePTFE precoating procedure for in vitro endothelial cell seeding. Biomaterials 26:5042–5047
13. Assmann A, Delfs C, Munakata H et al (2013) Acceleration of autologous in vivo recellularization of decellularized aortic conduits by fibronectin surface coating. Biomaterials 34:6015–6026
14. Avci-Adali M, Perle N, Ziemer G et al (2011) Current concepts and new developments for autologous in vivo endothelialisation of biomaterials for intravascular applications. Eur Cells Mater 21:157–176

15. Avci-Adali M, Paul A, Ziemer G et al (2008) New strategies for in vivo tissue engineering by mimicry of homing factors for self-endothelialisation of blood contacting materials. Biomaterials 29:3936–3945
16. Thomas LV, Lekshmi V, Nair PD (2013) Tissue engineered vascular grafts- preclinical aspects. Int J Cardiol 167:1091–1100
17. Sarkar S, Sales KM, Hamilton G et al (2007) Addressing thrombogenicity in vascular graft construction. J Biomed Mater Res B Appl Biomater 82B:100–108
18. de Valence S, Tille J-C, Chaabane C et al (2013) Plasma treatment for improving cell biocompatibility of a biodegradable polymer scaffold for vascular graft applications. Eur J Pharm Biopharm 85:78–86
19. Wong CS, Sgarioto M, Owida AA et al (2006) Polyethyleneterephthalate provides superior retention of endothelial cells during shear stress compared to polytetrafluoroethylene and pericardium. Heart Lung Circ 15:371–377
20. Andrews KD, Hunt JA (2008) Upregulation of matrix and adhesion molecules induced by controlled topography. J Mater Sci Mater Med 19:1601–1608
21. Isenberg BC, Backman DE, Kinahan ME et al (2012) Micropatterned cell sheets with defined cell and extracellular matrix orientation exhibit anisotropic mechanical properties. J Biomech 45:756–761
22. Chen J-H, Laiw R-F, Jiang S-F et al (1999) Microporous segmented polyetherurethane vascular graft: I. dependency of graft morphology and mechanical properties on compositions and fabrication conditions. J Biomed Mater Res 48:235–245
23. Blinder YJ, Mooney DJ, Levenberg S (2014) Engineering approaches for inducing blood vessel formation. Curr Opin Chem Eng 3:56–61
24. Ahmed M, Ghanbari H, Cousins BG et al (2011) Small caliber polyhedral oligomeric silsesquioxane nanocomposite cardiovascular grafts: influence of porosity on the structure, hemocompatibility and mechanical properties. Acta Biomater 7:3857–3867
25. Koens MJW, Faraj KA, Wismans RG et al (2010) Controlled fabrication of triple layered and molecularly defined collagen/elastin vascular grafts resembling the native blood vessel. Acta Biomater 6:4666–4674
26. Norotte C, Marga FS, Niklason LE et al (2009) Scaffold-free vascular tissue engineering using bioprinting. Biomaterials 30:5910–5917
27. Kang T-Y, Hong JM, Kim BJ et al (2013) Enhanced endothelialization for developing artificial vascular networks with a natural vessel mimicking the luminal surface in scaffolds. Acta Biomater 9:4716–4725
28. McNamara LE, Burchmore R, Riehle MO et al (2012) The role of microtopography in cellular mechanotransduction. Biomaterials 33:2835–2847
29. Biggs MJP, Richards G, Dalby MJ (2010) Nanotopographical modification: a regulator of cellular function through focal adhesions. Nanomed: Nanotechnol Biol Med 6:619–633
30. Vitte J, Benoliel AM, Pierres A et al (2004) Is there a predictable relationship between surface physical-chemical properties and cell behavior at the interface? Eur Cell Mater 7:52–63
31. Boccafoschi F, Mosca C, Ramella M et al (2013) Biological evaluation of materials for cardiovascular application: the role of the short-term inflammatory response in endothelial regeneration. J Biomed Mater Res A 101A:3131–3140
32. Filová E, Brynda E, Riedel T et al (2014) Improved adhesion and differentiation of endothelial cells on surface-attached fibrin structures containing extracellular matrix proteins. J Biomed Mater Res A 102A:698–712
33. Kirton JP, Xu Q (2010) Endothelial precursors in vascular repair. Microvasc Res 79:193–199
34. Li W, Wang H, Kuang C-Y et al (2012) An essential role for the Id1/PI3K/Akt/NFkB/surviving signaling pathway in promoting the proliferation of endothelial progenitor cell in vitro. Mol Cell Biochem 363:135–145
35. Hasan A, Memic A, Annabi N et al (2014) Electrospun scaffolds for tissue engineering of vascular grafts. Acta Biomater 10:11–25
36. Fullana MJ, Wnek GE (2012) Electrospun collagen and its applications in regenerative medicine. Drug Deliv Transl Res 2:313–322

37. Dargaville BL, Vaquette C, Rasoul F et al (2013) Electrospinning and crosslinking of low-molecular-weight poly(trimethylene carbonate-co-L-lactide) as an elastomeric scaffold for vascular engineering. Acta Biomater 9:6885–6897
38. Wong CS, Liu X, Xu Z et al (2013) Elastin and collagen enhances electrospun aligned polyurethane as scaffolds for vascular graft. J Mater Sci Mater Med 24:1865–1874
39. Hu J-J, Chao W-C, Lee P-Y et al (2012) Construction and characterization of an electrospun tubular scaffold for small-diameter tissue-engineered vascular grafts: a scaffold membrane approach. J Mech Behav Biomed Mater 13:140–155
40. Kuwabara F, Narita Y, Yamawaki-Ogata A et al (2012) Long-term results of tissue-engineered small-caliber vascular grafts in a rat carotid arterial replacement model. J Artif Organs 15:399–405
41. Andrews KD, Hunt JA, Black RA (2008) Technology of electrostatic spinning for the production of polyurethane tissue engineering scaffolds. Polym Int 57:203–210
42. Gui L, Niklason LE (2014) Vascular tissue engineering: building perfusable vasculature for implantation. Curr Opin Chem Eng 3:68–74
43. Do H, Owida AA, Yang W, Morsi YS (2011) Numerical simulation of the haemodynamics in end-to-side anastomoses. Int J Nume Methods Fluids 67:638–650. doi:10.1002/fld
44. Kouhi E, Morsi YS, Masood SH (2008) Haemodynamic analysis of coronary artery bypass grafting in a non-linear deformable artery and Newtonian pulsatile blood flow. Proc Inst Mech Eng H 222:1273–1287. doi:10.1243/09544119JEIM459
45. Dai G, Kaazempur-Mofrad MR, Natarajan S et al (2004) Distinct endothelial phenotypes evoked by arterial waveforms derived from atherosclerosis-susceptible and -resistant regions of human vasculature. Proc Natl Acad Sci U S A 101:14871–14876. doi:10.1073/pnas.0406073101
46. Cecchi E, Giglioli C, Valente S et al (2011) Role of hemodynamic shear stress in cardiovascular disease. Atherosclerosis 214:249–256. doi:10.1016/j.atherosclerosis.2010.09.008

Chapter 26
Preparation of Liposomes with Dual Fluorophores to Follow Real-Time Content Release In Vivo

Harmesh Singh Aojula

26.1 Introduction

Vasculature, a central barricade between blood and tissues, is essential for regulating most physiological processes. At diseased sites, the vascular permeability is often enhanced due to defects in the blood vessel as a result of rapid angiogenesis, chronic inflammation or local injury. The defects in endothelial lining in the locality of diseased tissues presents gaps through which nanometer-sized particles are able to escape the circulation and hence avoid rapid clearance from the body. This effect is especially pronounced in solid tumours where lymphatic drainage is also very poor to allow any further circulation of the extravasated particles. In this way, liposomes of size less than 200 nm diameter are able to form depots of drug reservoirs by the "enhanced permeability retention (EPR) effect", a phenomenon that is well exploited [1] in delivering drugs to tumours. While liposomes are yet to make significant clinical impact in the area of cardiovascular disease, such as in the case of ischemic myocardial tissue, the principles and practices that have made these versatile vehicles a commercial success for treatment of cancer are expected to expedite their utility in delivering active agents to infarcted areas [2]. Unlike the solid tumours, the widespread existence of the EPR effect in these applications is debatable, although can be applicable in some situations. For instance, given that in myocardial infarction blood vessels are leakier in the left ventricle and cell injury can result in chronic inflammation of blood vessels infers that passive targeting may have a role to play. In addition to this, active targeting of ligand sensitised liposomes constitutes an alternative to form depots of drug by accumulating liposomes through interaction with over-expressed cell receptors [3]. The aim of this chapter is not to dwell on the

H.S. Aojula (✉)
Manchester School of Pharmacy, University of Manchester,
Oxford Rd, Manchester M13 9PT, UK
e-mail: harmesh.aojula@manchester.ac.uk

M. Slevin, G. McDowell (eds.), *Handbook of Vascular Biology Techniques*,
DOI 10.1007/978-94-017-9716-0_26

details of any particular application. Suffice to say that actively targeted liposomal drug delivery systems are prepared by conjugating targeting moieties, including small-molecule ligands, peptides, receptors and monoclonal antibodies, on the liposomal surface [4, 5]. New targets are being identified to target stroke, atherosclerosis, neurodegenerative and many other inflammatory diseases [2, 3].

In this chapter, the intent is only to provide a starting point to a new researcher for acquaintance with the basic methodology of liposome preparation. It is assumed that the reader is already familiar with the structure of liposomes and lipids as well as the role transition phase, gel to liquid crystalline, plays in controlling bilayer permeability. Preparation of liposomes according to the procedure below should give initial liposome handling experience within a few weeks. One can then adventure and adapt these procedures to prepare liposomes of different lipid compositions to suit various research applications. Focus is on just one method of preparation, which we routinely use to yield liposomes from small (0.1 ml) to medium scale (200 ml) with excellent homogeneity and a range of payloads. This method is based on extrusion of a lipid suspension through polycarbonate membranes. The principle of encapsulation is solely based on passive loading.

26.2 Methodology

26.2.1 Materials

The following are available from Sigma-Aldrich: L-α-Phosphatidylcholine (P-3556), Cholesterol (C-8667), Disposable PD-10 desalting columns (54805), Iron (III) Chloride hexhydrate (31232), Ammonium Thiocyanate (431354). Millex-GP syringe filter unit (Z359904), Phosphate buffered Saline (P5493) and Calcein (C-0875), Triton X-100 (X-100). 5,5′-Ph2-DiIC18(3) lipophilic tracer dye was purchased from Invitrogen Netherlands (catalogue number D-7779). Whatman Nucleopore polycarbonate membranes (0.1 μm pore) and filter support (drain) discs can be purchased from the extruder manufactures as consumable items.

26.2.2 Equipment

General equipment; Rotary evaporator, Vortex mixer, Bench-top microfuge, Spectrophotometer, Flask shaker, Freeze drier.

Specialised equipment: Extrusion apparatus (see below for choice), Nanoparticle sizer (e.g. Malvern 4700 DLS system) and a Spectrofluorometer.

26.2.2.1 Extrusion Apparatus

(Lipex Biomembranes Inc., Vancouver, Canada), or Liposofast (push-pull) extruder (Avestin, Ottawa, Canada).

Fig. 26.1 Common types of research scale extruders used for producing liposomes. The mini-extruder (*A*) is a low cost option (Avanti Polar Lipids, Inc.) for producing small, under 1.0 ml, volumes and requires only manual pushes to drive the process. The high pressure extruders made of stainless steel (*B* and *C*) are driven by compressed gas. Extruder *B* (Lipix™, Northern Lipids Inc) is suitable for 1.0–10 ml scale while extruder *C* (Avestin Liposo-Fast™) has a 50 ml capacity. Both the latter devices have a thermobarrel for connection to a water circulating bath

Extrusion of liquid suspension containing large multi-lamellar liposomes is made through two (stacked) polycarbonate membranes of selected pore size, 0.2 μm or below, by applying high pressure in a closed system. There are a number of commercial suppliers of extrusion devices. For small scale preparation (<1.0 ml), the most common low cost option is to use the push-pull mini-extruders of the type (Avestin Inc or Avanti Polar lipids Inc) shown in Fig. 26.1A. With these devices, gas tight Hamilton syringe comprises the closed system and pressure is applied mechanically by hand. Mini extruders are better suited to prepare liposomes with low lipid concentrations and with the transition phase below room temperature. Although additional accessories are available, to maintain higher temperatures during preparation, it is preferable to use the larger extruders (Fig. 26.1B and C) as they are jacketed and extrusion is more readily controlled with high pressure gas (nitrogen) regulator.

26.3 General Procedure

The method of liposome preparation requires three key sequential steps. Firstly, the lipid solution is dried down to deposit a thin lipid film. This film is then hydrated with an aqueous solution containing the payload to be encapsulated. In the final step, sequential extrusions (Fig. 26.2) through polycarbonate membranes of defined pore size are carried out.

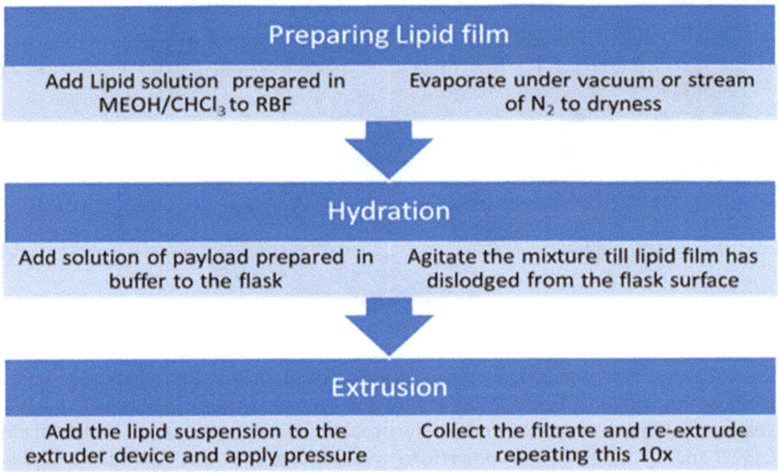

The assembly of polycarbonate filters and supports is a very straightforward task with all these extrusion devices. As an illustration, the 10.0 ml extruder (Northern Lipids Inc, Canada shown in Fig. 26.1B) device has a base plate onto which the drain discs and membranes are applied as illustrated in Fig. 26.3.

26.3.1 Preparing Blank Liposomes

The following protocol is used to prepare 2.0 mls of liposomes comprised from a 2:1 molar ratio of L-α-Phosphatidylcholine and cholesterol with an effective diameter of approximately 118 nm, using the 10 ml extruder shown in Fig. 26.1b (Northern Lipids Inc, Canada).

1. L-α-Phosphatidylcholine (25 mg) and cholesterol (7.0 mg) are dissolved in 4mls of 1:1 (v/v) chloroform/methanol mixture and poured into a 50.00 ml round-bottomed flask (RBF). The flask is placed on a rotary evaporator with water bath set at 40 °C and rotated (~100 r.p.m) under vacuum. A thin lipid film deposits on the side of the flask. The flask is then removed and the lipid film further dried by placing on a freeze-drier for 2 h to ensure all solvent has been removed. Alternatively a high vacuum pump may be attached to the rotary evaporator and the film left to dry under vacuum for further 2 h without rotation.

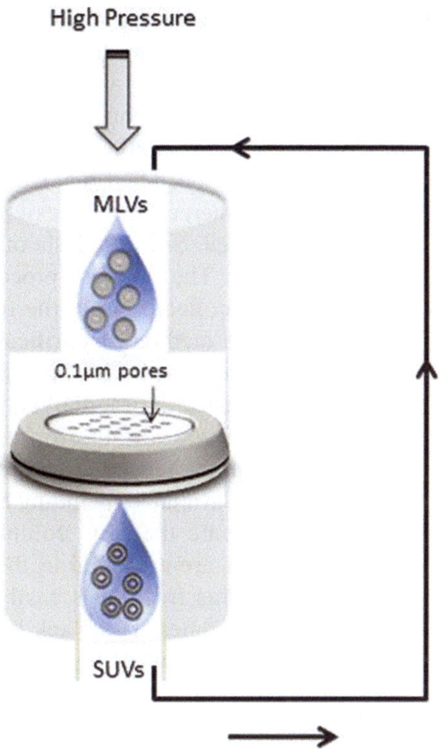

Fig. 26.2 Schematic diagram of the extrusion concept for preparing small unilamellar liposomes. Lipid suspension in aqueous buffer, containing large multilamellar vesicles (MLVs), is forced through 0.1 μm pore size filters by applying high pressure using nitrogen gas. The extrudate containing small unilamellar vesicles (SUVs) is reprocessed ten times to generate a homogenous preparation

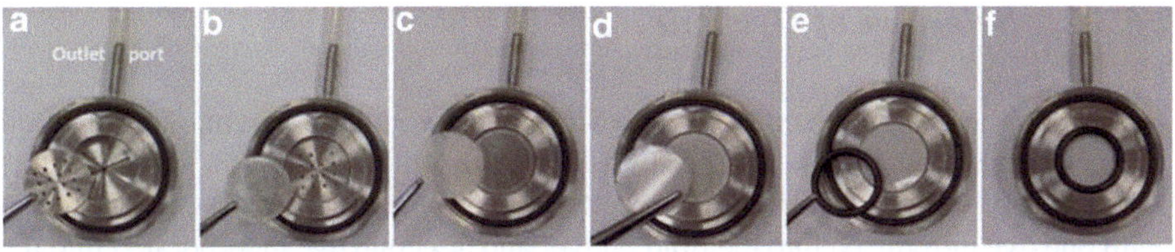

Fig. 26.3 Assembly of the filters and supports for extrusion. The filter support baseplate has a hole through which the extrudate escapes into the outlet port for collection. The following parts are inserted into the base plate sequentially; (**a**) support disc, (**b**) Stainless steel mesh, (**c**) Polyester drain disc, (**d**) two polycarbonate membrane (one at a time) and finally (**e**) an O-ring to seal. Both the Avestin™ and Lipix™ extruder bases are assembled in same way

2. To hydrate the lipid film, 2.0 ml of Phosphate buffered Saline pH 7.4 is added and the mixture shaken until a milky suspension is formed and the entire lipid has dislodged from the flask surface. This step may take an hour depending on the level of agitation and nature of the lipid film. See troubleshooting problem 1. The suspension contains multilamellar liposomes of a large size, which then need to be down-sized by extrusion.

3. The cloudy suspension is transferred into the extruder device with polycarbonate membranes already (Fig. 26.3) in place and washed with PBS. The extruder device is closed ensuring all fittings and seals are secured (as per manufacturer instructions), including connections to the nitrogen gas regulator. The high pressure gas regulator used should be capable of delivering nitrogen up to 500 psi. The gas line is opened and regulator pressure slowly increased until the extrudate begins to emerge from the exit port (Fig. 26.3). Liposomes are collected in a clean sterile vial. See troubleshooting problem 2. At this stage the liposomes are polydisperse. The extrusion procedure is thus repeated ten times.
4. The liposomes collected from the final extrusion cycle require some form of purification, the method of purification will vary depending on the payload. The most common methods of purification include size exclusion chromatography (SEC, gel-filtration), dialysis and ultracentrifugation. For payloads less than 3,000 Kda, gel filtration on a Sephadex G-25 column (PD-10, Sigma) offers a convenient and rapid method. For larger payloads, a choice of other gel matrices are available including Sepharose CL- 2B, 4B and 6B. The buffers used to equilibrate the SEC column should be isotonic. When purified by SEC, liposomes are collected in the fraction that elutes close to the column void volume. This fraction is easily noticed as it appears cloudy due to the presence of liposomes. The blank liposomes prepared above, after purification through G-25 column, using the PBS pH 7.4, are fairly homogenous with mean diameter around 120 nm and polydispersity value below 0.1 as measured by Dynamic Light Scattering (DLS).

26.3.2 Selection of Dye Pair

We will now prepare liposomes which encapsulate two fluorescent markers and find out how such preparations can be used to study payload release in vivo. For this purpose, we will make use of both the liposome compartments for encapsulation, with one dye introduced into the lipid bilayer and a second dye incorporated into the lumen (Fig. 26.4). The advantage of this is that the lipid bound dye becomes non-exchangeable with surrounding media while the lumen dye can be released. The lipophilic tracer, 1,1′-dioctadecyl-5,5′-diphenyl-3,3,3′,3′-tetramethyl indocarbocyanine (DiI) has E_x 576 nm and E_m 599 nm and is a suitable tracking dye, as it exhibits high fluorescence only when in association with the bilayer. Calcein, on the other hand, is a hydrophilic dye suitable for encapsulation in the aqueous lumen compartment. Its fluorescence is relatively insensitive to pH near the physiological value (pH 7.4 ± 1.0 unit). In addition, the fluorescence is highly dependent on concentration (Fig. 26.5), self-quenching above ~20 μM. Liposomes prepared with high (mM) Calcein concentration will thus have minimum fluorescence as it will be quenched inside the lumen. When liposomes leak, the released Calcein is diluted into surrounding media and quenching is relieved to produce an increase in fluorescence intensity; hence this dye is well suited to studying payload release. By knowing the relative fluorescence of the Calcein (green) and DiI (red), it is possible to track the

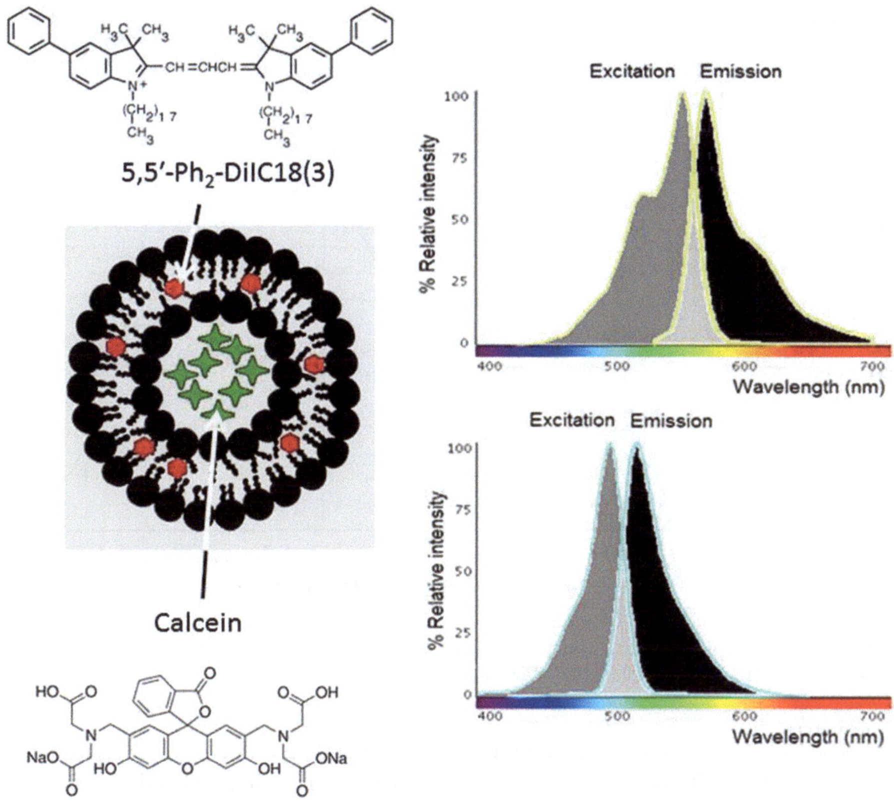

Fig. 26.4 Schematic diagram of the unilamellar liposome encapsulating Calcein in the lumen and DiI dye in the lipid bilayer. The excitation and emission spectra of the two fluorophores, generated with spectraview (*invitrogen*™), are shown on the right. The emission spectra for the dyes are sufficiently resolved to be monitored as green and red emission intensities in two channels using epifluorescence microscopy

liposomes in tissues and know their fate within the microvasculature, with respect to accumulation and payload release, monitored in vivo. A multiband-band filter (Chroma Technology corporation) for DAPI/FITC/Texas Red is ideally suited to collect and resolve green and red fluorescence signal for detection by epifluorescence microscopy.

26.3.3 Encapsulation of Dual Payload

The method is essentially the same as for blank liposomes with the following changes.

(a) The lipid film is prepared as before except that 50 μg of DiI 5,5′-Ph2-DiIC18(3), dissolved in 0.1 ml of ethanol is also added to the PC/Cholesterol (dissolved in 1:1 Chloroform/Methanol) mixture.

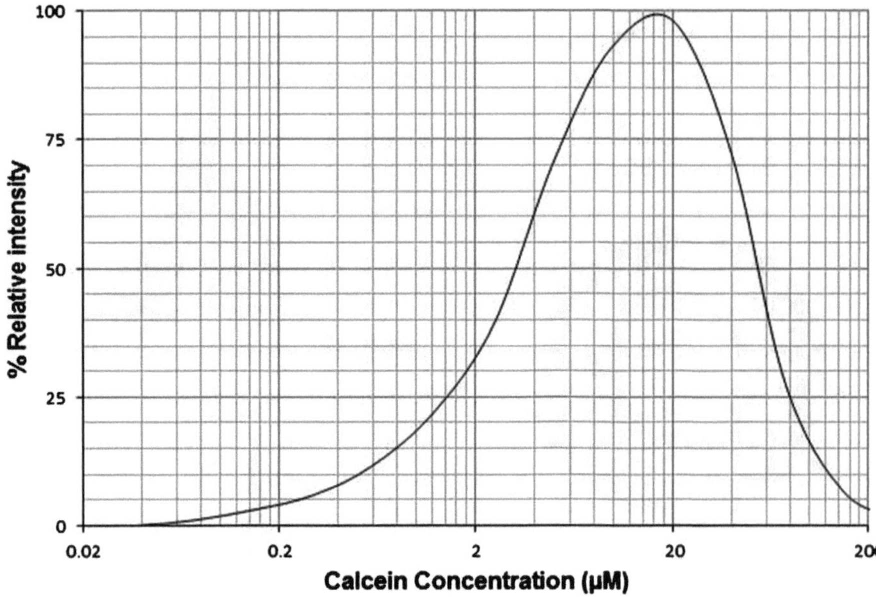

Fig. 26.5 Calcein fluorescence quenching curve. Fluorescence intensity was recorded using an excitation and emission wavelengths of 485 nm and 520 nm respectively

(b) The hydration solution used is prepared by suspending Calcein (0.748 g) in 5.0 ml of 20 mM NaCl, 10 mM sodium phosphate buffer pH 7.4. The pH of the suspension is adjusted by adding 5M NaOH with constant mixing, until the Calcein has completely dissolved. The pH is then readjusted to 7.4 using 1N HCl (see trouble-shooting problem 3) and the volume made up to 10.0 ml using 20 mM NaCl, 10 mM sodium phosphate buffer pH 7.4. A 2.0 ml aliquot of this solution is used to hydrate the lipid film as before. It is recommended to wrap the flask in metal foil while hydrating the lipid film to avoid photodegradation.

(c) Any non-encapsulated dyes are removed by gel-filtration on a PD-10 column in 0.5 ml lots. Alternatively a larger column (2 × 1.4 cm) packed with G-25 or Sepharose CL-4B can be used for a single run purification. Collect 0.5 ml fractions and pool those that emerge first and are orange and cloudy. Green colouration is an indication of contamination from free Calcein (see trouble-shooting problem 4).

26.4 Characterisation

The liposomes can be analysed by a number of techniques [6, 7]. At a minimum it is essential to know the lipid concentration, size of the liposomes, and assurance that the payloads have been encapsulated.

26.4.1 Lipid Assay

Several methods to measure the total amount of lipid are available [7]. Those methods which measure inorganic phosphate are incompatible with phosphate buffers. Stewart assay [8] uses ammonium ferrothiocyanate reagent (red) to bind with the phospholipid head groups in an emulsion of chloroform and water. The absorbance of the chloroform extracted lipid in complexation with the red reagent is then taken as proportional to the amount of lipid present. The extent of phospholipid binding to the reagent is different for different lipids, therefore a reference sample or a calibration curve is always necessary.

Ammonium ferrothiocyanate (0.1 M) reagent is prepared by dissolving ferric chloride hexahydrate (0.27 g) and ammonium thiocyanate (0.34 g) in 10 ml of deionised water. A standard solution of L-α-Phosphatidylcholine (2.0 mg/ml) in chloroform is prepared. In a microfuge tube (2.5 ml capacity) 1.0 ml of the ammonium ferrothiocyante reagent and 1.0 ml of chloroform are added. To this mixture is added an aliquot (10 μL) of test sample and the mixture vortexed for half a minute. The emulsion is then centrifuged at 1,000 r.p.m for 5 min. The absorbance of the lower layer (see troubleshooting problem 5) is recorded using a 1.0 ml glass cuvette in a spectrophotometer set at 485 nm wavelength. The concentration of phospholipid is calculated by comparing against the standard sample tested in the same way. The dual payload liposomes made above, gave 10.2 mg/ml lipid concentration. For further characterisation (described below) this can be diluted with PBS buffer to 3.0 mg/ml.

26.4.2 Size Distribution

The two key parameters for characterising liposome size are determination of the mean diameter and size distribution. To do this readily on a native sample, in colloidal form, photon correlation spectroscopy (PCS) based on dynamic light scattering (DLS), is most commonly used. While this technique is very sensitive and provides rapid results, cleanliness in sample preparation is of utmost importance, as the method of analysis is biased towards larger particle size.

Take a clean glass cuvette and wash it thoroughly with deionised water that has been passed through a 0.22 μm filter. Use clean ethanol to rinse further and dry by purging with nitrogen gas via an inline filter. Examine the cuvette to ensure it is clean and dust free.

Using a syringe filter, pass 2.0 ml of PBS buffer through a 0.22 μm filter (prewashed) and collect directly into the cleaned cuvette. Add a small aliquot (e.g. 10 μL) of liposomes and mix the contents. Place the cuvette into the cell holder (Malvern 4700 Zetasizer) thermostated at 25 centigrade and leave to equilibrate for 5 min. Perform the particle size analysis averaging at least ten runs. See troubleshooting problem 6. An example of the size distribution profile

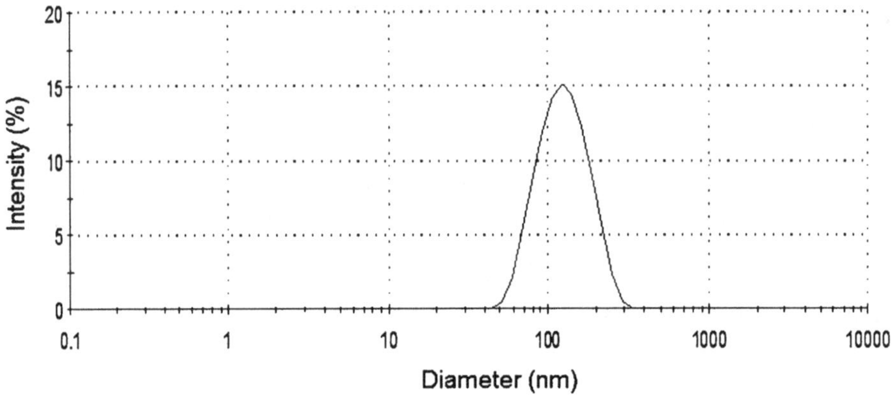

Fig. 26.6 Particle size distribution of liposomes as determined by dynamic light scattering analysis. The liposomes were found to have a mean size of 118 nm and PDI of 0.110

obtained for the dual payload liposomes shows (Fig. 26.6) a mean diameter of 118 nm with polydispersity index (PDI) value of 0.11. For an acceptable homogenous sample having narrow size distribution, PDI values should ideally not exceed the value of 0.15.

26.4.3 Encapsulation Dyes

To check if the DiI has been successfully incorporated into the liposomes, a fluorescence emission spectrum is recorded. Add 3.0 μl aliquot of liposomes to 2.0 mls of PBS pH 7.4 buffer in a clear, four-sided cuvette. Scan fluorescence emission from 580 to 680 nm using an excitation wavelength of 576 nm. An example of fluorescence emission spectrum with emission λ_{max} around 605 nm is observed (Fig. 26.7).

In order to check if the Calcein dye is encapsulated and can be released, a time course fluorescence assay is carried out. Spectrofluorometer (Perkin Elmer LS50B) is set at excitation and emission wavelength of 490 and 520 nm. A 3 μl aliquot of liposomes is added to 2 ml of PBS pH 7.4 buffer and the fluorescence intensity continually recorded until a stable baseline of low intensity (calcein being largely quenched) is observed. After a short period, the cuvette is quickly taken out and 10 μl of Triton X-100 (10 %) added. The contents are very quickly mixed and cuvette placed back in the cell holder to continue recording. Triton X-100 will cause immediate lysis of liposomes releasing the entrapped Calcein, which should become evident from a rapid rise (Fig. 26.8) in fluorescence intensity (see troubleshooting problem 7) as dequenching takes place. A good preparation will give greater than ten fold increase in fluorescence intensity upon 100 % lysis. By recording an absorbance value for the lysed solution on a spectrophotometer, at 495 nm wavelength and using an extinction coefficient of 80 mM^{-1} cm^{-1}, the actual amount of calcein in the liposome preparation can be calculated.

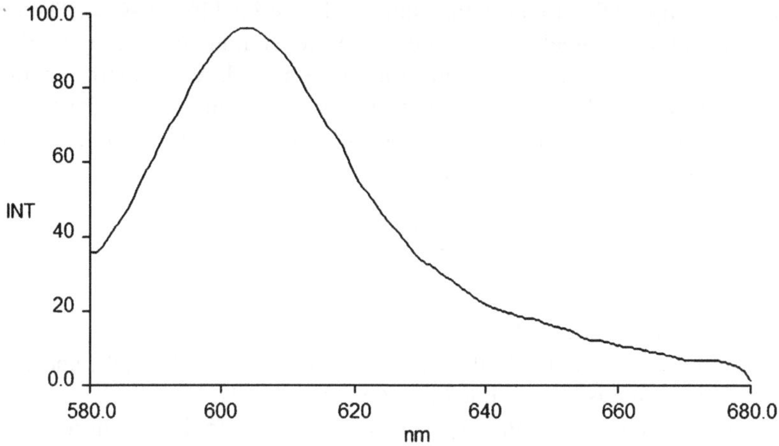

Fig. 26.7 Fluorescence emission spectrum of the liposomes containing the 5,5′-Ph2-DiIC18(3) red dye embedded in the lipid bilayer. Excitation wavelength was 576 nm

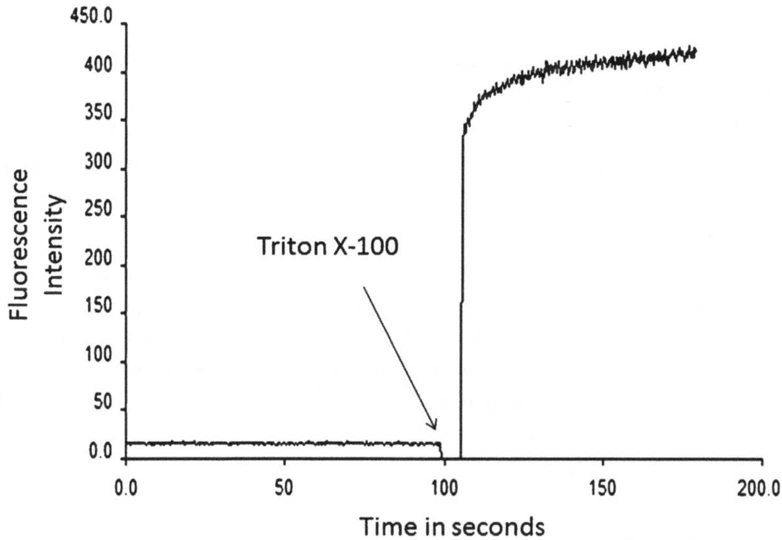

Fig. 26.8 Time course assay showing release of Calcein from liposomes. Fluorescence emission of liposome solution was recorded continually over 200 sec in a spectrofluorometer (Perkin Elmer LS50B) using an excitation and emission wavelengths of 485 and 520 nm. At approximately 100 sec (indicated by *arrow*) Triton X-100 was added to release the encapsulated Calcein

26.4.4 Judging the Efficiency of the Encapsulation Process

To judge the effectiveness of the encapsulation process, it is possible to calculate an expected level of the payload entrapped within the maximum possible entrapment volume and compare this to the actual amount determined by an assay. The internal volume (*VI*) of the unilamellar liposomes varies depending on the bilayer thickness, which in turn is dependent on the lipid composition used. As an approximation, the following equation will give a guideline indication of the encapsulation volume within the lumen, based on the assumption that a unilamellar vesicle is a perfect sphere of radius (r) with a bilayer thickness b.

$$V_i = \frac{4}{3}\pi(r-b)^3$$

The value of b is variable [9] although 4.1×10^{-3} cm can be used as an approximation. For the liposomes made above, with a diameter of 118 nm, the encapsulated volume per liposome is thus estimated as 6.93×10^{-16} mls.

The total entrapment volume V_t in mls at a given lipid concentration can be estimated:

$$V_t = \frac{(moles\ of\ lipid)(6.23\times10^{24})V_i}{(D-50)^2}$$

Where D is the outer diameter of liposomes in angstroms (Å).

In this example V_t=0.03 mls.

The amount of payload entrapped in this volume provides an indication of the maximum theoretical payload in the lumen. This can then be expressed as payload/ µmole of lipid.

Worked example

The liposomes of 118 nm diameter containing Calcein were diluted with PBS pH 7.4 to 3 mg/ml lipid solution. Calculate the expected payload per µmole of lipid.

V_i=6.93×10^{-16} mls, V_t=0.03 ml. As the concentration of Calcein in the hydration buffer was 120 mM. This equates to 0.91 µmoles Calcein/µmole lipid. This value is then compared to experimentally found value (Sect. 26.4.3) to assess if the encapsulation has been successfully achieved.

26.5 Application and Discussion

For animal studies the liposomes need to be sterile. Chemical and physical sterilization methods have limited value as they usually compromise liposome stability. Fortunately, for unilamellar liposomes under 200 nm diameter, the extrusion process is well suited for aseptic production. The entire method can be carried out in a laminar flow cabinet as the extruder devices and the polycarbonate membranes are designed to withstand autoclaving. Purification by gel-filtration can also be performed in a laminar flow cabinet. The sterility of liposomes produced by extrusion through

0.1 μm under laminar flow conditions can be further ensured by passing the preparation through a 0.22 μm sterile retentive filter. It is worth noticing that there will still remain a risk of smaller sized contaminants originating from any of the raw materials if they were not sterilised prior to extrusion.

The trafficking of liposomes into tumour interstitium by extravasation through the small pores in the endothelium has been studied in live tissues and in real-time by employing intravital fluorescence microscopy [10]. The technique has been particularly useful to understand how tumour microvasculature can be exploited for nanoparticle drug delivery [11–16] systems. A transparent window, implanted in the dorsal skin-fold and positioned over the tumour tissue, is imaged through a microscope objective. As an example, liposomes (100 μl) with dual dye payload are intravenously administered, via tail vein, to a tumour bearing mouse under the influence of anaesthetic. Time-lapsed epifluorescence fluorescence images are recorded, as described elsewhere [16] using a Texas Red filter set for DiI and fluorescein filter set for Calcein intensity measurements. By analysing signal intensities from such images as a function of time, the accumulation of liposomes can be followed based on DiI fluorescence while the release of contents (Calcein) computed using the equation below:

$$Normalised\,release = \frac{I_{cal}(t)/I_{cal}(t_0)}{I_{DiI}(t)/I_{DiI}(t_0)}$$

where I_{cal} and I_{DiI} are the calcein and DiI fluorescence intensities respectively, at a given time point. The DiI intensity represents the combined effects of liposome accumulation and clearance hence reflecting the proportion of liposomes at any particular point in time. Immediately following the injection, vascular filling of the tissue takes place, giving a rapid rise in both the Calcein and DiI intensities (Fig. 26.9a). Provided liposomes are sufficiently stable, up to this point in time, the intensities can be normalised and ratio set to one unit. Any increase in the ratio, following this initial phase, can then be taken to reflect release of liposome contents. Eventually, the ratio will decline well below one unit as contents are washed away. Normalised dual fluorescence ratiometric kinetic plots thus provides a handle on payload release within the tumour microvasculature. Figure 26.9b shows one such typical profile when fluorescence intensities were recorded for both the Calcein and the DiI signal. In this example there appears to be small gradual increase in ratio over time reflecting liposome destablisation. Using such plots release kinetics could, for instance, be quantitatively followed before and after a therapeutic intervention or application of a stimulus to assess the efficiency of a nanoparticle drug delivery system within microvasculature. It should be noted however that variables in local physiology and extracellular matrix content do affect microvasculature, with significant impact on vascular permeability. This will also influence liposome extravasation and retention, causing tissue to tissue variations in the observed release kinetics.

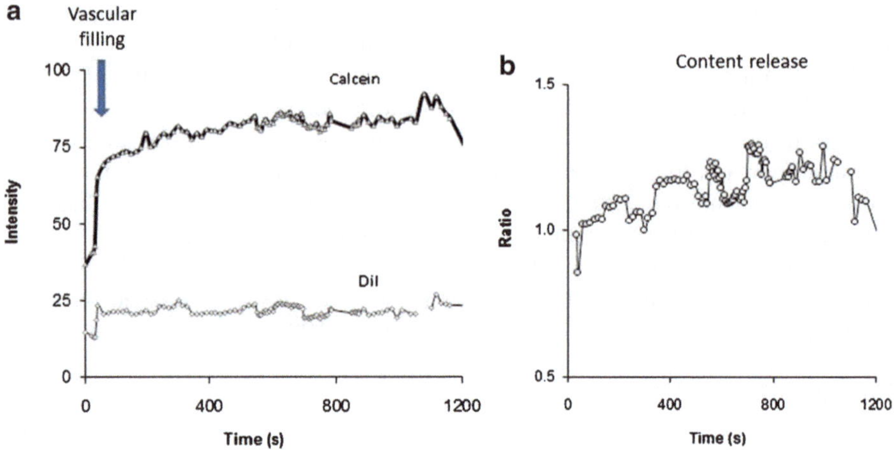

Fig. 26.9 Real-time monitoring of fluorescence intensities for each dye, following a tail vein injection of liposomes to a tumour-bearing mouse. RIF-1 (murine fibrosarcoma) tumour allograft in the window chamber was viewed with epifluorescence microscopy. (**a**) Intensity verses time for Calcein and DiI shows initial rise, indicated by an *arrow*, which is taken to represent vascular filling. (**b**) Ratiometric plot reflects changes in Calcein intensity not associated with liposome clearance or accumulation

26.5.1 Other Payloads

The method of liposome preparation described here is generally applicable to encapsulate a variety of payloads and may be extrapolated to other lipid compositions. For instance we have routinely used the method above to encapsulate the following payloads: Ciprofloxacin, Alkaline Phosphatase, Asparaginase and Glucose Oxidase Given below are conditions of encapsulation for some of these formulations.

Payload	Lipid composition	Hydration buffer	Purification & payload assay
Ciprofloxacin	PHPC: cholesterol	Ciporofloxicin.HCl in 10 mM sodium phosphate plus 20 mM NaCl pH 7.4 at 60 °C	Purify on PD-10 column
	2: 1 molar ratio (PHPC = Partially hydrogenated egg PC)		Assay: fluorescence dequenching
Alkaline phosphatase	PC: cholesterol	Alkaline phosphatase in 10 mM Tris–HCl pH 7.1	Purify: sepharose CL-6B.
	4: 1 molar ratio		Assay: colourimteric

(continued)

Payload	Lipid composition	Hydration buffer	Purification & payload assay
Biotinylated glucose oxidase	PC (40 mg), cholesterol (11 mg), dicetylphosphate (3.0 mg) and dipalmitoyl phosphatidylethanolamine-LC-biotin (DPPE-LC-biotin; 0.272 mg)	Glucose oxidase in 10 mM Tris–HCl pH 7.1	Purify: sepharose CL-6B
			Assay: colourimetric
Asparaginase	Egg yolk lecithin: cholesterol	Asparginase in 10 mM Tris–HCl pH 8.	Purify: G-200 column
	2: 1 molar ratio		Assay: conductometric

26.6 Troubleshooting

Problem	Remedy
1. The lipid film does not completely dislodge	The time taken to produce a good suspension is dependent on several factors such as lipid composition, surface area, shaking vigour and temperature. The dislodging of lipid film can be assisted by freeze-thawing the suspension in between shaking cycles. Doing this may also increase the encapsulation efficiency.
2. Extrudate emerges at a very slow rate	Check that the transition phase of the lipid being used is below the room temperature. If not then increase the temperature by connecting the thermo-barrel to a circulating bath set at higher temperature than the Phase transition.
	Blockage of polycarbonate membranes occurs due to large particulate matter in the suspension or when the lipid concentration is too high and the suspension viscous. Using course filters with larger pore size (0.4 μm) for the first extrusion cycle will remedy this.
3. Cannot achieve dissolution of Calcein	Solubility of the calcein is highly pH and salt form dependent. Use a magnetic flea to vigorously stir the solution while adding NaOH or HCL. Regularly check pH in between additions to keep it on the alkaline side.
4. The purified liposomes are contaminated by large amount of free Calcein	Liposomes can rupture though an osmotic shock. It is thus essential to use buffer of the same strength as the hydration mixture to equilibrate the SEC column.
	Column overloading can be a problem. If you are using PD-10 column, reducing the loading volume to 0.5 ml can significantly improve purity.
5. Stewart assay is reporting very high readings	The liposome concentration of the test sample should be lowered by diluting in PBS buffer and the analysis repeated.
	The spectrophotometer should be blanked against a sample prepared in the same manner but without any lipid.
	The lower layer (chloroform), after centrifugation step, needs to be carefully removed without contamination from the upper aqueous layer, which contains bulk of the red reagent. One way to do this is to lower a glass Pasteur pipette into the lower and expel a small bubble to empty the tip before sucking up the lower chloroform layer. Transfer this to a 1.0 ml glass cuvette and stopper to prevent evaporation.

(continued)

Problem	Remedy
6. Inconsistent mean diameter and PDI values between samples	Check the quality of filtered buffer in the cuvette before adding liposomes. Ideally the Kilicounts per second (Kcps) should be less than 50 for buffer alone. In the presence of liposomes it should ideally be above 300. Adjust the amount of sample to get a good count rate. If there are dust particles in the original liposome sample this will grossly affect results. To avoid this problem make sure liposome extrudate are collected in a dust free vial. Performing sample preparation in a laminar flow cabinet will help minimise contamination.
7. Background (baseline) to signal (triton X-100 lysed) ratio is low	This usually indicates poor encapsulation or unstable liposomes. A gradual rise in baseline fluorescence intensity indicates leaky liposomes. Steady baseline but not sufficient rise in signal intensity, when triton is added, indicates poor encapsulation. High background can also arise from poorly purified liposomes, in which case re-purification is recommended.

Finally a note to say that the design of liposome composition requires careful study to suit an application and thus factors such as the charge, size and lipid type will all influence its properties. Rational design approaches, based on careful selection of lipid compositions and controlling bilayer permeability with additional additives, are at the leading edge [17] of solving problems with undue drug retention once liposomes are administered in vivo. The future designs incorporates a feature of destabilising the lipid bilayer, using either a disease specific intrinsic trigger or externally applied localised stimuli to provide controlled drug release based on demand.

References

1. Maeda H, Wu J, Sawa T, Matsumura Y, Hori K (2000) Tumor vascular permeability and the EPR effect in macromolecular therapeutics: a review. J Control Release 65:271–284
2. Ruiz-Esparza GU, Flores-Arredondo JH, Segura-Ibarra V, Torre-Amione G, Ferrari M, Blanco E, Serda RE (2013) The physiology of cardiovascular disease and innovative liposomal platforms for therapy. Int J Nanomedicine 8:629–640
3. Dvir T, Bauer M, Schroeder A, Tsui JH, Anderson DG, Langer R, Liao R, Kohane DS (2011) Nanoparticles for targeting the infarcted heart. Nano Lett 11:4411–4414
4. Hermanson GT (2013) Bioconjugate techniques, 3rd ed Chapter 21. Liposome conjugates and derivatives. Academic Press, pp 921–949
5. Torchilin VP, Weissig V (2003) Liposomes: practical approach, 2nd edn. Oxford University Press, Oxford
6. Edwards KA, Baeumner AJ (2006) Analysis of liposomes. Talanta 68:1432–1441
7. New RRC (1990) Liposomes: a practical approach. IRL Press, Oxford
8. Stewart JC (1980) Colorimetric determination of phospholipids with ammonium ferrothiocyanate. Anal Biochem 104:10–14
9. Xu X, Khanb MA, Burgessa DJ (2012) Predicting hydrophilic drug encapsulation inside unilamellar liposomes. Int J Pharm 423:410–418
10. Sjoerd H, Reitan NK, Haraldseth O, de Lange Davies C (2010) Intravital microscopy in window chambers: a unique tool to study tumor angiogenesis and delivery of nanoparticles. Angiogenesis 13:113–130
11. Wu NZ, Braun RD, Gaber MH, Lin GM, Ong ET, Shan S, Papahadjopoulos D, Dewhirst MW (1997) Simultaneous measurement of liposome extravasation and content release in tumors. Microcirculation 4:83–101

12. Laschke MW, Vollmar B, Menger MD (2011) The dorsal skinfold chamber: window into the dynamic interaction of biomaterials with their surrounding host tissue. Eur Cell Mater 22:147–164
13. Djanashvili K, ten Hagen TLM, Blangé R, Schipper D, Peters JA, Koning GA (2011) Development of a liposomal delivery system for temperature-triggered release of a tumor targeting agent, Ln(III)-DOTA-phenylboronate. Bioorg Med Chem 19:112–1130
14. Pink DBS, Schulte W, Parseghian MH, Zijlstra A, Lewis JD (2012) Real-time visualization and quantitation of vascular permeability in vivo: implications for drug delivery. PLoS One 7:e33760. doi:10.1371/journal.pone.0033760
15. Gaber MH, Wu NZ, Hong K, Huang SK, Dewhirst MW, Papahadjopoulos D (1996) Thermosensitive liposomes: extravasation and release of contents in tumor microvascular networks. Int J Radiat Oncol Biol Phys 36:1177–1187
16. Offerman SC, Verma AK, Telfer BA, Berk DA, Clarke DJ, Aojula HS (2014) Ability of co-administered peptide liposome nanoparticles to exploit tumour acidity for drug delivery. RSC Adv 4:10779–10790
17. Oude BE, Mastrobattista E, Schiffelers RM (2013) Strategies for triggered drug release from tumor targeted liposomes. Expert Opin Drug Deliv 10:1399–1410

Chapter 27
Vascular Flow Modelling Using Computational Fluid Dynamics

Amir Keshmiri and Kirstie Andrews

27.1 Computational Fluid Dynamics

The equations governing fluid flows are a set of coupled, non-linear partial differential equations such as:

27.1.1 Continuity

$$\frac{\partial \rho}{\partial t} + \frac{\partial \rho U_i}{\partial x_i} = 0 \tag{27.1}$$

27.1.2 Momentum

$$\frac{\partial \rho U_i}{\partial t} + \frac{\partial \rho U_i U_j}{\partial x_j} = \frac{\partial P}{\partial x_j} + \frac{\partial}{\partial x_j}\left(\mu \frac{\partial U_i}{\partial x_j}\right) \tag{27.2}$$

A. Keshmiri (✉)
School of Engineering, Manchester Metropolitan University, Manchester M1 5GD, UK

School of Mechanical, Aerospace and Civil Engineering, The University of Manchester, Manchester M13 9PL, UK
e-mail: a.keshmiri@mmu.ac.uk

K. Andrews
Division of Mechanical Engineering, School of Engineering, Manchester Metropolitan University, Manchester M1 5GD, UK
e-mail: k.andrews@mmu.ac.uk

M. Slevin, G. McDowell (eds.), *Handbook of Vascular Biology Techniques*,
DOI 10.1007/978-94-017-9716-0_27

These equations are known as Navier–Stokes equations. Many real problems include additional terms and/or equations, governing heat transfer, chemical species, etc. Analytical solutions are known only for a few very simple flow cases. An alternative is to solve the governing equations numerically, on a computer. Computational Fluid Dynamics (CFD) is this process of obtaining numerical approximations to the solution of the governing fluid flow equations.

One could view CFD as *a numerical experiment*. In a typical fluids experiment, an experimental model has to be built and the flow interacting with that model needs to be measured using various measurement devices, and the results are then analysed. In CFD, the building of the model is replaced with the formulation of the governing equations and the development of the numerical algorithm. The process of obtaining measurements is replaced with running an algorithm on the computer to simulate the flow interaction. The analysis of the results is, however, the same for both techniques. There are several unique advantages of CFD over experiment-based approaches:

- **Relatively low cost:** Using physical experiments and tests to get essential engineering data for design can be expensive. CFD simulations are relatively inexpensive, and costs are likely to decrease as computers become more powerful.
- **Speed:** CFD simulations can be executed in a short period of time. Quick turnaround means engineering data can be introduced early in the design process.
- **Ability to simulate real conditions:** Many flow and heat transfer processes cannot be (easily) tested, e.g. hypersonic flow. CFD provides the ability to theoretically simulate any physical condition.
- **Ability to simulate ideal conditions:** CFD allows great control over the physical process, and provides the ability to isolate specific phenomena for study. For example extreme pressures or temperatures can easily be simulated.
- **Comprehensive information:** Experiments only permit data to be extracted at a limited number of locations in the system (e.g. pressure and temperature probes, heat flux gauges, etc.). CFD allows the analyst to examine a large number of locations in the region of interest, and yields a comprehensive set of flow parameters for examination.

However, despite its many advantages, CFD does not remove the need for experiments; numerical models need to be validated to ensure they produce reliable and accurate results. Particularly in clinical applications, there is a pressing need for rigorous model validation against detailed laboratory data. With the growth of available computing power and the advent of powerful user-friendly graphical user interfaces and automated options/features in commercial CFD codes, it has become possible for a wide range of users to apply CFD to even very complex flowfields, giving detailed information about the velocity field, pressure, temperature, etc. This sometimes results in producing solutions that are haemodynamically irrelevant and fail to capture even the most basic flow features. The key to successful use of CFD is an understanding of where the errors come from; their implications, and how to ensure they are small enough to be acceptable in a particular application.

27.2 Applications of CFD

27.2.1 Disease Research

In recent years, advances in vascular biology, biomechanics, medical imaging and computational techniques including CFD have provided the research community with a unique opportunity to analyse the progression of vascular diseases from a new angle and to improve the design of medical devices and develop new strategies for intervention. The increasing power-to-cost ratio of computers and the advent of methods for subject-specific modelling of cardiovascular mechanics have made the CFD-based modelling sometimes even more reliable than methods based solely on in vivo measurement.

Numerical simulations have played an important role in understanding the haemodynamics of several different areas including bypass grafting, cardiovascular treatment planning, cerebrovascular flow, the effects of exercise on aortic flow conditions, congenital heart disease and coronary stents. The patient-specific modelling of cardiovascular mechanics, however, have focused mainly on haemodynamic factors in atherosclerotic and aneurysmal disease.

Atherosclerosis, is the most widespread of the acquired cardiovascular diseases and affects various arteries supplying blood to brain, heart and other vital organs. It is a very focal disease, mainly affecting branches and bends of the arterial tree. There is currently extensive and increasing evidence, correlating the localisation of atherosclerosis and the patchiness of this disease to different local haemodynamic metrics [1] (see Sect. 27.4, below). Currently the majority of the research with an aim of correlating the haemodynamic parameters to the formation of atherosclerosis are based on the coronary arteries and the carotid bifurcation models of human, porcine, rabbit and murine [2]. Quantification of haemodynamic metrics in rodent models have recently become more feasible with the advent of small-animal imaging technology, advanced image-based modelling techniques and genetic manipulation. These models are particularly useful for studying atherosclerosis since the wall shear stress in the aorta of a mouse is more than 20-fold higher than that in humans [3].

In addition to atherosclerosis, aneurysm is another important disease which has been studied extensively in the literature. An aneurysm is defined as a focal dilatation of the arterial wall. Although the pathogenesis of aneurysms remains an enigma, the initial dilatation appears to be caused in part by degeneration of a portion of the arterial wall. There are two main types of aneurysms, namely Intracranial Aneurysms (IAs) and Abdominal Aortic Aneurysms (AAAs). IAs mostly occur at the circle of Willis, the major network of arteries that supplies blood to the brain. The major complication of IAs is their rupture, which causes subarachnoid haemorrhage. AAAs occur in the infrarenal aorta, the primary conduit that supplies blood to the legs. Similar to studying atherosclerosis, image-based computational modelling using CFD has recently been used extensively to calculate the haemodynamic forces and to correlate them to initiation, growth and rupture of aneurysms. Recent studies using CFD have shown the importance of haemodynamic metrics in understanding cerebral aneurysm rupture [4, 5]. Amongst several haemodynamic parameters calculated by CFD, Miura et al. [6] found the wall shear stress to be the most reliable parameter characterising the rupture status of middle cerebral artery. The review by Humphrey and Taylor [7] provides a thorough review of the role of computational mechanics in both types of aneurysms.

27.2.2 Predictive Medicine

In treating cardiovascular problems, interventional and surgical therapies are critical for restoring blood flow to affected organs and tissues. However, adverse and side effects from these therapies have always been a limitation. In addition, using a "trial-and-error" approach to surgical design is obviously not an option which limits testing alternate treatments in the patients. The recent developments in patient-specific computer simulations have provided a means to assess new surgeries and interventions at no risk to the patient. Also similar to other engineering fields such as aerospace and automotive, design optimisation is now possible and can be applied to predictive tools and methods to optimise surgeries for individual patients. Therefore, in the new paradigm of predictive medicine, the surgeons for example may use advanced imaging tools along with computational techniques such as CFD to create a patient-specific model and predict the outcome of a particular treatment for an individual patient. However, to be effective and attractive to the medical community, these simulation-based medical planning systems must be quick and efficient and should require minimum user intervention. They should also accurately calculate relevant haemodynamic variables, especially flow rate and pressure [8]. Such systems would require a number of key stages, which are briefly discussed in Sect. 27.3.

Another new and active area of research in which CFD and computational haemodynamic can play an important role is the design and clinical use of *artificial blood vessels*. While it is considered as a long-standing therapy, with the growth of multi-disciplinary fields such as biomaterials and tissue engineering, researchers are looking to provide clinicians and patients with enhanced options when repairing or replacing these structures. The artificial tissue analogues may be polymeric tubes (similar to the current "gold standards" of Dacron (i.e. polyethylene terephthalate, PET) or expanded polytetrafluoroethylene – ePTFE) or tissue engineered vessels ((degradable) polymeric tubes seeded with layers of vascular cells designed to regenerate into a fully formed structure that will mimic the natural healthy vessel) [9, 10]; whichever is chosen, the likely success of the replacement vessel needs to be analysed.

CFD has recently shown potential to provide researchers with valuable information regarding the forces, stresses and haemodynamic factors that will affect the specific anatomical location of the artificial vessel, for example the wall shear stress. It is critical to have this information as these variables will not only affect the material of the vessel wall, but also the contacting cells. As discussed in Chap. 25, the endothelialisation of artificial vessels is a key component of finding a bioengineered solution to the vascular problem of damaged and diseased blood vessels. As these cells form a critical layer between the circulating blood and the artificial underlying material layer(s), the haemodynamic factors they are exposed to, will directly induce cell responses ranging from preferential adaptation and alignment to inflammatory responses and detachment. The endothelial cells attachment and responses can be controlled to an extent through the alteration of the underlying artificial material structures; therefore, if the full environment of the artificial vessel, in its specific location, is understood beforehand then the forces, stresses and haemodynamic factors present can be factored into the design. The vessel substitutes not only need to withstand the applied forces/stresses but also need to produce and replicate the behaviour of natural healthy vessels as closely as possible to prevent anastomosis mismatch, subsequent cell loss, and poor haemodynamic performance [11, 12]. Hence, CFD provides a tool with vast potential for improving clinical therapeutic solutions, and is an area that, if more widely incorporated, can significantly contribute to the knowledge and development of artificial blood vessels. (Further details regarding the potential to directly affect the endothelialisation through the underlying material structural properties are discussed in Chap. 25.)

27.3 Computational Procedure

A typical CFD simulation consists of the following tasks:

1. Problem Identification
2. Pre-processing
3. Solving
4. Post-processing

These tasks are discussed below in more detail.

27.3.1 Task 1: Problem Identification

The first step in any CFD simulation is to define the modelling goals and subsequently identify the numerical domain that needs to be modelled.

In defining the modelling goal, the following points need to be considered:

- Decide if CFD is an appropriate tool
- Choose the modelling options

- The physical models needed for analysis (i.e. turbulence, heat transfer, compressibility, etc.)
- The simplification assumptions (i.e. steady-state, symmetry, periodicity, etc.)
- Unique modelling capability (i.e. Graphical User Interface, special features of a code, etc.)

- Choose the degree of accuracy
- Choose how quickly the results are needed

When identifying the flow domain, the following need to be taken into consideration:

- Decide how to choose the computational domain
  - Known boundary conditions and other input data
  - Extension of domain to enable comparison with data
- Decide if you can simplify the domain (e.g. 1D, 2D, etc.)

27.3.2 Task 2: Pre-processing

The pre-processing stage is normally the most time-consuming task in a CFD simulation, especially in vascular flow problems. The purpose of this task is essentially to prepare the flow problem before the governing equations of the flow can be solved computationally. The pre-processing typically consists of the following two main stages:

1. Definition of the geometry of the region of interest: *Computational Domain (or Geometry)*
2. Sub-division of the domain into a number of smaller, non-overlapping sub-domains: *Mesh (or Grid)*

Each stage is explained in more detail below.

27.3.2.1 Computational Domain

The computational domain which is usually based on a CAD (Computer-Aided Design) file, is a computer model to represent the fluid domain. In cardiovascular problems, there are different invasive and non-invasive techniques to acquire patient-specific anatomic and physiologic data. Common invasive methods include a combination of angiography and intravascular ultrasound (IVUS). Non-invasive imaging is usually carried out by Computerised Tomography (CT), Magnetic Resonance Imaging (MRI) and 3-Dimensional Ultrasound (3DUS). One of the advantages of MRI over other techniques is its ability to measure physiologic parameters such as blood flow and wall motion [8]. However, there are two major limitations in the computational imaging techniques which are namely

Table 27.1 List of current biomedical modelling codes (Adapted from [17])

		Capabilities		
Software	Ref.	Segmentation	Rendering	Simulation
ImageJ	[14]	✓	✓	×
SimVascular	[15]	✓	✓	✓
VMTK	[16]	✓	✓	✓

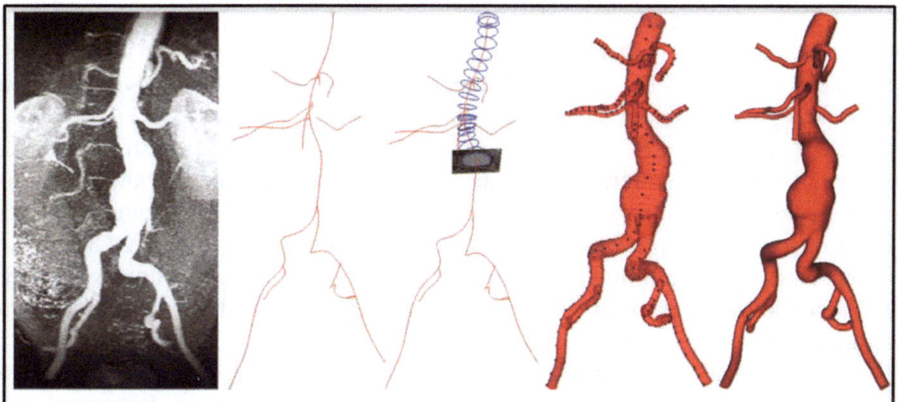

Fig. 27.1 Steps involved in creating a solid model from magnetic resonance angiography (MRA) for an abdominal aortic aneurysms (Adapted from [18])

the relatively low resolution of current imaging tools and geometrical variations during a cardiac cycle [13].

Following the acquisition of patient specific anatomic data, the computational domain can be created using a number of different software/codes. Table 27.1 lists three of the most common codes currently being used for creating a CAD model from medical images.

The solid model of, for example a blood vessel in these codes is typically created through the following steps [8]:

1. Obtaining a volume-rendered image of the area of interest usually by a contrast-enhanced magnetic resonance angiogram (MRA).
2. Creating centre-line paths along the vessels of interest.
3. Taking two-dimensional segmentations of vessel lumen perpendicular to the vessel path.
4. Lofting (or sweeping) the two-dimensional segmentations to form solid models for each vessel
5. Joining the different solid models of each vessel to form a complete three-dimensional solid model (which is ready to be meshed).

Figure 27.1 shows the application of the above steps for creating a typical abdominal aortic aneurysm (AAA) solid model.

27.3.2.2 Meshing

Once the computer model (or the CAD file) based on imaging data was created, the next step is to generate a mesh (or a grid). The mesh generation process consists of sub-dividing the domain into a number of smaller, non-overlapping sub-domains (known as 'cells', 'control volumes' or 'elements'). Meshing is needed since the numerical solution of the fluid flow governing equations cannot produce a continuous distribution of the variables (e.g. velocity, pressure, temperature etc.) over the whole computational domain, thus the aim instead becomes to produce a set of discrete values at a number of cells that cover the solution domain. In other words, the mesh designates the cells or elements on which the flow is solved and is a discrete representation of the geometry of the problem.

There are broadly two types of mesh: (1) structured and (2) unstructured. As shown in Fig. 27.2, structured grids are ideal for simple geometries and for simulations carried out using in-house CFD codes and therefore are rarely used in cardiovascular flow problems. Unstructured meshes (shown in Fig. 27.3) on the other hand, can accommodate completely arbitrary geometries; however, there are penalties to be paid for this flexibility, both in terms of the connectivity data structures and solution algorithms.

The mesh has a significant impact on the rate of convergence (i.e. reaching a solution), solution accuracy and the required computational time. In general, the

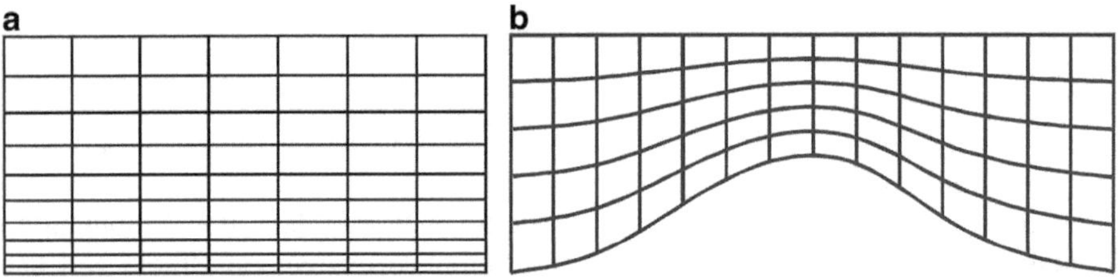

Fig. 27.2 Schematic of a typical structured mesh; (**a**) Cartesian and (**b**) Curvilinear

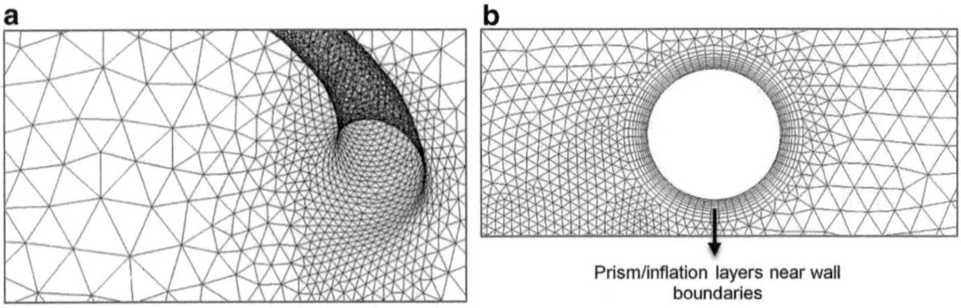

Fig. 27.3 Examples of typical unstructured grids; (**a**) tetrahedral mesh for an aorta with an intercostal branch, and (**b**) tetrahedral mesh with prism/inflation layers around a circular wall boundary

Table 27.2 List of popular mesh generation codes available in the market

Product(s) name	Developer	Mesh type
Gambit	Ansys	Finite volume
Ansys-meshing	Ansys	Finite volume
ICEM-CFD	Ansys	Finite volume
T-grid	Ansys	Finite element
STAR-CD+	CD-Adapco	Finite volume
HyperMesh	Altair	Finite element
PATRAN	MSC software	Finite element

larger the number of cells/elements, the better the solution accuracy. The number of cells in the mesh would have direct effects on the computational time and cost and therefore, the mesh resolution needs to be optimised to improve efficiency. As a general rule, the mesh needs to be finer in areas where large variations occur (e.g. near-wall regions) and coarser in regions with relatively little change. The mesh adjacent to the wall boundaries is particularly important as it needs to be fine enough to resolve the boundary layer flow. In addition, in unstructured grids, in near-wall regions, rectangular/quadrilateral cells (so called prism/inflation layers) are preferred over other shapes (see Fig. 27.3b).

It is difficult to create a comprehensive list of rules for generating an optimum mesh and hence, it is the skills and the experience of the user which play an important role. Nowadays, the majority of commercial mesh generation packages are equipped with semi-automatic mesh generation tools which enable even novice users to create acceptable grids within hours. Table 27.2 provides a list of common mesh generation programmes which are commonly used for biomedical and bioengineering CFD simulations and can create grids which are based on Finite Volume or Finite Element (explained in more detail in Sect. 27.3.3.2, below).

27.3.3 Task 3: Physical Definition and Solving

As the most crucial step in a CFD process, Task 3 is where the user is required to have expertise in a number of areas including fluid mechanics (e.g. mass and momentum conservation, incompressible flow, non-Newtonian flows, turbulence etc.), numerical methods (e.g. discretization, solution methods etc.), heat transfer (e.g. conduction, diffusion, advection etc.) and programming (e.g. C++, Fortran, Python etc.). The latter is nowadays probably not as important anymore thanks to the significant improvements made to the Graphical User Interfaces (GUI) in most CFD codes with an aim of making these tools available to a wider range of users.

Table 27.3 provides a list of popular CFD codes which are also commonly used for biomedical simulations.

One could generally divide Task 3 into the following two stages:

1. Defining the physical and boundary conditions: *Numerical Input*
2. Computing and monitoring the solution: *Solving*

Table 27.3 List of popular CFD codes currently available for biomedical CFD simulations

Product(s) name	Developer	Type
Fluent, CFX	Ansys	Commercial
STAR-CD, STAR-CCM+	CD-Adapco	Commercial
OpenFOAM	OpenCFD	Open-source
FLOW3D	Flow Science	Commercial
PHOENICS	CHAM	Commercial

Both of these stages (described further below) are carried out within the same code/package. The majority of the current commercial CFD codes make use of CAD-style graphical user interface to enable the user to import data from proprietary surface modellers and mesh generators and enter the numerical input. In addition to Navier–Stokes and energy equations, the CFD codes generally allow the user to also solve extra equations and models for special physical and chemical processes.

27.3.3.1 Numerical Input

Once the mesh is created or imported into the CFD code, the next step is to set up the physics of the flow through defining physical models, material properties, domain properties, solver settings and boundary conditions. In vascular flow modelling for instance, the latter usually encompasses most of the input information related to the patient including heart rate, blood pressure, blood flow velocity and temperature which may be obtained through invasive or/and non-invasive measurement techniques. The boundary conditions for a typical cardiovascular flow problem tend to incorporate upstream heart models and downstream microcirculation models too. The main types of boundary conditions in a typical CFD simulation include inlet, outlet, wall, symmetry[1] and cyclic (periodic).[2] Figure 27.4 shows an example of employing cyclic and symmetry boundary conditions in the simulation of a ribbed channel flow which results in the reduction of the computational domain i.e. less computational time and power required. The properties of the blood flow such as density, viscosity and temperature, etc. would also need to be defined as part of the material properties. Other numerical inputs include solver settings (e.g. numerical schemes, convergence controls, etc.), the discussion of which is beyond the scope of this chapter and the interested reader is referred to Peric and Ferziger [19] for further information.

[1] Symmetry boundary conditions are those that arise by viewing the computational domain as a sub-region of some larger domain which possesses planes or axes of symmetry.

[2] This type of boundary condition consists of pairs of geometrically identical boundaries at which all flow conditions are matched.

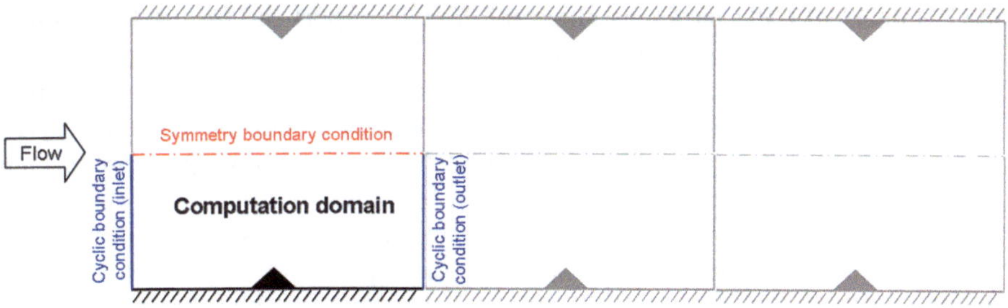

Fig. 27.4 An example of using cyclic and symmetry boundary conditions to reduce the computational domain

27.3.3.2 Solving

This is the step where the equations that govern the blood flow are solved. In summary, solving the main governing equations involves the following phases [20]:

1. *Approximation* of the unknown flow variables by means of simple functions.
2. *Discretisation* by substitution of the approximations into the governing flow equations and subsequent mathematical manipulations.
3. *Solution* of the algebraic equations.

A vast majority of the current CFD codes are based on one of the following three main numerical solution techniques: (1) Finite Difference, (2) Finite Element and (3) Finite Volume. The main differences between these techniques are associated with the way in which the flow variables are approximated and then discretised.

Finite difference is commonly used in in-house codes and simple structured CFD problems due to its simplicity and its application, therefore, is restricted to simple classical test cases (e.g. channel flows, pipes, etc.). While the finite element method is mainly used for structural stress analysis, it has some distinctive features such as its abilities to deal with arbitrary geometries which makes it somewhat attractive in vascular flow simulation problems. Finite element is not well-suited for turbulent flows and since the majority of haemodynamic problems involve laminar flow, it is often used in in-house codes which are developed mainly for biofluid/biomedical problems. However, it is the finite volume method (which was originally developed as a special finite difference formulation) that is currently the most popular technique in CFD. More detailed information about these three techniques can be found in several text books including Peric and Ferziger [19].

27.3.4 Task 4: Post-processing

The purpose of the post-processing task is to visualise the simulation results and to quantify the resulting physiological information. There are several different techniques and tools to enable the CFD user to extract desirable results. The majority of

of shear stress promotes the establishment of important physiologic characteristics of the artery wall promoting an anti-inflammatory, anti-thrombotic, anti-coagulative, profibrinolytic and anti-hypertrophic state [23].

As shown in Fig. 27.6a, for laminar flow of a Newtonian[3] fluid, the shear stress is simply proportional to the velocity gradient ($\partial u/\partial r$) and the fluid viscosity (μ):

$$\tau = -\mu \frac{\partial u}{\partial r} \tag{27.4}$$

and consequently the wall shear stress is the value of the shear stress evaluated at the wall:

$$\text{WSS} = \tau_w = \tau(R) = -\mu \left(\frac{\partial u}{\partial r} \right)_{wall} \tag{27.5}$$

In cardiovascular problems, due to the transient (pulsatile) nature of the blood flow, it is more useful to define the time-averaged WSS (TAWSS) which is calculated by integrating the WSS magnitude at each geometric part over the cardiac cycle:

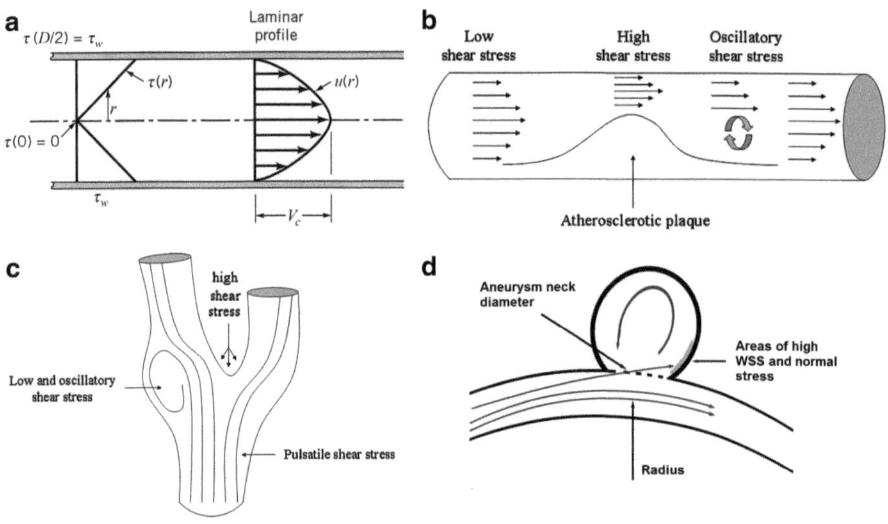

Fig. 27.6 (**a**) schematic of velocity profile and shear stress distribution within the blood flow in a typical artery [26], (**b**) distribution of shear stress in a straight arterial segment proximal to a lumen-protruding atherosclerotic plaque [27], (**c**) distribution of wall shear stress in a typical carotid artery bifurcation [28], and (**d**) schematic representation of an aneurysm originating on the outer curve of an intracranial artery; areas adjacent to the impact site experience high levels of wall shear stress [29]

[3] Fluids for which the shearing stress is *linearly* related to the rate of shearing strain (also referred to as rate of angular deformation) are designated as *Newtonian fluids*. On the other hand, fluids for which the shearing stress is not linearly related to the rate of shearing strain are designated as *non-Newtonian fluids*. Blood is a non-Newtonian fluid, even though for simplicity it is usually assumed to be a Newtonian fluid in most CFD simulations.

$$\mathrm{TAWSS} = \frac{1}{\mathrm{T}} \int_0^T |\vec{\tau}_w| \, \mathrm{dt} \tag{27.6}$$

where $\vec{\tau}_w$ is the instantaneous wall shear stress vector, t is the time and T is the duration of the cardiac cycle.

The relationships between the wall shear stress and distribution of atherosclerotic lesions in the arterial system has been the subject of several research papers in the past five decades with some contradicting conclusions. For instance, by studying hypercholesterolaemic animals, Fry [24] suggested that high WSS leads to endothelial damage and hence increases lipoprotein influx into the intima. On the other hand, through studying human post-mortem material, Caro et al. [25] proposed that high WSS was rather protective and found that atherosclerotic lesions instead occur in regions of low WSS. Recently, Peiffer et al. [2] carried out a systematic review of several papers and found that the majority of the studies support the low WSS theory. However, the presence of a direct correlation between lesions and WSS is probably too simplistic, hence suggesting a need for more complex haemodynamic metrics based on deeper understanding of arterial flow physics and its interaction with the endothelial cells and layers.

The significance of shear stress-based metrics in different vascular problems are depicted in Fig. 27.6b–d which show the expected shear stress distributions in a lumen protruding artery, a carotid artery bifurcation and an aneurysm, respectively.

An example of TAWSS magnitude calculated by CFD for rupture-predictions in intracranial aneurysms shown in Fig. 27.7. In Sect. 27.4.3, two more examples of TAWSS are given (Fig. 27.8) for the atherosclerosis localisation in the descending thoracic aorta.

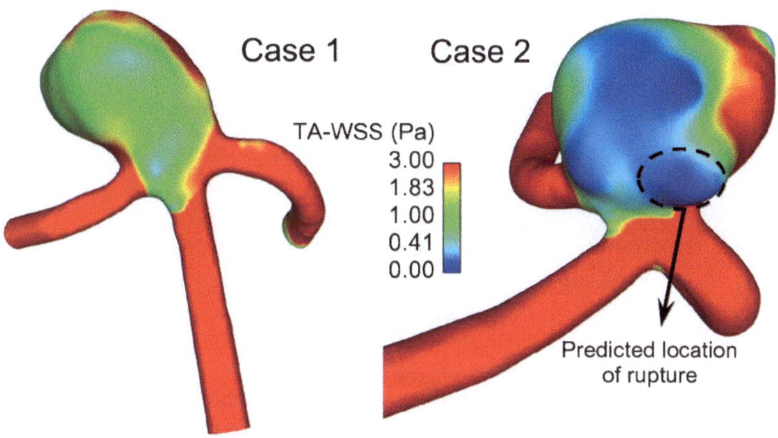

Fig. 27.7 Rupture prediction in two patient-specific intracranial aneurysms using predicted time-averaged wall shear stress magnitude. CFD simulations of both cases suggest that Case 2 is more likely to rupture based on low shear theory (the aneurysm models were provided by [30])

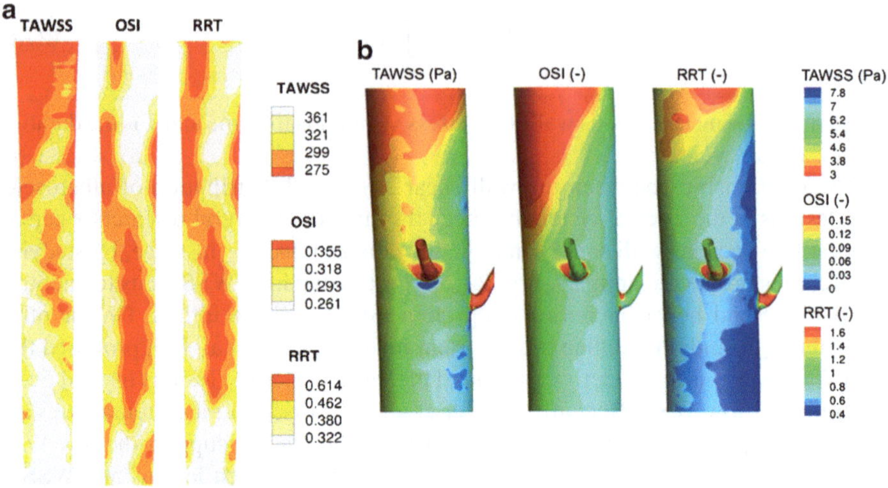

Fig. 27.8 Distribution of three haemodynamic metrics; (**a**) descending thoracic aorta of a mouse with aortic valve regurgitation (Modified from [34]); (**b**) close-up of the proximal left intercostal branch ostium in a rabbit aorta (Modified from [22])

27.4.2 Oscillatory Shear Index (OSI)

In addition to wall shear stress, in 1980s, Ku et al. [31] identified another haemodynamic metric known as Oscillatory Shear Index (OSI) which was associated with the flow reversal. A more common definition of OSI, generalised for three dimensional flows, was proposed by He and Ku [32] as

$$\mathrm{OSI} = \frac{1}{2}\left(1 - \frac{\left|\int_0^T \vec{\tau}_w \mathrm{dt}\right|}{\int_0^T |\vec{\tau}_w| \, \mathrm{dt}}\right) = \frac{1}{2}\left(1 - \frac{|\vec{\tau}_{mean}|}{\mathrm{TAWSS}}\right) \tag{27.7}$$

where

$$\vec{\tau}_{mean} = \frac{1}{\mathrm{T}}\int_0^T \vec{\tau}_w \, \mathrm{dt} \tag{27.8}$$

OSI represents a dimensionless number which measures the directional change of WSS during the cardiac cycle. It identifies regions where the WSS vector is reversed from its principal axial direction over each cardiac cycle. Its value changes between 0 and 0.5, with 0 corresponding to uni-directional flow while 0.5 represents purely oscillatory flow.

Note that as indicated in Eq. (27.7), high OSI values are associated with low mean shear stress (e.g. OSI = 0.5 when $\vec{\tau}_{mean} = 0$); this has led to the common use of the '*low and oscillatory shear*' concept in the literature (see Fig. 27.6b, c). It is also worth noting that even though OSI is useful in identifying regions with flow reversal, it is insensitive to the magnitude of shear stress. For instance, sites with low time-averaged shear stress do not necessarily indicate sites of significant OSI; low shear stress levels could be the result of flow expansion without any local flow reversal. Similarly, strongly oscillating flows and slow moving flows, both could exhibit the same OSI [33]. This suggests that while OSI is an important haemodynamic parameter, one should consider studying this index in combination with shear-based metrics to achieve a better understanding of the blood flow in vascular problems.

In the next section, examples of OSI calculated for descending thoracic aorta of a mouse and a rabbit aorta are shown in Fig. 27.8.

27.4.3 Relative Residence Time (RRT)

Relative Residence Time (RRT), initially introduced by Himburg et al. [33], is another common haemodynamic metric which has shown good spatial correlation for the localisation of atherosclerotic disease (see Hoi et al. [34], for example). The motivation behind the use of RRT is based on the importance of the 'residence time' of solutes and formed elements of the blood in the vicinity of vascular endothelium on the atherosclerotic process. At a particular site in a blood vessel, RRT is inversely proportional to a streamwise distance, Δx, that a fully entrained blood particle would travel at a small distance from the wall, y. In the near-wall regions where the viscous effects are significant, once could ignore the spatial variations in shear and therefore,

$$\Delta x = \left| \int_0^T u(y)\,dt \right| = \frac{y}{\mu} \left| \int_0^T \vec{\tau}_w\,dt \right| \tag{27.9}$$

where $u(y)$ is the streamwise velocity and μ is the blood dynamic viscosity.

Rearranging Eq. (27.7) to solve for $\vec{\tau}_{mean}$ and then substituting into Eq. (27.9) gives:

$$\Delta x = \left(\frac{Ty}{\mu} \right) \times \left[1 - (2 \times \mathrm{OSI}) \right] \times \mathrm{TAWSS} \tag{27.10}$$

For a small distance from the wall, (Ty/c) would be constant, therefore

$$\mathrm{RRT} \sim \frac{1}{\Delta x} \sim \frac{1}{\mathrm{TAWSS} \times \left[1 - (2 \times \mathrm{OSI}) \right]} \tag{27.11}$$

Equation (27.11) shows how RRT is affected by both TAWSS and OSI at any given site. It appears that small OSI values do not have significant effects on RRT, however, when OSI approaches its limit of 0.5, it would have important influence on this metric. Therefore, RRT would be a useful measure of the shear environment for correlative purposes that incorporates the level of the shear and its oscillatory character [33]. Note that since RRT is a relative concept, it is usually normalised using a reference value e.g. RRT at aortic inlet.

Figure 27.8 shows the distribution of the three hemodynamic metrics discussed above, for two different cases.

Acknowledgements The lead author would like to thank all colleagues at the University of Manchester and Manchester Metropolitan University, especially Dr Neil Ashton and Dr Alistair Revell at the University of Manchester who kindly provided some of the figures used in this chapter.

References

1. Steinman DA (2004) Image-based computational fluid dynamics: a new paradigm for monitoring hemodynamics and atherosclerosis. Curr Drug Targets Cardiovasc Haematol Disord 4:183–197
2. Peiffer V, Sherwin SJ, Weinberg PD (2013) Does low and oscillatory wall shear stress correlate spatially with early atherosclerosis? A systematic review. Cardiovasc Res 99:242–250
3. Suo J, Ferrara DE, Sorescu D et al (2007) Hemodynamic shear stresses in mouse aortas: implications for atherogenesis. Arterioscler Thromb Vasc Biol 27:346–351
4. Cebral JR, Mut F, Weir J, Putman C (2011) Quantitative characterization of the hemodynamic environment in ruptured and unruptured brain aneurysms. AJNR Am J Neuroradiol 32:145–151
5. Xiang J, Natarajan SK, Tremmel M et al (2011) Hemodynamic-morphologic discriminants for intracranial aneurysm rupture. Stroke 42:144–152
6. Miura Y, Ishida F, Umeda Y et al (2013) Low wall shear stress is independently associated with the rupture status of middle cerebral artery aneurysms. Stroke 44:519–521. doi:10.1161/STROKEAHA.112.675306
7. Humphrey JD, Taylor CA (2008) Intracranial and abdominal aortic aneurysms: similarities, differences, and need for a new class of computational models. Annu Rev Biomed Eng 10:221–246. doi:10.1146/annurev.bioeng.10.061807.160439.Intracranial
8. Taylor CA, Figueroa CA (2009) Patient-specific modeling of cardiovascular mechanics. Annu Rev Biomed Eng 11:109–134. doi:10.1146/annurev.bioeng.10.061807.160521
9. Andrews KD, Hunt J (2009) Developing smaller-diameter biocompatible vascular grafts. In: Di Silvio L (ed) Cell response to Biomater. Woodhead Publishing, Cambridge, pp 212–236
10. Menu P, Stoltz JF, Kerdjoudj H (2013) Progress in vascular graft substitute. Clin Hemorheol Microcirc 53:117–129
11. Andrews KD, Feugier P, Black RA, Hunt JA (2008) Vascular prostheses: performance related to cell-shear responses. J Surg Res 149:39–46
12. Vara DS, Salacinski HJ, Kannan RY et al (2005) Cardiovascular tissue engineering: state of the art. Pathol Biol 53:599–612
13. Lee B-K (2011) Computational fluid dynamics in cardiovascular disease. Korean Circ J 41:423–430. doi:10.4070/kcj.2011.41.8.423
14. http://rsb.info.nih.gov/ij/
15. https://simtk.org/home/simvascular

16. http://villacamozzi.marionegri.it/luca/vmtk/
17. Cannataro M, Guzzi PH, Tradigo G, Veltri P (2008) A tool for the semiautomatic acquisition of the morphological data of blood vessel networks. IEEE Int Symp Parallel Distrib Proc Appl 2008:837–840. doi:10.1109/ISPA.2008.120
18. Les AS, Shadden SC, Figueroa CA et al (2010) Quantification of hemodynamics in abdominal aortic aneurysms during rest and exercise using magnetic resonance imaging and computational fluid dynamics. Ann Biomed Eng 38:1288–1313. doi:10.1007/s10439-010-9949-x
19. Peric M, Ferziger JH (2002) Computational methods for fluid dynamics, 3rd edn. Springer, Berlin
20. Versteeg HK, Malalasekera W (1995) An introduction to computational fluid dynamics: the finite volume method. Longman Scientific & Technical, Harlow
21. Vincent PE, Plata AM, Hunt AAE et al (2011) Blood flow in the rabbit aortic arch and descending thoracic aorta. J R Soc Interface 8:1708–1719. doi:10.1098/rsif.2011.0116
22. Peiffer V, Sherwin SJ, Weinberg PD (2013) Computation in the rabbit aorta of a new metric – the transverse wall shear stress – to quantify the multidirectional character of disturbed blood flow. J Biomech 46:2651–2658. doi:10.1016/j.jbiomech.2013.08.003
23. Cunningham KS, Gotlieb AI (2005) The role of shear stress in the pathogenesis of atherosclerosis. Lab Invest 85:9–23
24. Fry DL (1969) Certain chemorheologic considerations regarding the blood vascular wall interface with particular reference to coronary artery disease. Circulation 40:38–59
25. Caro CG, Fitz-Gerald JM, Schroter RC (1971) Atheroma and arterial wall shear. Observation, correlation and proposal of a shear dependent mass transfer mechanism for atherogenesis. Proc R Soc Lond B Biol Sci 177:109–159
26. Munson BR, Young DF, Okiishi TH (2005) Fundamental of fluid mechanics. Wiley, New York
27. Koskinas KC, Chatzizisisa YS, Baker AB et al (2009) The role of low endothelial shear stress in the conversion of atherosclerotic lesions from stable to unstable plaque. Curr Opin Cardiol 24:580–590
28. Cecchi E, Giglioli C, Valente S et al (2011) Role of hemodynamic shear stress in cardiovascular disease. Atherosclerosis 214:249–256. doi:10.1016/j.atherosclerosis.2010.09.008
29. Nixon AM, Gunel M, Sumpio BE (2010) The critical role of hemodynamics in the development of cerebral vascular disease. J Neurosurg 112:1240–1253. doi:10.3171/2009.10.JNS09759
30. Janiga G (2013) Computational fluid dynamics challenge for rupture-prediction in intracranial aneurysms. http://www.ovgu.de/isut/LSS/CFD/CFD_Challenge.html
31. Ku DN, Giddens DP, Zarins CK, Glagov S (1985) Pulsatile flow and atherosclerosis in the human carotid bifurcation. Positive correlation between plaque location and low and oscillating shear stress. Arteriosclerosis 5:293–302
32. He X, Ku DN (1996) Pulsatile flow in the human left coronary artery bifurcation: average conditions. J Biomech Eng 118:74–82
33. Himburg HA, Grzybowski DM, Hazel AL et al (2004) Spatial comparison between wall shear stress measures and porcine arterial endothelial permeability. Am J Physiol Heart Circ Physiol 286:H1916–H1922
34. Hoi Y, Zhou Y-Q, Zhang X et al (2011) Correlation between local hemodynamics and lesion distribution in a novel aortic regurgitation murine model of atherosclerosis. Ann Biomed Eng 39:1414–1422

Chapter 28
Reverse Transcription Real-Time PCR Protocol for Gene Expression Analyses

M. Taliefar, S. Bradburn, G. Podda, and C. Murgatroyd

28.1 Introduction

This protocol describes the detailed experimental procedure for real-time reverse transcription (RT) PCR using SYBR Green. Gene expression differences between healthy and atherosclerotic arteries are important to understand cellular processes involved in the progression and development of atherosclerotic disease. The sequelae of atherosclerosis are leading causes of morbidity and mortality. Much is known about the pathogenesis of atherosclerosis [1], and many studies of gene expression in atherosclerotic lesions have been performed, both at the level of single gene analysis (e.g. [2]) and by global gene expression profiling technologies such as DNA microarray (e.g. [3]). However, there are still relatively little reproducible changes on gene expression patterns in atherosclerosis. This may relate to the fact that the heterogeneous progression of atherosclerotic disease in the peripheral arteries is currently not well understood [1]. For example, it has been hypothesized that the uneven onset and progression of atherosclerosis may be explained by artery specific transcriptomes [4]. Indeed, artery specific gene expression profiles, possibly a result of differences in angiogenesis and vasculogenesis, have previously been identified and may influence atherosclerotic disease susceptibility [5]. In sum, the need for detailed examination of gene expression patterns in atherosclerosis is crucial to aid understanding.

The procedure begins with reverse transcription of total RNA. The cDNA is then used as template for real-time PCR with gene specific primers. Optimization will be needed dependent on reagents or instruments for real-time PCR.

M. Taliefar • S. Bradburn • G. Podda • C. Murgatroyd (✉)
School of Healthcare Science, Manchester Metropolitan University, Manchester, UK
e-mail: c.murgatroyd@mmu.ac.uk

M. Slevin, G. McDowell (eds.), *Handbook of Vascular Biology Techniques*, DOI 10.1007/978-94-017-9716-0_28

28.1.1 Time Required

Time required for 20 samples:

DNA and RNA extraction:	3 h
cDNA synthesis:	3 h
Real-time PCR:	2 h
Data analysis:	1 h

28.2 Materials

- Oligonucleotide Primers *(Life Technologies™)*
- RNA
- SensiFAST™ SYBR Hi-ROX Kit *(Bioline Reagents)*
- Tetro cDNA Synthesis Kit *(Bioline Reagents)*
- RNaseZap® RNase Decontamination *(Ambion®)*
- DEPC-treated water *(Ambion®)*
- DNA LoBind 1.5 mL Tubes, PCR clean *(Eppendorf)*
- ART® self-sealing barrier pipette tips *(Sigma-Aldrich)*
- Eppendorf Mastercycler® PCR Cycler *(Eppendorf)*
- StepOnePlus™ Real-Time PCR *(Applied Biosystems®)*
- MicroAmp® Fast Optical 96-Well Reaction Plate *(Applied Biosystems®)*
- MicroAmp® 96 Well Optical Adhesive Film *(Applied Biosystems®)*
- 2100 bioanalyzer *(Agilent Technologies)*
- Gel Doc™ XR+ *(Bio-Rad Laboratories)*
- Multipurpose Centrifuge

28.2.1 Detailed Procedure

28.2.1.1 Analysis of RNA Quality and Quantity

Experiments using low-quality total RNA lead to poor results. Agarose gel electrophoresis is the conventional method for RNA analysis, but it is unreliable, labor intensive and slow. Several methods exist for quantifying RNA (Table 28.1). For RNA sample quality, the Agilent 2100 Bioanalyzer system provides sizing, quantitation and quality control of total RNA or mRNA on a single platform. This provides high quality digital data such as RNA Integrity Number (RIN) scale from 1 to 10 derived from Agilent 2100 expert software (Fig. 28.1) providing electropherogram, gel like images, histograms, dot plots and result table. RIN values range from 10 (intact) to 1 (totally degraded).

Table 28.1 Comparison of analytical and physical specification of various RNA quantification techniques

	NanoDrop 2000c	Qubit fluorometer	Bioanalyser 2100
Method	UV absorbance	Fluorescence-based dyes that bind specifically to nucleic acids	Microfluidics-based platform
Precision	3 % (0.74 at 350 nm)		10 % CV
Qualitative	NA		Nano 5–500 ng/μL; Pico 50–5,000 pg/μL; mRNA kit 25–250 ng/μL
Quantitative	0.4–15,000 ng/μL	>10 pg/μL	25–500 ng/μL
Sample amount required	2 μl	1 μl	1 μl
Number of samples per run	1	1	12 (+1 RNA ladder)
Total run time	<5 s	<5 s	30 min

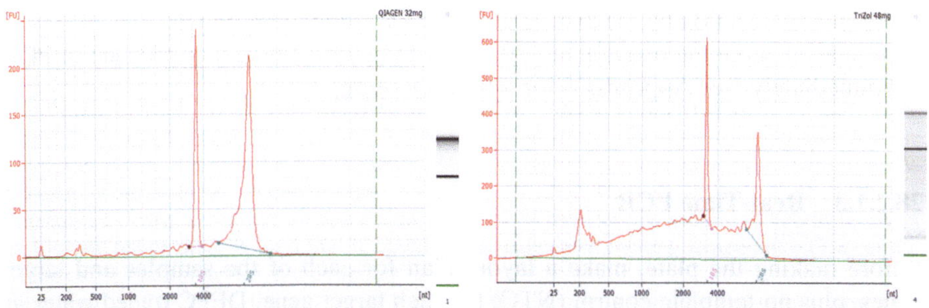

Fig. 28.1 Examples of bioanalysis electropherograms. Agilent 2100 expert software plots fluorescence intensity versus migration time and produces an electropherogram for each sample. This can also be displayed as a densitometry plot, creating a gel-like image. Intact mammalian total RNA shows two bands or peaks representing the 18S and 28S rRNA species. In general, the 28S rRNA is twice as bright and 18S rRNA. *Left Panel*; an electropherogram of a higher quality RNA preparations with a RIN value of 9.5. *Right panel*; an electropherogram of a lower quality RNA preparations with a RIN value of 5.4

28.2.1.2 Reverse Transcription

The reverse transcriptase (RT) is as critical to the success of qRT-PCR as the DNA polymerase. It is important to choose an RT that not only provides high yields of full-length cDNA, but also has good activity at high temperatures. High-temperature performance is also very important for denaturation of RNA with secondary structure. In one-step qRT-PCR, an RT that retains its activity at higher temperatures allows you to use a gene specific primer (GSP) with a high melting temperature (T*m*),

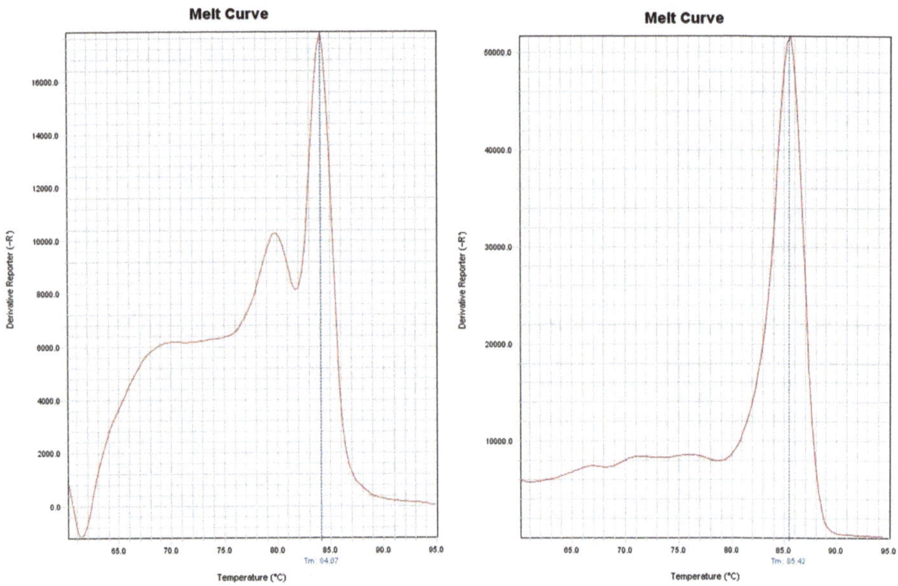

Fig. 28.2 Examples of melting curves resulting from real-time PCR reactions. *Left panel;* A melt curve plot representing unspecific primer binding resulting in multiple product formation. *Right panel;* An ideal melt curve produces one sharp peak indicating suitable primer specificity of the desired target

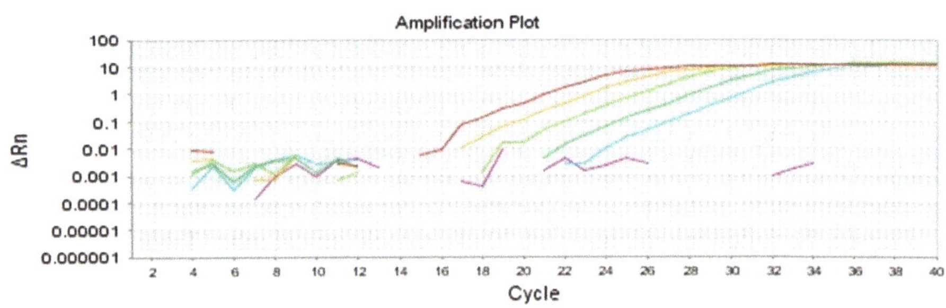

Fig. 28.3 A typical amplification plot from a set of real-time PCR reactions. Amplification plots are created when the fluorescent signal from each sample is plotted against cycle number. This can be used to extrapolate the Ct (cycle threshold) value. The incomplete plots at the start of the reaction represent background signals

also be detected as shown by additional peaks to the left of the peak for the amplified product in the melt curve. The PCR specificity can be further examined by running 5 μl of the PCR product on a 2.5 % agarose gel with a suitable DNA ladder.

Amplification plots are created when the fluorescent signal from each sample is plotted against cycle number. Therefore, amplification plots represent the accumulation of product over the duration of the real-time PCR experiment (Fig. 28.3). During the early cycles of the PCR reaction, there is little change in the fluorescent signal.

As the reaction progresses, with each cycle fluorescence begins to increase. A threshold is set and used to assign the threshold cycle, or Ct value, of each amplification reaction. Ct values are inversely related to the amount of starting template: the higher the amount of starting template in a reaction, the lower the Ct value for that reaction.

28.3 Quantitation of Results

Two strategies are commonly employed to quantify the results obtained by real-time RT-PCR; the standard curve method and the comparative threshold method.

The Standard Curve Method involves constructing a standard curve from an RNA of known concentration. This curve is then used as a reference standard for extrapolating quantitative information for mRNA targets of unknown concentrations. Though RNA standards can be used, their stability can be a source of variability in the final analyses. In addition, using RNA standards would involve the construction of cDNA plasmids that have to be in vitro transcribed into the RNA standards and accurately quantitated, a time-consuming process.

The Comparative Ct Method involves comparing the Ct values of the samples of interest with a control or calibrator such as a non-treated sample or cDNA from normal tissue. The Ct values of both the calibrator and the samples of interest are normalized to an appropriate housekeeping gene (see Appendix 2). The comparative Ct method is also known as the $2^{-\Delta\Delta CT}$ method:

$$\Delta\Delta C_{T} = \left(C_{T\text{Sample}} - C_{T\text{Sample housekeeping}}\right) - \left(C_{T\text{ Reference}} - C_{T\text{Reference housekeeping}}\right)$$

ΔC_T sample is the Ct value for any sample normalized to the housekeeping gene and ΔC_T reference is the Ct value for the calibrator also normalized to the housekeeping gene. To obtain tight data from RT-PCR experiment should ideally be prepared in replicates (triplicates), $2^{-\Delta\Delta CT}$ results used to perform independent *t*-test using IBM-SPSS or Microsoft Excel software.

Appendices

Appendix 1

Picking Random Primers and Oligo dT

Oligo dT primers are a favorite choice for two-step cDNA synthesis reactions because of their specificity for mRNA and because they allow many different targets to be studied from the same cDNA pool. However, because they always initiate reverse transcription at the 3′ end of the transcript, difficult secondary structure may lead to incomplete cDNA synthesis. Oligo dT priming of fragmented RNA, such as that isolated from Formalin-Fixed Paraffin-Embedded (FFPE) samples, may also be problematic. Nonetheless, as long as the primers are designed near the 3′ end of the target, premature termination downstream of this location is less of an issue.

Random primers are useful for synthesizing large pools of cDNA. They are also ideal for non-polyadenylated RNA, such as bacterial RNA, because they anneal throughout the target molecule. Degraded transcripts such as FFPE samples and secondary structure within the RNA do not pose as big a problem with random primers as they do with gene-specific primers and oligo dT primers. While increased cDNA yield is a benefit, data has shown that random primers can overestimate copy number when used in real-time RT-PCR experiments. Employing a combination of random and oligo dT primers can sometimes increase data quality by combining the benefits of both if used in the same first-strand cDNA synthesis reaction. Random primers are used only in two-step qRT-PCR reactions.

Appendix 2

Picking Housekeeping Gene

Relative quantification is a powerful technique that is commonly used to study RNA gene expression. In relative quantification the expression of a target gene is measured with respect to a stably expressed reference gene (so-called housekeeping gene); the two gene levels are expressed as a ratio.

Housekeeping genes must meet certain criteria before they can be effective reference genes. Housekeeping genes encode proteins that are essential for maintenance of cell function. For instance, housekeeping genes which code for components of the cytoskeleton (e.g., beta-actin, alpha-tubulin), components of the major histocompatibility complex (such as beta-2-microglobulin), enzymes of the glycolytic pathway (GAPDH (glyceraldehyde-3-phosphate dehydrogenase)) or ribosomal subunits appear to be expressed ubiquitously. However, several reports indicate that the expression of housekeeping genes is actively regulated; levels may vary across tissues and different types of cells, during cell proliferation and stages of development, or due to experimental treatment of cells.

Therefore, when choosing a housekeeping gene as a reference for relative quantification, one must identify a gene whose expression level remains relatively constant for a certain experimental setup. In fact, it is usually necessary to test a panel of housekeeping genes experimentally to find one that is not regulated in the investigated system. Since choosing an appropriate reference gene is critical for accurate quantitative RNA analysis, the behavior of candidate genes in different cell types and cell metabolic stages should be carefully examined.

Appendix 3

No-Template Control (NTC)

Controls in real-time PCR reactions prove that signal obtained from experimental samples represent the amplicon of interest, thereby validating specificity. All experiments should include a no-template control (NTC), and qRT-PCR reactions should also include a no-reverse transcriptase control (no-RT). NTC controls should contain all reaction components except the cDNA sample. Amplification detected in these wells is due to either primer-dimers or contamination with completed PCR reaction product. This type of contamination can make expression levels look higher than they actually are. No-RT reactions should contain all reaction components except the reverse transcriptase. If amplification products are seen in no-RT control reactions, this indicates that DNA was amplified rather than cDNA. This can also artificially inflate apparent expression levels in experimental samples

Appendix 4

Troubleshooting (Listed Some Major Causes for Real-Time PCR Failures)

Little or no PCR product: Poor quality of PCR templates, primers, or reagents may lead to PCR failures. First, please include appropriate PCR controls to eliminate these possibilities. Some genes are expressed transiently or only in certain tissues. In our experience, this is the most likely cause for negative PCR results. Please read literature for the gene expression patterns. One caveat is that microarrays are not always reliable at measuring gene expressions. After switching to the appropriate templates, we obtained positive PCR results in contrast to the otherwise negative PCRs (see our paper for more details).

Poor PCR amplification efficiency: The accuracy of real-time PCR is highly dependent on PCR efficiency. A reasonable efficiency should be at least 80 %. Poor primer quality is the leading cause for poor PCR efficiency. In this case, the PCR amplification curve usually reaches plateau early and the final fluorescence intensity is significantly lower than that of most other PCRs. This problem may be solved with re-synthesized primers.

Primer dimmer: Primer dimer may be occasionally observed if the gene expression level is very low. If this is the case, increasing the template amount may help eliminate the primer dimer formation.

Multiple bands on gel or multiple peaks in the melting curve: Agarose gel electrophoresis or melting curve analysis may not always reliably measure PCR specificity. From our experience, bimodal melting curves are sometimes observed for long amplicons (>200 bp) even when the PCRs are specific. The observed heterogeneity in melting temperature is due to internal sequence inhomogeneity (e.g. independently melting blocks of high and low GC content) rather than non-specific amplicon. On the other hand, for short amplicons (<150 bp) very weak (and fussy) bands migrating ahead of the major specific bands are sometimes observed on agarose gel. These weak bands are super-structured or single-stranded version of the specific amplicons in equilibrium state and therefore should be considered specific. Although gel electrophoresis or melting curve analysis alone may not be 100 % reliable, the combination of both can always reveal PCR specificity in our experience.

Non-specific amplicons: Non-specific amplicons, identified by both gel electrophoresis and melting curve analysis, give misleading real-time PCR result. To avoid this problem, please make sure to perform hot-start PCR and use at least 60 °C annealing temperature. We noticed not all hot-start Taq polymerases are equally efficient at suppressing polymerase activity during sample setup. The SYBR Green PCR master mix described here always gives us satisfactory results. If the non-specific amplicon is persistent, you have to choose a different primer pair for the gene of interest.

References

1. Slevin M (2011) Therapeutic angiogenesis for vascular disease: molecular mechanisms and targeted clinical approaches for the treatment of angiogenic disease. Springer, Dordrecht. ISBN 978-90-481-9494-0
2. Liu J, Sukhova GK, Sun JS, Xu WH, Libby P, Shi GP (2004) Lysosomal cysteine proteases in atherosclerosis. Arterioscler Thromb Vasc Biol 24:1359–1366
3. Eyster KM, Appt SE, Mark-Kappeler CJ, Chalpe A, Register TC, Clarkson TB (2011) Gene expression signatures differ with extent of atherosclerosis in monkey iliac artery. Menopause 18(10):1087–1095
4. Knowles JW, Assimes TL, Li J, Quertermous T, Cooke JP (2007) Genetic susceptibility to peripheral arterial disease: a dark corner in vascular biology. Arterioscler Thromb Vasc Biol 14(10):2068–2078
5. Burridge KA, Friedman MH (2010) Environment and vascular bed origin influence differences in endothelial transcriptional profiles of coronary and iliac arteries. Am J Physiol Heart Circ Physiol 299:H837–H846

Chapter 29
Oscillations, Feedback and Bifurcations in Mathematical Models of Angiogenesis and Haematopoiesis

Stephen Lynch and Jon Borresen

29.1 Introduction

Angiogenesis and haematopoiesis are processes that have been studied extensively in biomedical research where advances in cell culture have led to a deeper understanding of cell production and differentiation. Both are biological systems that are amenable to mathematical investigation and this chapter describes that approach. It is now understood that periodic, or oscillatory, behaviour is not confined to a limited number of physiological processes but is abundant in most biological systems. Periodicity in processes of the human body encompass phenomena such as genetic interactions, heartbeat rhythms, oscillating secretory, retina and muscle cells, cytoskeletal structures, bacterial oscillations, rhythmic behavior in growth and development, neuronal oscillations [1–3], and as shown in this chapter, angiogenesis and haematopoiesis. In 1952, whilst modeling neurons, Hodgkin and Huxley were able to accurately model the action potential in the giant squid axon [4]. Their nonlinear ordinary differential equations approximate electrical characteristics of excitable oscillatory cells such as cardiomyocytes and neurons. In 2009, Borresen and Lynch [5] published a paper suggesting a novel idea based on biological neural computing utilizing the Hodgkin-Huxley equations, and following this work, three years later UK and International patents were published [6, 7]. The patent illustrates how it is possible to construct binary logic gates from biological neurons grown on a chip. Utilizing excitatory and inhibitory neurons, it is proposed that these cells could be grown on a chip and that they could be trained to perform certain logical operations. Binary oscillator logic gates could then be used to test drugs that may halt or reverse symptoms of neurological disorders such as Alzheimer's, Parkinson's disease and epilepsy. Mathematical models of binary oscillator logic and memory are

S. Lynch (✉) • J. Borresen
School of Computing, Mathematics & Digital Technology, Manchester Metropolitan University, Manchester, UK
e-mail: s.lynch@mmu.ac.uk

M. Slevin, G. McDowell (eds.), *Handbook of Vascular Biology Techniques*,
DOI 10.1007/978-94-017-9716-0_29

remarkably robust [8], and results involving biological oscillating neurons will be published at a later date.

Following the primary creation of a network of vascular endothelial cells in the developing embryo (*vasculogenesis*), a process called *angiogenesis* remodels the network to form a child's circulatory system of new blood vessels or capillaries. Angiogenesis continues to be vital in growth and development and is involved in wound healing and in the formation of granulation tissue. Unfortunately, angiogenesis is also a fundamental factor in the transition of tumours from a benign state to a malignant one. Angiogenesis promotes cancer growth: cancerous tumour cells release molecules such as fibroblast growth factor (FGF), vascular endothelial growth factor (VEGF), VEGF receptors (VEGFR), angiopoietin 1 (Ang 1) and angiopoietin 2 (Ang 2). These molecules then signal certain genes in the host tissue to make proteins to encourage new blood cell formation supplying nutrients and oxygen and removing waste products. Current cancer research includes the use of *angiogenetic inhibitors*, and good functioning blood vessels may allow deeper penetration of anti-cancer drugs in to the tumour.

In 1971, Folkman [9] conjectured that all cancer tumours are angiogenesis-dependent. It was nearly a decade later when an application was discovered which helped in the treatment of macular degeneration. The Hahnfletd et al. [10] model of angiogenesis is regarded by many as the seminal work on this topic, however, this chapter will concentrate on the papers of Agur et al. [11] and Bodnar et al. [12, 13]. Note that there are lots of other mathematical models that can be used to model angiogenesis, see for example, the partial differential equations (PDEs) models in [14] and cellular automata based methods in [15] and [16].

This chapter will primarily be concerned with autonomous and non-autonomous ordinary differential equations (ODEs) and delay differential equations (DDEs) which model angiogenesis and haematopoiesis. The systems are nonlinear and a feedback mechanism is incorporated to make the models more realistic. It is shown that the models are history dependent and for certain parameter values they can undergo *hysteresis*, where the solutions obtained are dependent upon the history of the system. Hysteresis is not a new phenomenon, indeed it abounds throughout the realms of science. In biology, the author et al. have investigated hysteresis in blood cell population dynamics [17] and neuronal networks [18]. In physics, the author et al. have demonstrated hysteresis in a wide range of nonlinear optical resonators and microfiber ring resonators [19]. In chemical kinetics, simple models of Hopf bifurcation in a Brusselator model and hysteresis in an autocatalytic chemical reaction are demonstrated [20]. In electric circuits, the author and Borresen [21] have demonstrated hysteresis in Chua's electrical circuit for the first time. In mechanical engineering, the author and Christopher [22] have investigated surge in jet engines and wing rock phenomena and shown how hysteresis is possible between multiple stable limit cycles. It is shown that there can be several jumps between many different stable oscillations as parameter values are increased and decreased. In the engineering literature these are called *hard* or *dangerous bifurcations* and they are particularly relevant to angiogenesis where an oscillating tumour can suddenly jump from one size to another as a parameter is varied. This could indicate a change from benign to malignant behaviour in the tumour.

Section 29.2 will introduce basic mathematical definitions to be used in this chapter and a simple model of haematopoiesis will give the reader a gentle introduction to DDEs. Section 29.3 will list the differential equations to be used to model angiogenesis. Some Mathematica programs are listed in the text and the reader can download full working programs from the web. The conclusions will be presented in Sect. 29.4, a bibliography is listed in the Reference section and the URL to download the Mathematica notebook is given at the end of this chapter.

29.2 Definitions and a Simple Model of Haematopoiesis

This section is concerned with autonomous and non-autonomous ODEs and DDEs. For an introduction to autonomous and non-autonomous ODEs the reader is directed to references [23–25], for an introduction to DDEs with applications to the life sciences, the reader is directed to [26] and subsection 29.2.2 of this chapter. Note that since the present report concentrates on biological aspects, the mathematical analysis is omitted from this work. Interested readers should consult the research papers and mathematics textbooks in the reference section of this chapter for more information. For completeness, a list of commonly used definitions that will be used in this book chapter will now be listed.

29.2.1 Autonomous and Non-Autonomous ODEs

Definition 2.1 A system of the form

$$\frac{d}{dt}\underline{x} = \dot{\underline{x}} = f(\underline{x}), \tag{29.1}$$

where $\underline{x} = R^n$, is called an n-dimensional *autonomous system of ODEs*.

Definition 2.2 A system of the form

$$\frac{d}{dt}\underline{\mathrm{x}} = f(\underline{x},t), \tag{29.2}$$

where $\underline{x} \in R^n$, is called a *non-autonomous system of ODEs*.

Definition 2.3 For system (29.1), a *critical point* is a point that satisfies the equation

$$\frac{d}{dt}\underline{x} = f(\underline{x}) = 0.$$

Definition 2.4 A critical point, say $\underline{x}^*$, is *stable* if given $\varepsilon > 0$, there is a $\delta > 0$, such that for all $t \geq t_0$, $\left\|\underline{x}(t) - \underline{x}^*(t)\right\| < \varepsilon$, whenever $\left\|\underline{x}(t_0) - \underline{x}^*(t_0)\right\| < \delta$, where $\underline{x}(t)$ is a solution of (29.1). A critical point that is not stable is called *unstable*.

In applications, the stability of the critical points is usually determined by finding the *eigenvalues* and *eigenvectors* of the linearized system of (29.1), the reader is once more directed to references [23–25].

Definition 2.5 A system of the form (29.1) is said to be *multistable* if there exists at least two stable critical points or stable limit cycles.

Definition 2.6 A *limit cycle* is an isolated periodic solution of system (29.1).

Definition 2.7 A system of the form (29.1) is *structurally stable* if small perturbations leave the qualitative behavior unchanged. If small perturbations cause a change in the qualitative behavior, then system (29.1) is called *structurally unstable*.

Definition 2.8 A parameter value, μ_0 say, for which the system $\dot{\underline{x}} = f(\underline{x}, \mu_0)$, is not structurally stable is called a *bifurcation value*.

Definition 2.9 A *Hopf bifurcation* is a local bifurcation in which a critical point changes its stability as a pair of complex conjugate eigenvalues of the linearized system about the critical point cross the imaginary axis of the complex plane. The bifurcations can be *supercritical* resulting in *stable limit cycle* bifurcation or *subcritical* resulting in *unstable limit cycle* bifurcation.

Definition 2.10 A *Double-Hopf bifurcation* is another local bifurcation in which a fixed point of a discrete dynamical system has two pairs of purely imaginary eigenvalues. Generically two branches of torus bifurcations evolve leading to periodic and quasiperiodic behaviors, and this bifurcation can imply a local birth of chaos.

For a more detailed introduction to Hopf bifurcation the reader is directed to [27]. Hopf and double-Hopf bifurcations also occur in infinite dimensional DDEs to which the *Centre Manifold Theorem* applies, interested readers should consult textbooks such as [28–30].

29.2.2 An Introduction to DDEs

Many physiological processes in the human body are subject to time delays which can range from the cellular and molecular level up to the organ and inter-organ system level. These time lags are due to processes including maturation, regeneration, transmission, transport and threshold, to name but a few. As a result of the time lags, a delay term is introduced in to the differential equations as shown in Definition 2.11.

Definition 2.11 A system of the form

$$\frac{d}{dt}\underline{x} = f(\underline{x}, t, t - \tau_i), \tag{29.3}$$

where $\underline{x} \in R^n$, and $\tau_i, i = 1,2,3\ldots$, are constant delays, is called a *delay differential equation*.

Generally speaking, the dynamics involved in DDEs is much more complicated than that for autonomous ODEs. DDEs have infinitely many real and complex solutions and so are infinite dimensional. Unlike autonomous ODEs, the solution is not determined by its initial state but by an infinite set of initial conditions over an interval. For an introduction to DDEs with applications the reader is directed to [31], and for an introduction to numerical methods of DDEs, references [32] and [33] are recommended. Four well known examples of physiological delay models include haematopoiesis [34], the human respiratory control system and regulation of blood concentration of CO_2 [35], hormone regulation and maintenance in the menstrual cycle [36] and neural networks [37]. It transpires that haematopoiesis and angiogenesis are closely related which is unsurprising given that they have common ancestors in haematogenic endothelial cells or haemangioblasts. Indeed, some haematopoietic cells have been shown to be attracted to tumour tissue and contribute to angiogenesis and the growth of certain tumours [38]. Note that the DDEs considered in this chapter have constant delays, for systems with other types of delay the reader is directed to [39].

As a simple example of a DDE, consider the one dimensional Mackey-Glass model related to haematopoiesis [34] defined by

$$\frac{dx}{dt} = \frac{\beta x(t-\tau)}{1+x(t-\tau)^n} - \delta x(t), \tag{29.4}$$

where $x(t)$ is the blood cell population at time t, τ is a constant time lag and n is a constant. The first term in the right hand side of (29.4) represents the delayed production rate of the blood cell population and δ is a death rate of the cells. Figure 29.1a shows an interactive Mathematica solution plot of (29.4). Within the Mathematica notebook, the user can move the slider at the top of the figure left and right and the phase portrait will update automatically. As the parameter δ is increased, the periodic solution goes through period doubling and un-doubling routes to and from chaos. Figure 29.1b shows the power spectrum which displays the energy distribution over different frequencies of a time series.

By introducing a feedback mechanism and using the *second iterative method* [23–25], bifurcation diagrams may be plotted. A parameter is varied and the solution to the previous iterate is used as the initial condition for the next iterate. In this case, there is a history associated with the process and only one point is plotted for each value of the parameter (Fig. 29.3).

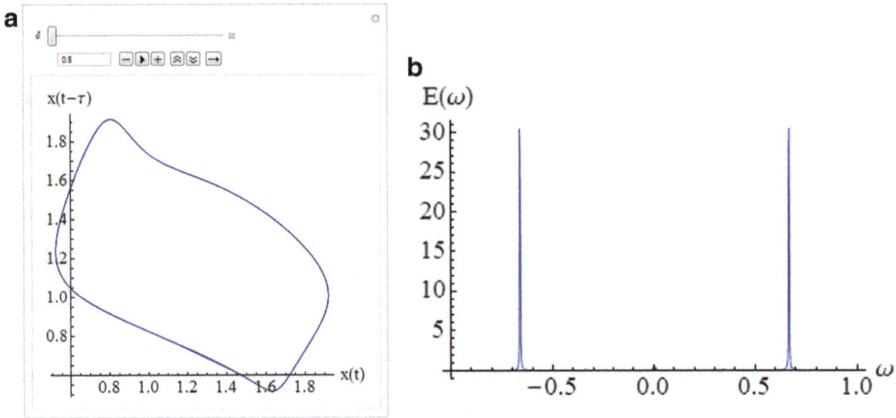

Fig. 29.1 Periodic behaviour in the blood cell population for (29.4) when $\beta = 2, n = 10,\ \tau = 2$ and $\delta = 0.5$. (**a**) Interactive phase portrait. (**b**) Power spectrum

```
(* Mathematica program to plot Figure 1(a) *)
β=2;τ=2;n=10;
Manipulate[sol=First[NDSolve[{x'[t]==β x[t-τ]/(1+x[t-τ]^n)-δ
x[t],x[t/;t≤0]==1/2},x,{t,0,200},MaxSteps→Infinity]];
p1=ParametricPlot[Evaluate[{x[t],x[t-τ]}/.sol],
{t,100,200}];Show[{p1},AxesLabel→{"x(t)","x(t-τ)"},
TextStyle→{FontSize→20}],{δ,0.5,1.5}]
```

```
(* Mathematica program to plot Figure 1(b) *)
τ=2;β=2;δ=0.5;n=10;
sol=First[NDSolve[{x'[t]==β x[t-τ]/(1+x[t-τ]^n)-δ
x[t],x[t/;t≤0]==1/2},x,{t,0,1000},MaxSteps→Infinity,Method →
{"StiffnessSwitching"}]];TS=Table[x[i],{i,301,1000,1}]/.sol;Nits=700;p1=List
LinePlot[Rest[Transpose[{Range[-
Nits/2+1,Nits/2]*(2/Nits),Abs[Fourier[TS]]^2}]],PlotRange→All,AxesOrigin→{1,
0},AxesLabel→{ω,"E(ω)"}];Show[{p1},TextStyle→20]
```

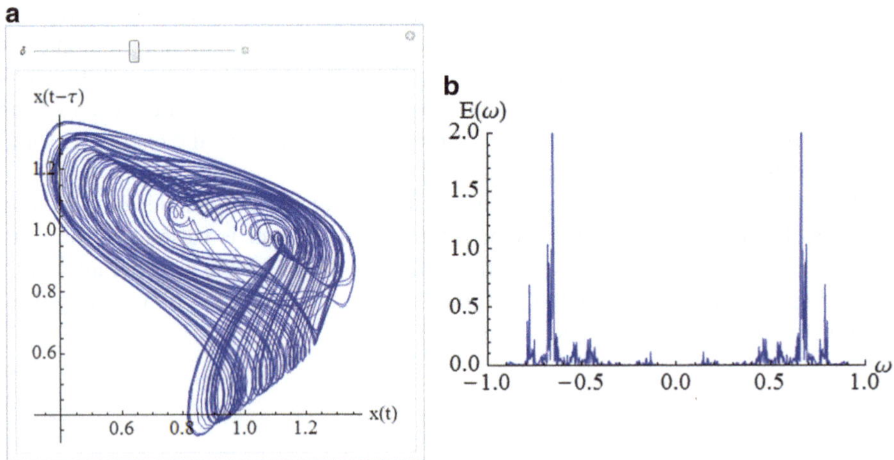

Fig. 29.2 Chaotic behaviour in the blood cell population for (29.4) when $\beta = 2, n = 10, \delta = 1$ and $\tau = 2$. (**a**) Interactive phase portrait. (**b**) Power spectrum

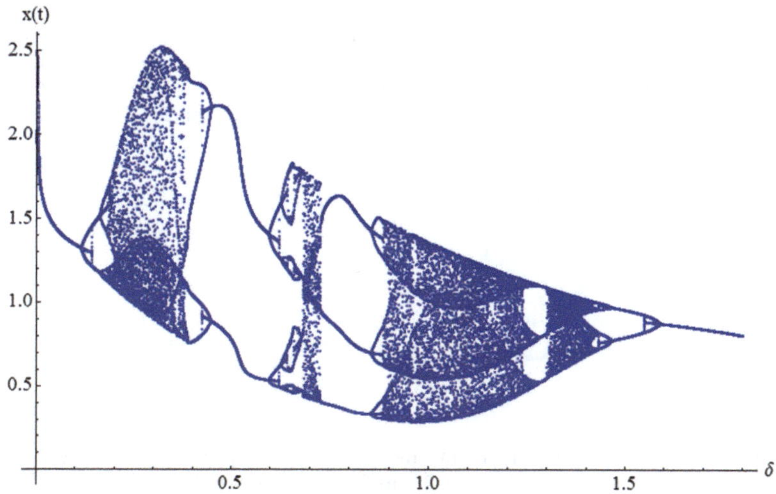

Fig. 29.3 Bifurcation diagram for system (29.4) as the parameter δ is increased linearly from $\delta = 0$ to $\delta = 1.8$, and then ramped back down again. Note that single branches are still depicting periodic solutions

```
(* Mathematica program to plot Figure 3 *)
Initδ=0;M=18000;step=0.0001;Maxδ=Initδ+M*step;τ=2;β=2;n=10;Tmax=τ;x0= 1/2;
(* Ramp the parameter up *)
For[i=1,i≤M,i++,δ=Initδ+step*i;MG=NDSolve[{x'[t]==β x[t-τ]*(1+x[t-τ]^n)
-δx[t],x[t/;t≤0]==x0},x,{t,0,Tmax},MaxSteps→Infinity];
x0=Flatten[x[Tmax]/.MG];rampup[i]=x0];
p1=ListPlot[Table[Flatten[{Initδ+m*step,rampup[m]}],{m,1,M}],PlotRange→All]
; (* Ramp the parameter down *)
For[i=1,i≤M,i++,δ=Maxδ-step*i;MG=NDSolve[{x'[t]==β x[t-τ]*(1+x[t-τ]^n)-δx[t],x[t/
;t≤0]==x0},x,{t,0,Tmax},MaxSteps→Infinity];x0=Flatten[x[Tmax]/.MG];rampdown[i]
=x0];p2=ListPlot[Table[Flatten[{Maxδ-m*step,rampdown[m]}],{m,1,M}],PlotRang
e→All];
Show[{p1,p2},PlotRange→All,AxesLabel→{"δ","x(t)"},AxesOrigin→{0,0},
TextStyle→{FontSize→20}]
```

Figure 29.3 shows the bifurcation diagram for system (29.4) obtained using the second iterative method with feedback. There are clearly regions of period doubling and un-doubling to and from chaos and there are also periodic windows.

To complete the introduction to DDEs consider the modified Mackey-Glass system given by

$$\frac{dx}{dt} = \frac{\beta\left(x\left(t-\tau_1\right)+x\left(t-\tau_2\right)\right)}{1+\left(x\left(t-\tau_1\right)+x\left(t-\tau_2\right)\right)^n} - \delta x\left(t\right), \tag{29.5}$$

where there are now two time delays, τ_1 and τ_2.

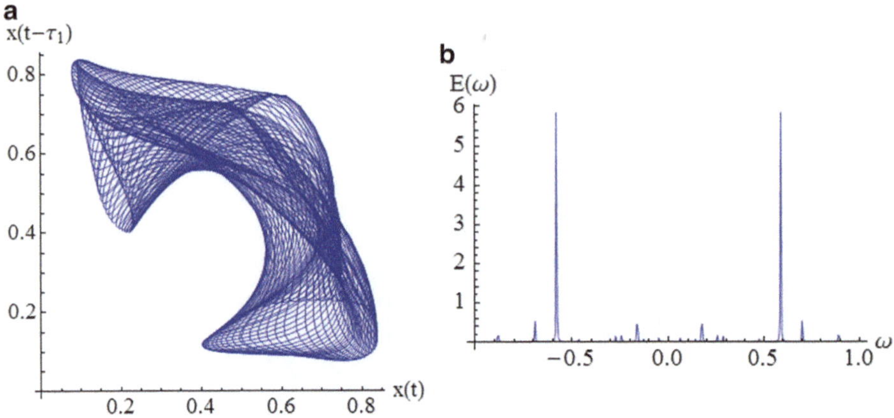

Fig. 29.4 Quasiperiodic behaviour in the blood cell population for (29.5) when $\beta = 2.4$, $n = 10$, $\delta = 2$, $\tau_1 = 2.4$ and $\tau_2 = 6.2$. (**a**) Phase portrait. (**b**) Power spectrum

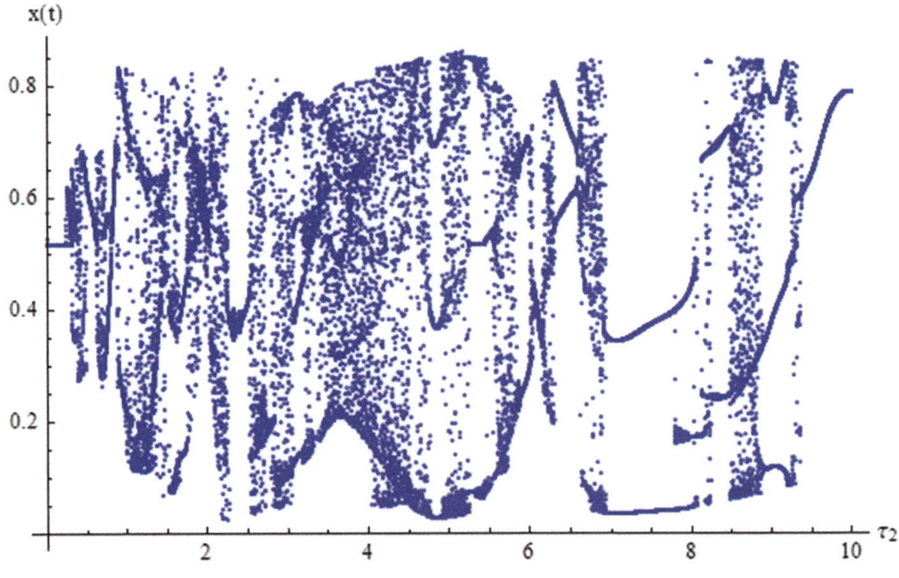

Fig. 29.5 Bifurcation diagram for system (29.5) when $\beta = 2.4$, $\delta = 2$, $n = 10$ and $\tau_1 = 2.4$ as the parameter τ_2 is ramped up from $\tau_2 = 0$ up to $\tau_2 = 10$, and then ramped down again. The system goes through double-Hopf bifurcations displaying periodic, quasiperiodic and possibly chaotic behaviours

Fix the parameters $\beta = 2.4$, $\delta = 2$, $n = 10$ and $\tau_1 = 2.4$. Figure 29.4 shows the phase portrait and corresponding power series spectrum when $\tau_2 = 6.2$.

Figure 29.5 shows the bifurcation diagram for system (29.5) as the parameter τ_2 is ramped up and down. The figure is a typical bifurcation diagram of double Hopf bifurcations displaying jumps between periodic, quasiperiodic and possibly chaotic behaviours.

29.3 Mathematical Models of Angiogenesis

29.3.1 *Results for the Model of Agur et al.*

Consider the model investigated by Agur et al. [10]:

$$\begin{aligned} \frac{dN}{dt} &= f_1(E)N \\ \frac{dP}{dt} &= f_2(E)N - \delta P \\ \frac{dE}{dt} &= f_3(P)E - f_1(E)E, \end{aligned} \tag{29.6}$$

where N is the number of tumour cells or tumour size, P is the quantity of growth factors known to be involved in supplying the tumour, and E represents the vessel density, where $E = \frac{V}{N}$, and V is the volume of blood vessels feeding the tumour. The functions f_1, f_2, f_3, model tumour cells proliferation rate, the protein production rate and the vessel growth rate, respectively. The functions satisfy the following conditions:

$$\begin{aligned} &\text{(i) } f_1 \text{ is an increasing function, } f_1(0) > 0, \lim_{E\to\infty} f_1(E) > 0; \\ &\text{(ii) } f_2 \text{ is a decreasing function, } f_2(x) > 0, \text{with } f_2(E) > 0; \\ &\text{(iii) } f_3 \text{ is an increasing function, } f_3(0) > 0, \lim_{E\to\infty} f_3(\mathrm{P}) > 0; \end{aligned} \tag{29.7}$$

Introducing protein consumption, system (29.6) becomes:

$$\begin{aligned} \frac{dN}{dt} &= f_1(E)N \\ \frac{dP}{dt} &= f_2(E)N - \delta P - f_c\left(f_3(P)EN\right)P \\ \frac{dE}{dt} &= f_3(P)E - f_1(E)E, \end{aligned} \tag{29.8}$$

where the function fc satisfies the additional conditions:

(iv) $f_c(x) > 0, \lim_{x\to\infty} f_c(x) > 0, f_c(0) \approx 0,$

The following theorem is proved in [10]:

Theorem 3.1 Systems (29.6) and (29.8) have no critical points that can undergo a Hopf bifurcation.

Now introduce time delays in to system (29.6) to obtain a DDE with constant delay terms:

$$\begin{aligned}\frac{dN}{dt} &= f_1\left(E\left(t-\tau_1\right)\right)N\left(t\right)\\ \frac{dP}{dt} &= f_2\left(E\left(t\right)\right)N\left(t\right)-\delta P\left(t\right).\\ \frac{dE}{dt} &= f_3\left(P\left(t-\tau_2\right)\right)E\left(t\right)-f_1(E\left(t-\tau_1\right)E\left(t\right),\end{aligned} \tag{29.9}$$

where τ_1 depicts a proliferation/death response to stimuli and τ_2 is the in vessel formation/regression response to stimuli. The following theorem is proved in [10]:

Theorem 3.2 For every $\left(\tau_1,\tau_2\right)\neq(0,0)$, system (29.9) has critical points that can undergo Hopf bifurcation and oscillatory behavior is possible.

29.3.2 Angiogenesis Models of Bodnar et al.

In 2009, Bodnar et al. [11] modified the ODEs (29.8) and DDEs (29.9) using some ideas from Hahnfeldt et al. [9], incorporating tumour carrying capacity depending on the volume of vessel endothelial cells. The new model is of the form:

$$\begin{aligned}\frac{dN}{dt} &= \alpha N\left(1-\frac{N}{1+f_1\left(E\right)}\right)\\ \frac{dP}{dt} &= f_2\left(E\right)N-\delta P.\\ \frac{dE}{dt} &= f_3\left(P\right)-\alpha\left(1-\frac{N}{1+f_1\left(E\right)}\right)E,\end{aligned} \tag{29.10}$$

where the functions satisfy the following conditions:

(i) f_1 is an increasing function, $f_1\left(0\right)=0, \lim_{E\to\infty} f_1\left(E\right)=b_1>0;$

(ii) f_2 is decreasing and convex, $f_2\left(0\right)=a_2>0, \lim_{E\to\infty} f_2\left(E\right)=0;$ (29.11)

(iii) f_3 is increasing, $f_3\left(0\right)=-a_3<0, f_3\left(c_3\right)=0, \lim_{P\to\infty} f_3\left(P\right)=b_3>0.$

As with the Agur et al. model, time delays are introduced to give the DDE:

$$\frac{dN}{dt} = \alpha N\left(1 - \frac{N}{1 + f_1\left(E\left(t - \tau_1\right)\right)}\right)$$
$$\frac{dP}{dt} = f_2\left(E(t)\right)N - \delta P(t). \tag{29.12}$$
$$\frac{dE}{dt} = f_3\left(P\left(t - \tau_2\right)\right) - \alpha\left(1 - \frac{N}{1 + f_1\left(E\left(t - \tau_1\right)\right)}\right)E(t),$$

where the parameters and functions are the same as those in system (29.10).

A numerical simulation is carried out using the following functions that satisfy conditions (29.11):

$$f_1(E) = \frac{b_1 E^2}{c_1 + E^2},\ f_2(E) = \frac{a_2}{1 + d_2 E},\ f_3(P) = \frac{\left(a_3 + b_3\right)P^2}{\frac{c_3^2 b_3}{a_3} + P^2} - a_3$$

The parameter values are:

$$c_1 = 5, d_2 = 0.8,\ a_3 = 2,\ b_3 = 1,\ c_3 = 2,\ \alpha = 1,\ \delta = 1.$$

The parameters b_1, a_2, τ_1 and τ_2 are varied in [11]. The parameters b_1 and a_2 alter the carrying capacity of the tumour and the angiogenic protein production, respectively. They show the co-existence of stable and unstable critical points in system (29.10), and in system (29.12) they are able to bifurcate both stable and unstable limit cycles by varying τ_1, τ_2, improving upon the results of Agur et al. [10].

In [12], Bodnar et al. vary the parameters m_3 and τ_1, τ_2 in systems (29.10) and (29.12) for the following function and parameter values:

$$f_1(E) = \frac{b_1 E^n}{c_1 + E^n},\ f_2(E) = \frac{a_2 c_2}{c_2 + E},\ f_3(P) = \frac{b_3\left(P^2 - m_3^2\right)}{\frac{m_3^2 b_3}{a_3} + P^2}. \tag{29.13}$$

The parameter values are:

$$n = 1 \text{ or } 2,\ a_2 = 0.4,\ a_3 = 1,\ b_1 = 2.3,\ b_3 = 1,\ c_1 = 1.5,\ c_2 = 1,\ \alpha = 1,\ \delta = 0.34$$

The parameter m_3 alters the stimulation of tumour vessel production. It is shown in that paper that it is possible to bifurcate a number of limit cycles for differing time delays and more complex dynamics are evident which match more closely with the physical models.

29.3.3 An Angiogenesis Model Subject to Feedback of Angiogenic Protein Production

Consider system (29.10) subject to the parameter values given in [11]. Fix the parameters $c_1 = 5,\ d_2 = 8,\ a_3 = 2,\ b_1 = 2.8,\ b_3 = 1,\ c_3 = 2,\ \alpha = 1,\ \delta = 1,$ and vary the parameter a_2, which is related to angiogenic protein production. The critical points of this system can be found using the NSolve command in Mathematica.

```
(* Mathematica program to obtain critical points *)
k1=c3^2*b3/a3;b1=2.8;a2=2.4;c1=5;d2=0.8;a3=2;b3=1;c3=2;α=1;δ=1;T
max=10;
NSolve[{x*α*(1-(x/(1+b1 z^2/(c1+z^2))))□0,a2*x/(1+d2*z)-δ*y□0,
(((a3+b3)*y^2/(k1+y^2)-a3)-α*(1-(x/(1+b1*z^2/(c1+z^2)))))*z□0},
{x,y,z},Reals]
```

When $a_2 = 2.4$, there are five critical points for $N, P, E \geq 0$, these are at $O = (0,0,0)$, $C_1 = (1,2.4,0)$, $C_2 = (1.067,2,0.351)$, $C_3 = (1.667,2,1.25)$ and $C_4 = (2.732,2,2.849)$. The stability of the critical points is determined using a linearization process, the theory is given in references [23–25]. It can be shown that the critical points C_2 and C_4 are stable, while the other three critical points are unstable. The behavior of the critical points can be summarized on a bifurcation diagram. By introducing a feedback mechanism it is possible for a steady state to jump from one stable critical point to another as a parameter varies. To demonstrate this phenomenon one has to use an iterative method with feedback.

Figure 29.6 shows bifurcation diagrams for system (29.10) as the parameter a_2 is linearly ramped up and down from $a_2 = 2$ to $a_2 = 3$, and then back down again. Hysteresis cycles are clearly visible in the plots. For example, in Fig. 29.6a, the steady state is on a critical point near C_2 as the parameter a_2 is increased and there is a jump to a critical point near C_4 at $a_2 \approx 2.5$, where it remains as a_2 is increased. As the parameter a_2 is decreased, the steady state drops back down near C_2 at $a_2 \approx 2.2$, where it remains until a_2 is once more back at $a_2 = 2$.

If the parameter a_2 oscillates about the value $a_2 \approx 2.35$, then it is possible to generate oscillatory behavior for system (29.10). In this way, system (29.10) now becomes a periodically forced non-autonomous system of ODEs and existence and uniqueness theorems no longer hold. The system can now display periodic behavior, quasiperiodic behavior and even chaos (Fig. 29.7 shows periodic behaviour).

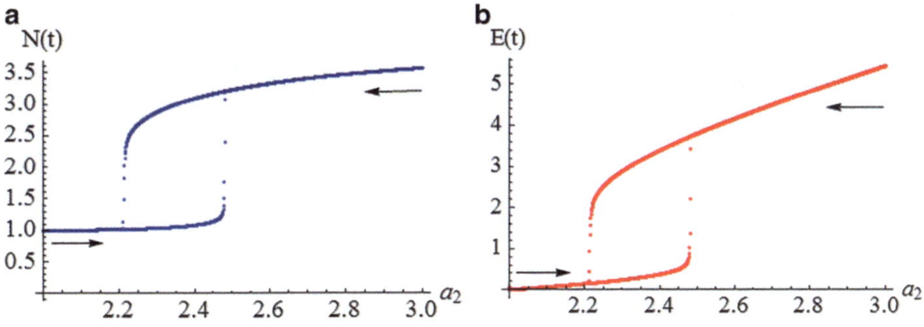

Fig. 29.6 Bifurcation diagrams for system (29.10) as the parameter a_2 is linearly increased and decreased from $2 \le a_2 \le 3$, and $c_1 = 5$, $d_2 = 0.8$, $a_3 = 2$, $b_3 = 1$, $c_3 = 2$, $\alpha = 1$, $\delta = 1$. (**a**) Hysteresis in $N(t)$, there is a counterclockwise bistable cycle. (**b**) Hysteresis in $E(t)$, again there is a counterclockwise bistable loop. Hysteresis is also present in P(t), however, the cycles are not as clear in the plots. Note that the steady states in this case are critical points

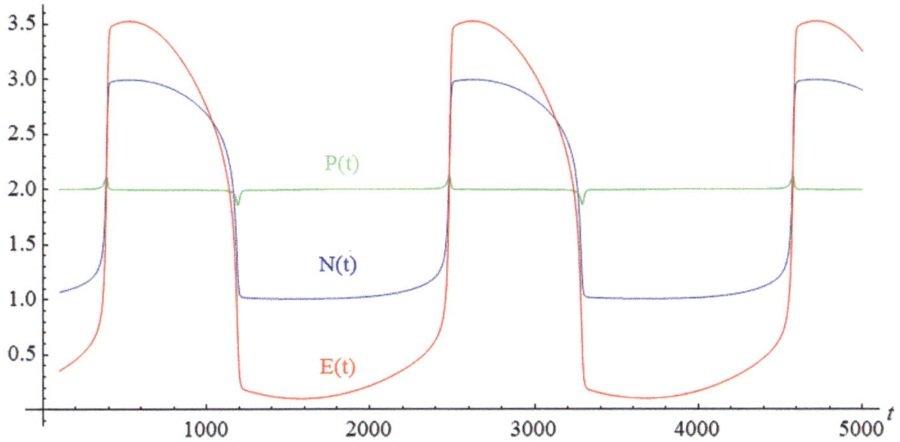

Fig. 29.7 Oscillatory behaviour of system (29.10) subject to periodic forcing $a_2(t) = 2.35 + 0.2\sin(0.003t)$ when $c_1 = 5$, $d_2 = 0.8$, $a_3 = 2$, $b_3 = 1$, $c_3 = 2$, $\alpha = 1$, $\delta = 1$

Figure 29.8 shows the bifurcation diagram for system (29.10) when $a_2 = 2.35 + 0.2\sin(0.003t)$ and the parameter b_1 is increased linearly from $b_1 = 2$ to $b_1 = 4$, and then decreased again. There is clearly a jump from one oscillator to another at $b_1 \approx 3$, however, in this case there is no hysteresis (Fig. 29.8).

To conclude this section, consider system (29.12) subject to conditions (29.13), where $m_3(t)$ is a periodic function given by $m_3(t) = 1.05 + 0.1\cos\left(\frac{t}{120}\right)$, and the two time delays vary. Figure 29.10 shows a gallery of time series plots for the quantities $N(t), P(t), E(t)$ for varying parameter values of τ_1 and τ_2. Figure 29.10a

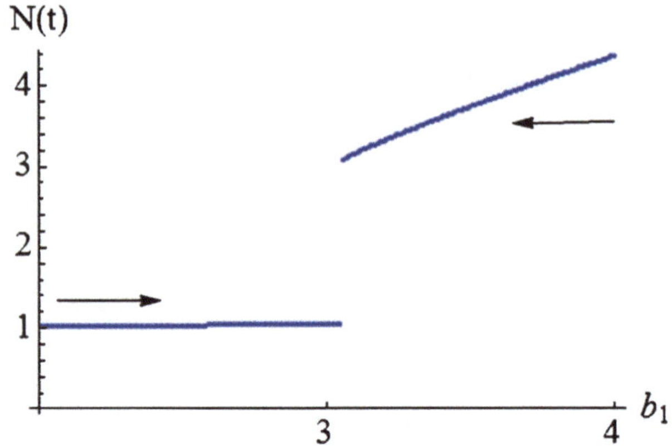

Fig. 29.8 Bifurcation diagram of system (29.10) when $a_2 = 2.35 + A\sin(0.003t)$, and the parameter b_1 is ramped up and down. The steady states in this case are stable oscillating limit cycles and there is no hysteresis

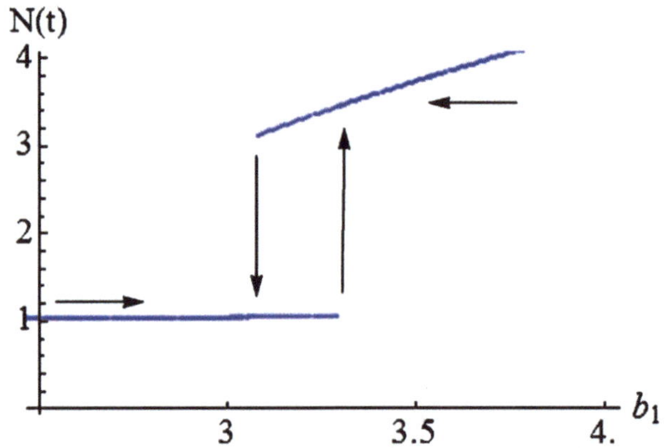

Fig. 29.9 Bifurcation diagram of system (29.10) when $a_2 = 2.35 + A\sin(0.003t)$, and the parameters A and b_1 are ramped up and down simultaneously. The steady states in this case are stable oscillating limit cycles and there is a counterclockwise hysteresis cycle

shows regular spiking, similar to the behavior of an oscillatory spiking neuron [4]. In Fig. 29.10b, the beginnings of bursting are evident and in Fig. 29.10c bursting is fully developed. Again this behavior has been observed for oscillating neurons [30]. Figure 29.10d displays amplitude modulation in all of the variables $N(t), P(t)$ and $E(t)$. In Fig. 29.10e, there is a different type of spiking behavior and in Fig. 29.10f there is a marked amplitude modulation in $E(t)$ but hardly any in $N(t)$ or $P(t)$.

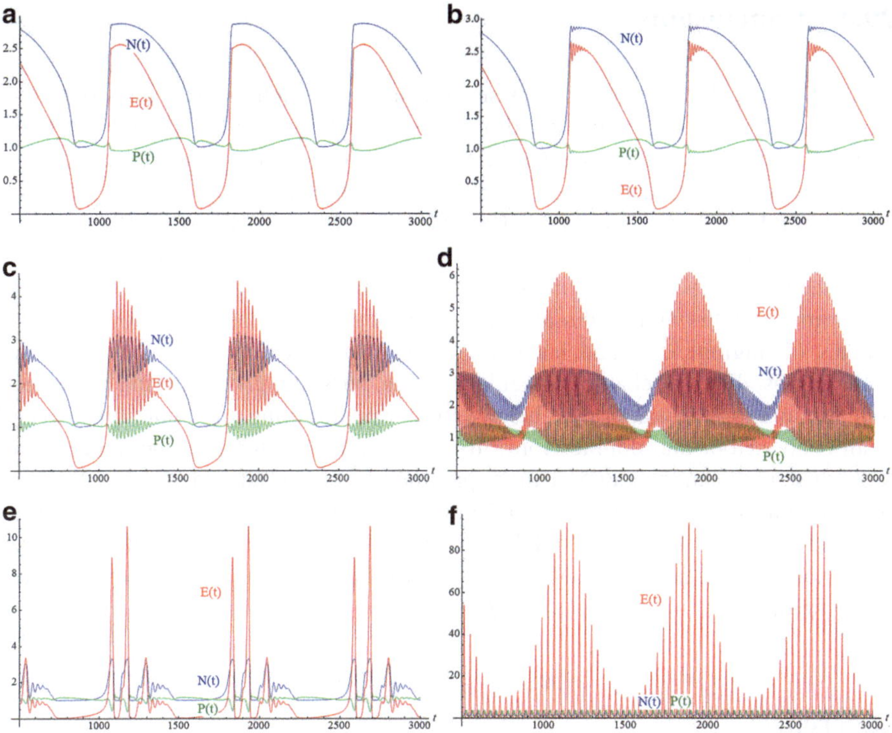

Fig. 29.10 Periodically forced angiogenesis model (29.12) subject to conditions (29.13) when $m_3(t) = 1.05 + 0.1\cos\left(\frac{t}{120}\right)$ and **(a)** $\tau_1 = \tau_2 = 1$; **(b)** $\tau_1 = 1, \tau_2 = 3$; **(c)** $\tau_1 = 1, \tau_2 = 5$; **(d)** $\tau_1 = 3, \tau_2 = 3$; **(e)** $\tau_1 = 1, \tau_2 = 15$; **(f)** $\tau_1 = 10, \tau_2 = 10$

```
(* Mathematica program to plot Figure 10(a) *)
a2=0.4;c1=1.5;b1=2.3;δ=0.34;Tmin=0;Tmax=3000;τ1=1;τ2=1;
Bodnar=NDSolve[{x'[t]==x[t]*(1-(x[t]/(1+b1 z[t-τ1]^2/(c1+z[t-τ1]^2)))),
   y'[t]==a2*x[t]/(1+z[t])-δ*y[t],z'[t]==(((y[t-τ2]^2-
(1.05+0.1*Cos[t/120])^2)/((1.05+0.1*Cos[t/120])^2+y[t-τ2]^2))-(1-
(x[t]/(1+b1*z[t-τ1]^2/(c1+z[t-τ1]^2)))))*z[t],
x[t/;t≤0]==0.4,y[t/;t≤0]==0,z[t/;t≤0]==0.1},{x[t],y[t],z[t]},
{t,0,Tmax},MaxSteps→Infinity,Method → {"StiffnessSwitching"}]
p1=ParametricPlot[Evaluate[{t,x[t]}
/.Bodnar],{t,Tmin,Tmax},PlotPoints→10000,PlotRange→All,
   AspectRatio→1/2];
p2=ParametricPlot[Evaluate[{t,y[t]}
/.Bodnar],{t,Tmin,Tmax},PlotPoints→10000,PlotRange→All,
   AspectRatio→1/2,PlotStyle→Green];
p3=ParametricPlot[Evaluate[{t,z[t]}
/.Bodnar],{t,Tmin,Tmax},PlotPoints→10000,PlotRange→All,
   AspectRatio→1/2,PlotStyle→Red];
Show[{p1,p2,p3},PlotRange→All,AxesLabel→{t,"
"},AxesOrigin→{Tmin,0},TextStyle→{FontSize→20}]
```

29.4 Conclusions

The cardiovascular system includes several functions such as the transportation of nutrients, blood, oxygen, carbon dioxide and other molecules around the body. Other functions include wound healing, tissue regeneration and reorganization, immunity, thermoregulation and developmental processes. All of these processes are inherently periodic and can display features as shown in this work. This chapter has concentrated on two mechanisms in this complex system, namely angiogenesis and haematopoieses. Mathematical models using autonomous ODEs, non-autonomous ODEs and DDEs have been investigated. It has been shown that the models of angiogenesis and haematopoiesis display a rich variety of dynamics including oscillatory behaviour (regular spiking, bursting, group spiking and amplitude modulation), quasiperiodic behaviour, period doubling and undoubling to and from chaos, Hopf and double-Hopf bifurcations, and hysteresis. The dynamics of these systems change as one or more parameters are varied and the systems have been shown to be history dependent. Mathematica programs have been listed in the text and more will be made available on the web. The numerical results displayed here should help clinical vascular biologists explain some phenomena encountered in the field.

References

1. Rapp PE (1979) An atlas of cellular oscillators. J Exp Biol 81:281–306
2. Hierlemann A, Frey U et al (2011) Growing cells atop microelectronic chips: interfacing electrogenic cells in vitro with CMOS-based microelectrode arrays introduction. Proc IEEE 99:249–251
3. Kruse K, Jülicher F (2005) Oscillations in cell biology. Opin Cell Biol 17:20–26
4. Hodgkin A, Huxley A (1952) A quantitative description of membrane current and its application to conduction and excitation in nerve. J Physiol 117:500–544
5. Borresen J, Lynch S (2009) Neuronal computers. Nonlinear Anal Theory Meth Appl 71:2372–2376
6. Lynch S, Borresen J (2012) Binary half adder using oscillators. International Publication Number WO 2012/001372 A1, International Patent, pp 1–57
7. Lynch S. Borresen J (2012) Binary half-adder and other logic circuits. UK Patent Number, GB 2481717 A, pp 1–57.
8. Borresen J, Lynch S (2012) Oscillatory threshold logic. PLoS One 7:e48498. doi:10.1371/journal.pone.0048498
9. Folkman J (1971) Tumor angiogenesis: therepeutic implications. N Engl J Med 285:1182–1186
10. Hahnfeldt P, Paigrahy D et al (1999) Tumor development under angiogenic signaling: a dynamical theory of tumor growth, treatment response, and postvascular dormancy. Cancer Res 59:4770–4775
11. Agur Z, Arakelyan L et al (2004) Hopf point analysis for angiogenesis models. Discrete Continuous Dynamical Syst B 4:29–38

12. Bodnar M, Forys U (2009) Angiogenesis model with carrying capacity depending on vessel density. J Biol Syst 17:1–25
13. Bodnar M, Piotrowska MJ et al (2013) Model of tumour angiogenesis – analysis of stability with respect to delays. Math Biosci Eng 10:19–35
14. Stamper IJ, Owen MR (2010) Oscillatory dynamics in a model of vascular tumour growth – implications for chemotherapy. Biol Direct 5:27
15. Bartha K, Riger H (2006) Vascular network remodelling via vessel co-option, regression and growth in tumors. J Theor Biol 241:903–918
16. Welter M, Bartha K et al (2008) Emergent vascular network inhomogeneities and resulting blood flow patterns in a growing tumor. J Theor Biol 250:257–280
17. Lynch S (2005) Analysis of a blood cell population model. Int J Bifurcation Chaos 15:2311–2316
18. Lynch S, Bandar Z (2005) Bistable neuromodules. Nonlinear Anal Theory Methods Appl 63:669–677
19. Lynch S, Steele AL (2011) Nonlinear optical fibre resonators with applications in electrical engineering and computing. In: Banerjee S, Mitra M, Rondoni L (eds) Applications of nonlinear dynamics and Chaos in engineering, vol 1. Springer, Berlin, pp 65–84
20. Lynch S (2011) MATLAB® programming for engineers and scientists. In: Banerjee S, Mitra M, Rondoni L (eds) Applications of nonlinear dynamics and Chaos in engineering, vol 1. Springer, Berlin, pp 3–35
21. Borresen J, Lynch S (2002) Further investigation of hysteresis in Chua's circuit. Int J Bifurcation Chaos 12:129–134
22. Lynch S, Christopher CJ (1999) Limit cycles in highly non-linear differential equations. J Sound Vib 224:505–517
23. Lynch S (2010) Dynamical systems with applications using Maple, 2nd edn. Springer-Birkhäuser, New York
24. Lynch S (2007) Dynamical systems with applications using Mathematica. Springer-Birkhäuser, New York
25. Lynch S (2014) Dynamical systems with applications using MATLAB, 2nd edn. Springer-Birkhäuser, New York
26. Smith H (2010) An introduction to delay differential equations with applications to the life sciences. Springer, New York
27. Kuznetsov Y (2010) Elements of applied bifurcation theory. Springer, New York
28. Marsden JE, McCracken M (1976) The Hopf bifurcation and its applications. Springer, New York
29. Hale J, Kocak H (1991) Dynamics and bifurcations. Springer, New York
30. Izhikevich EM (2000) Neural excitability, spiking and bursting. Int J Bifurcation Chaos 10:1171–1266
31. Erneux T (2009) Applied delay differential equations. Springer, New York
32. Dhooge A, Govaerts W, Kuznetsov YA (2003) MATCONT: a MATLAB package for numerical bifurcation analysis of ODEs. ACM Trans Math Software 29:141–164
33. Bellen A, Zennaro M (2013) Numerical methods for delay differential equations. Oxford University Press, Oxford, UK
34. Mackey MC, Glass L (1977) Oscillation and chaos in physiological control systems. Science 197:287–289
35. Carley DW, Shannon DC (1988) A minimal mathematical model of human periodic breathing. J Appl Physiol 65:1400–1409
36. Clark LH, Schlosser PM et al (2003) Multiple stable periodic solutions in a model for hormonal control of the menstrual cycle. Bull Math Biol 65:157–173
37. Baldi P, Atiya AF (1994) How delays affect neural dynamics and learning. IEEE Trans Neural Netw 5:612–621

38. Rafii S, Lyden D et al (2002) Vascular and haematopoietic stem cells: novel targets for anti-angiogenesis therapy? Nat Rev Cancer 2:826–835
39. Balachandran B, Kalmar-Nagy T et al (2009) Delay differential equations: recent advances and new directions. Springer, New York

Key web-address for accessing Mathematica notebook

http://library.wolfram.com/infocenter/Books/6881/

Chapter 30
Genomic Microarray Analysis

Stephen Hamlet, Eugen Petcu, and Saso Ivanovski

30.1 Introduction

New blood vessel formation i.e. angiogenesis, is a fundamental biological process involved in tissue development, maintenance, repair and disease. The use of high throughput approaches such as DNA microarray technology that enables the parallel analysis of multiple genes has identified an increasing number of genes that have been linked to angiogenesis in recent years [1–3]. New bioinformatic approaches, such as the meta-analysis of this publicly available microarray data, are now being investigated to potentially identify 'common higher-level' transcriptional profiles in different disease states. The inclusion of such large sample numbers promises to detect the differential expression of genes that may not be otherwise found in individual studies and to minimize the rate of false positive findings [4].

Gene expression profiling therefore provides a means to make valuable insights into the effect(s) of differential gene expression that are ultimately manifested by the phenotypic differences observed between normal and diseased states. Simply put, the 'expression' of the inherited set of instructions that an organism carries within its genetic code in response to stimuli or instruction, is facilitated by the transcription of messenger RNA (mRNA), which is subsequently translated into the

S. Hamlet (✉)
Griffith Health Institute, Gold Coast Campus, Griffith University,
Nathan, QLD 4222, Australia
e-mail: s.hamlet@griffith.edu.au

E. Petcu
Griffith Health Institute and School of Medicine, Gold Coast Campus, Griffith University,
Nathan, QLD 4222, Australia

S. Ivanovski
Griffith Health Institute and School of Dentistry and Oral Health, Gold Coast Campus,
Griffith University, Nathan, QLD 4222, Australia

M. Slevin, G. McDowell (eds.), *Handbook of Vascular Biology Techniques*,
DOI 10.1007/978-94-017-9716-0_30

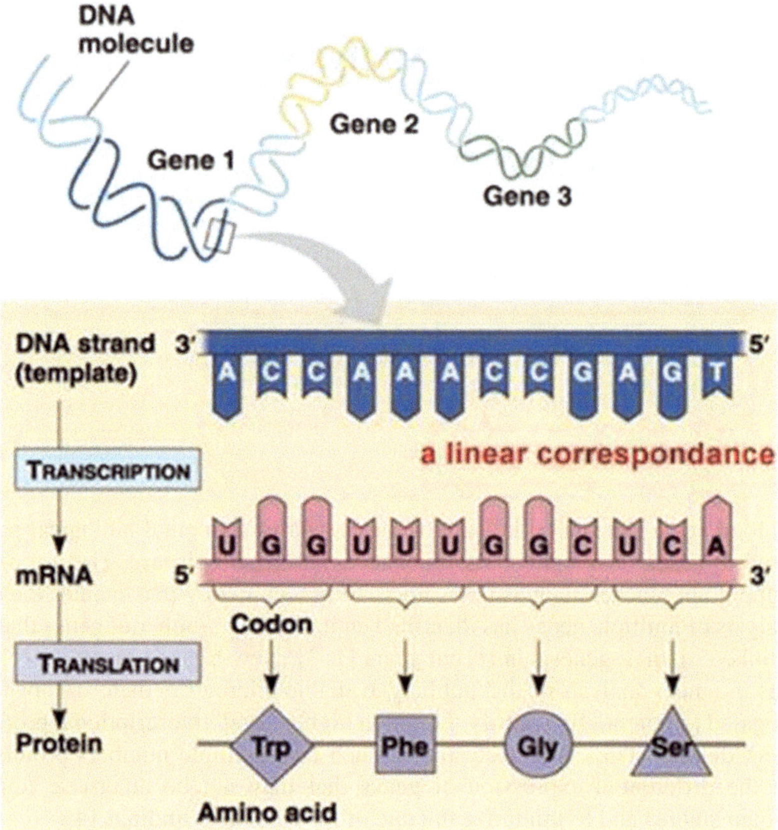

Fig. 30.1 The central dogma of genetics

proteins that ultimately determine the phenotype or observable characteristics and traits of an organism (Fig. 30.1).

Molecular methods of gene expression analysis based on the DNA sequence have evolved from the seminal work by Southern [5] and typically include hybridization techniques using DNA probes to identify and quantitate complementary DNA sequences. Microarray technology however is able to facilitate high throughput whole genome analysis of gene expression at both low cost and most importantly, increased sensitivity and specificity compared with these traditional methods.

In general terms, a microarray is a collection of microscopic features immobilized on a solid support surface. These features contain millions of immobilized identical probes that can hybridize with fluorescently labeled target molecules to produce an increase in fluorescence intensity over a background level that can be measured using a fluorescent scanner to produce both quantitative (gene expression) and qualitative (diagnostic) data. Microarrays can thus be categorized based upon the solid-surface support used, the nature of the probe and the methods used for probe addressing and/or target detection as illustrated in Table 30.1.

Table 30.1 Comparison of microarray platforms

Array type	Array support	Probe type	Probe density
Printed	Glass slides	cDNA, PCR amplicons 200–800 bp Oligonucleotides 25–80 bp	Moderate 10,000 – 30,000
In situ synthesized	Quartz wafer	Affymetrix – GeneChips® 20–25 bp	High
		Roche – NimbleGen® 60–100 bp	15,000 – >10^6
		Agilent – SurePrint® – 60 bp	
High-density bead arrays	Fiber optic bundles or silicon slides	Illumina – Sentrix BeadChip® 50-mer gene-specific probes are attached to beads, which are then assembled into the arrays	High 50,000 – >10^6
Electronic	Hydrogel permeation layer loaded with streptavidin over an electrode array	Nanogen – NanoChip® uses electronic addressing of charged biomolecules on the electrode array to separate and concentrate analyte targets	Low 400
Liquid-bead suspension	Microspheres	Non-specific capture sequences	Low 100

GeneChip® (Affymetrix, Santa Clara, CA) microarrays, arguably the most widely utilized for research purposes and therefore the focus of the methodology described in this chapter, use oligonucleotide probes which are synthesized using a semiconductor-based photochemical process directly onto the surface of the microarray, which is typically a 1.2 cm^2 quartz wafer (Fig. 30.2). These probes are typically short (20–25 base pairs) so multiple probes per target are included to improve sensitivity, specificity, and statistical accuracy. Classically, 11 probes are used per 600 bases being examined [6]. The gene sequence data for these probes is typically obtained from repositories of collected experimental data such as GenBank and the European Molecular Biology Laboratory (EMBL).

The use of 'probe sets' further increases the assay specificity. A probe set includes one perfect-match probe and one mismatch probe that contains a 1-bp difference in the middle position of the probe (i.e., position 13 of a 25-bp probe). Each member of the probe set is located in a separate feature, which allows the mismatch probe to act as a negative control to identify possible nonspecific cross-hybridization events. Affymetrix GeneChips typically have >10^6 features per microarray depending on the inter-feature distance [7, 6].

Technical issues (discussed in the methodology) aside, a poorly designed experiment is probably the next most common reason that microarray experiments fail to yield results that are interpretable. Success with microarray approaches therefore requires a sound experimental design and a coordinated and appropriate use of statistical tools (Fig. 30.3). Most microarray experiments for example can be classified into one of three types. The first is the comparison of the same sample type before or after some defined treatment. Issues such as the experimental time course or duration of the treatment and the dose, etc., that can have dramatic effects all need

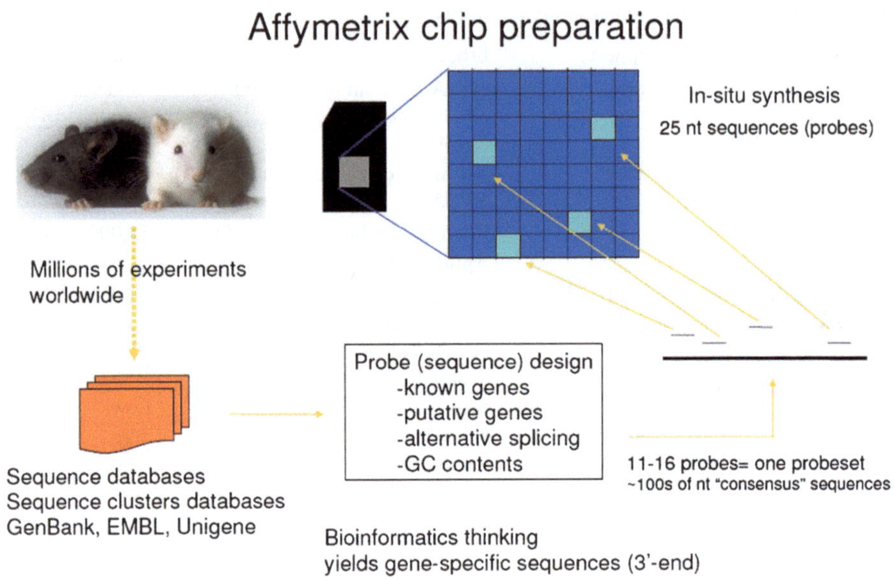

Fig. 30.2 Affymetrix GeneChip® preparation

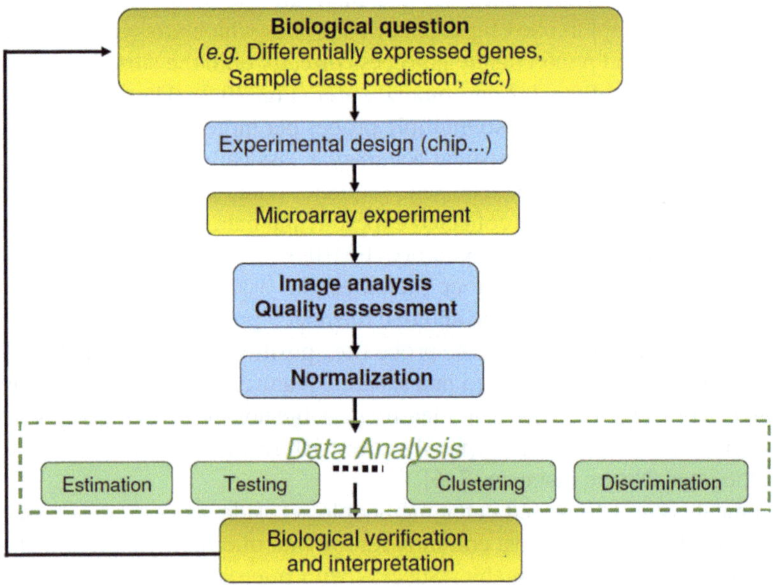

Fig. 30.3 Gene profiling workflow using microarray analysis

to be considered in these experiments. Second is a comparison of samples that are isogenic except for a limited number of genetic changes, such as a single over-expressed or mutated gene. Again, care in interpretation is needed as the loss of one gene may be compensated by the up-regulation of other genes. Finally, the third experiment type is the comparison of normal or diseased samples from multiple individuals. This type of study is particularly prone to problems with unbalanced cohorts.

30.2 Methodology

30.2.1 Double Stranded cDNA Synthesis

Notes

- SuperScript Double-Stranded cDNA Synthesis Kit (Life Technologies).
- The starting material should be 5–20 μg of *total RNA*
- The RNA quality should be checked using a spectrophotometer, a purity value of 1.9–2.1 is acceptable. It can also be checked on a 1–3 % agarose gel where there should be no smearing of the ribosomal bands or analyzed on the 'Lab Chip' from Agilent Technologies.

1st strand cDNA synthesis:

(a) In an RNase-free 1.5-mL microcentrifuge tube, add oligo-dT according to the table below. Incubate at 70 °C for 10 min then quick-chill on ice.

	μg of total RNA		
	5–8	8.1–16	16.1–20
T7-oligo dT	2 μL	2 μL	2 μL
RNA in DEPC water	9 μL	8 μL	7 μL
Total volume	*11 μL*	*10 μL*	*9 μL*

(b) Briefly centrifuge then add the following to the tube, mix well and incubate at 42 °C for 2 min to equilibrate sample.

Reagents	Volume
5× first strand buffer	4 μL
0.1 M DTT	2 μL
dNTP mix (10 mM)	1 μL

(c) Add SuperScript II RT according to the following table. Mix gently, and incubate at 45 °C for 1 h. The total volume should now be 20 μL. Place the tube on ice to terminate the reaction.

	μg of total RNA		
	5–8	8.1–16	16.1–20
SuperScript II RT	2 μL	3 μL	4 μL
Total volume	*20 μL*	*20 μL*	*20 μL*

2nd strand cDNA synthesis:

Notes

- Place the first strand reaction on ice; centrifuge briefly to bring down the condensation on the sides of the tubes.
- Transfer the first stand mixture to a larger tube and add:

Reagent	Volume (μL)
First strand cDNA synthesis	20
DEPC water	91
5× second strand buffer	30
10 mM dNTP mix	3
10U/μL *E. coli* DNA ligase	1
10U/μL *E. coli* DNA polymerase 1	4
2U/μL *E. coli* RNase H	1
Total volume	*150 μL*

(a) Gently mix the tube (don't vortex) then centrifuge to bring down any residue on the sides.
(b) Incubate at 16 °C for 2 h
(c) Add 2 μL of T4 DNA polymerase (5U/μL) and incubate at 16 °C for 5 min
(d) Add 10 μL of 0.5 M EDTA
(e) Then clean up the sample or freeze at −20 °C.

30.2.2 Clean Up of Double Stranded cDNA

Notes

- MinElute Reaction Cleanup Kit (Qiagen)
- All steps in this procedure are to be done at room temperature and the sample should be in a 1.5 or 2 mL eppendorf tube.

(a) Add 600 μL of cDNA binding buffer to the 162 μL final double stranded cDNA synthesis product. Mix by vortexing for 3 s. This should result in a solution that is slightly yellow, about the same colour as the cDNA binding buffer before the solution was added. If this is not the case and the solution is purple, then add 10 μL of 3 M sodium acetate to correct the pH.

(b) Apply 500 μL of the sample to the cDNA cleanup spin column. Have the column in a 2 mL collection tube, and centrifuge at 10,000 rpm for 1 min. Discard flow through.
(c) Reload the column with the remaining 262 μL and centrifuge as above and discard flow through.
(d) Transfer the column to a new 2 mL tube and wash with 750 μL of cDNA wash buffer then spin as above. Discard flow through.
(e) Open the cap of the spin column and centrifuge at 25,000 g and discard flow through and the tube. This is to dry the membrane.
(f) Transfer the spin column into a 1.5 mL collection tube and add 14 μL of cDNA elution buffer onto the membrane. Centrifuge at 10,000 rpm for 1 min. The elution volume is about 12 μL.
(g) It is not recommended to RNase treat the sample before cDNA labeling.
(h) Analyse the sample on a 1 % agarose gel (about 1 μL of sample) then stain with EtBr; this isn't to be done if the RNA is from total RNA.
(i) There is no need to try and quantify the amount of cDNA because the primers will also absorb at 260 nm.

30.2.3 Synthesis of Biotin Labeled cRNA

Notes

- BioArray HighYield RNA Transcript Labeling Kit (T7) (Enzo Life Sciences)
- All reagents should be briefly centrifuged to bring down the entire sample which may have collected on the sides of the tubes.
- If the sample is prepared correctly there will be enough RNA for at least three chips.

Total RNA (μg)	Volume of cDNA to use
5–8	10 μL
8.1–16	5 μL
16.1–20	3.3 μL

Add template cDNA and other reagents in order:

Reagent	Volume
Template cDNA	3.3–10 μL (see table above)
Distilled/deionised water	18.7–12 μL (see table above)
10× Hy reaction buffer	4 μL
10× biotin-labelled ribonucleotides	4 μL
10× DTT	4 μL
10× RNase inhibitor mix	4 μL

(continued)

Reagent	Volume
20× T7 RNA polymerase	2 μL
Total volume	*40 μL*

(a) Mix reagents then centrifuge to ensure everything is at the bottom of tube.
(b) Place in 37 °C water bath and incubate for 4–5 h gently mixing the tube every 30–45 min.
(c) Overnight incubation may produce shorter products which is less desirable.
(d) Store product at −70 or −20 °C.

30.2.4 Cleanup and Quantification of Biotin-labelled cRNA

Notes

- GeneChip® IVT cRNA Cleanup Kit (Affymetrix)
- All the nucleotides must be removed so the reading at 260 nm is accurate
- Don't use phenol/chloroform extraction as this lowers the yield (biotin is organic)
- Save aliquot to do gel electrophoresis analysis
- Work at room temperature

(a) Add 60 μL RNase free water to the in vitro transcription (IVT-step 2) reaction and mix well (vortex 3 s).
(b) Add 350 μL IVT cRNA binding buffer to sample and vortex.
(c) Add 250 μL ethanol (96–100 %) to the lysate and mix with a pipette.
(d) Apply sample (700 μL) to the IVT cRNA cleanup spin column sitting in a 2 mL tube. Spin for 15 s at 10,000 rpm. Discard the flow through.
(e) Transfer the spin column into a new 2 mL tube and wash with 500 μL wash buffer then spin for 15 s at 10,000 rpm. Discard the flow through.
(f) Put 500μI of 80 % ethanol onto the column then spin for 15 s at 10,000 rpm. Discard the flow through.
(g) Open the cap and spin for 5 min at maximum speed. Discard the flow through and the collection tube.
(h) Transfer the spin column into a new 1.5 mL tube and pipette 11 μL of RNase free water onto the middle of the membrane then centrifuge for I minute at maximum speed.

Notes

- When analysing use 1/100 or 1/200 dilutions
- To check the yield; Absorbance at 260 and 280 nm
- Adjusted cRNA yield = $RNA_{m\,-}$ (total RNA_i)(y)
- RNA_m = amount measured (μg)
- Total RNA_i = starting amount of RNA
- Y = fraction of cDNA in IVT

30.2.5 Fragmentation of cRNA

Notes

- GeneChip® 3′ IVT Express Kit Assay (Affymetrix)
- The minimum concentration of cRNA is 0.6 μg/mL

(a) Mix 2 μL of 5× fragmentation buffer per 8 μL of RNA and water. The final concentration can be between 0.5 and 2 μg/mL. It is important to use the adjusted cRNA concentrations.
(b) Incubate at 94 °C for 5 min then sit on ice.
(c) Save some of the sample to run on a gel (1 μg if using Ethidium bromide).
(d) Store at −20 °C until ready to hybridize.

Notes

- To check fragmentation, run 1 % of each sample on a 1 % agarose gel.

(a) Mix RNA with loading dye and heat to 65 °C for 5 min
(b) Stain after running
(c) Use a denaturing gel.

30.2.6 Hybridisation

Notes

- GeneChip® Hybridization, Wash, and Stain Kit (Affymetrix)
- The stock of 20× Genechip eukaryotic hybridisation control must be heated to 65 °C for 5 min to ensure cRNA is resuspended before aliquoting.
- The array must be allowed to equilibrate to room temperature before use.

Component	Test array	Standard array	Final concentration
Fragmented cRNA	5 μg	15 μg	0.5 μg/mL
Control oligonucleotide B2 (3nM)	1.7 μL	5 μL	50 pM
20× Eukaryotic hybridization controls (bio B, bio C, bio D, Cre)	5 μL	15 μL	1.5, 5, 25, 100 pM
Herring sperm DNA (10 mg/mL)	1 μL	3 μL	0.5 mg/mL
2× hybridization buffer	50 μL	150 μL	1×
Water	To 100 μL	To 300 μL	
Total volume	*100 μL*	*300 μL*	

(a) Heat the hybridization cocktail to 99 °C for 5 min.
(b) Fill the array with the maximal volume (100 or 300 μL depending upon the chip) of 1× hybridization buffer to wet the array.

(c) Incubate the array with buffer for 10 min at 45 °C rotating it in the oven.
(d) Transfer the hybridization cocktail from (b) to 45 °C and leave for 5 min.
(e) Spin cocktails (still in tube) at maximum speed for 5 min.
(f) Remove the buffer solution from the probe array cartridge and fill with appropriate volume of the clarified hybridization cocktail avoiding anything which has been deposited on the bottom.
(g) Place the probe array in the rotisserie box (make sure it is balanced) in the oven at 45 °C rotate at 60 rpm for 16 h.

30.2.7 Washing and Staining the Array

Notes

- An automated GeneChip® Fluidics Station (Fig. 30.4) is used for the wash and stain operation of Affymetrix GeneChip® arrays.
- After hybridization, remove the hybridization cocktail from the probe array and set the solution aside in a micro-centrifuge tube. Store the solution on ice at the moment, but for long term storage transfer to –80 °C.
- Fill the probe array completely with the appropriate volume of non-stringent wash buffer. Both high and low stringency buffers are used in the wash process.
- The probe array can be stored at 4 °C for a maximum of 3 h if essential, but it will need to be equilibrated to room temperature before processing on the fluidics station.

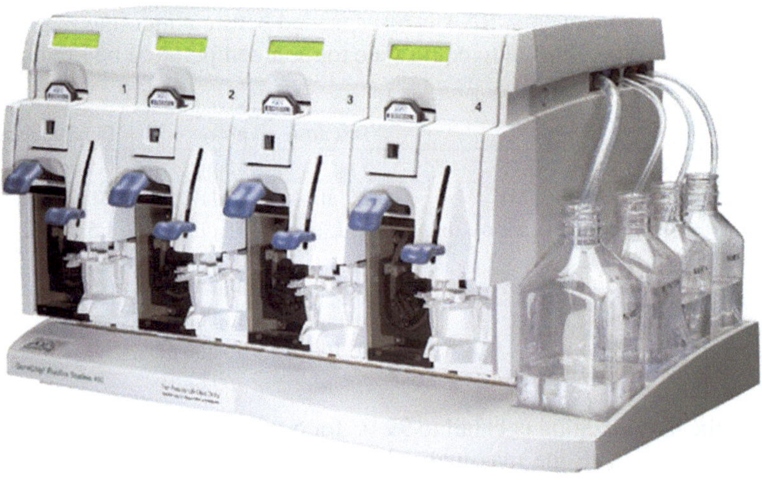

Fig. 30.4 GeneChip® Fluidics Station

- Don't freeze Streptavidin-phycoerythrin (SAPE); store in the dark wrapped in alfoil; prepare just before use.
- Follow prompts on fluidic station for wash/stain protocol

Staining Procedure 1 Eukaryotic targets (~75 min)

Components	Volume	Final concentration
2× MES stain buffer	300 μL	1×
50 mg/mL acetylated BSA	24 μL	2 mg/mL
1 mg/mL SAPE	6 μL	10 μg/mL
Water	270 μL	
Total volume	*600 μL*	

Staining Procedure 2 This protocol is recommended for use with probe arrays with probe cells of 24 μm or smaller (~90 min).

Components	Volume	Final concentrations
2× MES stain buffer	600 μL	1×
50 mg/mL acetylated BSA	48 μL	2 mg/mL
lmg/mL SAPE	12 μL	10 μg/mL
Water	540 μL	
Total volume	*1,200 μL*	

Antibody solution

Components	Volume	Final concentration
2× MES stain buffer	600 μL	1×
50 mg/mL acetylated BSA	24 μL	2 mg/mL
10 mg/mL normal goat IgG	6 μL	0.1 mg/mL
0.5 mg/mL biotinylated antibody	3.6 μL	3 μL/mL
Water	266.4 μL	
Total volume	*600 μL*	

30.2.8 Scanning the Array

Notes

- Affymetrix GeneChip® Scanner (Fig. 30.5)
- Warm up the laser up for at least 15 min
- For probe arrays with probe cells 24 μm or less – pixel value 3 μm, wavelength 570 nm
- For a 50 μm probe array – pixel value 6 μm, wavelength 570 nm

Fig. 30.5 GeneChip® scanner

30.3 Data Analysis

30.3.1 Data Extraction

Data is provided as .DAT files. These files are the output of the scanner and contain information about each pixel in the raw images. They are imported into the Affymetrix GCOS software to calculate the intensities. The libraries required to process the specific chip sets can be obtained from Netafix (Affymetrix). These intensity (.CEL) files may subsequently be used as input into image analysis software such as GeneSpring.

30.3.2 Importing Data Into GeneSpring

The intensity (.CEL) files are imported into GeneSpring and preprocessed with a normalisation algorithm. Robust Multiarray Averaging (RMA) adjusts the signal intensity for systematic errors introduced by differences in procedures and dye intensity effects. For example, the signal strength for the mismatched probes can often be larger than that of the perfect match probes, implying that the mismatched probe is detecting true signal as well as background signal resulting in nonsensical negative expression values.

RMA is a normalisation procedure for microarrays that background corrects, normalises and summarises the probe level information without the use of the information obtained in the mismatched probes. We routinely use the GeneChip RMA (GC-RMA) algorithm, which uses sequence-specific probe affinities of the

GeneChip probes to attain more accurate gene expression values as the preferred method for analyzing Affymetrix data. This method has been described in the literature to show the greatest consistency.

30.3.3 Normalisaton and Filtering

The preprocessed data is often further normalised per gene as the absolute level of expression among different genes varies dramatically. For example a small change in fluorescence of a high abundance transcript may represent a significant change in expression of a gene that is expressed at low levels. By default, most microarray data analysis programs calculate the mean expression level for each gene, then normalize each sample against that mean, or control value. Following this, the noise in the array can be filtered out by discarding any genes labeled as absent in any of the samples of the experiment, as well as discarding any gene which had a raw signal value under 100. Only genes that pass these criteria should be used onwards.

30.3.4 Finding Significant Genes

To determine genes that were significantly differentially regulated, i.e different between time points or between health and disease (Fig. 30.6), one way ANOVA is routinely performed on the filtered and normalized genes, followed by multiple testing correction to reduce the number of false positives. Post hoc analysis is then performed on the ANOVA results.

30.3.5 Functional Characterization

Significant genes as determined from the statistical analysis may then be further characterized functionally, by examining their over-representation in pathways and in gene ontology categories (Fig. 30.7).

1. Overrepresentation of genes in KEGG pathways:
   We use a script from the R integration package in GeneSpring (Bioscript library 2.2) to map to KEGG pathways. This script is designed to compare a user-defined gene list to known biological pathways, identifying genes in common and producing a measurement of confidence on the over-representation.
2. Overrepresentation of genes in Gene Ontology categories:
   Online tools such as The Database for Annotation, Visualization and Integrated Discovery (DAVID, http://david.abcc.ncifcrf.gov) inputs lists of significantly expressed genes to determine enriched functional-related gene groups based on Gene Ontology (GO) classifications.

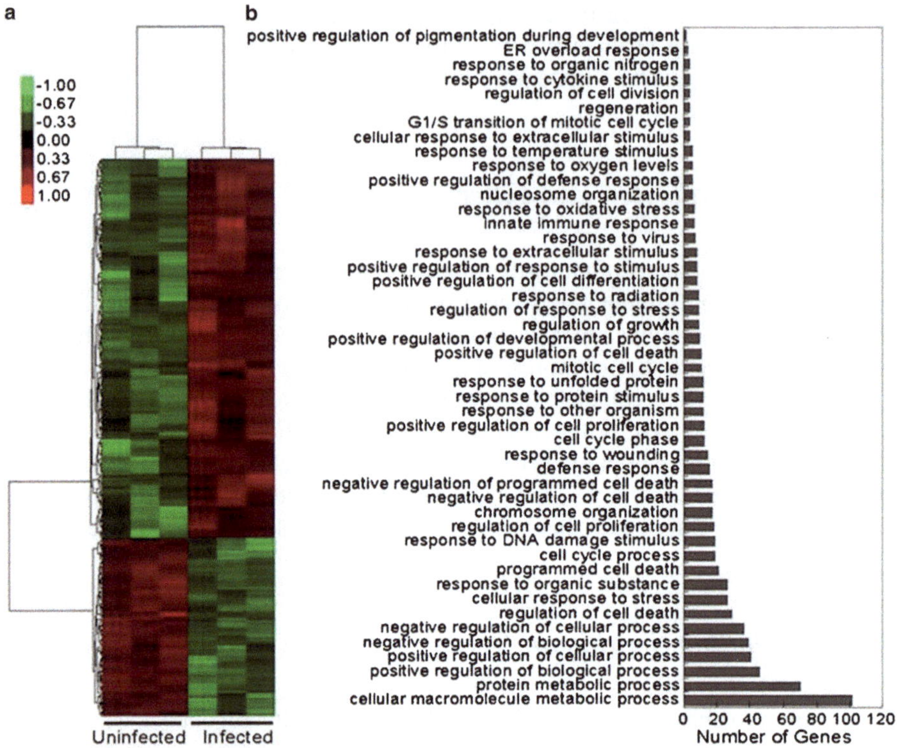

Fig. 30.6 Clustering analysis of significant genes

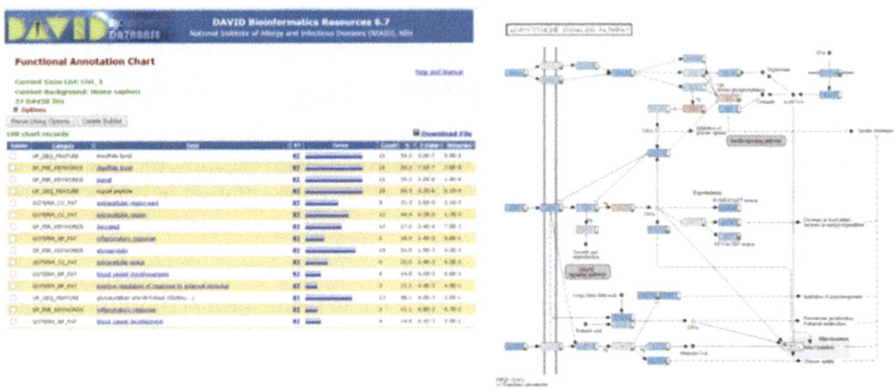

Fig. 30.7 'DAVID' and 'KEGG' functional analysis tools

References

1. Costouros NG, Libutti SK (2002) Microarray technology and gene expression analysis for the study of angiogenesis. Expert Opin Biol Ther 2(5):545–556
2. Toft JH, Lian IA, Tarca AL et al (2008) Whole-genome microarray and targeted analysis of angiogenesis-regulating gene expression (ENG, FLT1, VEGF, PlGF) in placentas from pre-eclamptic and small-for-gestational-age pregnancies. J Matern Fetal Neonatal Med 21(4):267–273
3. Anders M, Fehlker M, Wang Q et al (2013) Microarray meta-analysis defines global angiogenesis-related gene expression signatures in human carcinomas. Mol Carcinog 52:29–38
4. Choi JK, Choi JY, Kim DG et al (2004) Integrative analysis of multiple gene expression profiles applied to liver cancer study. FEBS Lett 565:93–100
5. Southern EM (1975) Detection of specific sequences among DNA fragments separated by gel electrophoresis. J Mol Biol 98:503–517
6. Dalma-Weiszhausz DD, Warrington J, Tanimoto EY et al (2006) The affymetrix GeneChip platform: an overview. Methods Enzymol 410:3–28
7. Fodor SP, Read JL, Pirrung MC, Stryer L, Lu AT, Solas D (1991) Light-directed, spatially addressable parallel chemical synthesis. Science 251:767–773

Chapter 31
Selection of Appropriate Housekeeping Genes for Quality Control

Stephen Hamlet, Eugen Petcu, and Saso Ivanovski

31.1 Introduction

Genetic differences inherent between individuals are often reflected in their susceptibility to disease. The determination of genetic differences between individuals can therefore allow insights into the role of genetic diversity in a given disease state. As such, traditional techniques to quantitate mRNA levels such as Northern blotting often require more RNA than is often available.

The development of the polymerase chain reaction (PCR), which allows the amplification of very small quantities of cDNA reverse transcribed from mRNA into amounts useful for further analysis, has subsequently revolutionised the analysis of gene expression. An insight into the role of specific genes in health and disease, or the effect of treatment on these genes for example, can now be determined simply by making comparisons between their levels of expression.

Unfortunately, at the heart of the PCR process, all downstream comparisons of the levels of amplified DNA are reliant upon their initial concentration in the reaction mix. This concentration is affected by both biological and technical factors including amongst others: the sample source, RNA quality and the reaction efficiencies etc. Clearly even small differences in the amount of starting template from sample to sample will affect subsequent amplified target gene values. To circumvent

S. Hamlet (✉)
Griffith Health Institute, Gold Coast Campus, Griffith University,
Nathan, QLD 4222, Australia
e-mail: s.hamlet@griffith.edu.au

E. Petcu
Griffith Health Institute and School of Medicine, Gold Coast Campus, Griffith University,
Nathan, QLD 4222, Australia

S. Ivanovski
Griffith Health Institute and School of Dentistry and Oral Health, Gold Coast Campus,
Griffith University, Nathan, QLD 4222, Australia

M. Slevin, G. McDowell (eds.), *Handbook of Vascular Biology Techniques*,
DOI 10.1007/978-94-017-9716-0_31

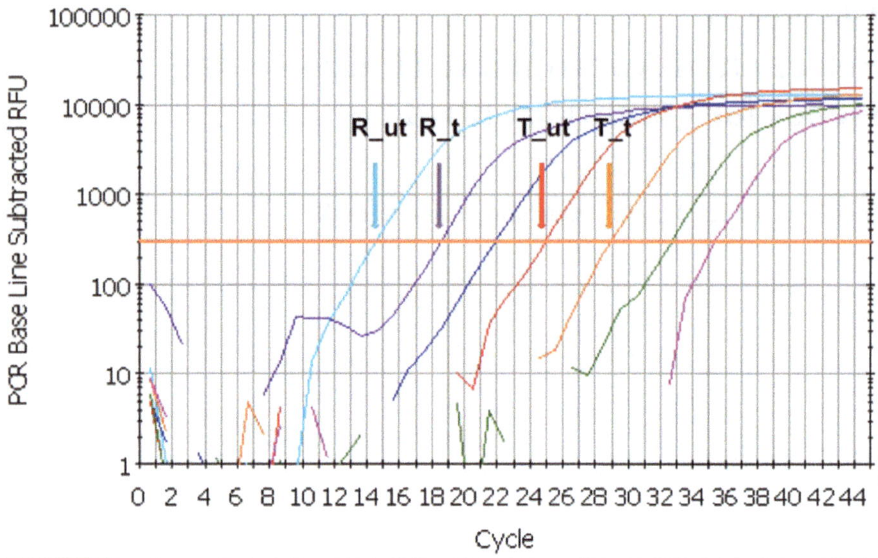

Fig. 31.1 *R* reference, *T* target, *ut* untreated, *t* treated

this problem, the level of expression of the target gene is compared with that of a gene that is expressed at a consistent level among the various sample types under study. This 'normalisation' process therefore allows changes in the level of gene expression for the target to be calculated irrespective of the starting mass of cDNA.

By way of example, if we were to consider only the target (T) gene's values illustrated in Fig. 31.1, we could conclude that its expression as a result of some treatment (ut vs. t) differed by 16-fold as there is a difference of 4 in their Ct values (i.e. 29 − 25 = 4Ct difference). However, as the expression of the reference gene also differs by four (19 c.f. 15), this difference tells us that there is actually a 16-fold difference in the concentration of starting cDNAs and the target's expression is actually equal in these two samples. Extending this argument further, clearly if the expression of the normalizer gene varies between the sample types under study, any final fold differences in expression between samples will be also be in error.

So how does the researcher select a normalization gene(s)? Fundamentally, the candidate gene has to fulfill two criteria: (*a*) it has to be stably expressed in the tissue of interest; and (*b*) it has to have an expression level above background [1]. A quest for a universal normalization gene with consistent expression across all possible tissue samples, cells, experimental treatments, and designs etc. is perhaps quixotic at best. However as suggested by Anderson et al. [1], most experimental designs are restricted to a few variables such that one or more genes are often constitutively expressed. However, the task of identifying these genes is not trivial. It is composed of two steps: first, to identify which genes are likely candidates; and second, to verify the stability of these candidates.

In trying to answer the first question, the 'Catch22' of reference gene selection arises i.e. how can you select a reference gene based on stability of expression without the use of further references? Several statistical algorithms discussed below have been developed to help identify the best reference genes to use under given experimental conditions [1, 2, 4]. Public microarray databases such as the Gene Expression Omnibus (GEO) from NCBI or ArrayExpress from EBI are also freely searchable repositories containing microarray gene expression data. A list of some of the most commonly used reference genes is also provided in the methodology section. Notwithstanding these publicly available lists, ultimate validation of any candidate reference genes needs to be performed by the researcher for each specific set of experimental conditions (Fig. 31.2).

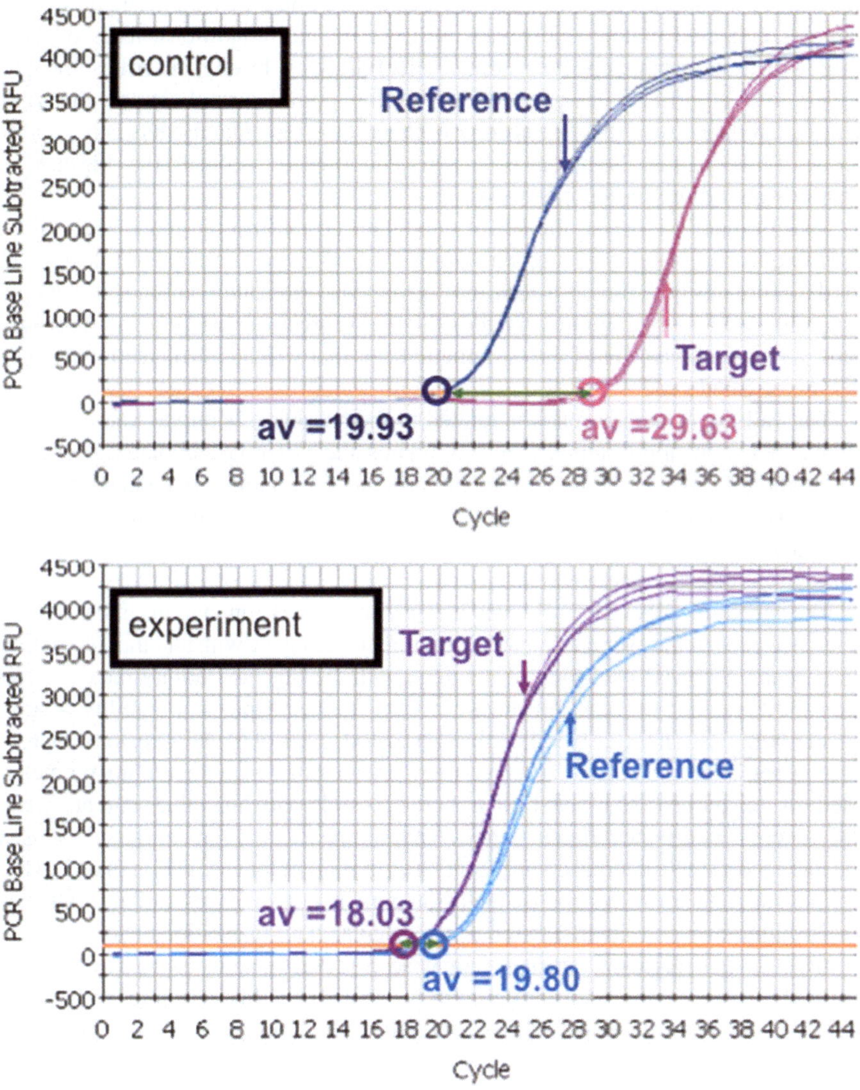

Fig. 31.2 Example data for a suitable reference gene

The top panel illustrates the level of expression for a target and reference gene prior to experimental treatment. Following treatment, the same amount of total RNA was used for reverse transcription of both RNAs, and the same amount of each reverse transcription reaction was used for real-time PCR (lower panel). As expected for a good reference gene, there is very little difference between the two RNA samples with regard to their levels of expression since the Ct values are very close suggesting that this is a good reference gene for this specific experiment.

More recently however, the use of multiple normalizers to minimize errors based on the choice of a single gene has become almost the norm in gene expression studies. This 'multiple normalizer' approach also alleviates the inherent problems of gene expression levels derived by averaging values obtained from large numbers of samples collected under widely varying experimental conditions, as is often the case with microarray repository data. The following briefly describes the most widely used algorithms designed to identify the optimal normalization gene from a set of candidates. As a full discussion of the various statistical models is beyond the scope of this chapter, the reader is strongly encouraged to consult the original publications for a more in-depth analysis.

'NormFinder' [1] ranks the set of candidate normalization genes according to their expression stability in a given sample set and given experimental design. The model estimates both the intra- and inter-group variation and combines the two into a stability value, which intuitively adds the two sources of variation and thus represents a practical measure of the systematic error that will be introduced when using the investigated gene.

The validity of the approach is related to the number of samples and candidates analyzed, i.e., the more samples and candidates, the better the estimates. The sample set should minimally contain 8 samples/group, and the number of candidates should be at least 3 for technical reasons, but 5–10 are recommended. It is a further requirement that the candidates are chosen from a set of genes with no prior expectation of expression difference between groups [1].

The 'geNorm' algorithm [4] calculates a gene expression stability measure for a reference gene as the average pairwise variation for that gene with all other tested reference genes. Stepwise exclusion of the gene with the highest stability measure value then allows ranking of the tested genes according to their expression stability. A subsequent normalization factor based on the expression levels of the best-performing reference genes is then calculated. The geometric mean rather than the arithmetic mean is used for this purpose as it better controls for possible outlying values and abundance differences between the different genes. Vandesompele et al. [4] recommend the use of the three most stable internal control genes as a minimum for calculation of an RT-PCR normalization factor with stepwise inclusion of more control genes until the extra gene has no significant contribution to the newly calculated normalization factor.

'BestKeeper' [2] uses repeated pair-wise correlation analysis to determine the optimal reference genes. This program uses cycle threshold (Ct) values of candidate reference genes instead of relative quantities. It employs pair-wise correlation analyses to calculate the Pearson correlation coefficient for each candidate reference gene pair, as well as the probability of correlation significance. Initial estimation of the data calculated variations: standard deviation (SD) and coefficient of variance

(CV) values for all the candidate reference genes show the overall stability in gene expression, from the most stable expression (with the lowest variation) to the least stable gene (with the highest variation). Any candidate gene with a SD value higher than one is considered inconsistent.

On a final note with regard to statistical algorithms, 'RefFinder' is a freely available web-based comprehensive tool (http://www.leonxie.com/referencegene.php) that integrates the currently available major computational programs i.e. geNorm, Normfinder, BestKeeper, and the comparative Ct method [3]. This tool has been developed to evaluate and screen reference genes from extensive experimental datasets. Based on the rankings from each program, it assigns an appropriate weight to an individual gene and calculates the geometric mean of their weights for the overall final ranking. An example data set is provided in Table 31.1.

31.2 Methodology

31.2.1 Choosing a Stable Normalizer for Gene Expression

1. Choose 3–4 representatives for each distinct sample type (treatment, tissue source, time point, etc.).
2. Isolate total RNA using a sample preparation method appropriate for your sample type(s).
3. DNase treat to remove potential contamination with genomic DNA.
4. Quantify RNA concentration (**see Note #1**).
5. Reverse transcribe equal mass amounts of RNA for each sample (**see Note #2**).
6. Amplify an equal volume of each cDNA (**see Note #3**) including the candidate normalizer gene(s) (**see Note #4**).
7. Assess the variability in Ct among the various sample types (**see Note #5**).

Note #1 Spectroscopic absorbance readings at 260 nm (A_{260}) measure not only RNA, but also genomic DNA and free nucleotides. Hence if the samples for analysis potentially contain unequal amounts, such that the A_{260} readings may prove inaccurate, the researcher should consider quantifying RNA levels using a fluorometer and an RNA-binding dye such as RiboGreen®.

Note #2 In order to minimize the possible effect of inhibitors on reverse transcription, use significantly less RNA than that recommended when using reverse transcription (RT) kits. The addition of one-tenth the suggested amount should still prove sufficient for testing a typical normalizer.

An appropriate choice of the input mass of RNA for reverse transcription reactions will ensure that a given reaction does not exhaust the reagents provided by the RT kit. More importantly however, the samples and genes under study also play critical roles in determining how much RNA can be added for reverse transcription. Converting all unknown samples into cDNA using the highest concentration of RNA with variable RT efficiencies among the samples, risks incorrect measurement of the normalizer gene(s) during real-time amplification.

Table 31.1 RefFinder analysis. *Top panel* Ct values of candidate reference genes, *Lower panel* final reference gene rankings according to algorithm

hBAct	hGAPDH	hSDHA	hTBCA	hTUBA1A	hRNU44	hU6	hRNU48	hRNU47	h18s
19.3112	22.28325	24.8479	22.9217	24.7194	17.5574	14.46205	19.4794	16.4062	18.99305
19.16265	22.63935	24.93535	22.8954	24.7734	17.58445	14.4329	19.5376	16.4733	19.33055
19.14815	22.3895	24.56275	22.51135	24.4619	17.93015	14.4635	19.73925	16.5504	19.5449
21.81065	24.6102	26.5362	23.36915	27.01725	18.1465	14.4691	20.0296	17.003	19.4468
21.1704	24.0964	26.02375	23.5005	26.0287	17.7986	15.0001	19.6619	16.43175	19.5778
23.4701	25.95015	27.0499	24.54845	28.30655	18.60915	16.04265	20.5171	17.3307	20.03305
19.27045	23.49115	25.0835	22.84805	24.67245	17.7206	14.336	19.8189	16.5204	19.30995
19.0253	22.8714	24.69045	22.7619	24.47635	17.8875	14.47215	19.87185	16.61655	20.05875
19.16015	22.9632	24.68925	22.5935	24.49845	18.026	14.72145	19.98605	16.76375	20.56225
20.23935	24.2292	25.4872	23.1425	25.45795	17.62315	14.73475	19.68395	16.3622	20.12155
20.6476	23.9726	25.84975	23.4667	25.92005	17.91115	15.0755	19.7871	16.47465	20.0937
22.8857	26.0722	27.2926	24.5212	27.9778	17.6749	15.2755	19.76915	16.386	20.35435
19.96615	22.7419	25.27745	22.9304	25.04025	18.04825	14.99655	20.29905	16.9748	20.3836
20.0786	22.61245	25.4461	22.79935	24.9942	17.74855	14.5316	20.155	16.67935	20.22445
20.7771	23.82425	25.7362	22.70535	25.11675	16.88815	13.50115	19.1055	15.6059	18.39635
21.58675	23.7839	26.3449	23.28645	26.0738	18.09565	15.0952	20.4421	17.02225	20.12955
22.15435	24.16015	26.665	23.533	26.52845	17.21855	14.51215	19.70135	16.02825	18.68725
24.07285	26.44245	27.4036	24.6452	29.01625	18.28	15.592	20.3794	16.7971	20.24645

Method	1	2	3	4	5	6	7	8	9	10
Delta CT	hTBCA	hU6	hRNU48	hRNU44	hSDHA	hRNU47	h18s	hGAPDH	hTUBA1A	hBAct
BestKeeper	hRNU47	hRNU44	hRNU48	hU6	hTBCA	h18s	hSDHA	hGAPDH	hTUBA1A	hBAct
Normfinder	hTBCA	hU6	hSDHA	hRNU48	hRNU44	hRNU47	hGAPDH	h18s	hTUBA1A	hBAct
Genorm	hRNU44 hRNU47		hRNU48	hU6	h18s	hTBCA	hSDHA	hGAPDH	hTUBA1A	hBAct
Ranking	hTBCA	hRNU47	hRNU44	hU6	hRNU48	hSDHA	h18s	hGAPDH	hTUBA1A	hBAct

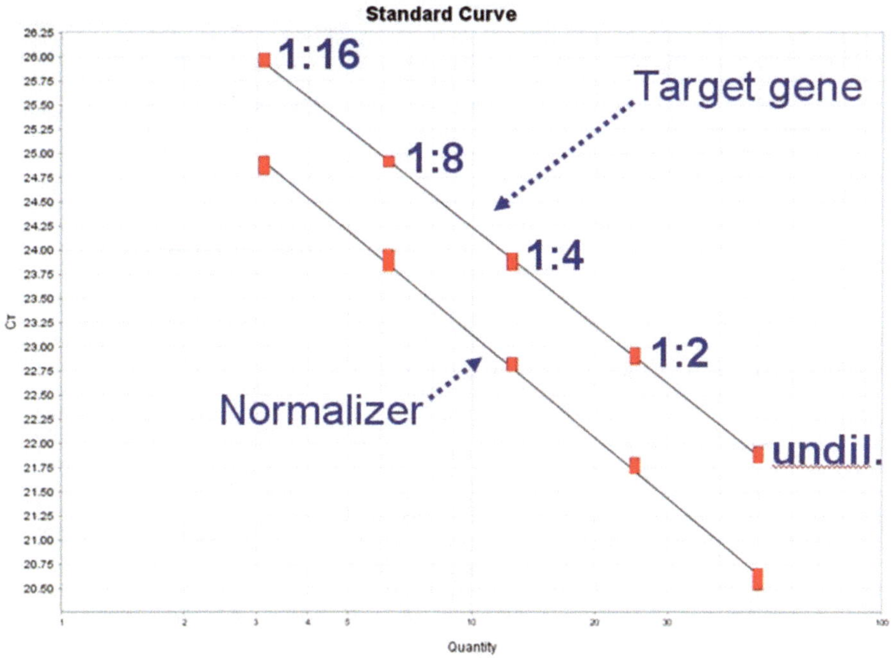

Fig. 31.3 RNA serial dilution plot

To test the dynamic range of RT:

(a) Select a representative RNA sample from among the unknowns or alternatively combine multiple samples and use the mixture as a starting point.
(b) Starting with 1 μg of RNA, serially dilute in nuclease-free water to obtain a five-point curve. Note: Depending on sample quality, a larger starting amount and/or dilution factor may be desirable or even necessary.
(c) Reverse transcribe each of these dilutions in the same reaction volume.
(d) Amplify in real-time an equal volume of each resulting cDNA with both the normalizer and target assay(s). A straight line connecting all dilution points indicates consistent RT efficiency across all starting concentrations of RNA (Fig. 31.3).
(e) Non-linear results often indicate the presence of RT inhibitors. (Note: higher concentrated gene assays are often inhibited to a much greater degree during RT compared to lower-expressed targets.)

Note #3 Comparative Ct (ΔΔCt) methodology relies on the assumption that the efficiencies of both the target and normalizer assays are identical (or at least very similar) i.e. if you dilute the sample, the difference between the target gene and the reference gene will remain the same, so that the delta Ct will be constant no matter how much you dilute the sample.

Assay efficiencies are easily determined by amplifying a dilution series of cDNA and examining the resulting slope of the regression line obtained after plotting the Ct values (Fig. 31.4a).

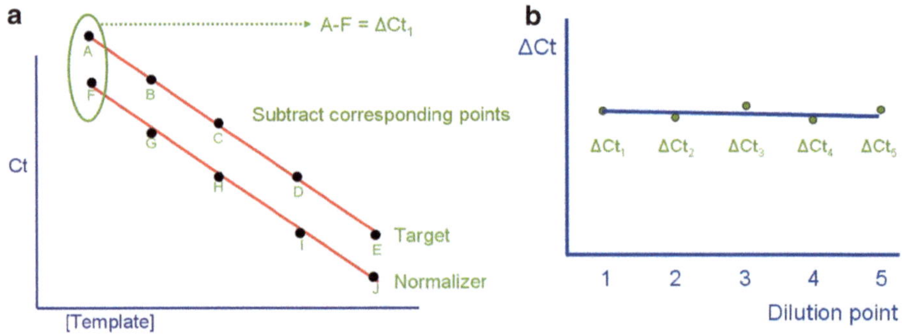

Fig. 31.4 (**a**) cDNA serial dilution plot; (**b**) ΔCt plot

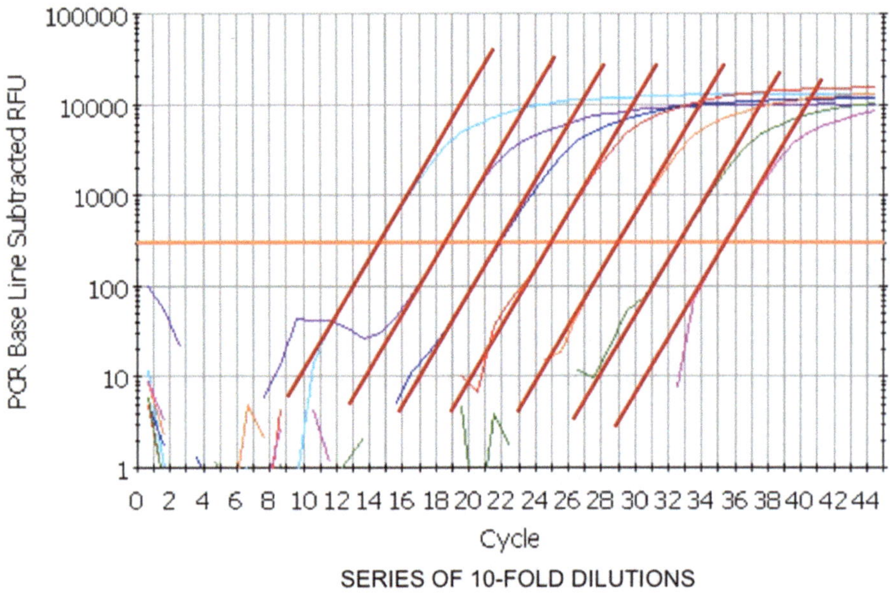

Fig. 31.5 Representative serial dilution amplification plot showing parallel geometric phases

Once the slope has been determined for your assay(s), the following calculation can be used to determine their respective efficiencies: $E = 10^{(-1/\text{slope})} - 1$. If the two dilution curves are parallel, clearly the ΔCt regression line pictured above will have a slope of zero (Fig. 31.4b). However, as long as the slope is greater than −0.1 and less than 0.1, the difference in efficiency between the two assays is not significant, and the ΔΔCt method is appropriate with this assay pair.

Another useful test of efficiency equality between two assays is a simple visual inspection of the amplification curves. When two assays produce amplification curves whose geometric phases are parallel, regardless of the slopes of their respective dilution curves, the efficiencies are either identical or very close (Fig. 31.5).

To help minimize errors:

(a) Generate the curve(s) from a dilution of an inhibitor-free nucleic acid e.g. concentrated cDNA or artificial template (plasmid) containing the assay-specific target in significant quantities.
(b) Use 1:10 dilutions over at least five points, this will cover four logs of dynamic range – large enough to minimize the effect of minor pipetting errors.
(c) Always dilute the template in a serial manner, rather than preparing each dilution from the stock tube.
(d) Mix each dilution point thoroughly by pipetting or vortexing before making the subsequent dilution.
(e) Always amplify at least three replicate wells for each dilution point.
(f) If the standard point representing the highest concentration does not fall in line with subsequent dilutions, the standard may contain inhibitors. Remove this point from consideration.

Note #4 There is no universal control gene, expressed at a constant level under all conditions and in all tissues, hence the choice of reference gene (Table 31.2) should be validated for each new study. The best way to choose the proper reference gene(s)

Table 31.2 Candidate reference genes

Gene symbol (s)	Gene class	Species
18SrRNA	18S ribosomal RNA	Human, mouse
ACTA1, ACTB	Actin	Chicken, cow, dog, hamster, horse, human, monkey, mouse, pig, rabbit, rat
B2M	Beta-2 microglobulin	Cow, dog, hamster, horse, human, monkey, mouse, pig, rabbit, rat, zebrafish,
CYP33	Cyclophilin-33	Fruit fly
GUSB	Glucuronidase, beta	Chicken, cow, dog, hamster, horse, human, monkey, mouse, pig, rabbit, zebrafish
GAPDH	Glyceraldehyde-3-phosphate dehydrogenase	Chicken, cow, dog, fruit fly, hamster, horse, human, monkey, mouse, pig, rabbit, zebrafish
HSP90AB1, HSP83, HSP90ab1	Heat shock protein	Cow, fruit fly, horse, human, monkey, mouse, rat, zebrafish
H6PD	Hexose-6-phosphate dehydrogenase	Chicken, cow, horse
HMBS	Hydroxymethylbilane synthase	Chicken, cow, rabbit
HPRT1	Hypoxanthine phosphoribosyltransferase 1	Cow, dog, horse, human, pig, rabbit, rat, zebrafish
LDHA, LDHAL6B	Lactate dehydrogenase	Chicken, dog, horse, human, mouse, pig, rabbit, rat, zebrafish
NONO	Non-POU domain containing, octamer-binding	Hamster, horse, human, monkey, rabbit, rat, zebrafish,

(continued)

Table 31.2 (continued)

Gene symbol (s)	Gene class	Species
PPIA, PPIH	Peptidylprolyl isomerase	Cow, dog, human, monkey, mouse, pig, rabbit, rat,
PGK1	Phosphoglycerate kinase	Cow, hamster, human, monkey, mouse, pig, rat, zebrafish,
RPL13A, RPL32, RPL4, RPLP0, RPLP1	Ribosomal protein	Chicken, cow, dog, fruit fly, human, monkey, pig, rabbit, rat, zebrafish,
SDHA	Succinate dehydrogenase A	Dog, fruit fly, hamster, horse, human, mouse, rat,
TBP	TATA binding protein	Cow, dog, fruit fly, horse, human, monkey, mouse, rabbit, rat, zebrafish,
TFRC	Transferrin receptor	Chicken, dog, hamster, human, monkey, mouse, pig, rabbit, rat, zebrafish
UBC	Ubiquitin C	Chicken, cow, dog, horse, mouse

is by running a panel of potential genes on a number of representative test samples. The most appropriate gene(s) for normalization are then chosen in each case.

Note #5 Whether or not the normalizer passes or fails validation depends on the agreement of Cts among the various samples. Ultimately, the researcher must decide if the observed variation is acceptable. While the statistical algorithms introduced earlier may aid in this decision, the answer ultimately depends on the experimental goals and expectations for the accuracy of the final data.

References

1. Andersen CL, Jensen JL, Orntoft TF (2004) Normalization of real-time quantitative reverse transcription-PCR data: a model-based variance estimation approach to identify genes suited for normalization, applied to bladder and colon cancer data sets. Cancer Res 64:5245–5250
2. Pfaffl MW, Tichopad A, Prgomet C, Neuvians TP (2004) Determination of stable housekeeping genes, differentially regulated target genes and sample integrity: BestKeeper–Excel-based tool using pair-wise correlations. Biotechnol Lett 26:509–515
3. Silver N, Best S, Jiang J, Thein SL (2006) Selection of housekeeping genes for gene expression studies in human reticulocytes using real-time PCR. BMC Mol Biol 7:33
4. Vandesompele J, De Preter K, Pattyn F, Poppe B, Van Roy N, De Paepe A, Speleman F (2002) Accurate normalization of real-time quantitative RT-PCR data by geometric averaging of multiple internal control genes. Genome Biol 3:RESEARCH0034

Chapter 32
Endothelial Transcriptomic Analysis

Dileep Sharma, Stephen Hamlet, Eugen Petcu, and Saso Ivanovski

32.1 Introduction

The genetic makeup of an individual is unique, termed genome and is stored in the nucleus of each cell. Genome is made up of deoxyribonucleic acid (DNA) a long, helical molecule that consists of instructions essential for formation and maintenance of cells. These instructions from the DNA are then transcribed into corresponding molecules of ribonucleic acid (RNA), termed as transcripts. A transcriptome essentially is a collection of all the transcripts within a given cell and the qualitative and quantitative evaluation of transcriptome is termed transcriptomics. Various "–omics" have been recently described in molecular biology including genomics or detection of total genes, transcriptomics or gene-expression profiling, proteomics or detection of proteins and metabolomics or detection of metabolites in a biological sample [1].

A cell contains various types of RNA including ribosomal (rRNA), transfer (tRNA), and messenger (mRNA). mRNA is the principal type and most abundant RNA involved in delivering the specific gene-transcribed message to ribosomes, the molecular factories located in the cytoplasm that produces (translates) the corresponding protein.

D. Sharma (✉)
Molecular Basis of Disease Group, Griffith Health Institute, School of Dentistry and Oral Health, Griffith University, Gold Coast Campus, Southport, QLD, Australia
e-mail: dileep.sharma@griffithuni.edu.au

S. Hamlet
Griffith Health Institute, Gold Coast Campus, Griffith University, Nathan, QLD 4222, Australia

E. Petcu
Griffith Health Institute and School of Medicine, Gold Coast Campus, Griffith University, Nathan, QLD 4222, Australia

S. Ivanovski
Griffith Health Institute and School of Dentistry and Oral Health, Gold Coast Campus, Griffith University, Nathan, QLD 4222, Australia

M. Slevin, G. McDowell (eds.), *Handbook of Vascular Biology Techniques*,
DOI 10.1007/978-94-017-9716-0_32

Genome-wide transcriptomics essentially focuses at the transcripts that possess the coding messages for all the proteins ultimately produced by the cell to perform its function in the tissue system. Thus, transcriptome reflects the genes that are actively expressed a given point in time. However, it can also include non-coding transcripts present within the cellular environment including, but not limited to maintaining its structural integrity and regulation of gene expression [2].

Endothelial cells activity can be studied targeting the mature cells within the tissues and blood vessels or it could be done using the precursor cells (EPC). For in vitro studies, pure EPC can be harvested from umbilical cord, peripheral blood and bone marrow [3]. However, detailed description of the endothelial cells is beyond the scope of this book. The following sections describe the methods of transcriptomic analyses of cells that can be employed to endothelial cells including its precursors.

32.2 Methodology

Three basic approaches are employed in transcriptomics including targeted gene expression analysis using real-time PCR, microarrays to detect total gene expression and sequencing (RNA-Seq) [4]. Microarrays and the related techniques have been discussed elsewhere in this book. This chapter essentially focuses on gene expression analyses using real time PCR technique (Fig. 32.1).

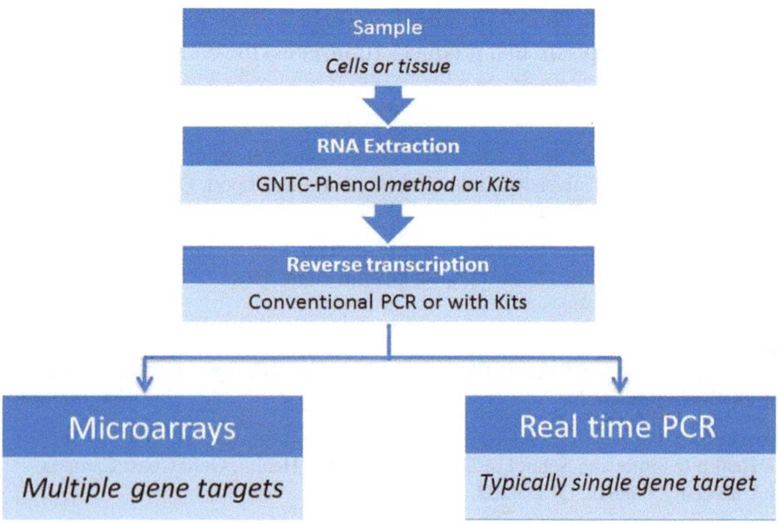

Fig. 32.1 Basic steps in gene expression analyses

32.2.1 *Steps Involved in Gene Expression Analyses*

1. Extraction of total RNA from cells/tissue
2. Reverse transcription of RNA to cDNA
3. Quantification of specific gene expression using gene specific primers.

32.2.1.1 RNA Extraction (GNTC/Phenol Method)

All the equipment, instruments and solutions should be sterile and RNAse free. Diethylpyrocarbonate (DPEC) treated solutions can be considered useful option. Personal protection equipment including gloves (to prevent RNA degradation), lab coat and safety glasses are recommended.

Reagents, Equipment and Supplies Required

1. Guanidinium thiocyanate/phenol (GNTCP) solution (TRIzol® Reagent, Life technologies) **Note 1**
2. Chloroform (Sigma)
3. Molecular grade Ethanol (Sigma)
4. Isopropyl alcohol (Sigma)
5. Diethylpyrocarbonate (DEPC)
6. DEPC-treated water
7. Phosphate Buffered Saline
8. Cell scraper
9. Refrigerated centrifuge
10. Microcentrifuge
11. Micropipettes
12. Filter tips
13. Vortex mixer
14. Powder-free gloves
15. Centrifuge tubes

Basic Protocol (Modified, Based on Ref. [5])

1. Homogenization: (**Note 2**)
   (a) Tissues: For optimal yield, tissue sample should be fresh or snap-frozen in liquid nitrogen and stored at −80 °C. Add at least 10 μl of GNTCP solution for every milligram of tissue and use a glass-Teflon homogenizer to homogenize the tissue.
   (b) Cell culture: Remove and discard the media from the plate and rinse with PBS (optional step). Addition of GNTCP solution directly to the culture dish/well will lyse the cells (minimum of 100 μl/cm^2). Scrape with cell scraper and resuspend the cell lysate 10–12 times through a pipette to fragment the DNA. Vortex thoroughly.

(c) Cells Grown in suspension: Spin cells for 5 min at 300 X g. Remove media and resuspend cells in ice cold PBS. Pellet cells by spinning at 300 X g for 5 min. Lyse cells with GNTCP solution by repetitive pipetting or by passing through syringe and needle. Use 1 ml of the reagent per $5–10 \times 10^6$ of animal cells.

2. Phase Separation:

   (a) To allow complete dissociation of the nucleoprotein complex, incubate the homogenized sample for 5 min at room temperature.
   (b) Add chloroform (200 μl/ml of GNTCP used) and manually shake tube vigorously (or vortex samples) or for 15 s. Let the tube incubate at room temperature for a couple of minutes at least.
   (c) Centrifuge the sample at $1.0–1.2 \times 10^4$ g for 15–20 min at 2–4 °C to obtain lower red, phenol-chloroform phase, an interphase, and a colourless upper aqueous phase. Transfer upper aqueous phase (RNA-containing, ~60 % of total initial volume) carefully without disturbing the interphase into RNase-free sterile tube. (**Note 3**)

3. RNA Precipitation:

   (a) Undiluted isopropyl alcohol is used (500 μl/ml of GNTCP used) to precipitate the RNA from the aqueous phase. Incubate samples at 15–30 °C for 10 min and centrifuge at $1.0–1.2 \times 10^4$ g for 10 min at 2–4 °C. Based on the quantity of RNA in the sample, precipitate is often invisible before centrifugation and usually a gel-like pellet can be noted on the side or bottom of the tube.

4. RNA Wash:

   (a) Remove the supernatant completely without disturbing the pellet. Wash the RNA pellet once with 75 % ethanol (1 ml/ml of GNTCP used).
   (b) Mix the samples by vortexing and centrifuge (2–4 °C) for 5 min, at $6.5–7.5 \times 10^3$ g.
   (c) Discard the wash and air-dry or vacuum dry RNA pellet 5–10 min (avoid complete drying that can decreases RNA solubility).

5. RNA Resuspension:

   (a) Dissolve RNA in DEPC-treated water by passing the solution a few times through a pipette tip to ensure complete solubilisation. Immediately store at −80 °C until its downstream use.

6. Spectrophotometric Analysis:
   After appropriate dilution, measure the OD (optical density) at 260 nm and 280 nm to determine RNA concentration and purity using spectrophotometer. The A_{260}/A_{280} ratio of 1.6 and above should be considered optimal.
7. DNAse treatment (Optional)
   Add specified volume of DNAse buffer followed by specified units of DNAse to each RNA preparation. Mix gently between additions. Incubate at 37 °C for 20–30 min. Add the DNAse inactivation reagent to each RNA preparation.

Incubate for 2–3 min. at room temperature, centrifuge for 2 min. at maximum speed in a microfuge and transfer the RNA to a fresh tube without touching the precipitate, if any.

32.2.2 cDNA Synthesis

Complimentary DNA is synthesised using the RNA and an enzyme termed reverse transcriptase which was first isolated from reverse-transcribing viruses. These viruses are termed retrovirus and common ones to used isolate reverse transcriptase are AMV (Avian Myeloblastosis Virus) and MMLV (Moloney Murine Leukemia Virus). The virus essentially reverse-transcribes its own RNA genomes into native DNA, which is then integrated into the host genome thus ensuing replication.

32.2.2.1 Reagents, Equipment and Supplies Required

1. RNA sample
2. Reverse transcriptase enzyme
3. RNase enzyme
4. Primers oligo(dT) and random
5. Thermo-cycler
6. PCR tubes
7. Microcentrifuge
8. Micropipettes
9. Filter tips
10. Vortex mixer
11. Powder-free gloves

32.2.2.2 Methodology

1. All the components must be mixed and centrifuged briefly before use.
2. Add the following components into a 200 μl PCR tube:
   (a) RNA sample (3–5 μg total RNA),
   (b) Primer 1 μl (oligo dT or gene-specific or random hexamers),
   (c) dNTP mix
   (d) DEPC-treated water to make up the required volume
3. Incubate 5–8 min at 62–65 °C & put on ice for 1 min;
4. Now add the cDNA synthesis solution which typically contains Reverse Transcriptase, A buffer, RNase Inhibitor (optional), $MgCl_2$ as per the requirement of the enzyme used.
5. Incubate for 50 min at 50 °C followed by 83–85 °C for 4–5 min, then chill on ice.
6. Store the cDNA at −20 °C or proceed to PCR step immediately.

32.2.3 Quantitative Real Time PCR for Gene Expression

The distinct feature of qPCR or real time PCR is its ability to monitor the DNA amplification process as it happens, in real time. Fluorescent reporter dyes in the reaction mixture emit fluorescence proportional to the amount of DNA present and the equipment detects the same with its optics and records it as a numerical value. This is based on the fact that fluorescent reporter dyes bind to the double-stranded DNA (i.e. SYBR® Green) or sequence specific probes (i.e. Molecular Beacons or TaqMan® Probes). As the quantification is in real-time, there is no need for the post PCR processing which saves both resources and time. Real time PCR assays are now easy to perform, have high sensitivity, more specificity, and provide scope for automation (**Note 4**).

32.2.3.1 Real Time PCR Methodology

Based on the molecule used for the detection, the real time PCR techniques can be of two categories using DNA Binding Dyes (e.g., SYBR® Green) and using probes specific for targets.

Detection using DNA Binding Dyes like SYBR® Green binds to the minor groove of the DNA double helix and is the most widely used double-strand DNA-specific reporter dye for detection in real time PCR. SYBR® Green, in the master mix solution, is present as the unbound dye which shows minimal fluorescence. However, as the PCR progresses and DNA double strands increase, the fluorescence is substantially and proportionally increases due to dye-binding. By quantifying the amount of fluorescence emitted by SYBR® Green at the end of each cycle, it is possible to monitor the PCR products in the exponential phase. One of the main drawbacks of this DNA binding dye-based PCR is that it detects both specific and non-specific products i.e., it detects all double stranded DNA in the reaction mixture.

Detection using Specific Probes employs specific oligonucleotide probes labelled with both a reporter fluorescent dye and a quencher dye. The reporter dye is "quenched" or "blocked" with the quencher until they contact a double stranded DNA Probes based on different chemistries are available for real time detection, including TaqMan® probes, FRET hybridization probes, molecular beacons, and scorpion® primers.

Preparation

1. Quantitative PCR detects the number of copies of transcribed mRNA based on the primer selected specifically targeting the gene of interest.
2. First, RNA isolated from samples is reverse transcribed to cDNA as described above. (**Note 5**)

3. Calculate the final volume of reaction mixture required for the intended genes of interest as per the equipment and qPCR SYBR Green kit instructions. Always run qPCR reaction on each sample in triplicates.
4. Final reaction tube or plate well would contain a "master primer mix" to which each cDNA sample is added to make up either 10, 20, 25, or 50 μl of total volume (dependant on the equipment and plate/tubes used)
5. Prepare a stock solution for each gene of interest called here as "master primer mix" containing primers specific for the genes of interest, and a SYBR Green mix which includes SYBR Green dye, Taq Polymerase, ROX, and dNTP (commercially available as qPCR SYBR Green kits under various brand names).
6. Here is a typical example of volumes of each component for a 25 μl reaction. SYBR Green Mix (2×) 12.5 μl, cDNA 0.2 μl, primer pair mix 1 μl (typically 0.5 μl each primer) and pure H_2O 11.3 μl.
7. The cycling conditions are optimised by the kit manufacturer and usually contain the steps as shown in Table 32.1.

Preparation

Primer Design-General Principles

To detect the desired target, accurate primer design selection is essential. Basic rules of designing a primer are

1. Forward and reverse primers sequences should not be complementary to each other, especially at their 3′ ends.
2. The 3′ region should not consist of secondary structure, repetitive sequences or palindromes.
3. They should ideally between 40 % and 70 % GC content. Try to avoid an unbalanced distribution of G/C and A/T-rich domains.
4. Design PCR primers that have annealing temperatures around 55–60 °C and use normal primer concentrations (e.g., 500 nM each).
5. Always use highly purified primers (e.g., HPLC), to prevent formation of primer dimers.
6. Ensure that primers are 18–30 bases long.
7. Amplicons (<300 bp) are optimal for most Real-Time PCRs.

Table 32.1 Typical cycling conditions in Real time PCR

Step	Temperature (°C)	Time	Number of cycles
Denaturation & enzyme activation	93–95	3–5 min	One
Denaturation	93–95	30–40 s	25–40
Annealing	60–65	20–30 s	
Extension	70–75	45–60 s	
Final extension	70–75	4–16 min	One

8. For selection of primers commercial software is available, e.g., Primer3 (http://frodo.wi.mit.edu/primer3/input.htm)
9. BLAST the primer sequences to ensure they are specific for the target species and gene (http://www.ncbi.nlm.nih.gov/BLAST)

32.2.4 Endothelial Specific Genes and Primers

Over past few years, there have been various reports of specific genes that are expressed by the endothelial cells and many others that are significant of metabolic state of the endothelial cell. Choice of gene targets (or primers) should be based on whether those genes are expected to be altered viz., over–expressed (up-regulated) or under-expressed (down-regulated) in the proposed experiment. Endothelial and angiogenesis focussed arrays are also commercially available that can be used to evaluate specific genes in these cell types. The targeted genes usually are related to the cytokines that the endothelial cells produce and/or possess receptors for. Also receptors or genes expressed on a particular form of stimulus can also be targeted based on the intended experimental protocol and the hypothesis. The list of the genes has been ever increasing since the role of angiogenesis in tumour enlargement has been explored (Table 32.2). However, a reference or housekeeping gene appropriate to the cell type and the intended modification is essential to compare the parameters in experimental as against control specimen. Further details and considerations in selection of reference gene have been included as a separate chapter in this book.

Note 1 Commercial kits are available for RNA extraction.

Note 2 If tissues contain fat, proteins, polysaccharides, or extracellular material, centrifuged at $1.0–1.2 \times 10^4$ g for few minutes at 4 °C to get a typical tri-phasic solution. Bottom pellet (tissue impurities including DNA) and top fatty liquid phase is discarded and the intermediate clear liquid (RNA rich) phase is transferred into a new clean tube for phase separation.

Note 3 Save the interphase and organic phenol-chloroform phase if isolation of DNA or protein is desired.

Note 4 Real time PCR is also referred to as **real time RT PCR** which has the additional cycle of reverse transcription that leads to formation of a DNA molecule from a RNA molecule. This is done because RNA is less stable as compared to DNA.

Note 5 The process of reverse transcription can be combined with QPCR (one step RT-PCR). However, it is not advisable as this will essentially use up all the cDNA synthesized in one reaction to detect one target gene.

Table 32.2 Target genes attributable to endothelial-related cells

Family	Description	Gene name
VEGF	Vascular endothelial growth factor A	VEGF, VPF
	Vascular endothelial growth factor C	Flt4-L, VRP
	Vascular endothelial growth factor receptor	FLT1
	Kinase insert domain receptor	FLK1, VEGFR2
	Vascular endothelial growth factor D	VEGF-D, VEGFD
EGF	Epidermal growth factor	HOMG4
FGF	Fibroblast growth factor 1 (acidic)	FGF-alpha, FGFA
	Fibroblast growth factor 2 (basic)	BFGF, FGFB
	Fibroblast growth factor receptor 3	ACH, CD333
HGF	Hepatocyte growth factor (hepapoietin A; scatter factor)	HGFB, SF
Interferon	Interferon, alpha 1	IFN, IFN-ALPHA
	Interferon, beta 1	IFB, IFNB
	Interferon, gamma	IFG, IFI
IGF	Insulin-like growth factor 1	IGF-I, IGF1A, IGFI
Interleukins	Interleukin 1, beta	IL-1, IL1-BETA
	Interleukin 6	IL-6
	Interleukin 8	CXCL8
Integrin	Integrin, alpha V (vitronectin receptor, alpha polypeptide, antigen CD51)	CD51
	Integrin, beta 3 (antigen CD61)	CD61
TGF	Transforming growth factor, alpha	TGFA
	Transforming growth factor, beta 1	TGFB, TGFbeta
	Transforming growth factor, beta 2	TGF-beta2
	Transforming growth factor, beta receptor 1	TGFR-1
PDGF	Platelet-derived growth factor alpha polypeptide	PDGF-A, PDGF1
PECAM	Platelet/endothelial cell adhesion molecule	CD31, PECAM-1
PF4	Platelet factor 4	CXCL4
PGF	Placental growth factor	PLGF
PLA	Plasminogen activator, urokinase	u-PA
TNF	Tumor necrosis factor	TNF-alpha, TNFA,
	Tumor necrosis factor, alpha-induced protein 2	B94
Angiopoietin	Angiopoietin-like 3	ANGPT5
	Angiopoietin-like 4	ANGPTL2
Chemokine motif	Chemokine (C-X-C motif) ligand 1	MGSA-a
	Chemokine (C-X-C motif) ligand 10	SCYB10
	Chemokine (C-X-C motif) ligand 3	SCYB3
	Chemokine (C-X-C motif) ligand 5	SCYB5
	Chemokine (C-X-C motif) ligand 6	SCYB6
	Chemokine (C-X-C motif) ligand 9	SCYB9

References

1. Horgan RP, Kenny LC (2011) 'Omic' technologies: genomics, transcriptomics, proteomics and metabolomics. Obstet Gynaecol 13:189–195
2. Eddy SR (2001) Non-coding RNA genes and the modern RNA world. Nat Rev Genet 2:919–929
3. Urbich C, Dimmeler S (2004) Endothelial progenitor cells: characterization and role in vascular biology. Circ Res 95:343–353
4. Dong Z, Chen Y (2013) Transcriptomics: advances and approaches. Sci China Life Sci 56:960–967
5. Chomczynski P, Sacchi N (1987) Single-step method of RNA isolation by acid guanidinium thiocyanate-phenol-chloroform extraction. Anal Biochem 162:156–159

Chapter 33
Protocol for Multiplex Amplicon Sequencing Using Barcoded Primers

S. Bradburn, J.S. McPhee, A. Williams, S. Heffernan, S. Lockey, S. Day, and C. Murgatroyd

33.1 Introduction

Recent advances in DNA sequencing technologies have created opportunities for sequencing at an unprecedented depth and breadth and multiplex sequencing has emerged as an important strategy for sequencing of many different samples in parallel. In multiplex sequencing, a unique multiplex identifier (MID), or "barcode" sequence, is added to the DNA that is to be sequenced. After sequencing, reads are sorted into sample libraries via detection of the appropriate barcode.

Atherosclerotic diseases, including coronary artery disease (CAD) and myocardial infarction (MI), are the leading causes of death in the world [1]. The genetic basis of CAD and MI, which are caused by multiple interacting endogenous and exogenous factors, has gained considerable interest in the last years as genome-wide association studies (GWASs) have identified many new susceptibility loci for CAD and MI (for review see [2]). The underlying genes provide new insights into the genetic architecture of these diseases.

This protocol describes the detailed experimental procedure for multiplex amplicon sequencing. The procedure begins with a PCR amplification of a sequence of interest. This is then followed by a second PCR with universal primers containing barcodes and sequences for the 454 sequencing reaction. Optimization will be needed dependent on reagents or instruments for PCR. The amplicons are then quantified prior to sequencing (Fig. 33.1).

S. Bradburn (✉) • J.S. McPhee • C. Murgatroyd
School of Healthcare Science, Manchester Metropolitan University, Manchester, UK
e-mail: s.bradburn@mmu.ac.uk

A. Williams • S. Heffernan • S. Lockey • S. Day
Department of Exercise and Sport Science, Manchester Metropolitan University, Crewe, UK

M. Slevin, G. McDowell (eds.), *Handbook of Vascular Biology Techniques*,
DOI 10.1007/978-94-017-9716-0_33

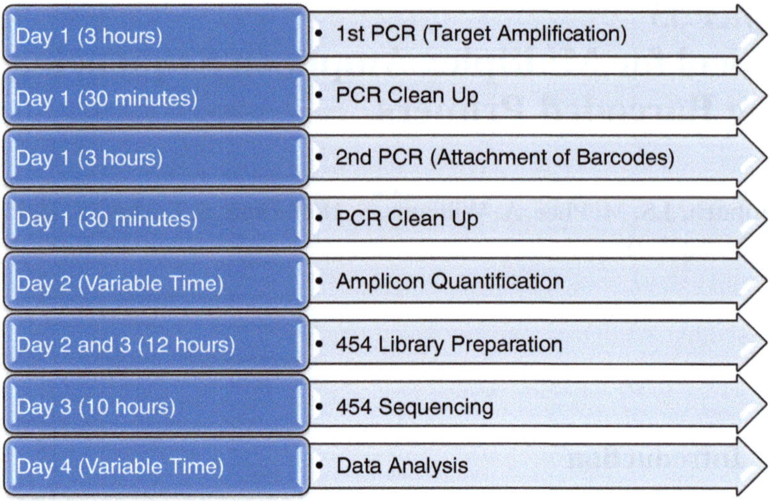

Fig. 33.1 Work flow for the sequencing procedure using the 454 GS Junior

33.2 Materials

33.2.1 First and Second PCR

Oligonucleotide primers	*Life technologies*
Genomic DNA (100 ng)	–
DNA polymerase and PCR reaction buffer	*Bioline reagents*
DNaseZap DNase decontamination	*Ambion*
RNase/DNase free water	*Ambion*
DNA LoBind 1.5 mL tubes, PCR clean	*Eppendorf*
ART self-sealing barrier pipette tips	*Sigma-Aldrich*
Mastercycler PCR thermocycler	*Eppendorf*
96-well PCR reaction plate	*Applied biosystems*
96-well adhesive film	*Applied biosystems*
Multipurpose tabletop centrifuge	–
Refrigerated centrifuge	*Sigma-Aldrich*
Vortex	–
[Optional] Ethidium bromide solution (0.5 μg/ml)	*Sigma-Aldrich*
[Optional] Agarose	*Sigma-Aldrich*
[Optional] 1X TBE buffer	*Thermo scientific*
[Optional] Gel Doc™ XR+	*Bio-rad laboratories*
[Optional] DNA hyperladder I and loading buffer	*Bioline reagents*
[Optional] Gel electrophoresis tanks, casts and power unit	*Bio-rad laboratories*

33.2.2 PCR Clean-Up

UltraClean PCR clean-up kit	*Mobio*
Multipurpose tabletop centrifuge	–
ART self-sealing barrier pipette tips	*Sigma-Aldrich*

33.2.3 Amplicon Quantification

2100 Bioanalyzer	*Agilent technologies*
DNA 1000 kit	*Agilent technologies*
Multipurpose tabletop centrifuge	–
DNA LoBind 0.5 mL tubes, PCR clean	*Eppendorf*
Vortex	–

33.2.4 454 Library Preparation, Sequencing and Data Analysis

454 GS Junior sequencer	*Roche*
GS Junior Titanium emPCR kit (Lib-A)	*Roche*
GS Junior Titanium PicoTiterPlate kit	*Roche*
GS Junior Titanium sequencing kit	*Roche*
GS Junior Titanium control bead kit	*Roche*

For a full list of all of the materials required for the sequencing procedures refer to the *GS Junior System Tables of Materials* document [3]

33.3 Detailed Procedure

33.3.1 First PCR (Target Amplification)

We implemented a two-step PCR procedure [4]. The first reaction involves conventional PCR primers (designed against the region of interest) to amplify the desired sequence form genomic DNA (Fig. 33.2). It is advisable to keep amplified products below 600 bp (see App. A.3). During primer design a 'universal tail' is also added to the 5′-end of each primer. The purpose of this modification is to allow the product gained from the first PCR to be used as a template for the second PCR.

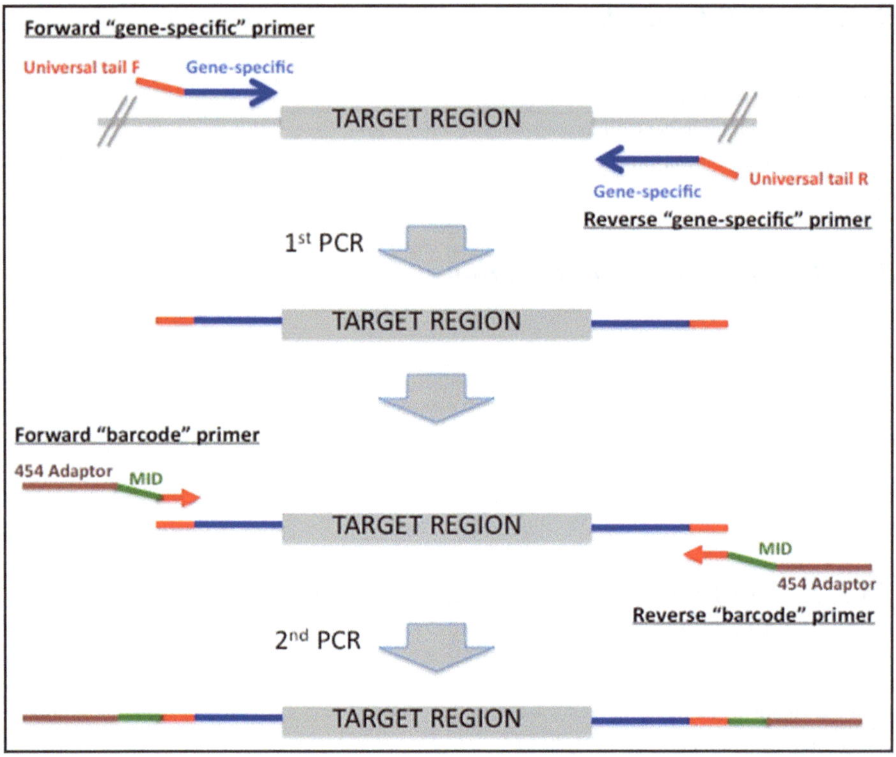

Fig. 33.2 Schematic representation of the two PCR reactions. The gene specific sequences for each primer are highlighted in *blue*. The universal tails which enable secondary primer binding are coloured *red*. The unique MID codes are highlighted in *green* and the 454 adaptor sequences are coloured *purple*

1. On ice prepare the following reaction mixture. Mix the reagents well before use with a vortex. Also allow an additional 10 % volume to account for pipetting inaccuracies:

Reagent	1 Reaction
5× PCR reaction buffer (3 mM $MgCl_2$ final concentration)	5 μl
10 μM forward primer	1 μl
10 μM reverse primer	1 μl
dNTPs (10 mM of each)	2.5 μl
1.25 U DNA polymerase	0.25 μl
Genomic DNA template (100 ng)	As required
RNase/DNase free water	Up to 25 μl
Final volume	25 μl

2. Pipette the above mixture into a 96 well plate whereby one well contains one DNA individual sample.

3. Attach the 96-well adhesive film tightly to the plate ensuring all of the wells are suitably covered and briefly pulse spin in a refrigerated centrifuge.
4. Place the plate in a thermocycler with the following cycle conditions applied:

Step		Temperature	Time
Initial denaturation		95 °C	10 min
30 cycles	Denaturation	95 °C	1 min
	Annealing	*User determined*	45 s
	Extension	72 °C	1 min
Final extension		72 °C	5 min
Hold		5 °C	Indefinitely

Additional information: A heated lid option is advisable to prevent condensation inside the wells. Also the ramp speed is recommended to be kept at standard

5. Once complete the plate can be stored at –20 °C.
   It is advisable to run some of the product on an agarose gel along with a DNA ladder to be confident of target amplification and primer specificity e.g. 5 μl of product on a 2 % agarose gel.
6. Before beginning the next PCR reaction all of the samples should be purified to remove any remaining primers and salts (see App. A.1).

33.3.2 *Second PCR (Attachment of Barcodes)*

A unique combination of forward and reverse barcode primers (Table 33.1) are then added to each sample in the second PCR (Fig. 33.2). These primers contain the same aforementioned 10 bp universal tails thus allowing attachment to the previously amplified product. Next to the universal tail is a 10 bp MID sequence which is used for sample identification during the sequencing. Adjacent to this is a 25 bp sequence which is able to hybridize to the DNA capture beads during the 454 sample library preparation.

1. On ice prepare the following reaction mixture, remembering to vortex all of the reagents well before use and to add an additional 10 % volume:

Reagent	1 Reaction
5× PCR reaction buffer (3 mM $MgCl_2$ final concentration)	5 μl
10 μM forward primer	1 μl
10 μM reverse primer	1 μl
dNTPs (10 mM of each)	2.5 μl
1.25 U DNA polymerase	0.25 μl
PCR product template	1 μl
RNase/DNase free water	14.25 μl
Final volume	25 μl

Table 33.1 A selection of primers required to be able to undertake multiplex sequencing. The forward and reverse universal tails are in red italics. The multiplex identifier (MID) is underlined in green and the 454 adaptors are in purple. For an extended list of MID primers available to use on the 454 GS Junior system refer to appropriate technical bulletin [9]

Primer name	Sequence (5′-3′)
Gene specific Forward	*GACACTATAG*-(Forward template specific sequence)
Gene specific Reverse	*CACTATAGGG*-(Reverse template specific sequence)
Forward barcode	
ForMID-1	CGTATCGCCTCCCTCGCGCCATCAG<u>ACGAGTGCGT</u>*GACACTATAG*
ForMID-3	CGTATCGCCTCCCTCGCGCCATCAG<u>AGACGCACTC</u>*GACACTATAG*
ForMID-4	CGTATCGCCTCCCTCGCGCCATCAG<u>AGCACTGTAG</u>*GACACTATAG*
ForMID-5	CGTATCGCCTCCCTCGCGCCATCAG<u>ATCAGACACG</u>*GACACTATAG*
ForMID-6	CGTATCGCCTCCCTCGCGCCATCAG<u>ATATCGCGAG</u>*GACACTATAG*
Reverse barcode	
RevMID-51	CTATGCGCCTTGCCAGCCCGCTCAG<u>AGCTCACGTA</u>*CACTATAGGG*
RevMID-52	CTATGCGCCTTGCCAGCCCGCTCAG<u>AGTATACATA</u>*CACTATAGGG*
RevMID-53	CTATGCGCCTTGCCAGCCCGCTCAG<u>AGTCGAGAGA</u>*CACTATAGGG*
RevMID-54	CTATGCGCCTTGCCAGCCCGCTCAG<u>AGTGCTACGA</u>*CACTATAGGG*

2. Pipette the above mixture into a 96 well plate. Each well should contain a unique combination of forward and reverse primers to allow for successful barcoding. This will allow the tracking of each sample when they are later pooled for sequencing.
3. Attach the film lid tightly to the plate ensuring all of the wells are suitably covered and briefly pulse spin in a refrigerated centrifuge.
4. Place the plate in a thermocycler with the following cycle conditions applied:

Step		Temperature	Time
Initial denaturation		95 °C	10 min
20–30 cycles	Denaturation	95 °C	1 min
	Annealing	60 °C	45 s
	Extension	72 °C	1 min
Final extension		72 °C	5 min
Hold		5 °C	Indefinitely

Additional information: A heated lid option is advisable to prevent condensation inside the wells. Also the ramp speed is recommended to be kept at standard

5. Once complete the plate can be stored at −20 °C.
   It is advisable to run some of the product on an agarose gel along with a DNA ladder to be confident of barcode attachment. If the attachment of the barcodes has been successful then the product size should be 70 bp larger than the product gained from the first PCR.
6. Finally, purify all of the newly gained PCR products (see App. A.1) to prepare for quantification.

33.3.3 *Amplicon Quantification*

Precise quantification of the PCR products gained in the previous reaction is essential to allow equal reads when it comes to multiplex sequencing. An inconsistently quantified amplicon library will ultimately lead to difficulties during data normalization. There are numerous methods available to quantify the purified PCR amplicons.

33.3.3.1 2100 Bioanalyzer

The 2100 Bioanalyzer is able to size and quantify PCR products from just 1 μl of sample via a microfluidic approach. Each sample produces an artificial gel image as well as an electropherogram (Fig. 33.3). In conjunction with the DNA 1,000 kit each chip can simultaneously analyze 12 samples ranging from 25 to 1,000 bp in size. One possible limitation to this assay, however, is the narrow quantitation range and low sensitivity available (Table 33.2).

33.3.3.2 Quant-iT PicoGreen dsDNA Kit

An alternative method for quantifying PCR products prior to sequencing is to use the commercially available Quant-iT PicoGreen dsDNA assay (*Invitrogen*). The fluorescent stain in the kit specifically binds to double stranded DNA (dsDNA), even in the presence of single stranded nucleic acids and other contaminants, and measured with a standard fluorometer. Quantification is extrapolated from an 8-point standard curve. The assay also supports 96- and 384-well plate formats to increase throughput capabilities.

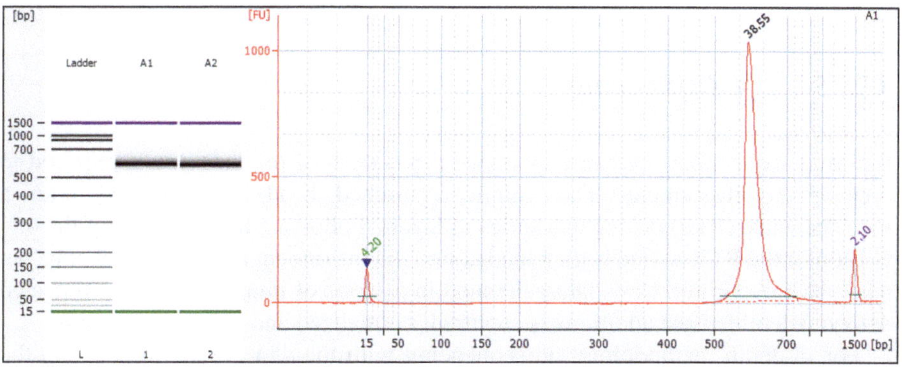

Fig. 33.3 *Left:* A gel representation of two samples generated with the 2100 Bioanalyzer on a DNA 1,000 chip. *Right:* The corresponding electropherogram for sample 'A1' denoting the concentration (ng/μl) of the desired amplicon and the two markers

Table 33.2 Comparison between the 2100 Bioanalyzer and the Quant-iT PicoGreen dsDNA methods for quantitation of amplions prior to sequencing. The values quoted for the 2100 Bioanalyzer are when using the DNA 1,000 kit. The values for the Quant-iT PicoGreen assay are based on a standard 2.0 ml reaction with a fluorometer

	2100 Bioanalyzer	Quant-iT PicoGreen kit
Quantitation range:	0.1–50 ng/μl	25 pg/ml–1,000 ng/ml
Sample volume required:	1 μl	Up to 1.0 ml
Number of samples (per run):	12	1
Qualitative analysis supported?:	Yes	No
Run time:	30 min	<5 s

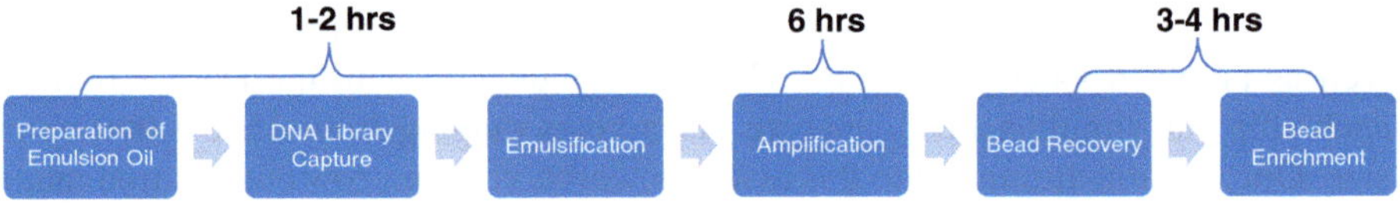

Fig. 33.4 Workflow of 454 amplicon library preparation prior to sequencing

33.3.4 454 Library Preparation

Prior to preparing the reagents required for sequencing it is first necessary to dilute and pool the amplicon library. This DNA pool is then processed in an emulsion-based clonal amplification (emPCR) reaction (Fig. 33.4). Finally the DNA capture beads containing the different amplicons are retrieved (approximately 500,000).

All of the above steps including amplicon dilution and emPCR procedures can be found in detail in the *Amplicon Library Preparation Manual* [5] and the *emPCR Amplification Method Manual – Lib-A* [6] respectively.

33.3.5 454 Sequencing

Before a sequencing run can commence a pre-wash of the machine is required to prepare the instrument. The amplicon-enriched beads are injected onto a PicoTiterPlate (PTP) device in a series of layers. Following a final prime of the 454 GS Junior the PTP device is then loaded into the instrument and the sequencing can take place overnight. Raw data generated in a series of images following repeated deoxynucleotide base additions is captured, normalized and collated into runs.

For in-depth methodology on sequencing with the 454 GS Junior refer to the *Sequencing Method Manual* [7].

33.3.6 Data Analysis

The benefit of next generation sequencing is having an immense data set containing thousands of reads therefore allowing the confidence to interpret results. However with such a vast supply of results provides difficulties during the analysis stage. Fortunately there is software available to aid in the presentation and interpretation of the experimental data.

A suitable program to analyze amplicon library reads is the GS Amplicon Variant Analyzer (*Roche*). This software is able to analyze both uni- and bi-directional sequencing reads (see App. A.2). Each amplicon sequenced is compared against a desired reference sequence to highlight any possible variants in the amplicon library (Fig. 33.5). The corresponding MIDs must first be selected on the software for each sample to differentiate between subjects. Flowgrams can also be created which are useful when presenting any novel variant found in a sequencing run (Fig. 33.6).

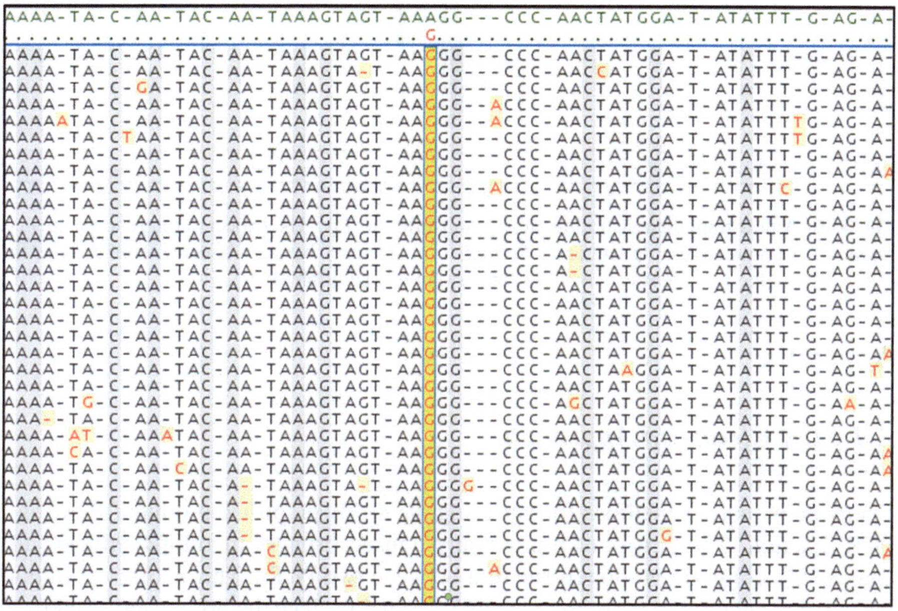

Fig. 33.5 Using the GS Amplicon Variant Analyzer to highlight single nucleotide polymorphisms. The top line (*green*) is the reference sequence. The second line is the sample sequence. Nucleotides highlighted (*red*) are those which differ from the expected sequence. The additional data represents individual reads for that sample

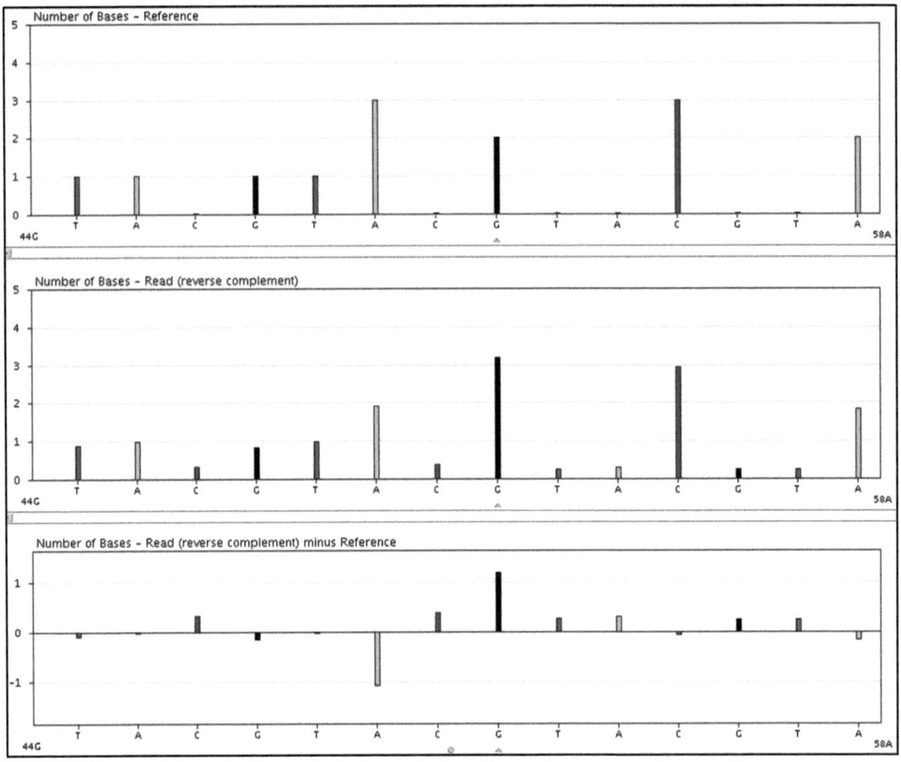

Fig. 33.6 A flowgram demonstrating an A/G substitution. *Top:* The expected reference sequence. *Middle:* An amplicon sample sequence. *Bottom:* The difference between both the amplicon and the reference. Notice a single A removal (−1) and a single G addition (+1)

Appendix

A.1 PCR Clean-Up Approach

There are numerous methods for purifying the newly amplified product following a PCR reaction. The aim is to remove any unused primers therefore reducing the chance of a chimeric product formation between two PCRs. This is of great importance when undertaking this approach to sequencing since the primer sets used from both of the aforementioned PCR reactions share certain sequence homology.

Column based PCR purification kits utilize an adapted silica membrane column to clean-up PCR products in three easy steps: bind, wash and elute. The products recovered are generally greater than 100 bp and depending on the kit used this can be up to 10 kb in size. This rapid approach requires very little laboratory equipment and is relatively cheap.

An alternative to the above post PCR clean-up process is the use of magnetic beads to purify the amplicon of interest. The overall process includes bead addition, separation, wash and elution. This approach is useful for high-throughput requirements. An addition benefit of this technique, in comparison to column based options, is the ability to size select the wanted product via altering the bead:DNA solution ratio. This is particularly useful when more than two products are present.

A third option is to run PCR products on a agarose gel before dissecting the wanted band and ultimately purifying the cut out with a gel purification commercial kit. This method also uses a silica membrane column to capture the amplicons prior to washing. This is mainly used as a last resort when there is unspecific binding present, such as when using bisulphite treated DNA, due to the lower recovery of DNA when compared to the above options.

A.2 Bidirectional v Unidirectional Sequencing

With the 454 GS Junior there possesses a capability to sequence an amplicon either bidirectionally or unidirectionally. As the name suggests, bidirectional sequencing includes reading the sequence in two directions (forward and reverse) and conversely unidirectional approaches produce reads in only one direction (either forward or reverse). Obviously when wanting to highlight possible variants in amplicons bidirectional sequencing provides a greater deal of confidence and accuracy when analyzing results. Alternatively, unidirectional approaches cannot read in both orientations however these will contain double the number of reads. It is therefore recommended to sequence bidirectionally when investigating rare single nucleotide polymorphisms (SNPs) and unidirectionally when analyzing a wide diversity of amplicons.

A.3 Amplicon Length Considerations

The average read length for the 454 GS Junior is 400 bp with longest reads being around 600 bp [8] including the two adaptor sequences. It should be noted that shorter than average amplicons (<300 bp) may result in excessive amplification during the emPCR stage when using the suggested conditions. This issue is avoidable by optimizing the emPCR conditions to contemplate for the desired shorter product.

References

1. WHO (2013) The top 10 causes of death. [Online]. http://who.int/mediacentre/factsheets/fs310/en/index.html. Accessed 8 Jan 2014
2. Maouche S, Schunkert H (2012) Strategies beyond genome-wide association studies for atherosclerosis. Arterioscler Thromb Vasc Biol 32(2):170–181. doi:10.1161/ATVBAHA.111.232652

3. 454 Life Sciences Corp. (2013) GS Junior system tables of materials. [PDF] USA. http://454.com/downloads/my454/documentation/gs-junior/system-wide-documents/GSJunior_TablesofMaterials_May2013.pdf. Accessed 7 Jan 2014
4. Berry D, Ben Mahfoudh K, Wagner MA, Loy A (2011) Barcoded primers used in a multiplex amplicon pyrosequencing bias amplification. Appl Environ Microbiol 77(21):7846–7849. doi:10.1128/AEM. 05220-11
5. 454 Life Sciences Corp. (2013) Amplicon Library Preparation Manual. [PDF] USA. http://454.com/downloads/my454/documentation/gs-junior/method-manuals/454SeqSys_AmpliconLibraryPrepMethodManual_Jun2013.pdf. Accessed 7 Jan 2014
6. 454 Life Sciences Corp. (2012) emPCR Amplification Method Manual – Lib-A. [PDF] USA. http://454.com/downloads/my454/documentation/gs-junior/method-manuals/GSJunioremPCRAmplificationMethodManualLib-A_March2012.pdf. Accessed 7 Jan 2014
7. 454 Life Sciences Corp. (2013) Sequencing Method Manual. [PDF] USA. http://454.com/downloads/my454/documentation/gs-junior/method-manuals/GSJuniorSequencingManual_Jan2013.pdf. Accessed 7 Jan 2014
8. Roche Diagnostics (2011) GS Junior System. [PDF] Germany. http://454.com/downloads/GSJuniorSystem_Brochure.pdf. Accessed 8 Jan 2014
9. 454 Life Sciences Corp. (2009) Using multiplex identifier (MID) adaptors for the GS FLX titanium chemistry – extended MID set. [PDF] USA. http://454.com/downloads/my454/documentation/technical-bulletins/TCB-09005_UsingMultiplexIdentifierAdaptorsForTheGSFLXTitaniumChemistry-ExtendedMIDSet.pdf. Accessed 7 Jan 2014

Chapter 34
Flow Cytometry Enumeration of Hematopoietic and Progenitor Stem Cells: Identification and Quantification

William Gilmore, Mayada Al Qaisi, and Nasser Al-Shanti

34.1 Introduction

Bone marrow (BM) was originally the primary source of hematopoietic progenitor stem cells for bone marrow transplantation marrow (BMT) which is used increasingly in the treatment of blood disorders, malignancies, and genetic abnormalities [1–3]. For a stem cell transplant, bone marrow aspiration will be performed at several sites on the body (generally from the back of the pelvic bone) to remove enough bone marrow cells for the transplant to work. Following the development of mobilization regimes (GCSF, GM-CSF, and chemotherapy), more rapid hematologic reconstitution has been achieved, hospitalization costs have been reduced, and the risks of general anesthesia and discomfort with a BM harvest avoided. Therefore, peripheral blood has become the preferred source of stem cells. Moreover, this technique shorten the time to engraftment and provide adequate numbers of stem cells for either autologous or allogeneic transplants [1, 2, 4–6]. Therefore, transplantation of hematopoietic progenitor cells obtained from the peripheral blood of a recipient patient or HLA-matched donor is used increasingly in the treatment of blood disorders, malignancies, solid tumours and genetic abnormalities. In cases of treatment of haematological malignancies, stem cell transplants allow the replacement of progenitor diseased bone marrow using high-dose chemo/radiotherapy and generate healthy progenitor cells that repopulate the marrow and produce normal blood cells more rapidly. Moreover, solid tumour patients can be salvaged from the lethal toxic effects caused by high dose chemotherapy using blood stem cell infusions. Recently, several findings revealed that progenitor cell therapy, employed in combination with high-dose chemotherapy led to clinically significant developments in the long-term survival rates among patients suffering from different types of cancer [1, 2, 7, 8].

W. Gilmore (✉) • M. Al Qaisi • N. Al-Shanti
Healthcare Science Research Institute, School of Healthcare Science,
Manchester Metropolitan University, Manchester, UK
e-mail: b.gilmore@mmu.ac.uk

M. Slevin, G. McDowell (eds.), *Handbook of Vascular Biology Techniques*,
DOI 10.1007/978-94-017-9716-0_34

Moreover, hematopoietic stem cell transplantation has been shown to be effective for many patients with non-malignant diseases such as primary immunodeficiency disorders, aplastic anemia, bone marrow failure syndromes, hemoglobinopathies, and inborn errors of metabolism [3, 9].

34.2 Methodology

34.2.1 Principle of the Procedure

The procedure described here is based on a signal platform by Keeney et al. [10]. The Beckton Dickinson (BD) cell enumeration kit is a single tube assay which is pereformed by staining the cells with the reagent in individual BD Trucount tubes for absolute CD34+ counts. This kit not only allows simultaneous enumeration of viable dual-positive CD45/CD34 hematopoietic stem cell population in CD34+ absolute counts (cells/μl) but also the percentage of the total viable leukocytes count that is CD34+ (%CD34 of viable $CD45^+$). Labelled antibodies (CD34 & CD45) in the reagent bind specifically to the cell surface, and the lyophilized pellet under the stainless steel retainer in the BD Trucount tube dissolves (Fig. 34.1), releasing a known number of fluorescent beads. The dye, 7-AAD, is then used to assess viability of the cells in which cells that are 7-AAD negative are viable. Ammonium chloride is added to lyse erythrocytes before the sample is acquired on a flow cytometer. During analysis of the sample, the concentration of viable CD34+ cells and viable CD45+ cells, and the percentage of viable CD34+ cells in the viable CD45+ cell population, are calculated using the fluorescent bead count which is indicated on the pouch (Fig. 34.1).

34.3 Materials

In this protocol, BD stem cell enumeration kit (cat # 344563) was used to measure the absolute the CD34 cells. This kit was sufficient for 50 tests. This kit was solely for *in-vitro* diagnostic single tube, single platform assay, and plays an important role in stem cell transplantation by enumerating CD34 cells in many samples including mobilized and normal peripheral blood, bone marrow and cord blood. This kit comes with pre-mixed antibodies, and the BD Trucount tubes loaded with beads which reduces the time for reagent preparation and analysis.

34.4 Reagents

BD Stem Cell reagent (CD45/CD34) provided in phosphate buffered saline containing bovine serum albumin and 0.1 % sodium azide. The pre-mixed antibodies contains CD45 fluorescein isothiocyanate (FITC), clone 2D1, and anti-CD34 phycoerythrin (PE), clone 8G12.

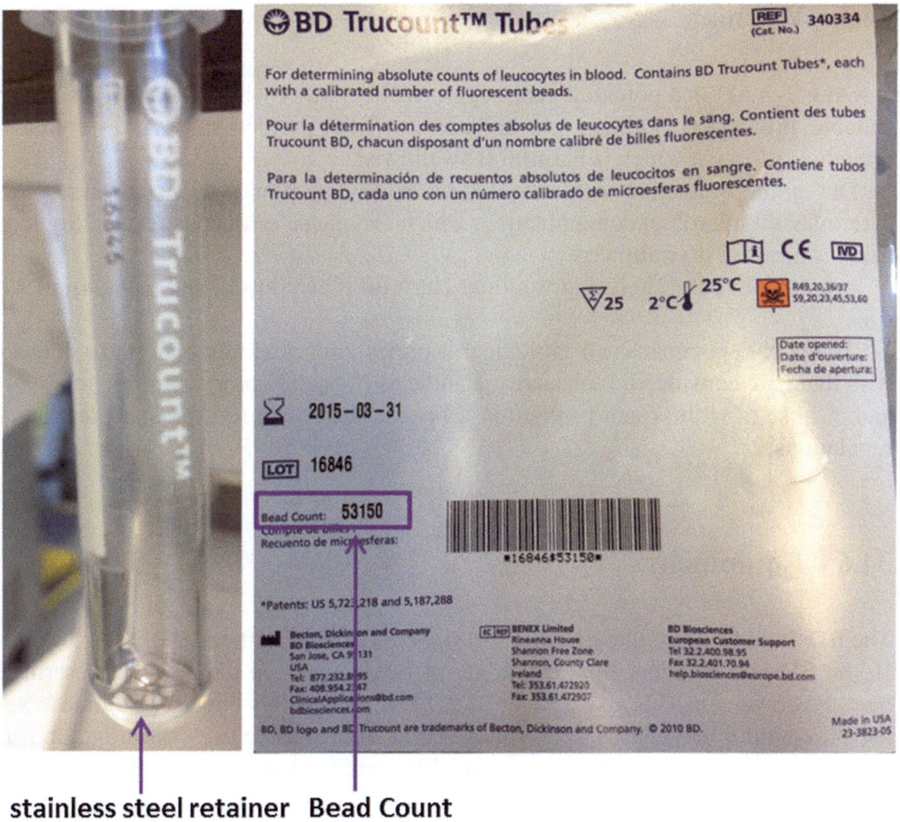

Fig. 34.1 Image of the Trucount tube. *Left panel* image shows stainless steel retainer to keep the fluorescence bead pellet is intact within the metal retainer at the bottom of BD Trucount tube. *Right panel* image shows the bead count shown on the Box/lot of BD Trucount tubes, which is, varies by lot of BD Trucount tubes

- 7-aminoactinomycin-D (7-AAD) reagent 7-AAD is a nucleic-acid dye used to identify non-viable cells
- 10× ammonium chloride lysing solution The ammonium chloride solution is a fixative-free solution for red blood cell lysis.
- 50 BD Trucount tubes, each single-use tube contains a freeze-dried pellet of APC-fluorescent beads.

The CD45 antigen is detected on all human leucocytes but is weakly expressed on hematopoietic progenitor cells. The CD34 antigen is present on immature hematopoietic precursor cells and all hematopoietic colony-forming cells in bone marrow and blood, including unipotent and pluripotent progenitor cells [4, 6, 11, 12]. CD45 and CD34 antibodies are each composed of mouse IgG1 heavy chains and kappa light chains. BD Trucount tubes are used with this reagent. By adding the reagent and the sample directly to the BD Trucount tube, the absolute count of the cell population of interest can be directly determined.

34.5 Warning

7-AAD: 7-AAD is a potential carcinogen Avoid contact with skin and eyes. Wear suitable protective clothing, gloves, and eye/face protection. 7-AAD contains dimethyl sulfoxide which is harmful if swallowed.

The antibody reagent: contains sodium azide as a preservative; however, take care to avoid microbial contamination, which can cause erroneous results. Sodium azide is harmful if swallowed.

Ammonium chloride lysing solution: Harmful if swallowed and irritating to the eyes.

Biological specimens: all biological samples and materials coming in contact with them are considered biohazards. Handle as if capable of transmitting infection and dispose of with proper precautions in accordance with federal, state, and local regulations.

34.6 Equipment

The BD Stem Cell Enumeration kit is designed for use on either the BD FACSCanto II, the BD FACSCalibur flow cytometer or FACSVerse. In the protocol described here, FACSVerse will be used. Refer to BD FACSVerse System User's Guide to set the photomultiplier tube (PMT) voltages and fluorescence compensation, and to check instrument sensitivity before acquisition and analysis.

34.7 Basic Protocols

For an overview of the staining process, see below the staining summary (Fig. 34.2).

1. For each patient specimen, label a BD Trucount tube with the sample identification.
2. Pipette 20 μL of the BD Stem Cell reagent into the bottom of the tube.
   Pipette just above the stainless steel retainer of the BD Trucount tube. Do not touch the bead pellet.
3. Add 20 μL of 7-AAD reagent into the BD Trucount tube.
4. Add 100 μL of a well-mixed specimen onto the side of the tube just above the retainer.
   Note: Avoid smearing the specimen down the side of the tube. If any specimen remains on the side of the tube, it will not be stained with the reagent but can be re-suspended by the lysing solution and therefore results may be affected. Accurate pipetting is critical when using a BD Trucount tube. Use the reverse pipetting technique, or a positive displacement pipettor, to pipette specimen onto the side of the tube just above the retainer.

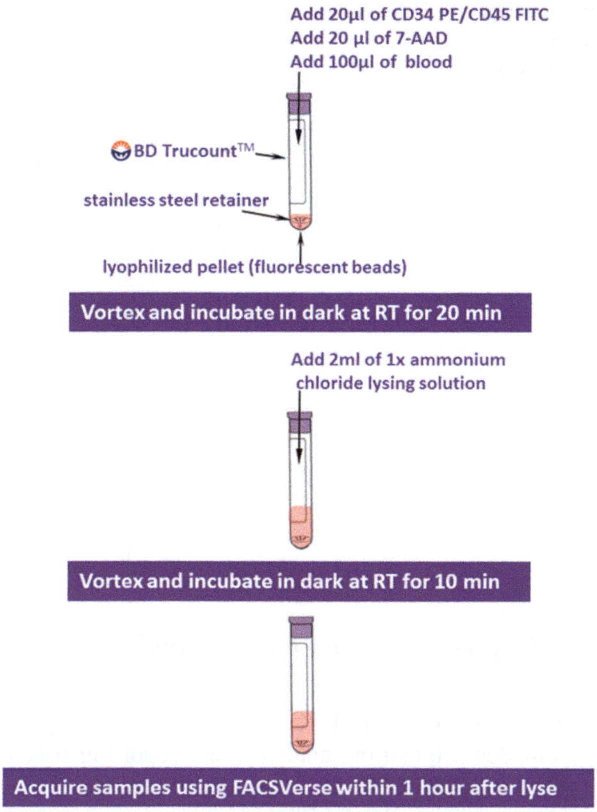

Fig. 34.2 Basic staining summary for CD34+ stem cell. This diagram summaries the staining protocol for CD34+ cells. Briefly, BD Stem Cell reagent, 7-AAD reagent and blood were added above the stainless steel retainer, mixed and incubated for 20 min at RT in dark. Subsequently, ammonium chloride lysing solution (1×) was added and Trucount tube was well-vortexed and incubated for further 10 min at RT in dark. Following this incubation period, cells were ready for analysis

5. Cap each tube and vortex gently to mix.
6. Incubate for 20 min in the dark at room temperature.
7. Add 2 mL of 1× ammonium chloride lysing solution to each tube to lyse red blood cells.
8. Cap each tube and vortex gently to mix.
9. Incubate for 10 min in the dark at room temperature.
10. Immediately, place tubes on wet ice in the dark until ready to acquire samples.

Acquire samples on flow cytometer within 1 h of lysing to minimise cells susceptibility to cell death. For information on setup, acquisition, and data analysis, see the BD Stem Cell Enumeration Application Guide for FACSVerse flow cytometer.

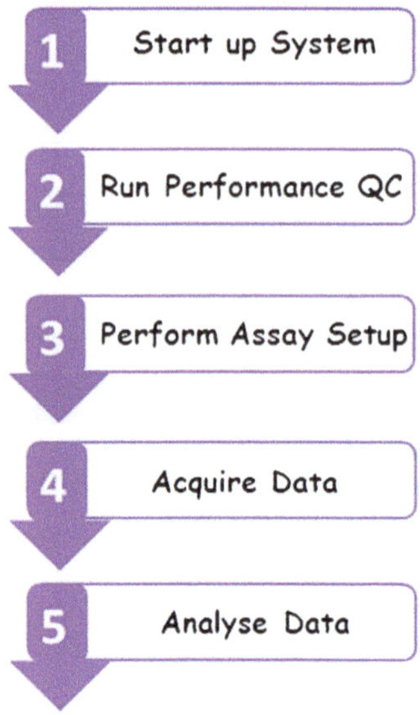

Fig. 34.3 Workflow for BD FACSverse setup. This diagram summaries the steps for using BD FACSuite™ software with the BD FACSVerse™ flow cytometer using the experiment workflow for staining CD34+ cells

34.7.1 Setup, Acquisition, and Data Analysis

For setup and acquisition on a BD FACSCVerse system using BD FACSuite™ software which provides acquisition, analysis, and reporting functions to CD34+ absolute count, refer to BD FACSVerse System User's Guide.

The basic workflow for the BD FACSVerse instrument setup, acquisition and data analysis is shown in Fig. 34.3. (1) **Start-up system**: log in to BD FACSuite software, check the fluid levels, connection and fluidics status. (2) **Run performance QC**: performance quality control (PQC) is performed using BD FACSuite CS &T research beads as outlined in the BD FACSVerse System User's Guide (Open the Setup & QC workspace, Run the BD FACSuite™ CS&T Research beads, View the Setup and QC Report and QC Tracking tabs). (3) **Perform assay setup**: Select the Setup & QC tab, Select the Tube Settings tab, Run the BD FACSuite CS&T Research beads. (4) **Acquire data**: Open the Experiment workspace, Create plots, gates, statistics, and a hierarchy needed for acquiring data, install a sample tube onto the cytometer. Click Preview, Click Stop; Double-click the tube to modify the tube properties, Click next in the data sources panel to create additional tubes, as required, finally click acquire. (5) **Analyse data**: Create any additional plots, gates, and statistics in the worksheet, or create a report, Perform quality control of the analysis, select File > export to PDF to create a PDF of the active worksheet [13].

During the settings of the instrument, the following values, bead count which is labelled on the BD Truncount pouch, blood dilution factor and blood sample volume need, to be entered manually into gate statistics of BD FACSuite software, so that the results (viable CD34 cells/μl, viable CD45 cells/μl, total CD34 cells/μl, CD34 % viability and viable CD34 % of viable CD45) can be automatically calculated by the BD FACSuite software.

34.8 Gating Strategy

The following steps explain the template gating strategy for enumeration of viable CD34+ hematopoietic stem cells

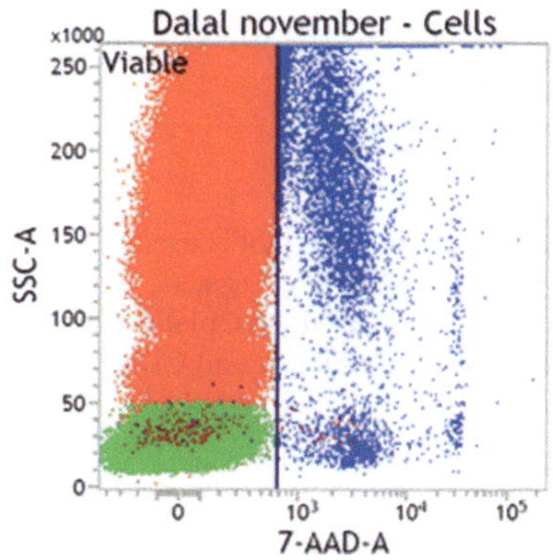	1. **7-ADD vs. SSC**: this plot is used to create the first gate to identify viable cells (negative 7-ADD)
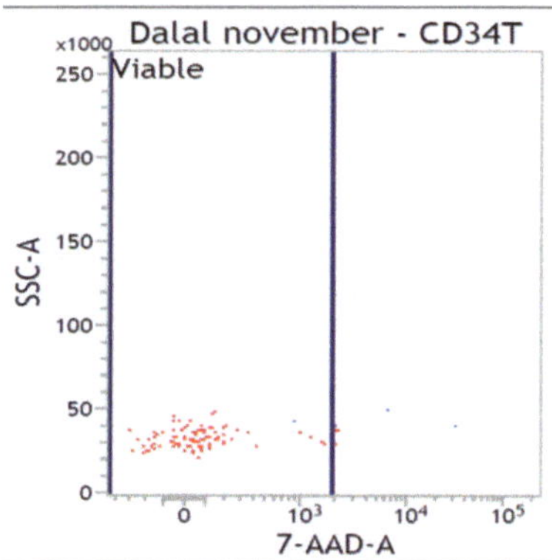	2. **7-ADD vs. SSC:** this plot is used to gate the total CD34+
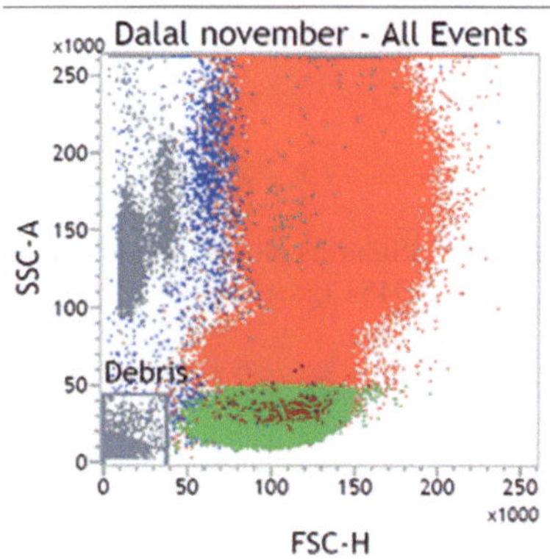	3. **FSC-H vs. SSC:** this plot is used to adjust the debris gate to exclude debris (no gate)

(continued)

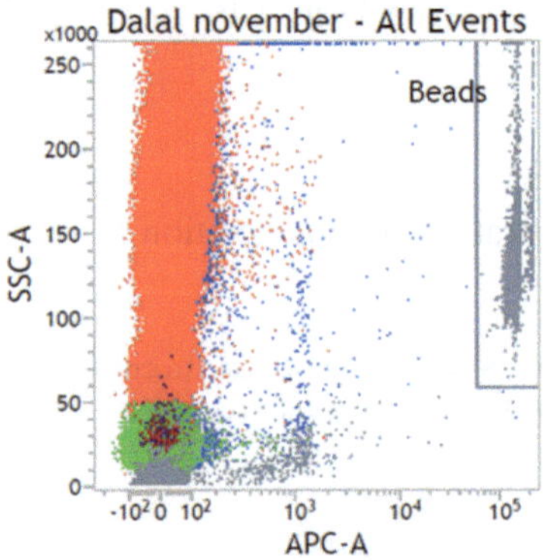

4. **APC-A vs. SSC:** this plot is used to adjust the bead gate to catch all beads including any multiples (no gate)

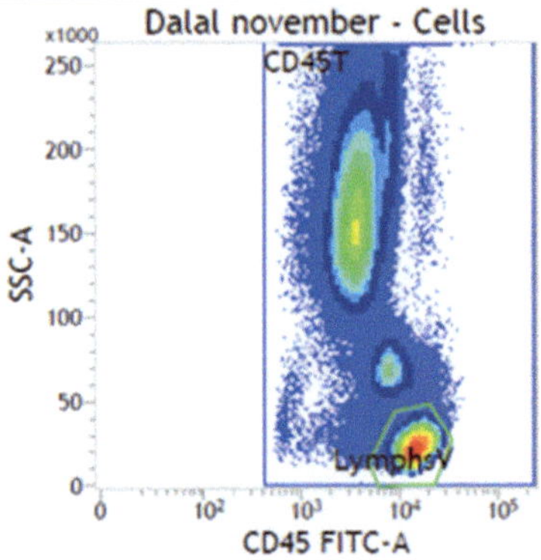

5. **CD45 FITC vs. SSC:** this plot is used to include all CD45^{dim} to CD45^{bright} (gated lymphocytes) and excludes debris, platelets and unlysed red blood cells which are all CD45 negative

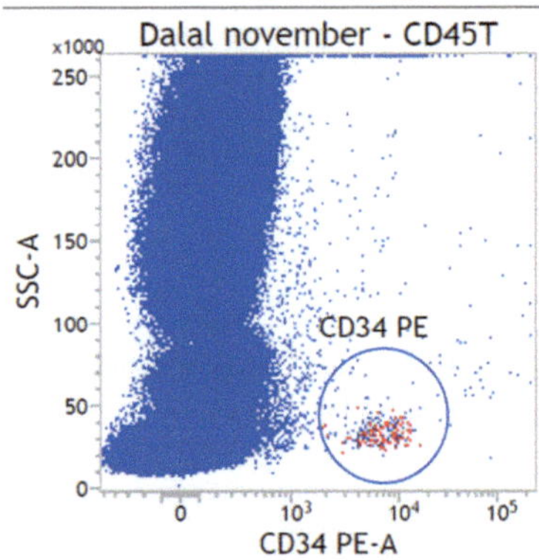

6. **CD34 PE vs. SSC:** this plot is used to display CD45 cells and is to identify CD34+ cells (gated cells)

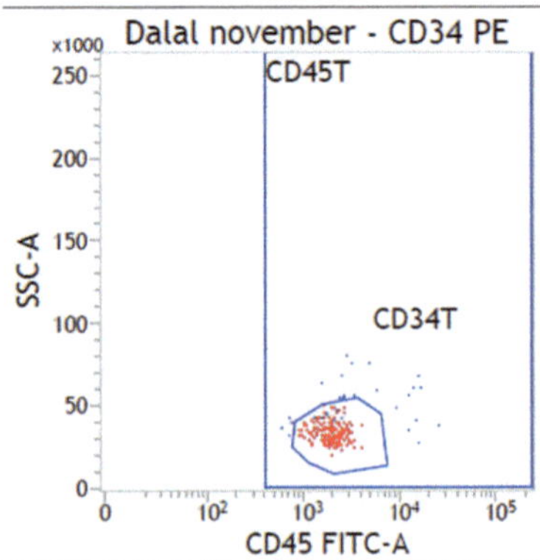

7. **CD45 FITC vs. SSC:** this plot is used to define the viable CD34+ CD45+ (gated cells)

(continued)

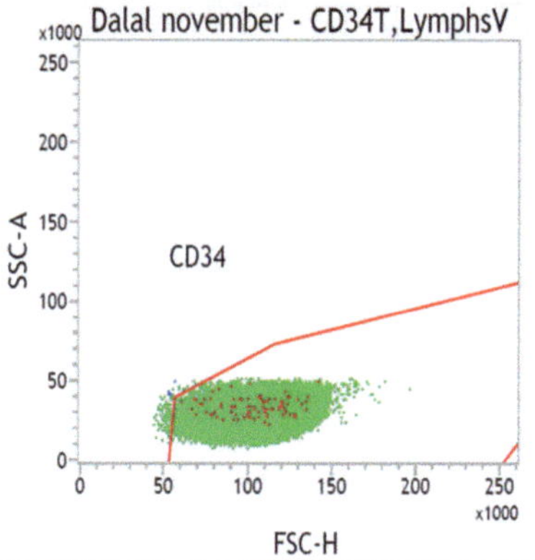

8. **FSC vs. SSC:** this plot is used to define the viable CD34+CD45+ (red-gated cells)

34.9 Quality Control

It is highly recommended in each run, to run low and high process controls with known assayed CD34+ absolute count values using the BD Stem Cell Control kit (Catalog No. 340991) to confirm staining and system integrity. These controls are fixed cells therefore run without the addition of the 7-AAD dye (positive for 7-AAD).

34.10 Analysing the Data Using BD FACSuite Software

Results of viable CD34 cells/μl, viable CD45 cells/μl, total CD34 cells/μl, CD34 % viability and viable CD34 % of viable CD45 will be automatically calculated by the BD FACSuite software. These results can also be calculated manually as follow (Figs. 34.4 and 34.5):

$$\text{Viable CD34 cells}/\mu\text{l} = \frac{\text{viable CD34} \times \text{BD Trucount} \times \text{dilution factor}}{\text{Beads} \times \text{sample volume}}$$

$$\text{Viable CD45 cells}/\mu\text{l} = \frac{\text{viable CD45} \times \text{BD Trucount} \times \text{dilution factor}}{\text{Beads} \times \text{sample volume}}$$

$$\text{Total CD34 cells}/\mu\text{l} = \frac{\text{Total CD34} \times \text{BD Trucount} \times \text{dilution factor}}{\text{Beads} \times \text{sample volume}}$$

$$\text{CD34 \% Viability} = \frac{\text{Viable CD34} \times 100}{\text{Total CD34}}$$

$$\text{Viable CD34 \% of viable CD45} = \frac{\text{Viable CD34} \times 100}{\text{Viable CD45}}$$

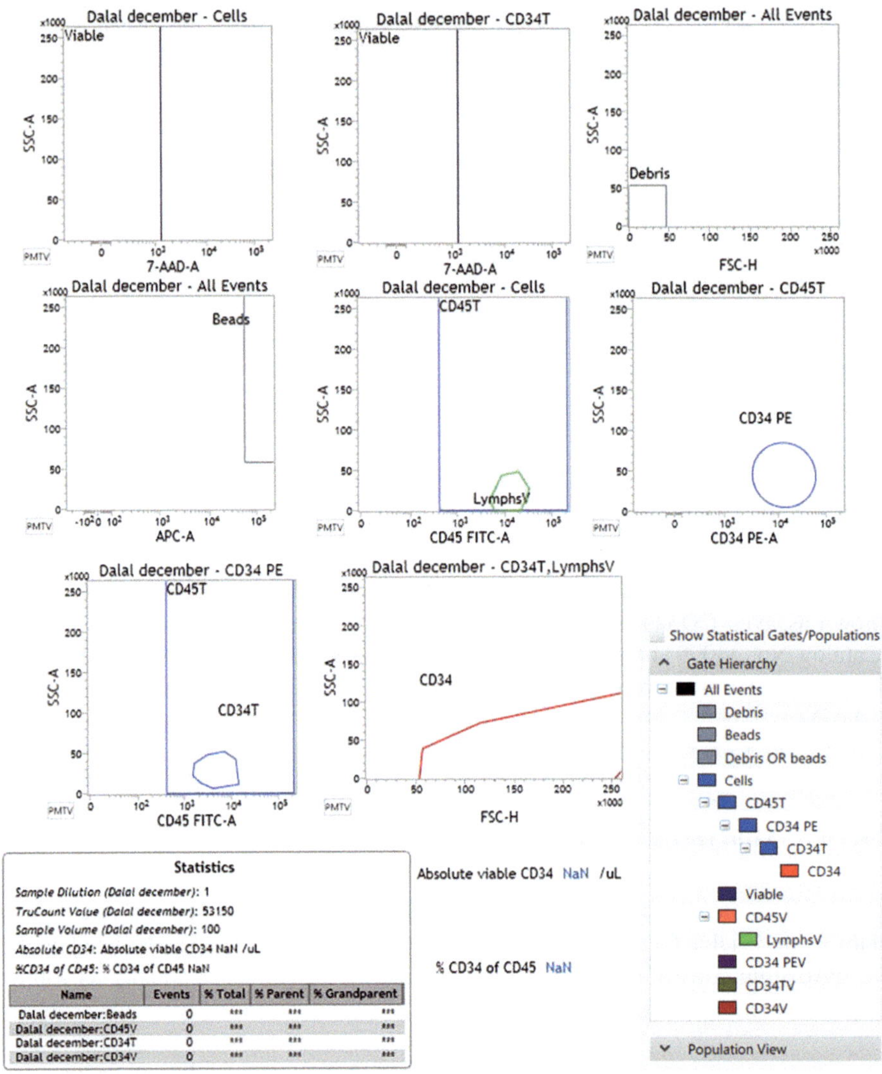

Fig. 34.4 BD FACSVerse CD34 enumeration template without data

34.11 Notes and Troubleshooting Tips

34.11.1 Notes

Note 1 Before start, make sure that fluorescence bead pellet is under the metal retainer at the bottom of the BD Trucount tube, otherwise, do not use it and replace it with another.

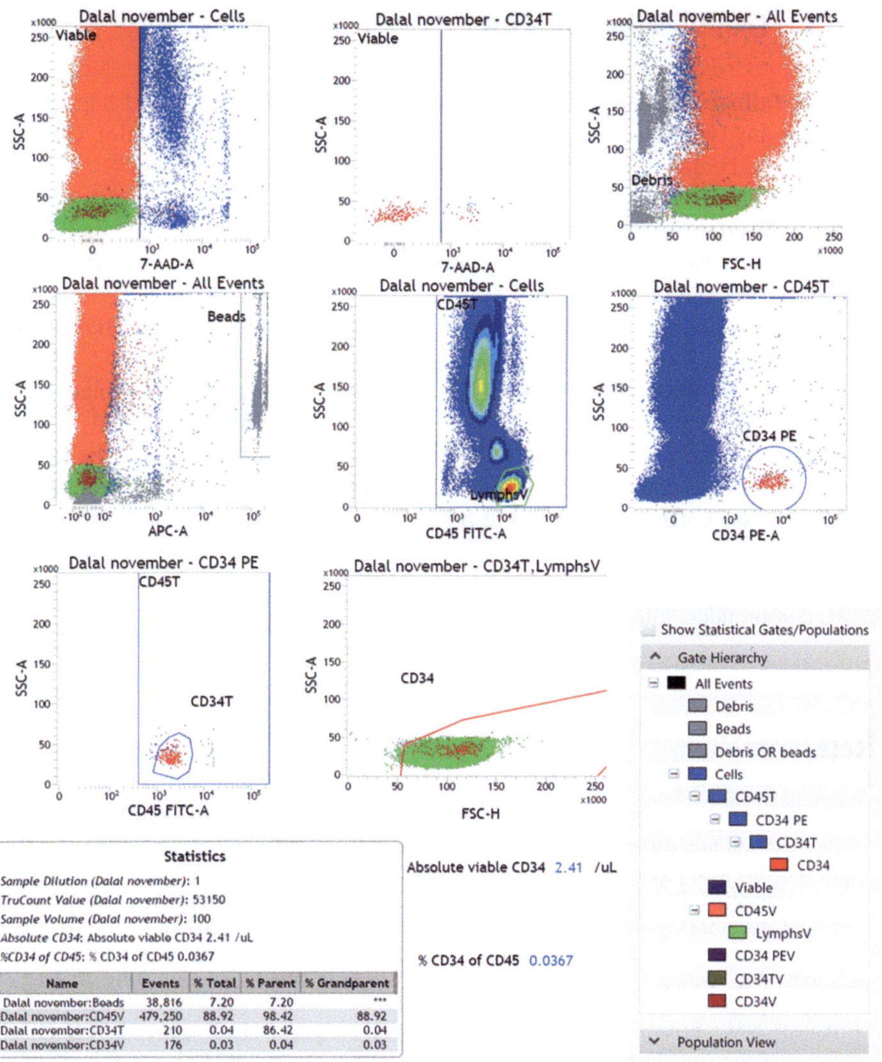

Fig. 34.5 BD FACSVerse CD34 enumeration template with data

Note 2 Do not transfer beads to another tube.

Note 3 Always change tips between tubes.

Note 4 Each day, prepare enough 1× ammonium chloride lysing solution for use, 1 volume of 10× ammonium chloride solution with 9 volume of distilled water. Store and use at room temperature.

Note 5 Acquire samples on flow cytometer within 1 h of lysing. Highly manipulated or processed samples can be more susceptible to increased cell death after preparation

34.12 Troubleshooting Tips

1. **Gate failure** can cause the results not to be generated, this could happened in gating any type of cells (lymphocytes, CD45$^+$, CD34+, CD34 viable). Gate failure could be due to insufficient number of the cells in the sample.
   Recommended solution: manually re-gate the sample
2. **No beads detected or low bead number detected** (<1,000 beads) could lead to incomplete results report, this could be possibly due to missing bead pellet from the BD Trucount tube.
   Recommended solution: repeat the sample and make sure using the BD Trucount tube
3. **High white blood cells count** ($>45 \times 10^3$ cells/μl) can cause incomplete results report
   Recommended solution: Dilute the blood sample, and then repeat the samples but you should change the dilution factor accordingly in the stats equation.
4. **Inaccurate Gating**, which could happened in gating any cell population. This gate suspect results in inaccurate gate location but still results completed and reported
   Recommended solution: Manually re-gate the sample

References

1. Shpall EJ, Jones RB, Bearman SI et al (1994) Transplantation of enriched CD34+ autologous marrow into breast cancer patients following high-dose chemotherapy: influence of CD34+ peripheral blood progenitors and growth factors on engraftment. J Clin Oncol 12(1):28–36
2. Langenmeyer I, Weaver C, Buckner CD et al (1995) Engraftment of patients with lymphoid malignancies transplanted with autologous bone marrow, peripheral blood stem cells or both. Bone Marrow Transplant 15:241–246
3. Zander AR, Lyding J, Bielack S (1991) Transplantation with blood stem cells. Blood Cells 17:301–309
4. Bensinger WI, Longin K, Appelbaum F et al (1994) Peripheral blood stem cells (PBSCs) collected after recombinant granulocyte colony stimulating factor (rhG-CSF): an analysis of factors correlating with the tempo of engraftment after transplantation. Br J Haematol 87:825–831
5. van der Wall E, Richel DJ, Holtkamp MJ et al (1994) Bone marrow reconstitution after high-dose chemotherapy and autologous peripheral blood progenitor cell transplantation: effect of graft size. Ann Oncol 5:795–802
6. Bender JG, To LB, Williams S et al (1992) Defining a therapeutic dose of peripheral blood stem cells. J Hematother 1:329–341
7. Civin CI, Trischman TW, Fackler MJ et al (1989) Report on the CD34 cluster workshop. In: Dorken B, Gilks WR, Knapp W et al (eds) Leucocyte typing IV: white cell differentiation antigens. Oxford University Press, Oxford, pp 818–829
8. Berenson RJ, Bensinger WI, Hill RS et al (1991) Engraftment after infusion of CD34+ marrow cells in patients with breast cancer or neuroblastoma. Blood 77(8):1717–1722
9. Pelehach L (1996) The story of the stem cell. Lab Med 27(9):588–599

10. Keeney M, Chin-Yee I, Weir K, Popma J, Nayar R, Sutherland DR (1998) e platform flow cytometric absolute CD34+ cell counts based on ISHAGE guidlines. Cytometry 34:61–70
11. Loken MR, Shah VO, Hollander Z, Civin CI (1988) Flow cytometric analysis of normal B lymphoid development. Pathol Immunopathol Res 7:357–370
12. Sutherland DR, Anderson L, Keeney M et al (1996) The ISHAGE guidelines for CD34+ cell determination by flow cytometry. J Hematother 5:213–226
13. BD FACSuite Software Quick Reference Guide Workflow. http://www.bdbiosciences.com/documents

Chapter 35
A Scheme for the Development and Validation of Enzyme Linked Immunosorbent Assays (ELISA) for Measurement of Angiogenic Biomarkers in Human Blood

Garry McDowell, Richard Body, Cliona Kirwan, Ged Byrne, and Mark Slevin

35.1 Introduction

This handbook of methods in vascular biology reports a number of analytical methods for the quantification of angiogenesis, both *in vitro* and *in vivo*. In this chapter the authors discuss a scheme for the development and validation of ELISA for the quantification of circulating vascular biomarkers. The principles we discuss, can be applied to the majority of ELISA techniques, irrespective of the biomarker being measured. We will illustrate our method using Vascular Cell Adhesion Molecule-1 (VCAM-1) as an example.

The ELISA is an example of a non-competitive sandwich assay. The components of the ELISA consist of a capture antibody, secondary detection antibody and detection reagent. Briefly, an analyte specific capture antibody is bound to an ELISA plate, forming the solid phase. The analyte of interest in samples, or standards is then incubated with the solid phase antibody, capturing the analyte in the solid phase, due to the antibody-antigen reaction. The plate is washed to remove unbound analyte. Following the wash step, a secondary biotin-conjugated antibody is added. The secondary antibody binding to different epitopes on the antigen to the capture antibody. Following a wash step to remove unbound secondary antibody, a Streptavidin-horseradish peroxidase enzyme conjugate is added. Streptavidin binds biotin with high affinity. Again following a wash step to remove unbound enzyme

G. McDowell, Ph.D., F.R.C.Path., F.E.S.C. (✉) • M. Slevin, Ph.D., F.R.C.Path., F.A.H.A.
School of Healthcare Science, Manchester Metropolitan University, Manchester, UK
e-mail: g.mcdowell@mmu.ac.uk

R. Body, Ph.D., F.C.E.M.
Institute of Cardiovascular Science, University of Manchester, Manchester, UK

C. Kirwan, Ph.D., F.R.C.S. • G. Byrne, M.D., F.R.C.S.
Department of Academic Surgery, University of Manchester and University Hospital of South Manchester, Manchester, UK

M. Slevin, G. McDowell (eds.), *Handbook of Vascular Biology Techniques*,
DOI 10.1007/978-94-017-9716-0_35

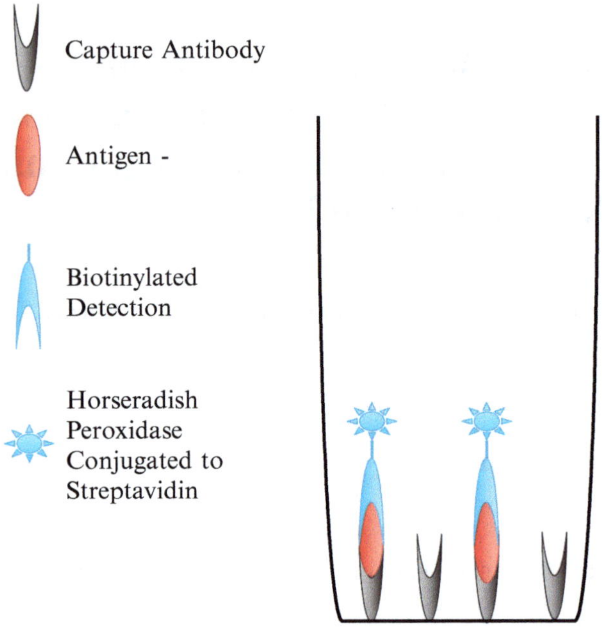

Fig. 35.1 The analytical principle of the ELISA

conjugate a substrate solution is added, which changes colour in the presence of the enzyme. The reaction can be stopped and the colour intensity (optical density) measured. In the non-competitive format, optical density is directly proportional to concentration. Figure 35.1 illustrates the analytical principle of the ELISA.

A number of manufacturers supply readymade ELISA kits which have been fully validated providing convenience and ease of use, however, these may be impractical especially on a restricted budget. In this Chapter the authors will illustrate how it is possible to use commercially available reagents to generate and validate an in-house assay, custom designed to meet your research needs.

35.2 Materials and Methods

35.2.1 Materials and Reagents

1. VCAM-1 DuoSet® Elisa Development System kit. (R & D systems; Abingdon, Oxford. Catalogue Number DY809). The kit includes the basic reagents required to establish the ELISA
   (a) Capture Antibody (Part 841127 1 vial) – 360 μg/ml of mouse monoclonal anti-human VCAM-1, when reconstituted with 1 ml buffer without protein. The manufacturer recommends a working concentration of 2.0 μg/ml. Note 1

(b) Detection antibody (Part 841128 1 vial) – 36 µg/ml of polyclonal biotinylated sheep anti human VCAM-1 when reconstituted. The manufacturer recommends a working concentration of 200 ng/ml. Note 1
(c) Standard solution. (Part 841129 1 vial) – 80 ng/ml of recombinant human VCAM-1, when reconstituted with 0.5 ml of reagent buffer. A upper standard concentration of 1,000 pg/ml is recommended for the standard curve.
(d) Streptavidin-HRP (Part 890803 2 vial) – Streptavidin conjugated to horse – radish peroxidase.

2. Sodium dihydrogen orthophosphate-1-hydrate ($NaH_2PO_4.H_2O$) Molecular Weight = 137.99 g. BDH Analar®, BDH Ltd, Poole, England.
3. Di-Sodium Hydrogen orthophosphate-2-hydrate ($Na_2HPO_4.2H_2O$) Molecular Weight = 177.99 g. BDH GPR™, BDH Ltd, Poole, England.
4. Albumin-Bovine (BSA): Sigma-Aldrich, Poole, Dorset. Fraction V (Initial fractionation by heat shock) Minimum 98 % (Electrophoresis).
5. Sodium azide (NaN_3). Molecular weight = 65.01 g. BDH GPR™, BDH Ltd, Poole, England. Warning COSHH: Very toxic if swallowed; contact with acids liberates a very toxic gas. Avoid contact with skin. After contact with skin, wash immediately with plenty of water. Heating may cause an explosion.
6. Tetramethylbenzidine (TMB): obtained as a commercial pre-prepared solution with substrate (DAKO, Cambridgeshire, UK). TMB and Hydrogen peroxide (substrate) in a 1:1 mixture for use as the Substrate-chromogen reagent.
7. 1 M Sulphuric acid – H_2SO_4. Sulphuric acid Analar® specific. gravity. 1.84; Molecular Weight = 98.07. BDH Ltd, Poole, England. Warning COSHH: Causes severe burns. Never add water directly to this reagent.
8. De-ionised water.

35.3 Buffers Required

35.3.1 Capture Antibody Buffer

0.05 M Phosphate Buffer, pH 7.4 (4.83 g of $NaH_2PO_4.H_2O$ and 38.28 g of $Na_2HPO_{4.}2H_2O$ made up to 5 L).

35.3.2 Wash Buffer

0.05 % Tween 20 in 0.05 M phosphate buffer (125 µl into 250 ml 0.05 M Phosphate Buffer, pH 7.4).

35.3.3 *Reagent Buffer*

0.1 % BSA in 0.05 M Phosphate Buffer, pH7.4 containing 0.01 % sodium azide. Note 2.

35.3.4 *Sample/Standard Buffer*

1 % BSA in 0.05 M phosphate buffer, pH 7.4 containing 0.1 % sodium azide.

35.3.5 *Equipment*

1. 96 well Microtitre polystyrene base immunoassay plates.
2. Plate reader, software, and printer:
3. Sterile 30 ml universal containers.
4. Gilson eight-channel multi-pipette.
5. Plate shaker; Perkin Elmer Ltd, Milton Keynes, UK.
6. Assorted tubes to prepare dilutions and standard curve.
7. pH meter.

35.4 Plate Coating

1. Prepare the working dilution of capture antibody in capture antibody buffer without BSA. Note 1
2. Add 100 μl of diluted antibody to each well on a 96 well microtitre plate.
3. Seal and incubate overnight at room temperature.
4. Following overnight incubation, wash the plate with wash buffer (x3), completely filling each well using a wash bottle.
5. Remove wash solution completely between washes.
6. Following the final wash remove any residual wash buffer by blotting on clean absorbent towels.
7. Block plates by adding 300 μl of reagent buffer to each well.
8. Incubate at room temperature for 1 h.
9. Repeat washing steps 5 and 6 above.
10. The plates are now ready for use. Based on our experience plates can be dried at room temperature, sealed with an adhesive plate cover and stored in at 4 °C for up to 1 week.

35.5 Preparation of a Typical Standard Curve

The standard curve details the relationship between the instrument response (optical density, OD) and known concentrations of the analyte. The standard curve should contain a sufficient number of points to determine the relationship between response and concentration to enable determination of the linear range of the assay. Ideally the curve height should be >1 and preferable between 1 and 3 OD units.

1. Prepare a range of concentrations of standards using recombinant VCAM-1 as supplied.
2. Standards should be diluted in reagent buffer and an upper concentration of standard of 1,000 pg/ml is recommended.
3. Prepare a working stock standard of 1,000 pg/ml from the reconstituted standard supplied in the DuoSet ELISA kit.
4. Use the stock standard to prepare a range of standard concentrations as shown below using doubling dilutions (Table 35.1) using doubling dilutions:

Table 35.1 Table showing the preparation of VCAM-1 standard curve by doubling dilutions method using recombinant VCAM-1 standard as supplied.

Tube number	Dilution factor	Volume of diluted standard added (ml)	Volume of reagent buffer added	VCAM-1 [pg/ml]
1		1	0	1,000
2	1:2	1	1	500
3	1:4	1	1	250
4	1:8	1	1	125
5	1:16	1	1	62.5
6	1:32	1	1	31.3
7	1:64	1	1	15.6
8	1:128	1	1	7.8
9	1:256	1	1	3.9
10	1:512	1	1	1.95
11	1:1024	1	1	0.98
12	Blank	0	1	Blank

35.6 Assay Procedure

The typical assay procedure is detailed below. In the following sections the authors will describe a scheme to validate the assay to ensure optimal performance and determine the optimal performance characteristic of the assay.

1. The following procedures are carried out in duplicate.
2. Add 100 μl of standard or unknown, diluted as appropriate in reagent buffer.
3. Cover the plate with an adhesive plate cover and incubate plate for 2 h at room temperature. Note 3
4. Aspirate and wash each well with wash buffer, Using a wash bottle fill each well with wash buffer to ensure unbound VCAM-1 in standard or sample has been removed.
5. Repeat step 4 above a further two times (three washes in total).
6. Add 100 μl of detection antibody diluted in reagent buffer to each well.
7. Cover the plate with an adhesive cover and incubate for 2 h at room temperature. Note 3
8. Repeat wash steps 4 and 5 above.
9. Add 100 μl of the streptavidin-HRP conjugate to each well.
10. Cover the plate with an adhesive strip and incubate at room temperature for 20 min, avoiding exposure to direct sunlight.
11. Repeat was steps 4 and 5 above.
12. Add 100 μl of Tetramethylbenzidine (TMB): obtained as a commercial pre-prepared solution with substrate to each well.
13. Incubate the plate for 20 min at room temperature.
14. Add 50 μl of 1 M H_2SO_4 to each well to stop the reaction.
15. Determine the optical density of each well immediately using a microplate reader set a 450 nm. If wavelength correction is available set this to 540 nm or 570 nm. If wavelength correction is not available subtract readings at 540 nm or 570 nm from readings at 450 nm, to correct for imperfections in the microtitre plate.

35.7 Calculation of Results

Plot a calibration curve of optical density against concentration of standard. The optical density of the unknown can then be used to determine the concentration.

A number of software packages are available which will do this automatically, however it is always advisable to visually inspect the calibration curve to check assay range and linearity.

35.8 Assay Validation Methodology

Our previous validation work has utilised the Guidance for Industry: Bioanalytical Method Validation protocol as a reference guide. Please also refer to the Manufacturers ELISA Development Guide (see link below). The material that follows is in addition to the optimization that this recommended in this guide

35.8.1 Use of a Secondary Calibrator in Biological Matrix

According to the Guidance for Industry: Bioanalytical Method Validation, calibrators should be prepared in the same biological matrix as the biological samples of unknown concentration. In addition, calibrators of higher concentration may facilitate an extended calibration curve.

1. Use pooled human serum from healthy volunteers and subjects suffering from pathological conditions of interest as a secondary calibrator. Note 4
2. Determine from the literature the concentrations of VCAM-1 expected.
3. Aliquot and freeze pooled serum in 1 above.
4. Prepare a standard curve of a range of dilutions of recombinant VCAM-1 as supplied in the DuoSet ELISA Development kit.
5. Prepare a range of dilutions of pooled human plasma or serum, which parallel the range of dilutions in 4 above.
6. Run an assay as detailed in Steps 1–15 under assay procedure.
7. Use the standard curve (4) to determine the concentration of VCAM-1 in the pooled human serum.
8. To assign a VCAM-1 concentration to your secondary calibrator, calculate the mean concentration from dilutions on the linear point of the calibration curve by visual inspection of curves prepared in 4 an 5 above.
9. To view the calibrations curves, plot both the standard curve of the kit standards and the pooled human serum on the same graph. They should parallel each other.

35.9 Establishing a Quality Control Pool

1. Use pooled human serum as described above. Note 4
2. Pool serum from healthy individuals and patients with the pathological condition of interest. This will enable the analyst to quality control the assay at normal and pathological concentrations.
3. Measure the VCAM-1 concentration across a range of dilutions and a minimum of 10 times (in duplicate) within the same assay run, and across 10 separate days (in duplicate).
4. Calculate the mean and SD for both the intra- and inter-assay results.
5. Incorporate QC material into each assay run and use to monitor assay performance over time.

35.10 Determine the Intra- and Inter-assay Co-efficient of Variation (CV)

The precision if an analytical method describes the closeness of individual measures of an analyte when the procedure is repeated to multiple aliquots of a single concentration in the biological matrix of interest. Precision should be measured a minimum of 5 times (10 preferred) at each concentration.

1. Within each assay, add 100 μl of pooled QC material at three separate locations across the assay plate.
2. The reminder of the assay is performed as usual using the procedure detailed above.
3. The QC material simply replaces the unknown sample.
4. Calculate the inter-assay CV from the mean and SD of the QC material included in at least five individual assays.
5. The CV = (SD/Mean)*100 and is expressed as a percentage.
6. The Intra-assay CV may be determined by repeating the process above, however the measurements of mean and SD must be determined within one assay run (i.e. within one ELISA plate).
7. The CV = (SD/Mean)*100 and is expressed as a percentage.

35.11 Determination of Detection Limit

This is the minimum amount of VCAM-1 that can be detected with the assay.

1. Perform the assay as described previously.
2. Set up the standard curve by adding 100 μl of calibrator in duplicate to each well.
3. Into the remaining wells of the plate, add 100 μl of blank calibrator.
4. Perform the remainder of the assay as described earlier.
5. Calculate the mean and SD of the blank readings.
6. The absorbance of the minimum detectable VCAM-1 equates to the ***mean blank absorbance reading + 2SD of the mean blank absorbance reading.***
7. The absorbance reading determined in 7 above can then be read from the standard curve to give the concentration of VCAM-1 corresponding to the minimum concentration that can be determined by the assay.

35.12 Final Assay Format

The format of the VCAM assay is shown in the diagrammatic representation of a 96 well plate in Fig. 35.2. This incorporates analytical standards and QC samples prepared in biological matrix and unknown samples for the determination of VCAM.

	1	2	3	4	5	6	7	8	9	10	11	12
A	Blank	Blank	QC Low	QC Low	Unknown Sample	Unknown Sample	Unknown Sample	Unknown Sample	Unknown Sample	Unknown Sample	Unknown Sample	Unknown Sample
B	Std 1	Std 1	QC High	QC High	Unknown Sample	Unknown Sample	Unknown Sample	Unknown Sample	Unknown Sample	Unknown Sample	Unknown Sample	Unknown Sample
C	Std 2	Std 2	Unknown Sample	Unknown Sample	Unknown Sample	Unknown Sample	Unknown Sample	Unknown Sample	Unknown Sample	Unknown Sample	Unknown Sample	Unknown Sample
D	Std 3	Std 3	Unknown Sample	Unknown Sample	Unknown Sample	Unknown Sample	QC Low	QC Low	Unknown Sample	Unknown Sample	Unknown Sample	Unknown Sample
E	Std 4	Std 4	Unknown Sample	Unknown Sample	Unknown Sample	Unknown Sample	QC High	QC High	Unknown Sample	Unknown Sample	Unknown Sample	Unknown Sample
F	Std 5	Std 5	Unknown Sample	Unknown Sample	Unknown Sample	Unknown Sample	Unknown Sample	Unknown Sample	Unknown Sample	Unknown Sample	Unknown Sample	Unknown Sample
G	Std 6	Std 6	Unknown Sample	Unknown Sample	Unknown Sample	Unknown Sample	Unknown Sample	Unknown Sample	Unknown Sample	Unknown Sample	QC Low	QC Low
H	Std 7	Std 7	Unknown Sample	Unknown Sample	Unknown Sample	Unknown Sample	Unknown Sample	Unknown Sample	Unknown Sample	Unknown Sample	QC High	QC High

Fig. 35.2 Format of ELISA plate showing position of standards, low and high quality control samples and unknown samples

35.13 Sample Results

In our hands, using the approach above we developed and validated an assay using the DuoSet kits with the following characteristics:

Detection Limit: 25 pg/ml.
Assay Range using Extended Matrix based Secondary Calibrator: 25–3,000 pg/ml.
Inter-assay CV: 9 % at 550 pg/ml and 12 % at 1,100 pg/ml.
Intra-assay CV: 7 % at 550 pg/ml and 9 % at 1,100 pg/ml.

35.14 Applications

We have used the above methodology to develop and validate ELISA's, using commercially available reagents, for determination of serum and/or plasma VCAM, P-selectin and E-selectin. The assays have been applied to our clinical investigations in rapid diagnosis of acute coronary syndromes [1] and in predicting venous thromboembolism in breast cancer [2–4].

35.15 Notes

Note 1 We used the antibody concentrations specified by the manufacturer as these gave an acceptable standard curve during assay validation. Please see R&D systems ELISA Development Guide Below which contains details of how to perform a grid experiment to determine the most favourable concentrations of capture and detection antibody combination.

Note 2 To avoid foaming when making up BSA buffer, sprinkle BSA onto the liquid surface and leave to stand for 10–15 min, until BSA dissolves. **Do not** stir or mix. We used a 0.1 % BSA buffer as on validation for our purpose this improved assay sensitivity.

Note 3 We used manufacturers instructions for incubation time, however we validated incubation times of 1, 2 and 3 h. A 1 h incubation lowered assay sensitivity significantly, whilst 3 h increased background with minimal improvement in sensitivity.

Note 4 Depending on National Laws and procedures for research governance, this may require approval of a Research Ethics Committee and/or other Institutional Committee overseeing research governance.

References

1. Body R, Pemberton P, Ali F, McDowell G, Carley S, Smith A, Mackway-Jones K (2011) Low soluble P-selectin may facilitate early exclusion of acute myocardial infarction. Clin Chim Acta 412:614–618
2. McDowell G, Temple I, Li C, Kirwan CC, Bundred NJ, McCollum CN, Burton IE, Kumar S, Byrne GJ (2005) Alteration in platelet function in patients with early breast cancer. Anticancer Res 25:3963–3966
3. Byrne GJ, McDowell G, Agarawal R, Sinha G, Kumar S, Bundred NJ (2007) Serum vascular endothelial growth factor in breast cancer. Anticancer Res 27:3481–3487
4. Kirwan CC, McDowell G, McCollum CN, Kumar S, Byrne GJ (2008) Early changes in the haemostatic and procoagulant systems after chemotherapy in breast cancer. Br J Cancer 99:1000–1006

Helpful Websites Including Troubleshooting Guides

DuoSet ELISA Development Systems, R&D Systems. http://www.rndsystems.com/product_detail_objectname_duoset.aspx

R&D Systems ELISA Development Guide. http://www.rndsystems.com/resources/images/5670.pdf

Chapter 36
Analysis of Phosphorylated Protein Kinases in Endothelial Cells by Flow Cytometry

Nina C. Dempsey-Hibbert

36.1 Introduction

The use of flow cytometry for both diagnostic and research purposes has increased dramatically over the past decade. The versatility and high sensitivity of this technique allow the analysis of multiple parameters in many different sample types, including cells, nuclei and chromosomes. As a result, flow cytometry has become a fundamental tool in the fields of immunology, cell biology and haematology.

36.1.1 The Flow Cytometer

Flow cytometry utilises the principles of light scattering, light excitation and light emission as a stream of cells in single file move past multiple beams of light of single wavelength. As each cell within the sample intercepts the light source, the light is deflected or 'scattered' allowing each cell to be analysed individually. Forward-scattered light (FSC) is proportional to the size or surface area of the corresponding cell, while side-scattered light (SSC) is proportional to the granularity of the cell. The deflection of the light after interaction with the cells is detected using photodetectors (PMTs). An electrical pulse is generated from these photodetectors, which is then processed and analysed in real-time. The use of fluorochrome-conjugated antibodies allows specific proteins inside the cell or on the cell surface to be detected and so, in addition to FSC and SSC, the fluorescent signal intensity of the cell is also analysed.

N.C. Dempsey-Hibbert (✉)
Cell and Molecular Biology Group, Manchester Metropolitan University, Oxford Road, Manchester, UK
e-mail: n.dempsey-hibbert@mmu.ac.uk

M. Slevin, G. McDowell (eds.), *Handbook of Vascular Biology Techniques*,
DOI 10.1007/978-94-017-9716-0_36

Many cell characteristics including the presence of cell surface receptors and intracellular proteins can be analysed simultaneously by the use of various combinations of fluorochrome-conjugated antibodies. This allows the identification of homogenous sub-populations within a heterogeneous cell population, which can be 'gated' to allow further analysis to be made on these specific populations. This has made flow cytometry the preferred method over alternative approaches such as ELISA, Western blot and spectrofluorometry.

36.1.2 Analysis of Cell-Signalling Pathway Activity by Flow Cytometry

Activation of cell signalling pathways via a cascade of protein kinase activation is a crucial process in endothelial cells during angiogenesis. PI3K/Akt signalling, in particular, is activated by many angiogenic growth factors, such as VEGF [1–3], TGF [4] and TNF [5], and regulates downstream target molecules that are involved in blood vessel growth [6]. Activation of the PI3K/Akt pathway is involved in a wide variety of both normal and pathological cellular processes and, consequently, analysis of Akt activity is integral to the study of many human diseases. Akt is activated by phospholipid binding and activation loop phosphorylation at Thr308 by PDK1 and by phosphorylation within the carboxy terminus at Ser473 [7]. Phosphorylation of Akt, leads to phosphorylation of downstream protein kinases, which ultimately results in a cellular response, such as proliferation, cell migration or cell survival in conditions that would otherwise be lethal to the cell [2, 8, 9].

Fluorochrome-conjugated antibodies that can differentiate between Akt phosphorylated at Ser473 and Akt phosphorylated at Thr308 are commercially available. This is particularly useful as different stimuli can promote the phosphorylation of Akt at distinct residues [10]. Total Akt within the cell (phosphorylated and non-phosphorylated) can also be measured using a fluorochrome-conjugated Pan-Akt antibody.

It is well-established that the Vascular Endothelial Growth Factor Receptors (VEGFRs) regulate numerous, if not all, facets of endothelial cell function during active angiogenesis. This is achieved by regulation of many key signalling pathways including PI3K/Akt [1, 11, 12]. Indeed, ligation of VEGFR-2 by VEGF induces PI3K/Akt activation in a wide variety of endothelial cells, positively regulating angiogenesis and promoting cell survival [3, 11, 13]. Studies have shown increased levels of Akt following VEGFR ligation and have also shown that chemical inhibition of Akt can prevent cellular morphogenesis [14]. The interplay between VEGF and the PI3K/Akt pathway in-vivo is complex, because PI3K/Akt activation, in turn, promotes VEGF production in tissues [15].

Activation of the NF-κB signalling pathway is also pro-angiogenic and pro-oncogenic, [16–18]. Activation of this pathway, can occur as a direct result of PI3K/Akt activation, due to phosphorylation and consequent degradation of the NF-κB inhibitory protein IκK [19]. It is therefore important to be able to examine the interplay between signalling pathways by measuring several kinases

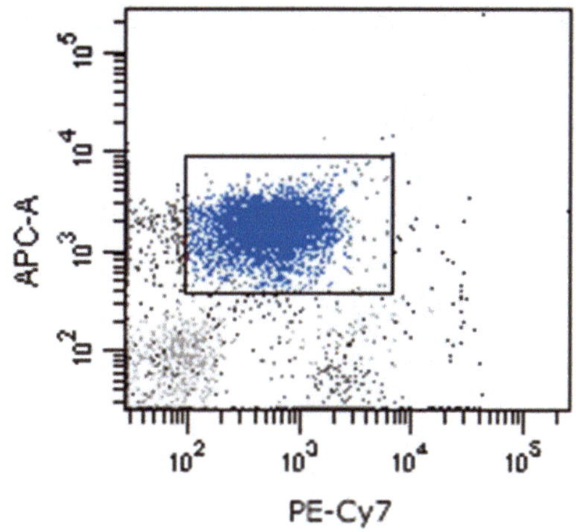

Fig. 36.1 Double staining of HUVECs with the PE-Cy7- and APC-conjugated antibodies. Cells of interest are recognised by double-fluorescence and 'gated' to distinguish them from the remaining cell population on the histograms. This is particularly important when analysing endothelial cells within a heterogeneous population

simultaneously. By using multiple antibodies with differing fluorochromes, flow cytometry allows the analysis of several protein kinases in one sample.

Identification of endothelial cells in heterogeneous populations is typically determined by the presence of the cell surface markers CD31, CD146 and vWF [20]. Recognition of the cells by flow cytometry involves simple labelling of the sample with fluorochrome-conjugated antibodies, targeted towards these cell surface markers. Evidence suggests that the cell surface expression of CD105 could be a more useful endothelial marker in the study of angiogenesis. CD105 is expressed on activated endothelial cells and so it is suggested that it may better reflect the process of neoangiogenesis [21, 22]. Indeed, cancer-associated blood vessels express high levels of CD105, while adjacent normal mucosa do not [23]. Conversely, CD31 expression is universally expressed in blood vessels of carcinoma lesions as well as normal mucosa vessels [23].

Simultaneous analysis of CD105 expression and protein kinase activity in endothelial cells is therefore critical for study of neoangiogenesis in primary samples. The following methods describe the simultaneous analysis of CD105, CD31 (Fig. 36.1) and Akt (both pS437 and pT308) (Figs. 36.2 and 36.3) in Human Umbillical-Cord Vascular Endothelial Cells (HUVEC). The methods may be adapted for analysis of primary tissues (see notes), other cell lines or other phosphorylated kinases.

36.2 Materials

36.2.1 Preparation of HUVECs for Cell Staining

1. Human Umbilical-Cord Vascular Endothelial Cell (HUVEC) (ATCC, CRL-1730)
2. 25 cm^2 plastic tissue culture flasks
3. F-12K Medium, supplemented with 0.1 mg/ml heparin, 0.05 mg/ml endothelial cell growth supplement (ECGS) and 10 % foetal bovine serum.

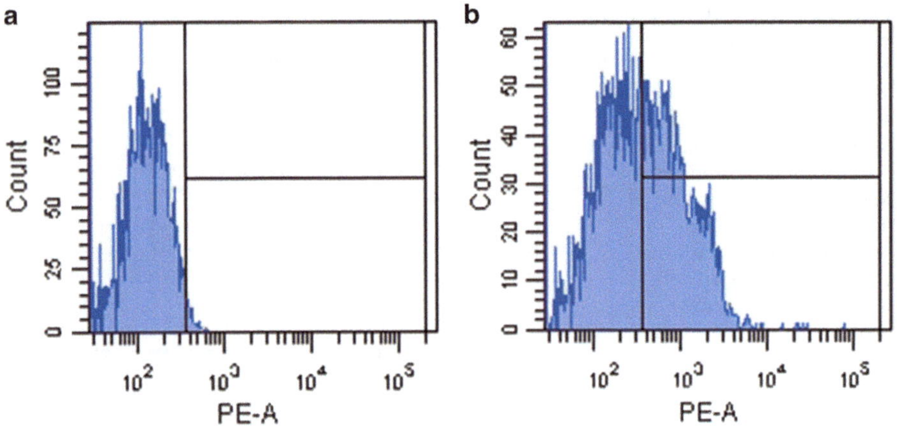

Fig. 36.2 Phosphorylated Akt (pT308) expression in CD31+/CD105+ HUVECS. CD31+/CD105+ events were gated to exclude other events. (**a**) Cells were stained with a PE-conjugated mouse IgG1 isotype control antibody (**b**) Cells were stained with a PE-conjugated mouse phosphorylated Akt (pT308) antibody

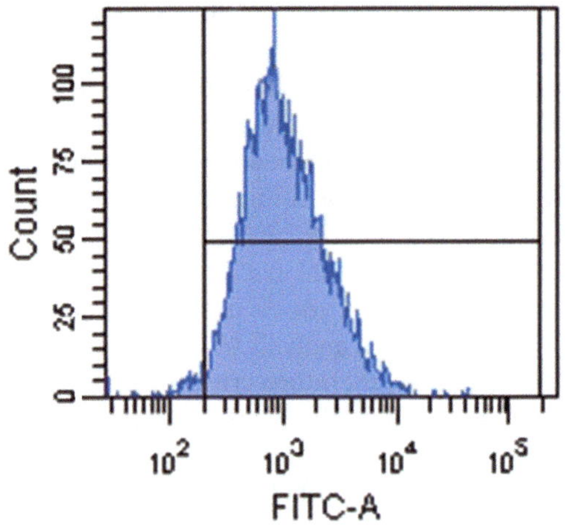

Fig. 36.3 Phosphorylated Akt (pS473) expression in CD31+/CD105+ HUVECS. CD31+/CD105+ events were gated to exclude other events

4. 0.25 % Trypsin/EDTA (Lonza)
5. Trypan Blue (Sigma)
6. Haemocytometer (Sigma)

36.2.2 Detection of Phosphorylated-Kinase in HUVEC

1. Binding Buffer; 5 % FBS (Lonza) in DPBS (Lonza). Make up fresh each time.
2. Antibodies:

   APC-conjugated Anti-human CD105 (BD Biosciences, 562408)
   Store at 4 °C. Protect from the light.

PE-Cy7-conjugated Anti-human CD31 (BD Biosciences, 563651)
Store at 4 °C. Protect from the light.

Alexa-Fluor 488®-conjugated Anti-human Akt pS473 (BD Biosciences, 560404)
Store at 4 °C. Protect from the light.
PE-conjugated Anti-human Akt pT308 (BD Biosciences, 558275)
Store at 4 °C. Protect from the light.

3. 4 % Paraformaldehyde (PFA) (Sigma).
   Store at room temperature. Toxic.
4. Perm Buffer III Solution (BD Biosciences, 558050).
   Store at room temperature. Toxic.
5. DPBS (Lonza)
6. 12×75 mm plastic flow cytometry tubes
7. Flow Cytometer (FACScanto II, Becton Dickinson, San Diego, CA, USA)

36.3 Methods

36.3.1 Preparation of HUVECs for Cell Staining

1. Remove and discard culture medium from the tissue culture flask.
2. Briefly rinse the adherent cell layer with 1 ml of 0.25 % Trypsin/EDTA, to remove all traces of the serum, containing trypsin inhibitor.
3. Discard the 0.25 % Trypsin/EDTA.
4. Add 2 ml 0.25 % Trypsin/EDTA to the tissue culture flask.
5. Incubate the tissue culture flask at 37 °C for 5 min to allow detachment of the cells.
6. Observe the cells under a light microscope to ensure the cell layer is observed. HINT: the 0.25 % Trypsin/EDTA can remain on the cells for 15 min if necessary but should not be left any longer.
7. Add 8 ml of F-12 K culture medium to the flask and gently mix to disperse aggregates into single cells.
8. Count cells by Trypan Blue exclusion method using a haemocytometer – HINT: The experiment should only proceed when the percentage of dead cells <1 %
9. Aliquot volumes of sample containing 1×10^6 cells into 1.5 ml microcentrifuge tubes (one aliquot per test)
10. Centrifuge the samples at 500 g for 5 min at RT
11. Discard the supernatant and proceed to sub-heading 3.2

36.3.2 Cell Surface Double-Labelling of HUVECs with CD31 and CD105

Cell surface proteins are detected on cells using fluorochrome-conjugated antibodies to specific antigens without prior cell fixation or permeabilisation. Each antibody is conjugated to a different fluorochrome to ensure the distinction can be made between each antibody (see notes section for information on selecting the best combination of fluorochromes).

1. Wash the cells by adding 100 μl binding buffer to each sample, incubate at RT for 5 min, then centrifuge at 500 g for 5 min. This limits non-specific binding of antibodies in subsequent steps. Remove the supernatant.
2. Make up the cocktail of each 'panel' of antibodies to be used by multiplying the number of samples by the volume of antibody required;

   20 μl of anti-human CD31-PE-Cy7 and 20 μl of anti-human CD105-APC are required per sample. Therefore, for example, for four replicates of HUVEC samples, mix 80 μl of anti-human CD31-PE and 80 μl of anti-human CD105-APC in a foil-wrapped microcentrifuge tube.

   Add 40 μl of antibody cocktail to each cell pellet and mix.
3. Incubate at 4 °C for 30 min in the dark.
4. Add 100 μl binding buffer to each sample to wash any unbound antibodies and centrifuge at 500 g for 5 min at RT.
5. Remove the supernatant. At this point, cells can be resuspended in 500 μl of DPBS and analysed immediately on the flow cytometer if only detection of surface markers is wanted. Otherwise, proceed to Sect. 36.3.3.

36.3.3 Detection of Phosphorylated Akt in HUVECs

Intracellular proteins are detected on cells using fluorochrome conjugated antibodies to specific antigens following fixation and permeabilisation of the cells using a methanol-based permeabilisation solution.

1. Fix the cells by adding 100 μl/sample of 4 % Paraformaldehyde and incubating at 37 °C for 10 min.
2. Add 100 μl/well binding buffer to each sample to wash the cells and centrifuge at 500 g for 5 min at RT. Remove the supernatant.
3. Permeabilise the cells by adding 100 μl/sample of Perm Buffer III solution and incubate for 20 min at 4 °C. This allows any subsequent antibodies to diffuse into the cytosol and nucleus of the cells and allow intracellular detection.
4. Add 100 μl/well binding buffer to each sample to wash the cells and centrifuge at 500 g for 5 min at RT. Remove the supernatant.

5. Make up the cocktail of each 'panel' of intracellular antibodies to be used by multiplying the number of samples by the volume of antibody required; 20 μl of anti-human Akt (pS473) –Alexa-Fluor 488® and 20 μl of anti-human Akt (pT308)-PE are required per sample. Therefore, for four replicates of HUVEC samples, mix 80 μl of anti-human CD31-PE and 80 μl of anti-human CD105-APC in a foil-wrapped microcentrifuge tube.
6. Add 40 μl of antibody cocktail to each cell pellet and mix
7. Incubate for 60 min at 4 °C in the dark.
8. Add 100 μl binding buffer to each sample and leave for 5 min to allow any unbound antibodies to diffuse out of the cells.
9. Centrifuge at 500 g for 5 min at RT and remove the supernatant.
10. Resuspend the cell pellet in 500 μl DPBS.
11. Transfer the cell suspension to 12×75 mm tubes ready for analysis on the flow cytometer. (see notes)

36.4 Notes

1. Analysis of phosphorylated protein kinases in endothelial cell lines is a relatively simple, yet sensitive technique. The homogenous cell population allows for easy identification of the cells of interest. However, the benefits of flow cytometry become apparent when attempting to study these kinases in primary samples with heterogeneous cell populations. The Akt detection methods described above may also be followed to analyse Akt phosphorylation in non-adherent cells and also primary tissue culture samples. Primary tissue samples must be in single cell suspension prior to staining. This can be achieved by enzymatic digestion using the following method (optimised for isolating cells from primary colorectal tissue):

   (a) Immediately submerge the tissue in 1.5 ml of Hanks Balanced Salt Solution (HBSS) media (Lonza) following surgical removal and transport to the laboratory within 30 min
   (b) All subsequent steps should be performed under sterile conditions
   (c) Wash the sample three times with 2 ml of 10 % antibiotic (Lonza) in HBSS. Samples must be incubated in the antibiotic solution for 1–2min between each wash with F12K Medium
   (d) Dissect the tissue into 1×1×1 mm pieces using sterile scalpel and forceps
   (e) Centrifuge the sample at 500 g for 8 min and remove the supernatant
   (f) Resuspend the cell pellet in 3 ml of trypsin and incubate for 100 min at 37 °C, agitating regularly
   (g) Centrifuge the sample at 500 g for 8 min and resuspend in 2 ml of 10 % antibiotic in HBSS
   (h) Pass the sample through a nylon cell strainer (BD Biosciences)
   (i) Centrifuge at 500 g for 8 min and remove supernatant

(j) If traces of red blood cells are observed, the cells can be incubated in 3 ml of lysing solution (BD Biosciences) for 6 min at RT before another centrifugation step at 500 g for 8 min.
(k) Remove supernatant and resuspend cells in 1 % antibiotic solution, made up in the optimal culture media for your chosen cell.
(l) Culture cells at 37 °C in 5 % CO_2 in air.

2. It is critical to select the right combination of fluorchromes for the experiments and this will largely depend on the type and number of lasers and detectors on the specific flow cytometer. The idea is to choose the stain with maximum 'brightness' but with minimal 'spectral overlap' (otherwise known as 'spill-over'). The more fluorochromes used on one sample, the more 'spectral overlap' will have to be dealt with. Although this can largely be achieved through the process of 'compensation' (see note 4), reagents should be selected so that the potential for spectral overlap is minimized.
3. The Alexa-Fluor 488® fluorochrome will be detected in the FITC channel on the FACS Canto II, so it is important to be aware that you cannot stain the same samples with Alexa-Fluor 488®-conjugated antibodies and FITC-conjugated antibodies. This is also the case for the APC fluorochrome and the Alexa-Fluor 647® fluorochrome which are both detected in the APC channel.
4. A number of control samples must be analysed alongside the test samples in order to ensure accurate results.

   (a) A no-stain control sample which has not been probed with any antibodies is used as a negative control. This sample is analysed prior to the stained samples in order to adjust the PMT voltages and to place the negative population within the first log decade of the histogram (Fig. 36.2a).
   (b) Isotype control samples must also be analysed. Isotype control antibodies have no relevant specificity and help to distinguish background fluorescence. Isotype controls must be matched to the specific test primary antibody by both species and isotype. For example, a mouse IgG1-APC can be used as an isotype control against the mouse monoclonal anti-human CD105-APC.

5. Compensation experiments must be performed prior to running the test samples when more than one fluorochrome-conjugated antibody is added to a single sample. If the wavelengths of two fluorochromes overlap, the fluorescent signal detected by the flow cytometer may not be the actual signal emitted by the cell. This results in the incorrect assessment of cell phenotype. This overlap can be corrected by running a number of samples stained with a single antibody from each of the 'panel' of antibodies used in the experiment. For example, compensation controls for the present method would include the use of samples stained with Anti-CD31, Anti-CD105, Anti-Akt (pS473) and Anti-Akt (pT308) in isolation. The need to run compensation experiments can be avoided by carefully selecting fluorochromes that do not have spectral overlap.
6. The use of a fixative reagent in Sect. 36.3.3 prevents any further biochemical reactions occurring within the cells. Immediate analysis on the flow cytometer

following staining is therefore not required and the cells can be stored in DPBS at 4 °C protected from the light for up to 2 weeks before analysis. If only surface staining is being performed (Sect. 36.3.2), 100 µl of 4 % PFA can be added after step 9 and the cells can be incubated at 37 °C for 10 min before washing off the PFA and resuspending the samples in 500 µl DPBS. Again, this gives the user the opportunity to perform the flow cytometry analysis at a later date.

7. When labelling intracellular proteins in cells, it is important to choose the correct permeabilisation solution. Saponin-based permeabilisation solutions, such as Cytofix/Cytoperm (BD Biosciences) are useful for staining proteins that are restricted to the cytoplasm. However, if the proteins of interest may also be located in the nucleus, a methanol-based solution is required to permeabilise the nuclear membrane.
8. Samples can be stained in microcentrifuge tubes and transferred to 12×75 mm flow cytometry tubes before analysis. Alternatively, the cells may be transferred to 96-well v-bottom plates prior to surface staining so that all subsequent steps and flow cytometry are performed in the wells. The analysis of samples in the plate requires the use of a high throughput sampler (HTS) attachment on the flow cytometer. Staining in plates not only saves time at washing stages but also allows the use of 3 times lower concentrations of antibody as cell concentrations can be adjusted to 3×10^5 cells/well.

References

1. Abid MR, Guo S, Minami T, Spokes KC, Ueki K, Skurk C, Walsh K, Aird WC (2004) Vascular endothelial growth factor activates PI3K/Akt/forkhead signaling in enothelial cells. Arterioscler Thromb Vasc Biol 24:294–300
2. Fujio Y, Walsh K (1999) Akt mediates cytoprotection of endothelial cells by vascular endothelial growth factor in an anchorage-dependent manner. J Biol Chem 274:16349–1635
3. Gerber HP, McMurtrey A, Kowalski J, Yan M, Keyt BA, Dixit V, Ferrara N (1998) Vascular endothelial growth factor regulates endothelial cell survival through the phosphatidylinositol 3′-kinase/Akt signal transduction pathway. J Biol Chem 273:30336–30343
4. Viñals F, Pouysségur J (2001) Transforming growth factor 1 (TGF-1) promotes endothelial cell survival during in vitro angiogenesis via an autocrine mechanism implicating TGF-signaling. Mol Cell Biol 21:7218–7230
5. Wang L, Chopp M, Teng H, Bolz M, Francisco MA, Aluigi DM, Wang XL, Zhang RL, Chrsitensen S, Sager TN, Szalad A, Zhang ZG (2011) Tumor necrosis factor a primes cerebral endothelial cells for erythropoietin-induced angiogenesis. J Cereb Blood Flow Metab 31:640–647
6. Shiojima I, Walsh K (2002) Role of Akt signaling in vascular homeostasis and angiogenesis. Circ Res 90:1243–1250
7. Sarbassov DD, Guertin DA, Ali SM, Sabatini DM (2005) Phosphorylation and regulation of Akt/PKB by the rictor-mTOR complex. Science 307:1098–1101
8. Morales-Ruiz M, Fulton D, Sowa G, Languino LR, Fujio Y, Walsh K, Sessa WC (2000) Vascular endothelial growth factor–stimulated actin reorganization and migration of endothelial cells is regulated via the serine/threonine kinase Akt. Circ Res 86:892–896
9. Sinnberg T, Sauer B, Holm P, Spangler B, Kuphal S, Bosserhoff A, Schittek B (2012) MAPK and PI3K/AKT mediated YB-1 activation promotes melanoma cell proliferation which is counteracted by an autoregulatory loop. Exp Dermatol 21:265–270

10. Vincent EE, Elder DJE, Thomas EC, Phillips L, Morgan C, Pawade J, Sohail M, May MT, Hetzel MR, Tavare JM (2011) Akt phosphorylation on Thr308 but not on Ser473 correlates with Akt protein kinase activity in human non-small cell lung cancer. Br J Cancer 104:1755–176
11. Dellinger MT, Brekken RA (2011) Phosphorylation of Akt and ERK1/2 is required for VEGF-A/VEGFR2-induced proliferation and migration of lymphatic endothelium. PLoS One 6:e2894–28947
12. Gliki G, Wheeler-Jones C, Zachary I (2002) Vascular endothelial growth factor induces protein kinase C (PKC)-dependent Akt/PKB and phosphatidylinositol-3′-kinase-mediated PKCδ phosphorylation: role of PKC in angiogenesis. Cell Biol Int 26:751–759
13. Chen J, Somanath PR, Razorenova O, Chen WS, Hay N, Bornstein P, Byzova TV (2005) Akt1 regulates pathological angiogenesis, vascular maturation and permeability in vivo. Nature 11:1188–1196
14. Kaneko Y, Kitazato K, Basaki Y (2004) Integrin-linked kinase regulates vascular morphogenesis induced by vascular endothelial growth factor. J Cell Sci 117:407–415
15. Jiang BH, Zheng JZ, Aoki M, Vogt PK (2000) Phosphatidylinositol 3-kinase signaling mediates angiogenesis and expression of vascular endothelial growth factor in endothelial cells. Proc Natl Acad Sci U S A 97:1749–1753
16. Jiang L, Song L, Wu J, Yang Y, Zhu X, Hu B, Cheng SY, Li M (2013) Bmi-1 promotes glioma angiogenesis by activating NF- kB signaling. PLos One 8:e55527
17. Iosef C, Alastalo TP, Hou Y, Chen C, Adams ES, Lyu SC, Cornfield DN, Alvira CM (2012) Inhibiting NF-κB in the developing lung disrupts angiogenesis and alveolarization. Am J Physiol Lung Cell Mol Physiol 302:L1023–L1036
18. Tabruyn SP, Mémet S, Avé P, Verhaeghe C, Mayo KH, Struman I, Martial JA, Griffioen AW (2009) NF-κB activation in endothelial cells is critical for the activity of angiostatic agents. Mol Cancer Ther 8:2645–2654
19. Bai D, Ueno L, Vogt PK (2009) Akt-mediated regulation of NFjB and the essentialness of NFjB for the oncogenicity of PI3K and Akt. Int J Cancer 125:2863–2870
20. Mesri M, Birse C, Heidbrink J, McKinnon K, Brand E, Bermingham CL, Feild B, FitzHugh W, He T, Ruben S, Moore PA (2013) Identification and characterization of angiogenesis targets through proteomic profiling of endothelial cells in human cancer tissues. PLos One 8:e78885
21. Dales JP, Garcia S, Bonnier P, Duffaud F, Andrac-Meyer L, Ramuz O, Lavaut MN, Allasia C, Charpin C (2003) CD105 expression is a marker of high metastatic risk and poor outcome in breast carcinomas. Am J Clin Pathol 119:374–380
22. Miyata Y, Sagara Y, Watanabe SI, Asai A, Matsuo T, Ohba K, Hayashi T, Sakai H (2013) CD105 is a more appropriate marker for evaluating angiogenesis in urothelial cancer of the upper urinary tract than CD31 or CD34. Virchows Arch 463:673–679
23. Minhajat R, Mori D, Yamasaki F, Sugita Y, Satoh T, Tokunaga O (2006) Endoglin (CD105) expression in angiogenesis of colon cancer: analysis using tissue microarrays and comparison with other endothelial markers. Virchows Arch 448:127–134

Index

M. Slevin, G. McDowell (eds.), *Handbook of Vascular Biology Techniques*,
DOI 10.1007/978-94-017-9716-0

MIX
Papier aus verantwortungsvollen Quellen
Paper from responsible sources
FSC® C105338

If you have any concerns about our products,
you can contact us on
ProductSafety@springernature.com

In case Publisher is established outside the EU,
the EU authorized representative is:
Springer Nature Customer Service Center GmbH
Europaplatz 3, 69115 Heidelberg, Germany

Printed by Libri Plureos GmbH
in Hamburg, Germany